A CHECKLIST OF THE ORCHIDACEAE OF INDIA

A CHECKLIST OF THE ORCHIDACEAE OF INDIA

André Schuiteman, B. Ramamurthy Kailash,
and Uttam Babu Shrestha

MISSOURI BOTANICAL GARDEN PRESS

ISBN 978-1-935641-25-4
Library of Congress Control Number: 2021939534
Monographs in Systematic Botany from the Missouri Botanical Garden, Volume 139
ISSN 0161-1542

Publisher: Liz Fathman
Managing Editor: Allison M. Brock
Editor: Lisa J. Pepper
Press Coordinator: Amanda Koehler
Cover design: Mary Shocklee

This monograph was printed on 26 January 2022.

Front cover photo: *Coelogyne nitida* (Wall. ex D. Don) Lindl. Back cover photos (clockwise from below): *Paphiopedilum spicerianum* (Rchb. f.) Pfitzer, *Dendrobium brymerianum* Rchb. f., *Bulbophyllum ornatissimum* (Rchb. f.) J. J. Sm., *Spathoglottis ixioides* (D. Don) Lindl. All photos by André Schuiteman.

Contents

Acknowledgments

André Schuiteman wishes to thank Christopher Ryan and Bala Kompalli (Tropical Nursery, Kew) for helping to locate and extract type specimens in the Kew Herbarium, and Amy Martin for compiling information on type specimens from selected literature and for preparing some statistics. Leonid Averyanov commented on an early draft of the checklist. Daniel Geiger reviewed the treatment of *Oberonia*. Paul Ormerod drew our attention to nomenclatural issues and overlooked names. Lisa Thoerle and Lisa Pepper corrected many imperfections while proofreading. We thank Peter H. Raven and Kamal Bawa for their encouragement and patience and two anonymous reviewers for helpful suggestions and corrections. The early part of the work in generating the data was funded by the Department of Biotechnology, Government of India, New Delhi. The final compilation and publication were supported by the Sehgal Foundation, and in part by the Chris Davidson Fund and a grant from the Office of the Principal Scientific Advisor to the Prime Minister, Government of India, for developing India's National Mission on Biodiversity and Human Well-Being.

INTRODUCTION

Indian orchids have long attracted the attention of botanists, even before Linnaeus, and the history of taxonomic research on these fascinating plants parallels that of Indian systematic botany as a whole. We find here largely the same names of collectors and scientists as for most other plant groups, so that it seems superfluous to describe in any detail the gradual accumulation of our knowledge since the time of Linnaeus. Suffice it to mention that it was Linnaeus himself who, in 1753, formally named the first Indian orchids according to the system of binomial nomenclature that we still use today. Some Indian orchids were described and illustrated before Linnaeus in the famous *Hortus Malabaricus* by Hendrik Adriaan van Rheede tot (or van) Drakestein (also spelled Rhede, Reede, and Dra[a]ke[n]stein), published between 1678 and 1703. Linnaeus established five Indian orchid species based on Rheede's work.

If we had to choose the three most important publications on Indian orchids prior to the 20th century, we would suggest John Lindley's *Contributions to the Orchidology of India*, published in two parts (Lindley, 1857, 1858); Joseph Hooker's treatment of the Orchidaceae for the *Flora of British India* (Hooker, 1890a, 1890b); and George King and Robert Pantling's monumental *The Orchids of Sikkim-Himalaya* (King & Pantling, 1898). By 1900, about 750 orchid species were known from India.

Since then, relatively little work of major importance on Indian orchids was published until late in the 20th century, apart from Duthie's *The Orchids of the Northwestern Himalaya* (1906). Arguably, the first milestone after that was the publication by Udai C. Pradhan of his two-volume *Indian Orchids* (Pradhan, 1976, 1979), which was also one of the first large-scale orchid works written by an Indian botanist. Pradhan treated ca. 825 species in 135 genera, which represented a relatively modest increase since the time of Hooker and King and Pantling. Another important work is *The Orchid Flora of North West Himalaya* (Deva & Naithanai, 1986). Since 1986, however, the number of publications on Indian orchids has risen exponentially. Too numerous to mention are the books on orchids of individual states, the flora treatments of states and districts, and the never-ending stream of local checklists and publications of new taxa and records. Until recently, a few nationwide checklists were available (Jain & Mehrotra, 1984; Karthikeyan et al., 1989, Kumar & Manilal, 1994; Sarkar, 1995a). The most detailed of these is the one by Kumar and Manilal; they list 1141 species in 166 genera and indicate for each species if it occurs in Peninsular India, the Himalaya region, or the Andaman and Nicobar Islands. None of these four checklists provide information on the occurrence by state or on type specimens, and they give no or very limited synonymy. In addition, since they were published before DNA analyses were applied to orchid systematics, their taxonomy is outdated.

While the present work was in its final stages, an important new publication appeared, *Orchids of India: A Pictorial Guide* by Singh et al. (2019), published by the Botanical Survey of India. This is essentially an illustrated checklist and so covers much of the same ground as our work. However, there are some significant differences. Singh et al. do not list type specimens and have not aimed for completeness in synonymy; therefore, many names based on Indian types are unaccounted for. Their taxonomic treatment differs significantly from ours in places, as they have not adopted recently modified concepts of several genera (Pridgeon et al., 1999–2014; Chase et al., 2015; Ng et al., 2018) (Table 1). They list a few species not included in our list (and vice versa) as well as numerous state-level records that differ from ours (see Table 5 under Distribution). It is not clear whether these are based on published sources that we have not seen or on unpublished data. Surprisingly, no mention is made of conservation in Singh et al.'s book, and there is no identification key.

An even more recent publication of note is *Orchids of India—A Handbook* by Sarat Misra (2019). This is mainly a compendium of genera together with a checklist of accepted names, with distribution data limited to occurrence in the Peninsula, Himalayas, and Andaman and Nicobar Islands. It has an extensive bibliography and for each species there is a reference to a published source. There is also a key to the identification of genera, which unfortunately has many problems. For example, *Dendrobium* Sw. and *Eria* Lindl. (the latter in the broad, now obsolete sense) are distinguished as follows:

124a. Flowers brightly colored, usually large and showy, glabrous . . . . *Dendrobium*
124b. Flowers not brightly colored, medium- to small-sized; ovary and sepals outside softly hairy. . . . . *Eria*

To arrive at this point in the key one had to assume that the inflorescences are short and borne on a leafy shoot from the upper internodes, which in both genera is not always the case. Furthermore, *Dendrobium* can have small, dull-colored flowers (e.g., *D. parcum* Rchb. f.) and *Eria* s.l. may have brightly colored, glabrous flowers (e.g., *E. coronaria* Rchb. f.). The most reliable

Table 1. Genera in Singh et al. (2019) not recognized here.

Amitostigma Schltr. (= *Ponerorchis* Rchb. f.)
Androcorys Schltr. (= *Herminium* L.)
Ascocentrum Schltr. (= *Vanda* Jones ex R. Br.)
Bhutanthera Renz (= *Herminium*)
Bulleyia Schltr. (= *Coelogyne* Lindl.)
Cephalantheropsis Guillaumin (= *Calanthe* R. Br.)
Dickasonia L. O. Williams (= *Coelogyne* Lindl.)
Dienia Lindl. (= *Crepidium* Blume p.p. & *Malaxis* Sol. ex Sw. p.p.)
Dithrix (Hook. f.) Schltr. ex Brummitt (= *Gennaria* Parl.)
Epigeneium Gagnep. (= *Dendrobium* Sw.)
Esmeralda Rchb. f. (= *Arachnis* Blume)
Flickingeria A. D. Hawkes (= *Dendrobium*)
Geodorum Jacks. (= *Eulophia* R. Br.)
Neogyna Rchb. f. (= *Coelogyne* Lindl.)
Odisha S. Misra (= *Habenaria* Willd.)
Otochilus Lindl. (= *Coelogyne* Lindl.)
Panisea (Lindl.) Lindl. (= *Coelogyne* Lindl.)
Phaius Lour. (= *Calanthe* R. Br.)
Pholidota Lindl. ex Hook. (= *Coelogyne* Lindl.)
Plectoglossa (Hook. f.) K. Prasad & Venu (= *Habenaria*)
Vandopsis Pfitzer p.p. (= *Cymbilabia* D. K. Liu & Ming H. Li)

distinguishing character, four versus eight pollinia, is not mentioned.

The differences between Misra's checklist and ours are too numerous to discuss in detail, as should be clear from the fact that this author recognizes 192 genera (against our 148) and 1430 species (against our 1234). Almost all the additional genera and species found in Misra's book are here considered synonyms.

The aim of the current checklist is fivefold:

(1) To align the nomenclature with the latest insights from molecular phylogenetic studies, mainly following Pridgeon et al. (1999–2014) and papers published since.
(2) To provide distribution data at state level.
(3) To give information on type specimens.
(4) To list the relevant synonymy, but mainly limited to names that may be encountered in literature on Indian orchids.
(5) To provide a key for the identification of genera.

The first three aims are discussed in more detail in separate chapters below.

Work on this checklist was begun by B. Ramamurthy Kailash at ATREE, who compiled an initial version based on a survey of the literature. Uttam B. Shrestha at Harvard University Herbaria added many more species and checked the list sent by Kailash as part of the Plant Checklist of India project directed by K. N. Ganeshaiah, Peter H. Raven, Kanchi N. Gandhi, J. Rich-

ard Abbott, and K. S. Bawa. The initial list also received some valuable comments from Leonid Averyanov from the Komarov Institute at Saint Petersburg, Russia. André Schuiteman joined the project in 2011 and over the following years revised the checklist as the taxonomy of several groups underwent considerable change based on molecular phylogenetic research. New species and records published since 2011 have also been incorporated and much information on type specimens was added.

Our sources include most regional floras and manuals on Indian orchids, as well as numerous incidental publications, the majority of which are listed in the bibliographies of Kumar and Manilal (1994), Sarkar (1995b), Misra (2019), and on Rudolf Jenny's *Orchilibra* website (Jenny, 2021). The references at the end are only of works explicitly mentioned in the text; they do not comprise a full bibliography.

SYSTEMATICS

According to our checklist, 1234 species in 148 genera have been recorded from India. Following Pridgeon et al. (1999–2014) and Chase et al. (2015), with some modifications discussed below, they are classified as in the following list, where numbers of species recorded from India are indicated for each genus.

All five subfamilies of Orchidaceae are represented in India. The only endemic genera here recognized are *Aenhenrya* Gopalan and *Smithsonia* C. J. Saldanha.

I subfamily APOSTASIOIDEAE
Apostasia Blume 3
II subfamily CYPRIPEDIOIDEAE
Cypripedium L. 4
Paphiopedilum Pfitzer 9
III subfamily VANILLOIDEAE
III.1 tribe Vanilleae
Cyrtosia Blume 4
Erythrorchis Blume 1
Galeola Lour. 2
Lecanorchis Blume 2
Vanilla Mill. 7
III.2 tribe Pogonieae
Pogonia Juss. 1
IV subfamily ORCHIDOIDEAE
IV.1 tribe Cranichideae
  IV.1.a subtribe Goodyerinae
  Aenhenrya Gopalan 1
  Anoectochilus Blume 7
  Chamaegastrodia Makino & F. Maek. 3
  Cheirostylis Blume 9
  Erythrodes Blume 2
  Goodyera R. Br. 16
  Herpysma Lindl. 1

Pinalia Buch.-Ham. ex Lindl. 22
Podochilus Blume 5
Porpax Lindl. 20
Strongyleria (Pfitzer) Schuit., Y. P. Ng & H. A. Pedersen 1
Thelasis Blume 4
Trichotosia Blume 3
V.11 tribe Vandeae
  V.11.a subtribe Adrorhizinae
Sirhookera Kuntze 2
  V.11.b subtribe Polystachyinae
Polystachya Hook. 3
  V.11.c subtribe Aeridinae
Acampe Lindl. 3
Aerides Lour. 9
Arachnis Blume 4
Biermannia King & Pantl. 4
Brachypeza Garay 1
Chiloschista Lindl. 5
Cleisocentron Brühl 1
Cleisomeria Lindl. ex D. Don 1
Cleisostoma Blume 22
Cottonia Wight 1
Cymbilabia D. K. Liu & Ming H. Li 1
Diplocentrum Lindl. 2
Diploprora Hook. f. 2
Gastrochilus D. Don 19
Grosourdya Rchb. f. 2
Holcoglossum Schltr. 3
Luisia Gaudich. 22
Macropodanthus L. O. Williams 2
Micropera Lindl. 4
Papilionanthe Schltr. 5
Pelatantheria Ridl. 1
Pennilabium J. J. Sm. 3
Phalaenopsis Blume 15
Pomatocalpa Breda 4
Pteroceras Hasselt ex Hassk. 5
Renanthera Lour. 1
Rhynchostylis Blume 2
Robiquetia Gaudich. 8
Saccolabiopsis J. J. Sm. 1
Sarcoglyphis Garay 1
Schoenorchis Blume 7
Seidenfadeniella C. S. Kumar 2
Smithsonia C. J. Saldanha 3
Smitinandia Holttum 2
Stereochilus Lindl. 4
Taeniophyllum Blume 10
Taprobanea Christenson 1
Thrixspermum Lour. 13
Trachoma Garay 1
Trichoglottis Blume 4
Uncifera Lindl. 3
Vanda Jones ex R. Br. 17

In the last 20 years, DNA-based phylogenetic studies have led to numerous changes in generic circumscription while greatly increasing our insight into the relationships between genera. It is instructive to compare the list of genera presented in Kumar and Manilal (1994) with the one above. Twenty-eight genera accepted by these authors—most by many others before them—are not recognized here (Table 2).

Conversely, 22 genera here accepted (Table 3) are not found in Kumar and Manilal's list.

On balance, the number of genera recognized has decreased somewhat, from 166 in Kumar and Manilal (1994) to 148 in this publication. Singh et al. (2019) enumerate 155 genera with 1256 infrageneric taxa. Two generic names not yet mentioned but frequently encountered in Indian orchid literature, *Cirrhopetalum* Lindl. and *Rhytionanthos* Garay, Hamer & Siegerist, are here included in *Bulbophyllum*.

Research in orchid phylogeny is ongoing and some changes compared with the classification of Chase et al. (2015), as far as relevant for Indian orchids, are as follows: *Androcorys* and *Bhutanthera* have been merged with *Herminium* (Raskoti et al., 2016); *Dienia* is to be included in *Crepidium* (Tang et al., 2015), necessitating the conservation of the name *Crepidium*; and in tribe Podochileae the genera *Bambuseria*, *Cylindrolobus*, *Dendrolirium*, and *Strongyleria* have been added while *Conchidium* Griff. has been merged with *Porpax* (Ng et al., 2018).

Coelogyne is polyphyletic (Gravendeel in Pridgeon et al., 2005). This is addressed in a paper by Chase et al. (2021), with the main implication that the genera *Bulleyia*, *Dickasonia*, *Neogyna*, *Otochilus*, *Panisea*, and *Pholidota*, as well as others not occurring in India, are to be merged with *Coelogyne*.

In subtribe Malaxidinae, the large genus *Liparis* is clearly polyphyletic (Tang et al., 2015) but it appears difficult to divide it into smaller genera that are all morphologically recognizable.

Work on Eulophiinae (Bone et al., 2015) strongly suggested that *Geodorum* should be included in *Eulophia*, as it is deeply nested within that genus. This change has been effected in a recent paper (Chase et al., 2021).

In Collabieae, *Calanthe* and *Phaius* as traditionally circumscribed are polyphyletic and their species are to some extent interdigitated (Xiang et al., 2014; Zhai et al., 2014), with *Cephalantheropsis* also nested in the clade that contains these two well-known genera. Splitting *Calanthe* s.l. into the genera *Calanthe*, *Preptanthe* Rchb. f., and *Styloglossum* Breda (Yukawa & Cribb, 2014) has been proposed, but if this is accepted it then is inevitable to split up *Phaius* as well, leading to a classification with morphologically poorly defined genera. The alternative, to merge the genera concerned into

Table 2. Genera in Kumar and Manilal (1994) not recognized here.

Acrochaene Lindl. (= Bulbophyllum)
Amitostigma Schltr. (= Ponerorchis)
Androcorys Schltr. (= Herminium)
Aorchis Verm. (= Galearis)
Archineottia S. C. Chen (= Neottia)
Bulleyia Schltr. (= Coelogyne)
Cephalantheropsis Guillaumin (= Calanthe)
Dickasonia L. O. Williams (= Coelogyne)
Didiciea King & Prain (= Tipularia)
Diphylax Hook. f. (= Platanthera)
Doritis Lindl. (= Phalaenopsis)
Epigeneium Gagnep. (= Dendrobium)
Esmeralda Rchb. f. (= Arachnis)
Evrardianthe Rauschert (= Odontochilus)
Flickingeria A. D. Hawkes (= Dendrobium)
Geodorum Jacks. (= Eulophia)
Hygrochilus Pfitzer (= Phalaenopsis)
Jejosephia A. N. Rao & K. J. Manilal (= Bulbophyllum)
Kingidium P. F. Hunt (= Phalaenopsis)
Listera R. Br. (= Neottia)
Malleola J. J. Sm. & Schltr. (= Robiquetia)
Mischobulbum Schltr. [as "Miscobulbon"] (= Tainia)
Monomeria Lindl. (= Bulbophyllum)
Neogyna Rchb. f. (= Coelogyne)
Neottianthe (Rchb.) Schltr. (= Ponerorchis)
Ornithochilus (Lindl.) Wall. ex Benth. (= Phalaenopsis)
Otochilus Lindl. (= Coelogyne)
Panisea (Lindl.) Lindl. (= Coelogyne)
Phaius Lour. (= Calanthe)
Pholidota Lindl. ex Hook. (= Coelogyne)
Rhinerrhiza Rupp (an Australian genus mistakenly reported from India)
Staurochilus Ridl. (= Trichoglottis)
Sunipia Buch.-Ham. ex Sm. (= Bulbophyllum)
Trias Lindl. (= Bulbophyllum)
Trudelia Garay (= Vanda)
Vandopsis Pfitzer p.p. (= Cymbilabia)
Xenikophyton Garay (= Schoenorchis)

Table 3. Genera here accepted but not in Kumar and Manilal (1994); those marked with * are also absent from Singh et al. (2019).

Aenhenrya
Ania
*Bambuseria
*Brachypeza
*Bryobium
*Callostylis
Chamaegastrodia
Cleisomeria
Crepidium
*Cylindrolobus
*Cymbilabia
*Dendrolirium
Galearis
*Gennaria
Holcoglossum
*Hsenhsua
Lecanorchis
*Mycaranthes
Odontochilus
*Pinalia
Rhomboda
*Strongyleria

a single genus, has been implemented recently (Chase et al., 2020) and is followed here.

Goodyerinae are the subject of active research (Chen et al., 2019), and a stable generic delimitation has not yet been established. Pace (2020) has proposed that *Goodyera* should be split in at least five genera, which in our view leads to an unworkable classification with genera distinguished by minute, not to say trivial, differences. At the same time, we are aware that some of the genera here listed, such as *Myrmechis*, will have to be merged with others in a future classification, but this will require making new combinations. At present, this seems premature.

The huge genus *Habenaria* is distinctly polyphyletic in its current circumscription (Jin et al., 2017), but it seems nearly impossible to split it into morphologically recognizable monophyletic entities. It could be merged with related genera like *Peristylus* and *Herminium* to create an even larger and even more diffuse genus. Unfortunately, *Herminium* is an older name than *Habenaria*, so this option would require nomenclatural conservation of the name *Habenaria*. *Pecteilis* is nested in a clade of *Habenaria* but is here provisionally retained until a well-founded proposal is made to reclassify *Habenaria* s.l.

In subtribe Aeridinae, the alliance of *Cleisostoma*, *Sarcoglyphis*, *Stereochilus*, and several other genera is phylogenetically unresolved, and more changes will probably be required once enough species have been subjected to molecular analyses.

The two Indian endemic genera *Aenhenrya* and *Smithsonia* have not been included in phylogenetic studies and it remains to be seen if they can be maintained.

TYPIFICATION

It can be remarkably difficult to establish the nomenclatural type of a taxon. In the past, before typification was fully codified, authors of a species, subspecies, or variety often provided only scant information about the specimens they studied. In many cases, they would mention the names of the collector(s) of the specimens, without giving the collecting numbers and without explicitly designating one specimen as the type. Conversely, some collectors, such as Wallich and Pantling, often used the same collection number for specimens collected at different dates in different localities, sometimes even for material gathered by different collectors in different countries. Even if both a single collector and number as well as a date and a locality are mentioned in the protologue, this does not always pin down a holotype specimen, because there may be duplicate specimens carrying the same label data. To make matters worse, sometimes this information was not transferred to the herbarium sheet carrying the specimen. It is also not uncommon to find that the published data (e.g., dates, localities, elevations) and specimen data do not match. On the other hand, in many cases where the collector's number is not mentioned in the protologue it is present on the specimen label, frequently with other unpublished information, such as collecting dates and localities. Such collector's numbers that are absent from the protologue are here added between square brackets, often along with unpublished information taken from the collector's label. We have not tried to give the modern equivalents of each locality mentioned, although many obvious cases have been silently corrected (e.g., Bootan to Bhutan, Darjiling to Darjeeling, Neilgherries to Nilgiri Hills).

Under the current rules, unless the author explicitly designated one particular specimen as the holotype, duplicate specimens are all to be considered syntypes, even those that may not have been seen by the author (McNeill, 2014). Until recently, the specimen that was deemed most plausible to be the one used by the author was routinely designated as the "holotype." As McNeill (2014) points out, in many cases such specimens should in fact be considered lectotypes.

In this checklist we have often copied the provenance and status of types from revisions and monographs, and we are aware that these often use "holotype" where "lectotype" or "syntype" would have been correct. However, since lectotypification under the current rules of botanical nomenclature (Turland et al., 2018) is a formal process, we could not simply turn all the misapplied "holotypes" into lectotypes. *It is not our intention to typify any name in this work*, as we believe this is best done in the context of a taxonomic revision by a specialist who has examined all the relevant material. The only exception here is in the case of *Bulbophyllum sterile* (Lam.) Suresh, where we have designated an epitype.

In view of the above, the user of this checklist should keep in mind that many "holotypes" here listed are in fact isotypes or syntypes and may or may not be formally designated as lectotypes in future.

We have not aimed at completeness in listing the herbaria that hold type specimens; we have used whatever information was at hand. Only in the collections at the Royal Botanic Gardens, Kew, did we actively search for type specimens. Even here, we have undoubtedly overlooked some. Where we believe that the type specimen should be in Kew, but failed to find it, we have added "not found." Where we suspect that a type specimen may still be in some herbarium, but has not been located by us, we have not given an indication of the whereabouts of the specimen(s). We use the standard herbarium abbreviations of *Index Herbariorum* (Thiers, 2021).

DISTRIBUTION

For its size, India is not unusually rich in orchid species, considering that Orchidaceae are a predominantly tropical plant family with about 27,000 known species. Although 1234 is a respectable number, it is relatively low in terms of species per area compared with many other orchid-rich tropical countries, as Table 4 demonstrates (in part based on data from Kurzweil & Lwin, 2014; Zhou et al., 2016; Gale et al., 2018; Hassler, 2019).

However, the table also shows that certain states within India do have a relatively high orchid diversity, as here exemplified by Arunachal Pradesh and Kerala (Table 5). Evidently, the species richness is quite unevenly distributed. This is true as well for other large countries, such as Brazil, China, and Mexico, where orchid diversity is similarly concentrated in hotspots. In India, such hotspots are clearly correlated with mountainous areas with high rainfall, especially the northeastern Himalaya ranges and the Western and Eastern Ghats.

The figures presented here are necessarily approximations only. In many states there are areas that have not yet been thoroughly explored; new records, and

Table 4.   Orchid richness of selected countries and states.

Region	Number of orchid species / 1000 km²
India	0.38
Kerala	6.99
Arunachal Pradesh	7.09
China	0.16
Brazil	0.32
Mexico	0.67
Myanmar	1.18
Thailand	2.37
Laos	2.89
Vietnam	3.21
Papua New Guinea	4.89
Ecuador	14.76
Costa Rica	29.45

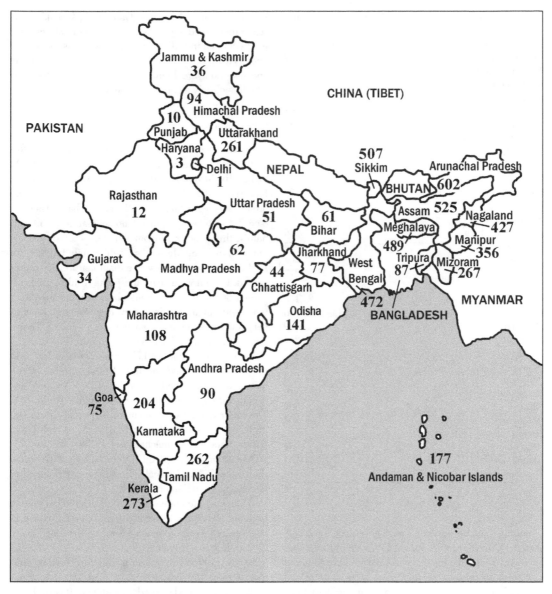

Map of India, showing number of species recorded from each state.

even species new to science, still turn up regularly. As we have not searched in Indian herbaria, unpublished records to be found there have not been included. Another reason our numbers may be off is that the delimitation of states has changed over time, so that, for example, what used to be called Assam may refer to present-day Meghalaya as well as Assam. Old records from "Sikkim" could refer to West Bengal, Bhutan, or present-day Sikkim. "Sylhet" could be in Bangladesh, Meghalaya, or Assam. Uttarakhand was split off from Uttar Pradesh in 2000; earlier records from Uttar Pra-

desh mostly refer to Uttarakhand (high elevation species are almost certainly from there). It is not always easy or even possible to tell in such cases what the actual range of the species is. Finally, species may have been recorded through mistaken identifications. (Note added in proof: The papers by Gogoi et al. [2015] and Rao & Kumar [2018] came too late to our attention; these authors record 398 species from Assam and 386 species from Manipur respectively.)

Ideally, each record should be backed up by a voucher specimen in a public herbarium, so that iden-

Table 5.   Number of orchid species recorded from each state.

Andaman & Nicobar Islands	177
Andhra Pradesh	90
Arunachal Pradesh	602
Assam	525
Bihar	61
Chhattisgarh	44
Dadra & Nagar Haveli (not shown on map)	7
Daman & Diu (not shown on map)	2
Delhi	1
Goa	75
Gujarat	34
Haryana	3
Himachal Pradesh	94
Jammu & Kashmir	36
Jharkhand	77
Karnataka	204
Kerala	273
Madhya Pradesh	62
Maharashtra	108
Manipur	356
Meghalaya	489
Mizoram	267
Nagaland	427
Odisha	141
Punjab	10
Rajasthan	12
Sikkim	507
Tamil Nadu	262
Tripura	87
Uttar Pradesh	51
Uttarakhand	261
West Bengal	472

tifications can be verified and, where necessary, corrected. A future orchid flora of India should be based on verified records; for this checklist we have followed the literature unless we had grounds to mistrust and reject identifications. Where records are unlikely to be due to misidentification but do not appear to be verifiable, we have added a question mark to the name of the state. In our checklist, distribution data outside India are limited to occurrence in neighboring countries, except for non-endemic species that are not found in any of the neighboring countries. For full, country-level distribution data we refer to *Plants of the World Online* (POWO, 2021) and WCSP (2021).

At species level and below, endemism in the Indian orchid flora is about 24%, with 292 endemic species, one endemic variety, and one endemic subspecies. This is a relatively high percentage compared with most other continental Asian countries. For example, in Myanmar the percentage of endemic orchid species is ca. 10% of ca. 800 species (Kurzweil & Lwin, 2014) and in Laos it

is only 3% of 683 species (Gale et al., 2018). As already mentioned above, two orchid genera are considered endemic in India: *Aenhenrya*, which is monotypic and terrestrial, and *Smithsonia*, with three epiphytic species. Both are known from southwest India only, which is clearly a center of endemism.

CONSERVATION

Relatively few Indian orchid species have been formally assessed using IUCN Red List criteria (IUCN, 2012). By November 2019, the IUCN Red List database (IUCN Red List, 2021) contained only 40 Indian orchid species, which had all been globally assessed (see Table 6, with updated nomenclature). Unfortunately, this is not a random sample, as all slipper orchids (Cypripedioideae) were deliberately targeted for Red Listing, so that these are disproportionally represented. Even if we discard these, the resulting list still cannot be considered a random sample, because relatively widespread species are over represented, whereas very few Indian endemic species are included. Therefore, not many conclusions can be drawn from it.

We note that almost all slipper orchids in India, in the genera *Cypripedium* and *Paphiopedilum*, fall in one of the threatened categories (CR, EN, and VU), except for *Cypripedium tibeticum* King ex Rolfe (LC). The latter is relatively common in parts of China and Tibet, but less so in India, where it is only recorded from Uttarakhand and Arunachal Pradesh. Two species, *Paphiopedilum druryi* (Bedd.) Stein and *P. fairrieanum* (Lindl.) Stein, are Critically Endangered (CR), the former being an Indian endemic.

As most of the non-slipper orchids in the list tend to be more widespread species, there is probably a bias towards relatively common species, with only about 20% falling in a threatened category.

Sarkar (1995c), based on a literature review, enumerated ca. 550 Indian orchid species that he categorized as rare, endangered, or endemic. It is probably safe to assume that most of the endemic species would fall in one of the threatened categories, if only because the majority have a restricted range. Therefore, as a rough estimate it may well be the case that at least about 500 Indian orchid species, or 40%, fall in one of the globally threatened categories according to the IUCN Red List criteria. This is likely to be an underestimate, considering that Bhattacharjee et al. (2018) report that in the Barak Valley in southern Assam out of 61 recorded species only five were found to be still common, all others being rare or endangered within the area.

Unfortunately, primary forest destruction in India is ongoing and shows no signs of abating (Deforestation Statistics, 2021). Levels of primary forest loss were at

Table 6. Indian orchid species globally assessed according to IUCN Red List criteria as of 29 November 2019.

Brachycorythis wightii Summerh.	DD	*Neottia acuminata* Schltr.	LC
Bulbophyllum cauliflorum Hook. f.	LC	*Paphiopedilum charlesworthii* (Rolfe) Pfitzer	EN
Bulbophyllum delitescens Hance	LC	*Paphiopedilum druryi* (Bedd.) Stein	CR
Bulbophyllum leopardinum (Wall.) Lindl. ex Wall.	LC	*Paphiopedilum fairrieanum* (Lindl.) Stein	CR
Bulbophyllum macranthum Lindl.	LC	*Paphiopedilum hirsutissimum* (Lindl. ex Hook.) Stein	VU
Bulbophyllum restrepia (Ridl.) Ridl.	LC	*Paphiopedilum insigne* (Wall. ex Lindl.) Pfitzer	EN
Coelogyne rigida C. S. P. Parish & Rchb. f.	LC	*Paphiopedilum spicerianum* (Rchb. f.) Pfitzer	EN
Cryptochilus acuminatus (Griff.) Schuit., Y. P. Ng & H. A. Pedersen	LC	*Paphiopedilum venustum* (Wall. ex Sims) Pfitzer	EN
		Paphiopedilum villosum (Lindl.) Stein	VU
Cypripedium cordigerum D. Don	VU	*Paphiopedilum wardii* Summerh.	EN
Cypripedium elegans Rchb. f.	EN	*Peristylus aristatus* Lindl.	LC
Cypripedium himalaicum Rolfe	EN	*Phreatia plantaginifolia* (J. Koenig) Ormerod	LC
Cypripedium tibeticum King ex Rolfe	LC	*Podochilus khasianus* Hook. f.	LC
Dendrobium aphyllum (Roxb.) C. E. C. Fisch.	LC	*Pomatocalpa maculosum* (Lindl.) J. J. Sm.	LC
Dendrobium bensoniae Rchb. f.	LC	*Porpax pusilla* (Griff.) Schuit., Y. P. Ng & H. A. Pedersen	LC
Epipactis veratrifolia Boiss. & Hohen.	LC		
Erythrorchis altissima (Blume) Blume	LC	*Sirhookera lanceolata* (Wight) Kuntze	LC
Eulophia recurva (Roxb.) M. W. Chase, Kumar & Schuit.	LC	*Spiranthes sinensis* (Pers.) Ames s.l. [including *S. australis* (R. Br.) Lindl.]	LC
Habenaria dichopetala Thwaites	EN	*Taprobanea spathulata* (L.) Christenson	VU
Luisia volucris Lindl.	VU	*Vanda tessellata* (Roxb.) Hook. ex G. Don	LC
Malaxis muscifera (Lindl.) Kuntze	VU	*Zeuxine strateumatica* (L.) Schltr.	LC

their highest for over a decade in 2016 and 2017, at around 30,000 hectares/year. Orchids mainly occur in old growth ("primary") forest and other habitats with low levels of disturbance; they are slow to colonize planted and secondary forests, and this is often limited to a relatively small subset of species (Schuiteman, pers. obs. in Southeast Asia). Therefore, preserving the remaining primary forests is far more important for orchid conservation than planting new forests, although the latter may of course be beneficial in terms of climate and hydrology. In 2014, protected areas covered only 5% of India's territory (Pande & Arora, 2014). A few reserves were established specifically to protect orchids, for example the Sessa Orchid Sanctuary in Arunachal Pradesh, which harbors approximately 175 orchid species (Pande & Arora, 2014).

There are several orchid collections in botanic gardens in the country and the Botanical Survey of India established two orchidaria, at Shillong (Meghalaya) and at Yercaud (Tamil Nadu).

In view of the large number of threatened orchid species, conservation efforts need to be stepped up dramatically to decrease the risk of extinction. Measurable targets, in line with the Global Strategy for Plant Conservation (CBD, 2021), must be set and policies should be implemented to meet those targets.

KEY TO THE GENERA

For the terminology used in the following key see Beentje (2016). Parts of the key are adapted and modified from Ng et al. (2018; tribe Podochileae) and Schuiteman (2014; subtribe Aeridinae). Please note that the key may not apply to species from outside India.

A few genera have been included of which the occurrence in India has not been confirmed but which have sometimes been reported from there. These are marked with *. Names of some recently synonymized genera are given between square brackets.

24b. Plant without roots but with an underground coralloid rhizome; lip rounded at apex; pollinia 4, waxy . . . . . . *Corallorhiza*

25a(7). Pollinia mealy or sectile; plant without pseudobulbs or aerial roots 26

25b. Pollinia waxy or hard, plastic-like; plant with or without pseudobulbs and with or without aerial roots 65

26a. Pollinia inserted in 2 pockets, not covered by a detachable or hinged anther cap; sectile 27

26b. Pollinia or pollen covered by a single, detachable or hinged anther cap; sectile or mealy 42

27a. Lateral sepals spurred . . . .*Disperis*

27b. Lateral sepals not spurred 28

28a. Lip with 2 spurs . . . .*Satyrium*

28b. Lip with a single spur or not spurred 29

29a. Plant with underground rhizome, lacking tubers or finger-like swollen roots . . . . *Galearis*

29b. Plant without rhizome, with tubers or finger-like swollen roots 30

30a. Tubers lobed, claw-like or hand-like 31

30b. Tubers simple, possibly forming a cluster of elongate finger-like, but separate tubers 32

31a. Lip purple and spotted or lip brownish; spur broadest at the base, tapering towards the apex; stigma a single concave surface . . . . *Dactylorhiza*

31b. Lip not spotted, pink or purple; spur narrowest at the base, not tapering towards the apex; stigma 2-lobed . . . *Gymnadenia*

32a. Floral bracts large and broad, leaf-like . . . . *Brachycorythis*

32b. Floral bracts small or narrow 33

33a. Plant with a single, usually spotted leaf; lip purple. . . . . *Hemipilia*

33b. Not this combination of characters, leaves not spotted 34

34a. Tuber tapering to a root-like tip; lip simple . . . . *Platanthera*

34b. Tuber rounded at apex; lip simple or lobed. 35

35a. Petals much broader and longer than the dorsal sepal; inflorescence 1- or 2-flowered . . . . *Diplomeris*

35b. Petals not both much broader and longer than the dorsal sepal; inflorescence 1- to many-flowered 36

36a. Flowers pink or purple, lip with spots or stripes; stigma concave or slightly raised, not stalked. . . . . *Ponerorchis*

36b. Flowers green, white, or yellowish, if with pink or purple colors then lip not spotted and stigma lobes on 2 distinct stalks 37

37a. Inflorescence 1-flowered; flowers pale yellow; stigma lobes not stalked . . . .*Hsenhsua*

37b. Inflorescence 1- to many-flowered; if 1-flowered then flowers white and stigma lobes stalked 38

38a. Stigma lobes on slender stalks that are not adnate to the lip; petals lobed or entire . . . . *Habenaria*

38b. Stigma lobes not stalked (though possibly raised and projecting) or, if stalked, adnate to the base of the lip; petals entire 39

39a. Flowers more than 3 cm diam., white; lateral lobes of lip entire to deeply lacerate-fimbriate, not filiform; spur long and slender. . . . . *Pecteilis*

39b. Flowers less than 1.5 cm diam., white, green or yellowish, lip sometimes red; lateral lobes of lip entire, sometimes filiform; spur absent or present, short to long and slender 40

40a. Column on either side with an antenna-like appendage . . . .*Gennaria*

40b. Column without such lateral appendages 41

41a. Stigma lobes shortly stalked or knob-like, attached to base of lip; spur present, short to elongate . . . . *Peristylus*

41b. Stigma lobes pulvinate or stalked but then not attached to base of lip; spur absent or present, short to elongate . . . . *Herminium*

42a(26). Leaves and inflorescences on separate growths, usually not present simultaneously, arising from an underground tuber; leaves single, not longer than wide, in outline cordate, reniform or angular . . . . *Nervilia* (p.p.)

42b. Leaves and inflorescence on the same growth, present simultaneously (leaves sometimes withered at flowering time in *Cheirostylis* and *Zeuxine*), plant with or without underground tuber; leaves possibly single, variously shaped, but not angular in outline 43

43a. Leaves plicate, without distinctly reticulate venation 44

43b. Leaves not plicate, i.e., without rib-like veins and without folds along the veins; often with distinctly reticulate venation 48

44a. Inflorescences lateral or terminal; flowers crowded; rostellum well developed, pointing towards the column apex; pollinia attached to a viscidium 45

44b. Inflorescence terminal; flowers well-spaced; rostellum an inconspicuous transverse ridge or absent; pollinia without viscidium 46

45a. Lip spatulate, not strongly concave, not spurred. . . . .*Corymborkis*

45b. Lip boat-shaped, strongly concave, spurred or not . . . . *Tropidia*

46a. Leaves 2, (sub)opposite; lip bifid or entire at apex . . . . *Neottia* (p.p.)

46b. Leaves 2 or more, not (sub)opposite; lip entire at apex 47

47a. Flowers whitish; pedicel-with-ovary of almost equal thickness throughout . . . . *Cephalanthera*

47b. Flowers variously colored in combinations of green, yellow, and brown; pedicel-with-ovary clavate . . . . *Epipactis*

48a(43). Leaves 2, (sub)opposite; pollinia mealy . . . . *Neottia* (p.p.)

48b. Leaves 1 to several, not (sub)opposite; pollinia mealy or sectile or pollen not arranged in pollinia 49

49a. Flowering stems 1-leaved 50

49b. Flowering stems with 2 or more leaves or a leafless inflorescence arising from a basal cluster of leaves. 52

50a. Leaf longer than wide, tapering to the narrow base, without reticulate venation; pollen loose, not arranged in pollinia . . . . *Pogonia*

50b. Leaf about as long as wide or wider than long, with a broad, cordate base and with reticulate venation; pollen arranged in pollinia 51

51a. Flower pitcher-shaped; dorsal sepal much broader than lateral sepals; column without appendage below the stigma
....*Corybas*
51b. Flower spidery; dorsal sepal similar to the very narrow lateral sepals and petals; column with an appendage below
      the stigma...*Stigmatodactylus*
52a. Flower with the lip positioned above the column; sepals and petals very narrow; pollinia mealy ....*Cryptostylis*
52b. Flower with the lip positioned above or below the column; sepals and petals not very narrow; pollinia mealy or sectile
     53
53a. Flowers arranged in a spiral; pollinia mealy..*Spiranthes*
53b. Flowers not arranged in a spiral; pollinia sectile 54
54a. Lip spurred...55
54b. Lip not spurred...58
55a. Lip on either side with a dentate to long-fimbriate flange ....*Anoectochilus*
55b. Lip without lateral flanges..56
56a. Inflorescence lax, with a long peduncle; spur without glands or warts inside ....*Erythrodes*
56b. Inflorescence dense, peduncle short; spur with glands or warts inside 57
57a. Spur with 2 drumstick-shaped glands inside; lateral sepals porrect.....................................*Vrydagzynea*
57b. Spur with several small warts inside; lateral sepals spreading ....*Herpysma*
58a(54). Sepals connate ....*Cheirostylis*
58b. Sepals free 59
59a. Flowers with the lip positioned above the column ....*Hetaeria*
59b. Flowers with the lip positioned below the column 60
60a. Lip in the concave basal part with a patch of hair-like papillae ....*Goodyera*
60b. Lip in the concave basal part with 2 warts or with 2 small clusters of papillae.................................61
61a. Lip on either side of the long and narrow middle part with a dentate to long-fimbriate flange (sometimes reduced to
      1 or 2 teeth) ....*Odontochilus*
61b. Lip without dentate or fimbriate flange on the sides of the narrow middle part, or without a long narrow middle part
      between the apical lobules and the basal concave part of the lip 62
62a. Lip with a long and narrow tubular part below the apical lobule(s); inflorescence 1- to few-flowered..............63
62b. The middle part of the lip, between the concave basal part and the apical lobule(s), short, not tubular; inflorescence
      many-flowered 64
63a. Leaves marble-like variegated; column on the ventral side with an elongate appendage; lip at base with 2 rows of
      hair-like papillae ....*Aenhenrya*
63b. Leaves uniformly green, sometimes with pink or silvery midvein; column without an appendage on the ventral side;
      lip at base with 2 warts..*Myrmechis*
64a. Lip inside at the base with a low crest that terminates in a bilobed lamella; leaves with silvery white or pink midvein
      ....*Rhomboda*
64b. Lip inside at the base without a crest; leaves without differently colored midvein ....*Zeuxine*
65a(25). Plant sympodial 66
65b. Plant monopodial 149
66a. Plant terrestrial...67
66b. Plant epiphytic or lithophytic 98
67a. Pollinia 2 or 4...68
67b. Pollinia 8 86
68a. Leafy stems elongate, erect, many-leaved, cane-like 69
68b. Leafy stems short or, if elongate and many-leaved, creeping, not erect.......................................70
69a. Inflorescences lateral; column with a foot...*Dendrobium* (p.p.)
69b. Inflorescences terminal; column without a foot ....*Thunia* (p.p.)
70a. Inflorescences terminal from an exposed (not subterranean) leafy growth.......................................71
70b. Inflorescences lateral or basal, or terminal from a subterranean corm, or terminal from a leafless growth 74
71a. Flowers large and showy, 4 cm or more across; inflorescence 1- or 2-flowered ....*Pleione* (p.p.)
71b. Flowers minute to small, less than 2 cm across; inflorescence many-flowered 72
72a. Flower with the lip positioned below the column ....*Liparis* (p.p.)
72b. Flower with the lip positioned above the column 73
73a. Leaves plicate; lip with or without teeth along the margin ....*Crepidium* (p.p.)
73b. Leaves not plicate; lip without teeth along the margin ....*Malaxis*
74a(70). Flowers large and showy, 4 cm or more across; inflorescence 1- or 2-flowered (inflorescence arises from a young,
      eventually leafy shoot and is actually terminal); column at apex wing-like flattened, forming a hood which overtops
      the anther ....*Pleione* (p.p.)
74b. Flowers small to large and showy; inflorescence with 3 or more flowers; column not forming an apical hood which
      overtops the anther 75
75a. Sepals connate, forming a basal tube; flower with the lip positioned above the column ....*Anthogonium*
75b. Sepals free; flower with the lip positioned below the column 76
76a. Pollinia 4 77
76b. Pollinia 2 81
77a. Inflorescence arising from an exposed (not subterranean) pseudobulb 78
77b. Inflorescence arising from a subterranean corm 79

78a. Leaves conduplicate; inflorescence glabrous; lip concave due to the raised lateral margins or erect side-lobes, not
    mobile. . . . . *Cymbidium* (p.p.)

78b. Leaves plicate; inflorescence finely densely pubescent; lip convex, without side-lobes or raised margins, actively
    mobile (snaps upward when touched) . . . . *Plocoglottis*

79a. Lip at base saccate or spurred . . . . *Tipularia*

79b. Lip at base not saccate or spurred 80

80a. Flowers facing forwards, widely opening; sepals less than 15 mm long . . . . *Oreorchis*

80b. Flowers bent downwards, not widely opening; sepals more than 20 mm long . . . . *Cremastra*

81a(76). Growths few- to several-leaved; inflorescences arising laterally from the leafy growth 82

81b. Growths 1-leaved; inflorescences arising terminally from specialized leafless growths. 84

82a. Leaves conduplicate, plant without underground corms . . . . *Cymbidium* (p.p.)

82b. Leaves plicate or, if not plicate, plant with underground corms 83

83a. Inflorescence nodding at apex; lip bowl-shaped, not spurred, although sometimes with a short conical depression
    at the base. . . . . *Eulophia* (p.p.) [*Geodorum*]

83b. Inflorescence straight and erect at apex; lip with a distinct spur or, if not spurred, not bowl-shaped . . . . *Eulophia* (p.p.)

84a. Flower not spurred . . . . *Diglyphosa*

84b. Flower spurred 85

85a. Lip with strongly pleated margins at base, mobile; spur formed by the lip . . . . *Chrysoglossum*

85b. Lip without pleated margins at base, not mobile; spur formed by the column-foot with the lateral sepals . . . . . . *Collabium*

86a(67). Plant with cane-like, many-leaved stems and terminal inflorescences 87

86b. Stems not cane-like or inflorescences lateral 88

87a. Petals much broader than the sepals; leaves long-lived, not annually deciduous . . . . *Arundina*

87b. Petals as wide as the sepals; leaves annually deciduous. . . . . *Thunia* (p.p.)

88a. Plant with underground corms or fleshy rhizome. 89

88b. Plant with exposed stems or pseudobulbs 92

89a. Lip with a pair of calli in the center just below the mid-lobe . . . . *Spathoglottis*

89b. Lip without a pair of calli below the mid-lobe. 90

90a. Leaf bases forming an aerial pseudostem; floral bracts deciduous; column glabrous . . . . *Bletilla*

90b. Leaf bases not forming an aerial pseudostem; floral bracts persistent; column pubescent 91

91a. Flowers ca. 5 cm diam., widely opening; inflorescence 1- to few-flowered, not secund. *Ipsea*

91b. Flowers ca. 2 cm diam., not widely opening; inflorescence few- to many-flowered, secund. . . . . *Pachystoma* (p.p.)

92a(88). Sepals connate, forming an urn-shaped flower . . . . *Acanthophippium*

92b. Sepals free, flower not urn-shaped 93

93a. Column for its entire length connate to the lip . . . . *Calanthe* (p.p.)

93b. Column entirely free from the lip or connate to it only in the basal part 94

94a. Plant with cane-like stem with several to many leaves; inflorescence a lateral raceme with unspurred flowers . . . .
    . . . . *Calanthe* (p.p.) [*Cephalantheropsis*]

94b. Plant not with cane-like stem or flowers spurred. 95

95a. Flower with the lip positioned above the column; leaves marbled or with darker spots . . . . *Nephelaphyllum*

95b. Flower with the lip positioned below the column; leaves uniformly green or with light spots 96

96a. Growths with more than 1 leaf . . . . *Calanthe* (p.p.) [*Phaius*]

96b. Growths with 1 leaf only 97

97a. Pseudobulbs of more than 1 internode; inflorescence arising laterally from the leaf-bearing pseudobulb . . . . *Ania*

97b. Pseudobulbs of 1 internode; inflorescence arising on a specialized leafless shoot . . . . *Tainia*

98a(66). Pollinia 2 or 4. 99

98b. Pollinia 6 or 8. 124

99a. Leaves bilaterally flattened or subterete; inflorescence a terminal raceme of many minute flowers (less than 4 mm
    diam.) without a column-foot . . . . *Oberonia*

99b. Leaves dorsiventrally flattened or, if bilaterally flattened or subterete, not with many minute flowers and column-foot
    present 100

100a. Stems elongate (as long as or longer than the leaves), with many leaves and at least as many internodes. 101

100b. Stems short (shorter than the leaf), with 1 to many leaves, with 1 or more internodes. 103

101a. Inflorescence terminal; flowers large and showy; column without a foot . . . . *Thunia* (p.p.)

101b. Inflorescences lateral or, if terminal, flowers small and inconspicuous; column with a foot 102

102a. Leaves less than 1 cm long; pollinia attached to a funnel-shaped appendage; flowers minute (less than 4 mm long);
    stems thin, not thick and fleshy. . . . . *Podochilus*

102b. Leaves more than 1 cm long; pollinia naked, without any appendages; flowers minute to large; stems thin to thick
    and fleshy. . . . . *Dendrobium* (p.p.)

103a. Plant with pseudobulbs of 1 internode (sometimes small and button-shaped in *Bulbophyllum*), with 1 to 3 non-
    sheathing leaves at apex; inflorescence not branching 104

103b. Plant without pseudobulbs or with pseudobulbs of more than 1 internode, or with branching inflorescence; leaves
    sheathing or not 119

104a. Inflorescences arising from the base of the pseudobulb or from the rhizome. 105

104b. Inflorescence arising from the apex of the pseudobulb (possibly from a young growth that later develops into a
    pseudobulb with leaves; in that case, immature leaves are present surrounding the base of the inflorescence and
    older pseudobulbs will have remains of an inflorescence at the apex). 110

105a. Lip hinged with the column-foot, mobile . . . . *Bulbophyllum*
105b. Lip rigidly attached to the column or the column-foot 106
106a. Column at apex with 2 arm-like appendages; column-foot prominent, tubular; pollinia 2 . . . . *Thecostele*
106b. Column without apical appendages; column foot absent or not tubular; pollinia 4 107
107a. Lip adnate to the column over more than half the length of the column . . . . *Coelogyne* (p.p.) [*Dickasonia*]
107b. Lip largely free from the column 108
108a. Lip divided into a concave hypochile without side-lobes and a flat epichile . . . .*Coelogyne* (p.p.) [*Pholidota* (p.p.)]
108b. Lip not clearly divided into 2 parts 109
109a. Lip with erect side-lobes that clasp the column or lip with erect basal margins . . . . *Coelogyne* (p.p.)
109b. Side-lobes, if present, not clasping the column, basal margins of the lip not erect . . . . . *Coelogyne* (p.p.) [*Panisea* (p.p.)]
110a(104). Pseudobulbs annual, plant not forming a rhizome; leaves plicate; inflorescence 1- or 2-flowered; flowers large
         and showy . . . . *Pleione* (p.p.)
110b. Pseudobulbs longer-lived, attached to a rhizome; leaves plicate or conduplicate; inflorescence 1- to many flowered;
         flowers small to large and showy 111
111a. Lip divided into a concave hypochile and a flat epichile 112
111b. Lip not clearly divided into 2 parts 113
112a. Column short, not much longer than wide . . . . *Coelogyne* (p.p.) [*Pholidota* (p.p.)]
112b. Column long and slender, much longer than wide . . . . *Coelogyne* (p.p.) [*Otochilus*]
113a. Lip saccate or spurred at the base 114
113b. Lip not saccate or spurred at the base 115
114a. Lip with a distinct incurved spur; lateral sepals not saccate at base . . . . *Coelogyne* (p.p.) [*Bulleyia*]
114b. Lip shortly saccate at base, the sac not incurved; lateral sepals saccate at base . . . . *Coelogyne* (p.p.) [*Neogyna*]
115a. Inflorescence with many densely arranged, laterally flattened floral bracts in 2 opposite rows, with the flowers ap-
         pearing in succession over a long period . . . . *Stichorkis*
115b. Inflorescence without densely arranged, laterally flattened bracts; flowers opening largely simultaneously or inflo-
         rescence 1-flowered 116
116a. Column with a distinct foot; pollinia tightly cohering in 2 pairs . . . .*Dendrobium* (p.p.)
116b. Column without a foot; pollinia not cohering 117
117a. Lip with erect side-lobes that clasp the column or lip with erect basal margins . . . . *Coelogyne* (p.p.)
117b. Side-lobes, if present, not clasping the column, basal margins of the lip not erect 118
118a. Lip at least 1 cm long; petals lanceolate; inflorescence 1- to 8-flowered; rostellum projecting, table-like . . . .
         . . . . *Coelogyne* (p.p.) [*Panisea* (p.p.)]
118b. Lip usually less than 5 mm long, rarely up to ca. 1 cm long; petals linear; inflorescence few- to many-flowered;
         rostellum a simple rim . . . . *Liparis* (p.p.)
119a(103). Inflorescence branching 120
119b. Inflorescence not branching 122
120a. Lateral sepals entirely connate, forming a single synsepal . . . . *Acriopsis*
120b. Lateral sepals free or only connate at their base 121
121a. Pseudobulbs of 1 internode, hidden between the scale leaves; flower with the lip positioned below the column;
         column without a foot . . . . *Sirhookera*
121b. Pseudobulbs with more than 1 internode, exposed; flower with the lip positioned above the column; column with a
         foot . . . . *Polystachya*
122a. Lateral sepals forming a chin-like or spur-like structure with the column-foot; pollinia without a viscidium . . . . . .
         . . . .*Dendrobium* (p.p.)
122b. Column-foot absent, or, if present, no chin-like or spur-like structure formed by the lateral sepals; pollinia with or
         without a viscidium 123
123a. Flowers medium-sized to large (2.5–12 cm diam.); lip with erect sidelobes that more or less clasp the column;
         pollinia with a viscidium . . . . *Cymbidium* (p.p.)
123b. Flowers minute to small, less than 2 cm diam.; lip without erect sides, not clasping the column; pollinia without a
         viscidium . . . . *Liparis* (p.p.)
124a(98). Leaves and/or leaf sheaths covered with persistent, brownish or yellowish hairs . . . .*Trichotosia*
124b. Leaves and leaf sheaths glabrous or leaf sheaths or scale leaves covered with whitish hairs when young but becom-
         ing glabrous later on 125
125a. Stem elongate (as long as or longer than the leaves), with > 6 leaves 126
125b. Stem short (shorter than the leaf) or, if elongate, with < 7 leaves 131
126a. Flowers large and showy, > 10 cm diam. when spread . . . .*Thunia* (p.p.)
126b. Flowers small, less than 2 cm diam. 127
127a. Inflorescences very short, arranged in a head-like cluster at the stem apex . . . . *Agrostophyllum* (p.p.)
127b. Inflorescences lateral or terminal, solitary when short, sometimes fascicled when elongate-racemose, not forming
         head-like clusters 128
128a. Lip in basal half with a broad, rounded, blade-like appendage; peduncle glabrous; pollinia 6 . . . . *Appendicula*
128b. Lip in basal half without blade-like appendage; peduncle glabrous or pubescent; pollinia 8 129
129a. Inflorescences much shorter than the leaves, 1- to several-flowered . . . .*Cylindrolobus* (p.p.)
129b. Inflorescences as long as or longer than the leaves, many-flowered 130
130a. Inflorescences arising in terminal fascicles (2 or more simultaneously from the stem apex); lip with short, raised,
         farinose calli . . . . *Mycaranthes* (p.p.)

130b. Inflorescences arising from the upper internodes of the stem, 1 per internode, not in terminal fascicles; lip with a few low keels along its length, without farinose calli . . . . *Bambuseria*

131a(125). Inflorescences very short, arranged in a head-like cluster at the stem apex; stem with 3 or more leaves . . . . *Agrostophyllum* (p.p.)

131b. Inflorescences lateral or terminal, solitary when short, or, if in head-like clusters, stem 1-leaved 132

132a. Lip spurred; column entirely adnate to the lip . . . . *Calanthe* (p.p.)

132b. Lip not spurred, or, if spurred, column largely free from the lip 133

133a. Inflorescence tall, arising from the base of an ovoid pseudobulb with 2 to 3 apical leaves, often branched; flowers medium-sized (ca. 2.5 cm diam.), widely opening . . . . *Eriodes*

133b. Inflorescence short or tall, not branched, if arising from the base of a 2- to 3-leaved pseudobulb then flowers very small (< 5 mm diam.) and not widely opening. 134

134a. Inflorescence arising from the base of a pseudobulb, with many very small (< 5 mm) flowers 135

134b. Plant without pseudobulb or inflorescence arising above the base of a pseudobulb or, if inflorescence basal, not with many very small flowers 136

135a. Column with a foot; flower with a chin-like mentum . . . . *Phreatia* (p.p.)

135b. Column without a foot; flower without a mentum . . . . *Thelasis*

136a. Leaves 2 to 4 per stem, bilaterally flattened, subterete. . . . . *Strongyleria*

136b. Leaves not bilaterally flattened, if subterete then stems 1-leaved. 137

137a. Pseudobulbs club-shaped, laterally flattened; column-foot with a cushion-like callus . . . . *Callostylis*

137b. Plant with or without club-shaped pseudobulbs but these not laterally flattened; column-foot without a cushion-shaped callus 138

138a. Pseudobulbs flattened, broader than tall (button-shaped), or pseudobulbs about as broad as tall and truncate at apex (barrel-shaped), with the truncate surface broader than the leaf scar; pseudobulbs often covered with net-like, persistent veins. . . . . *Porpax*

138b. Pseudobulbs, if present, cylindrical to ovoid or subglobose, apex not truncate, except for the leaf scar; pseudobulbs not covered with net-like, persistent veins 139

139a. Inflorescences terminal (arising from the axil of the uppermost leaf of the pseudobulb or stem). 140

139b. Inflorescences all arising laterally or near-basally from the pseudobulb or stem (sometimes close to the apex, but then next to the leaf, i.e., on its abaxial side, not from the leaf axil) 144

140a. Inflorescences 1-flowered (sometimes fascicled); stems 1-leaved, stalk-like or rudimentary, not swollen into distinct pseudobulbs, consisting of 1 internode; stigma-lobes often on arm-like projections at column-apex . . . . *Ceratostylis*

140b. Not this combination of character states; stigma-lobes not on arm-like processes 141

141a. Leaves with long-sheathing bases, distributed along the stem, plant without pseudobulbs; lip with farinose calli . . . . *Mycaranthes* (p.p.)

141b. Leaves not sheathing at the base, or with short sheaths, all clustered (or singly) at the apex of the stem or pseudo-bulb; calli on lip, if any, not farinose. 142

142a. Floral bracts scarious (dead at anthesis), or bracts inconspicuous, much shorter than the pedicel-with-ovary, scale-like, appressed to the pedicel, not broader than the ovary; flowers not both numerous and secund . . . . *Eria* (p.p.)

142b. Floral bracts herbaceous, usually conspicuous, either as long as or longer than the pedicel-with-ovary, or, if shorter, bracts broader than ovary and patent to reflexed, or flowers numerous and secund 143

143a. Pseudobulb not completely enveloped in a sheath, largely exposed; sepals free or connate . . . . *Cryptochilus*

143b. Pseudobulb completely enveloped in a sheath; sepals free. . . . . *Pinalia* (p.p.)

144a(139). Inflorescences arising from the basal internode of a pseudobulb with more than 1 internode . . . . *Dendrolirium*

144b. Plant without pseudobulb or inflorescences arising from the upper internodes of a pseudobulb or from the apex of a pseudobulb with a single internode 145

145a. Inflorescence arising from a fully-grown shoot, with short rachis; floral bracts large, patent, usually with some sterile bracts present on the rachis. . . . . *Cylindrolobus* (p.p.)

145b. Inflorescence with elongate rachis, or, if with short rachis, floral bracts inconspicuous or inflorescence arising from young developing shoot; no sterile bracts on the rachis 146

146a. Plant without swollen stems or pseudobulbs, leaves more or less arranged in a fan; inflorescence glabrous. . . . . *Phreatia* (p.p.)

146b. Plant with swollen stems or pseudobulbs; leaves not arranged in a fan; inflorescence glabrous or pubescent 147

147a. Pseudobulbs with a single internode. . . . . *Eria* (p.p.)

147b. Pseudobulbs with several internodes 148

148a. Pseudobulbs 1-leaved. . . . . *Bryobium*

148b. Pseudobulbs or stems 2- or more-leaved. . . . . *Pinalia* (p.p.)

149a(65). Leaves terete or subterete. 150

149b. Leaves dorsiventrally flattened or plant leafless 155

150a. Leaves on a flowering plant much longer than the stem . . . . *Holcoglossum* (p.p.)

150b. Leaves on a flowering plant about as long as or shorter than the stem 151

151a. Lip not spurred. . . . . *Luisia*

151b. Lip spurred. 152

152a. Inflorescence 1- to few-flowered; flowers 2–10 cm diam. . . . . *Papilionanthe*

152b. Inflorescence 1- to many-flowered; flowers < 1.5 cm diam. (excluding the spur) 153

153a. Lip (excluding spur) longer than dorsal sepal; spur without ornaments on back wall; pollinia 4 in 2 unequal pairs . . . . *Schoenorchis* (p.p.)

183a. Inflorescence finely muricate . . . . *Grosourdya*
    [Remark: a few species currently in *Pteroceras* will key out here; it is not clear if they really belong in *Grosourdya*.]
183b. Inflorescence glabrous . . . . *Brachypeza*
184a. Inflorescence with floral bracts arranged in 2 ranks or quaquaversal; column at most slightly longer than wide,
    without a distinct lateral constriction below stigma . . . . *Thrixspermum* (p.p.)
184b. Inflorescence with quaquaversal floral bracts; column longer than wide, or, if shorter, with a distinct constriction
    below stigma (column pandurate in front view) . . . . *Phalaenopsis* (p.p.)
185a(181). Lip articulate with column foot; anther with 2 wiry lateral appendages . . . . *Chiloschista* (p.p.)
185b. Lip not articulate with column foot; anther without wiry lateral appendages . . . . *Phalaenopsis* (p.p.)
186a(180). Lip with 2 spurs . . . . *Diplocentrum*
186b. Lip with a single spur or sac 187
187a. Floral bracts almost extending to tip of dorsal sepal, finely pubescent . . . . *Cleisomeria* (p.p.)
187b. Floral bracts at most reaching half the length of dorsal sepal, glabrous 188
188a. Mouth of spur or sac largely closed by 1 or 2 calli 189
188b. Mouth of spur or sac unobstructed 195
189a. Rostellum raised and crest-like, extending across clinandrium . . . . *Sarcoglyphis*
189b. Rostellum tooth-like or beak-like, not raised, not extending across clinandrium 190
190a. Interior of back wall of spur with an obliquely erect lamella or an uncinate callus 191
190b. Interior of back wall of spur without ornaments or with a thick, cushion-shaped callus 192
191a. Interior of back wall of spur with an obliquely erect lamella inserted well inside the spur . . . . *Pomatocalpa* (p.p.)
191b. Interior of back wall of spur with an uncinate callus inserted near the mouth . . . . *Robiquetia* (p.p.)
192a. Interior of back wall of spur without a callus 193
192b. Spur on the back wall with a thick, cushion-shaped callus 194
193a. Leaves linear; column foot absent or indistinct . . . . *Vanda* (p.p.)
193b. Leaves elliptic; column foot well-developed . . . . *Phalaenopsis* (p.p.)
194a. Rostellum beak-like, almost as long as column . . . . *Stereochilus* (p.p.)
194b. Rostellum tooth-like, much shorter than column . . . . *Cleisostoma* (p.p.)
195a(188). Lip (excluding spur) longer than dorsal sepal; rostellum bifurcate, abruptly curved upwards, often surpassing
    anther; pollinia 4 . . . . *Schoenorchis* (p.p.)
195b. Lip (excluding spur) as long as or shorter than dorsal sepal; rostellum entire or bidentate, not abruptly curved up-
    wards, not surpassing anther; pollinia 2 or 4 196
196a. Mouth of spur situated near middle of lip . . . . *Phalaenopsis* (p.p.)
196b. Mouth of spur at base of lip 197
197a. Interior of back wall of spur with an obliquely erect lamella . . . . *Pomatocalpa* (p.p.)
197b. Interior of back wall of spur without lamellae 198
198a. Lip clearly divided into a blade-like epichile and a cup- or pitcher-like hypochile without distinct lateral lobes;
    front rim of hypochile sometimes with 2 hornlike extensions 199
198b. Lip otherwise 200
199a. Rachis of inflorescence not much longer than pedicel-with-ovary . . . . *Gastrochilus* (p.p.)
199b. Rachis of inflorescence clearly longer than pedicel-with-ovary . . . . *Smithsonia* (p.p.)
200a. Leaves linear to elliptic, flat, usually twisted at base through 90° . . . . *Saccolabiopsis*
200b. Leaves linear, strongly V-shaped in cross-section, not twisted at base through 90° 201
201a. Leaf tip bilobed or premorse . . . . *Vanda* (p.p.)
201b. Leaf tip narrowly obtuse to subacute . . . . *Holcoglossum* (p.p.)
202a(172). Pedicel arising from a longitudinal concavity delimited by sharp edges on rachis 203
202b. Rachis with smooth or not sharply delimited concavities 206
203a. Column with foot about as long as column or longer; lip mobile; pollinia 2, partly cleft, or 4 204
203b. Column foot absent or not as long as column; lip rigidly attached; pollinia 2, entire 205
204a. Lip continuing the line of column foot; rostellum tooth-like . . . . *Pteroceras* (p.p.)
204b. Lip bent upwards relative to column foot; rostellum beak-like . . . . *Macropodanthus* (p.p.)
205a. Stipe strongly spatulate; rostellum proboscis-like; spur usually much longer than wide . . . . *Pennilabium* (p.p.)
205b. Stipe suborbicular, linear or oblong; rostellum tooth-like; spur not much longer than wide . . . . *Trachoma* (p.p.)
206a(202). Floral bracts almost extending to tip of dorsal sepal, finely pubescent . . . . *Cleisomeria* (p.p.)
206b. Floral bracts at most reaching half the length of dorsal sepal, glabrous 207
207a. Interior of spur mouth with 1 or 2 horizontal calli that largely obstruct entrance 208
207b. Interior of spur mouth without calli that largely obstruct entrance, but possibly with a thin lamella on back wall or
    with calli lower down 216
208a. Back wall of spur without callus 209
208b. Back wall of spur with a callus 212
209a. Column bent downwards, with rostellum projecting into mouth of spur; stipe with a hook-like extension that carries
    the pollinia . . . . *Uncifera* (p.p.)
209b. Column not with rostellum projecting into mouth of spur; stipe without a hook-like extension 210
210a. Plant often well over 1 m tall, sometimes shorter; mid-lobe of lip tooth-like; pollinia 4 . . . . *Micropera* (p.p.)
210b. Plant < 1 m tall; mid-lobe of lip not tooth-like but suborbicular to flabellate, sometimes linear; pollinia 2 or 4 211
211a. Flowers ca. 5 mm diam.; column without a foot; pollinia 4 . . . . *Smitinandia*
211b. Flowers > 1 cm diam.; column with a (sometimes indistinct) foot; pollinia 2, incompletely cleft . . . . *Aerides* (p.p.)
212a(208). Rostellum beak-like 213

CHECKLIST

ORCHIDACEAE Adans., Fam. Pl. (Adanson) 2: 68. 1763, nom. cons.

TYPE: *Orchis* L.

———

Acampe Lindl., Fol. Orchid. 4(*Acampe*): 1. 1853, nom. cons.

TYPE: *Acampe multiflora* (Lindl.) Lindl. (*Vanda multiflora* Lindl. = *Acampe praemorsa* (Roxb.) Blatt. & McCann).

Acampe carinata (Griff.) S. G. Panigrahi, Taxon 34(4): 689. 1985. *Saccolabium carinatum* Griff., Not. Pl. Asiat. 3: 354. 1851. *Gastrochilus carinatus* (Griff.) Schltr., Repert. Spec. Nov. Regni Veg. 12: 314. 1913. TYPE: India, West Bengal, Serampore, 22 Nov. 1841, *Griffith s.n.* (not found). NEOTYPE: India, Assam, Cachar, Borail Wildlife Sanctuary, near Bihara, epiphytic on a mango tree, 12 Dec. 2010, *Barbhuiya 112206* (neotype, ASSAM, designated by Barbhuiya et al., 2017).

Acampe cephalotes Lindl., Fol. Orchid. 4(*Acampe*): 3. 1853. *Gastrochilus cephalotes* (Lindl.) Kuntze, Revis. Gen. Pl. 2: 661. 1891. *Saccolabium cephalotes* (Lindl.) Hook. f., Fl. Brit. India [J. D. Hooker] 6(17): 63. 1890. TYPE: India or Bangladesh, above Sylhet, 18 Nov. 1851, *Hooker & Thomson s.n.* [197] (holotype, K-LINDL; isotype, K).
Acampe excavata Lindl., Fol. Orchid. 4(*Acampe*): 3. 1853. TYPE: India, Concan, *Hooker (leg. Law) s.n.* (holotype, K-LINDL; isotype, K).
Acampe papillosa Lindl. var. *flava* A. P. Das, Katham & Nirola, Pleione 4: 155. 2010. *Acampe carinata* (Griff.) S. G. Panigrahi var. *flava* (A. P. Das, Katham & Nirola) S. Misra, Orchids India: 484. 2019. TYPE: India, West Bengal, Jalpaiguri, Duars, Nagrakata, 17 Nov. 2009, *Das & Katham 4193* (holotype, NBU).

Distribution. Andhra Pradesh, Arunachal Pradesh, Assam, Bihar, Jharkhand, Karnataka, Madhya Pradesh, Manipur, Meghalaya, Mizoram, Nagaland, Odisha, Tripura, Uttar Pradesh, Uttarakhand, West Bengal [Bangladesh, Bhutan, China, Myanmar, Nepal].

Habit. Epiphytic herb.

Acampe ochracea (Lindl.) Hochr., Bull. New York Bot. Gard. 6: 270. 1910. *Saccolabium ochraceum* Lindl., Edwards's Bot. Reg. 28 (Misc.): 2. 1842. *Gastrochilus ochraceus* (Lindl.) Kuntze, Revis. Gen. Pl. 2: 661. 1891. TYPE: Sri Lanka, *Loddiges s.n. (leg. Wilmot Horton s.n.)* (holotype, K-LINDL).

Acampe dentata Lindl., Fol. Orchid. 4(*Acampe*): 3. 1853. TYPE: Sikkim, *Hooker s.n.* [179] (holotype, K-LINDL). *Acampe griffithii* Rchb. f., Flora 55: 277. 1872. TYPE: Bhutan, *Griffith s.n.*[547] (holotype, W; isotype, K).

Distribution. Andhra Pradesh, Arunachal Pradesh, Assam, Karnataka, Kerala, ?Maharashtra, Manipur, Meghalaya, Mizoram, Nagaland, Odisha, Sikkim, Tripura, West Bengal [Bangladesh, Bhutan, Myanmar, Nepal, Sri Lanka].

Habit. Epiphytic herb.

Acampe praemorsa (Roxb.) Blatt. & McCann, J. Bombay Nat. Hist. Soc. 35(3): 495. 1932. *Epidendrum praemorsum* Roxb., Pl. Coromandel 1: 34, t. 43. 1795. *Cymbidium praemorsum* (Roxb.) Sw., Nova Acta Regiae Soc. Sci. Upsal. 6: 75. 1799. *Aerides undulata* Sm. in A. Rees, Cyclop. 39: Aerides no. 13. 1820. *Sarcanthus praemorsus* (Roxb.) Lindl. ex Spreng., Syst. Veg. (ed. 16) [Sprengel] 3: 721. 1826. *Sarcochilus praemorsus* (Roxb.) Spreng., Syst. Veg. (ed. 16) [Sprengel] 3: 721. 1826. *Saccolabium papillosum* Lindl., Edwards's Bot. Reg. 18: t. 1552. 1833 (excl. descr. et icon.), nom. superfl. *Rhynchostylis papillosa* (Lindl.) Heynh., Alph. Aufz. Gew. 2: 594. 1846, nom. superfl. *Acampe papillosa* (Lindl.) Lindl., Fol. Orchid. 4(*Acampe*): 2. 1853, nom. superfl. *Gastrochilus papillosus* (Lindl.) Kuntze, Revis. Gen. Pl. 2: 661. 1891. *Saccolabium praemorsum* (Roxb.) Hook. f., Fl. Brit. India [J. D. Hooker] 6(17): 62. 1890. TYPE: India, icon. "Thalia maravara" in Rheede, Hort. Malab. 12: t. 4 (lectotype, designated by Seidenfaden in Matthew, 1983).

Aerides rigida Buch.-Ham. ex Sm., Cycl. (Rees) 39: Aerides no. 12. 1818 (as "*rigidum*"). *Acampe rigida* (Buch.-Ham. ex Sm.) P. F. Hunt, Kew Bull. 24(1): 98. 1970. *Acampe praemorsa* (Roxb.) Blatt. & McCann var. *rigida* (Buch.-Ham. ex Sm.) Barbhuiya, D. Verma & Vik. Kumar, Phytotaxa 303: 274. 2017. TYPE: Nepal, icon. *Buchanan-Hamilton s.n.* (holotype, LINN).
Vanda multiflora Lindl., Coll. Bot. (Lindley) t. 38. 1826. *Acampe multiflora* (Lindl.) Lindl., Fol. Orchid. 4(*Acampe*): 1. 1853. TYPE: China, *cult. Cattley s.n.*, July 1822 (not found).
Vanda longifolia Lindl., Gen. Sp. Orchid. Pl.: 215. 1833. *Acampe longifolia* (Lindl.) Lindl., Fol. Orchid. 4(*Acampe*): 1. 1853. *Saccolabium longifolium* (Lindl.) Hook. f., Fl. Brit. India [J. D. Hooker] 6(17): 62. 1890. *Gastrochilus longifolius* (Lindl.) Kuntze, Revis. Gen. Pl. 2: 661. 1891. TYPE: Myanmar, Tavoy [Dawei], 11 Oct. 1827, *Wallich 7322 (leg. W. Gomez 411)* (holotype, K-WALL; isotype, K-LINDL, icon.).

Vanda congesta Lindl., Edwards's Bot. Reg. Misc. 61. 1839. *Acampe congesta* (Lindl.) Lindl., Fol. Orchid. 4(*Acampe*): 2. 1853. *Gastrochilus congestus* (Lindl.) Kuntze, Revis. Gen. Pl. 2: 661. 1891. *Saccolabium congestum* (Lindl.) Hook. f., Fl. Brit. India [J. D. Hooker] 6(17): 63. 1890. *Acampe wightiana* (Lindl. ex Wight) Lindl. var. *longepedunculata* Trimen, Syst. Cat. Fl. Pl. Ceylon: 90. 1885. *Acampe praemorsa* (Roxb.) Blatt. & McCann var. *longepedunculata* (Trimen) Govaerts, Skvortsovia 4(3): 75. 2018. SYNTYPES: Madras, *Wight s.n.* (syntype, K-LINDL); Ceylon, *cult. Loddiges s.n.* (syntype, K-LINDL).

Vanda wightiana Lindl. ex Wight, Icon. Pl. Ind. Orient. [Wight] 5(1): 9, t. 1670. 1851. *Acampe wightiana* (Lindl. ex Wight) Lindl., Fol. Orchid. 4(*Acampe*): 2. 1853. *Saccolabium wightianum* (Lindl. ex Wight) Hook. f., Fl. Brit. India [J. D. Hooker] 6(17): 62. 1890. TYPE: India, Tamil Nadu, Iyamally Hills, *Wight s.n.* (holotype, K-LINDL).

Acampe intermedia Rchb. f., Berliner Allg. Gartenzeitung 24: 217. 1856. TYPE: Calcutta, *cult. Schiller s.n.* (holotype, W).

Distribution. Andaman & Nicobar Islands, Andhra Pradesh, Arunachal Pradesh, Assam, Bihar, Dadra & Nagar Haveli, Daman & Diu, Goa, Gujarat, Jharkhand, Karnataka, Kerala, Madhya Pradesh, Maharashtra, Manipur, Meghalaya, Mizoram, Nagaland, Odisha, Sikkim, Tamil Nadu, Tripura, Uttarakhand, West Bengal [Bangladesh, Bhutan, China, Myanmar, Nepal, Pakistan, Sri Lanka].

Habit. Epiphytic herb.

Remarks. We interpret this as a widespread and variable taxon. Studies throughout the range of the species are needed in order to establish if the variation warrants formal recognition, as suggested by Barbhuiya et al. (2017). If *Acampe rigida* is recognized as a variety of *A. praemorsa* the correct combination would be *A. praemorsa* var. *longepedunculata* (Trimen) Govaerts. The name *A. praemorsa* has often been misapplied to specimens of *A. carinata*. We consider the African *A. pachyglossa* Rchb. f. to be conspecific with *A. praemorsa.*

As noted by Seidenfaden (1977b), when Lindley described and illustrated *Saccolabium papillosum* he included *Epidendrum praemorsum* in the synonymy, rendering the name *S. papillosum* superfluous and making it a homotypic synonym of *Acampe praemorsa.* However, the species he illustrated as *S. papillosum* is what is now called *A. carinata.* His later combination *A. papillosa* was still superfluous, as he explicitly stated that it had the same synonyms as *S. papillosum* and the combination *A. praemorsa* was still available at that time.

———

Acanthophippium Blume, Bijdr. Fl. Ned. Ind. 7: 353. 1825.

TYPE: *Acanthophippium javanicum* Blume.

Acanthophippium bicolor Lindl., Edwards's Bot. Reg. 20: t. 1730. 1835. TYPE: Sri Lanka, *cult. Horticultural Society (leg. Watson) s.n.* (holotype, K).

Distribution. Karnataka, Kerala, Madhya Pradesh, Odisha, Tamil Nadu [Sri Lanka].

Habit. Terrestrial herb.

Acanthophippium striatum Lindl., Edwards's Bot. Reg. 24(Misc.): 41. 1838. TYPE: Nepal, *cult. Kew (leg. Bateman) s.n.* (holotype, K-LINDL).

Distribution. Assam, Arunachal Pradesh, ?Meghalaya, Nagaland, Sikkim, West Bengal [Bhutan, China, Myanmar, Nepal].

Habit. Terrestrial herb.

Acanthophippium sylhetense Lindl., Gen. Sp. Orchid. Pl.: 177. 1833. TYPE: Bangladesh or India, Sylhet, *Wallich s.n.* (holotype, K-LINDL).

Distribution. Arunachal Pradesh, Assam, Manipur, Meghalaya, Mizoram, Nagaland, Odisha, Sikkim, West Bengal [Bangladesh, China, Myanmar].

Habit. Terrestrial herb.

———

Acriopsis Blume, Bijdr. Fl. Ned. Ind. 8: 376. 1825.

TYPE: *Acriopsis javanica* Blume.

Acriopsis indica Wight, Icon. Pl. Ind. Orient. [Wight] 5(1): 20, t. 1748-1. 1851. TYPE: India, *Wight s.n.* (holotype, probably lost); Myanmar, Moulmein [Mawlamyine], *Parish 76* (neotype, K, designated by Minderhoud & de Vogel, 1986).

Distribution. Andaman & Nicobar Islands, Assam, Tripura [Myanmar].

Habit. Epiphytic herb.

Acriopsis liliifolia (J. Koenig) Ormerod, Opera Bot. 124: 58. 1995 var. **liliifolia**. *Epidendrum liliifolium* J. Koenig, Observ. Bot. (Retzius) 6: 61. 1791. TYPE: Not designated.

Acriopsis javanica Reinw. ex Blume, Bijdr. Fl. Ned. Ind. 8: 377. 1825. TYPE: Indonesia, Java, *Blume s.n.* (HLB 902-322-65) (lectotype, L, designated by Minderhoud & de Vogel, 1986).
Acriopsis harae Tuyama, J. Jap. Bot. 39(5): 129. 1964 (as "harai"). TYPE: India, Sikkim, near Gangtok [see note below], Apr. 1960; *Togashi s.n.* (holotype, TI).

Distribution. Andaman & Nicobar Islands, ?Sikkim [Myanmar].

Habit. Epiphytic herb.

Remarks. The only record for the Himalayan region (Sikkim) is the type of *Acriopsis harae*; as this was a cultivated specimen in a Japanese botanical garden, we are skeptical about its origin. This species has never been found since in Sikkim.

――――

Aenhenrya Gopalan, J. Bombay Nat. Hist. Soc. 90(2): 270. 1993 [1994].

TYPE: *Aenhenrya agastyamalayana* Gopalan (= *Aenhenrya rotundifolia* (Blatt.) C. S. Kumar & F. N. Rasm.).

Aenhenrya rotundifolia (Blatt.) C. S. Kumar & F. N. Rasm., Novon 7: 81. 1997. *Odontochilus rotundifolius* Blatt., J. Bombay Nat. Hist. Soc. 32: 521. 1928. *Anoectochilus rotundifolius* (Blatt.) N. P. Balakr., J. Bombay Nat. Hist. Soc. 63: 330. 1967. TYPE: India, Tamil Nadu, Madura Dist., High Wavy Mountain, May 1917, *Hallberg & Blatter 25802* (holotype, K).

Aenhenrya agastyamalayana Gopalan, J. Bombay Nat. Hist. Soc. 90: 271. 1993 [1994]. TYPE: India, Tamil Nadu, Tirunelveli Kattabomman District, Poonkulam, Agastyamalai Hills, 3600 ft, 24 Apr. 1991, *Gopalan 93224* (holotype, CAL; isotype, MH).

Distribution. Kerala, Tamil Nadu. Endemic.

Habit. Terrestrial herb.

――――

Aerides Lour., Fl. Cochinch. 2: 525. 1790.

TYPE: *Aerides odorata* Lour.

Aerides crassifolia C. S. P. Parish ex Burb., Garden (London 1871–1927) 3: 461. 1873. TYPE: Myanmar, Moulmein, *Parish 146* (holotype, K; isotype, W).

Distribution. Assam, Nagaland [Myanmar].

Habit. Epiphytic herb.

Aerides crispa Lindl., Gen. Sp. Orchid. Pl.: 239. 1833 (as "*crispum*"). TYPE: India, Courtallum, *Wallich 7319 (leg. Mitchell)* (holotype, K-LINDL; isotype, K-WALL).

Aerides brookei Bateman ex Lindl., Edwards's Bot. Reg. 27(Misc.): 55. 1841 (as "*brookeii*"). TYPE: *cult. Brooke s.n.* (not found).

Aerides lindleyana Wight, Icon. Pl. Ind. Orient. [Wight] 5(1): 9, t. 1677. 1851. TYPE: India, Nilgiri Hills, Kartairy falls below Kaitie, Apr., *Wight s.n.* [*2985*] (holotype, K).

Distribution. Dadra & Nagar Haveli, Goa, Gujarat, Karnataka, Kerala, Maharashtra, Tamil Nadu. Endemic.

Habit. Epiphytic herb.

Aerides emericii Rchb. f., Gard. Chron., n.s., 18: 586. 1882. TYPE: India; *Berkeley s.n.* (holotype, W).

Distribution. Andaman & Nicobar Islands. Endemic.

Habit. Epiphytic herb.

Aerides falcata Lindl. & Paxton, Paxton's Fl. Gard. 2: 142. 1851 (as "*falcatum*"). TYPE: "West Indies." June 1847, *cult. Larpent s.n.* (holotype, K-LINDL).

Distribution. Arunachal Pradesh, Assam, Madhya Pradesh [Myanmar].

Habit. Epiphytic herb.

Aerides maculosa Lindl., Edwards's Bot. Reg. 31: t. 58. 1845 (as "*maculosum*"). TYPE: Origin unknown, June 1844, icon. *cult. Rollissons s.n.* (holotype, K-LINDL).

Saccolabium speciosum Wight, Icon. Pl. Ind. Orient. [Wight] 5(1): t. 1674–1675. 1851. *Gastrochilus speciosus* (Wight) Kuntze, Revis. Gen. Pl. 2: 661. 1891. TYPE: India, Malabar District, Paulghaut, July–Aug., *Wight s.n.* [?*2986*] (syntypes, AMES, C, G, K [2×], L, M, NY, W).

Distribution. Andhra Pradesh, Goa, Gujarat, Karnataka, Kerala, Madhya Pradesh, Maharashtra, Odisha, Rajasthan, Tamil Nadu. Endemic.

Habit. Epiphytic herb.

Remarks. Specimens at Kew collected by Wight (2986 and s.n.) have Iyamalay or Iamally, Coimbatore (July 1847 & 1849) as collecting locality. It is probable that the published locality in the protologue, "about Paulghaut," refers to the same place.

Aerides multiflora Roxb., Pl. Coromandel 3: 68, t. 271. 1820 (as "*multiflorum*"). SYNTYPES: Bangladesh, Sylhet and India, Meghalaya, Garrow [Garo] Hills and vicinity, icon. *Aerides multiflorum* [sic], *Roxburgh 2346* [?origin] (syntypes, CAL, K).

Aerides affinis Wall. ex Lindl., Gen. Sp. Orchid. Pl.: 239. 1833. SYNTYPES: India, Sylhet, Purrarooah, Jan. 1829, *Wallich 7316A (leg. De Silva & Bruce)* (syntypes, K, K-LINDL, K-WALL); Nepal, Toka, Feb. 1821, *Wallich 7316B* (syntypes, K, K-LINDL, K-WALL).

Distribution. Andaman & Nicobar Islands, Andhra Pradesh, Arunachal Pradesh, Assam, Chhattisgarh, Hi-

machal Pradesh, Jharkhand, Madhya Pradesh, Maharashtra, Manipur, Meghalaya, Mizoram, Nagaland,
Odisha, Rajasthan, Sikkim, Tamil Nadu, Tripura, Uttar
Pradesh, Uttarakhand, West Bengal [Bangladesh, Bhutan, Myanmar, Nepal].

Habit.   Epiphytic herb.

Aerides odorata Lour., Fl. Cochinch. 2: 525. 1790.
   Epidendrum odoratum (Lour.) Poir., Encycl. (La-
   marck) Suppl. 1. 385. 1810. TYPE: Vietnam, Hue,
   Loureiro s.n. (holotype, BM; isotype, K).

Aerides cornuta Roxb., Fl. Ind. ed. 1832, 3: 472. 1832 (as
   "*cornutum*"). SYNTYPES: Bangladesh, Dacca [Dhaka]
   and India, eastern frontier of Bengal, icon. *Aerides cornutum* [sic], *Roxburgh 2093* [?origin] (syntypes, CAL, K).

Distribution.   Andaman & Nicobar Islands, Andhra
Pradesh, Arunachal Pradesh, Assam, Bihar, Chhat-
tisgarh, Jharkhand, Karnataka, Madhya Pradesh, Ma-
harashtra, Manipur, Meghalaya, Mizoram, Nagaland,
Odisha, Sikkim, Tamil Nadu, Tripura, Uttar Pradesh,
Uttarakhand, West Bengal [Bangladesh, Bhutan, China,
Myanmar, Nepal].

Habit.   Epiphytic herb.

Aerides ringens (Lindl.) C. E. C. Fisch., Fl. Madras
   3: 1442. 1928. *Saccolabium ringens* Lindl., Gen.
   Sp. Orchid. Pl.: 221. 1833. *Gastrochilus ringens*
   (Lindl.) Kuntze, Revis. Gen. Pl. 2: 661. 1891.
   TYPE: India, Madras, *Wallich 7313 (leg. Wight
   s.n.)* (holotype, K-WALL; isotype, K-LINDL, icon.).

Saccolabium lineare Lindl., Numer. List [Wallich] n. 7312.
   1832, nom. inval.
Saccolabium wightianum Lindl., Gen. Sp. Orchid. Pl.: 221.
   1833. *Gastrochilus wightianus* (Lindl.) Kuntze, Revis.
   Gen. Pl. 2: 661. 1891. SYNTYPES: East Indies, *Wallich
   7303 (leg. Wight)* (syntype, K-LINDL); *Wallich 7303
   (leg. Heyne)* (not found).
Aerides radicosa A. Rich., Ann. Sci. Nat., Bot. sér. 2, 15: 65.
   1841 (as "*radicosum*"). TYPE: India, Nilgiri Hills, Otacamund, May, *Perrottet s.n.* [*425 & 1096*] (syntypes,
   G, P).
Saccolabium rubrum Wight, Icon. Pl. Ind. Orient. [Wight]
   5(1): 9, t. 1673. 1851 (non Lindl.). TYPE: India, Nilgiri
   Hills, *Wight(?) s.n.* (synypes, K, K-LINDL).
Saccolabium paniculatum Wight, Icon. Pl. Ind. Orient. [Wight]
   5(1): 9, t. 1676. 1851. TYPE: India, Tamil Nadu, Iyamally Hills, Sep.–Oct., *Wight 2982* (holotype, K; isotype, LE, W).
Aerides lineare Hook. f., Fl. Brit. India [J. D. Hooker] 6(17):
   47. 1890. SYNTYPES: India Deccan Peninsula, Ghats
   from Canara, alt. 5000–7000 ft, *Wight s.n.*; Sri Lanka,
   Doombera District, *Thwaites s.n.*; *Wallich 7312 (leg.
   Heyne)* (syntypes, K-LINDL, K-WALL).

Distribution.   Andhra Pradesh, Goa, Gujarat, Karnataka, Kerala, Maharashtra, Odisha, Tamil Nadu [Sri
Lanka].

Habit.   Epiphytic herb.

Aerides rosea Lodd. ex Lindl. & Paxton, Paxton's Fl.
   Gard. 1: t. 60. 1850 (as "*roseum*"). TYPE: India,
   Loddiges 1530 (holotype, K-LINDL).

Aerides fieldingii B. S. Williams, Orch.-Grow. Man., ed. 2: 39.
   1862 (as "*fieldingi*"). TYPE: Not designated.
Aerides williamsii R. Warner, Select. Orchid. Pl. ser. 1, t. 21.
   1862. TYPE: Not designated.

Distribution.   Arunachal Pradesh, Assam, Manipur,
Meghalaya, Mizoram, Nagaland, West Bengal [Bhutan,
China, Myanmar].

Habit.   Epiphytic herb.

———

Agrostophyllum Blume, Bijdr. Fl. Ned. Ind. 8: 368.
   1825.

TYPE: *Agrostophyllum javanicum* Blume.

Agrostophyllum brevipes King & Pantl., Ann. Roy.
   Bot. Gard. (Calcutta) 8: 156, t. 213. 1898. TYPE:
   India, Sikkim, tropical valleys, May 1893, *Pant-
   ling 34* (holotype, K; isotype, AMES, BM, CAL,
   K, L, W).

Distribution.   Arunachal Pradesh, Assam, Meghalaya, Nagaland, Sikkim, West Bengal [Bhutan].

Habit.   Epiphytic herb.

Agrostophyllum callosum Rchb. f., Fl. Vit. [Seeman]
   296. 1868. TYPE: India, Sikkim, *Hooker s.n.* [*O.
   201*] (holotype, K-LINDL).

Distribution.   Arunachal Pradesh, Assam, Manipur,
Meghalaya, Mizoram, Nagaland, Sikkim, West Bengal
[Bhutan, China, Myanmar, Nepal].

Habit.   Epiphytic herb.

Agrostophyllum flavidum Phukan, J. Indian Bot.
   Soc. 69: 209. 1990. TYPE: India, Meghalaya,
   Khasi Hills, Shillong Peak, 29 May 1985, *S. Phu-
   kan 68259* (holotype, CAL; isotype, ASSAM).

Distribution.   Meghalaya. Endemic.

Habit.   Epiphytic herb.

Agrostophyllum myrianthum King & Pantl., Ann.
   Roy. Bot. Gard. (Calcutta) 8: 155, t. 211. 1898.
   TYPE: India, Sikkim, Teesta Valley, 1200 ft, Aug.
   1895, *Pantling 459* (holotype, K; isotype, BM,
   CAL, G).

Distribution. Arunachal Pradesh, Manipur, Sikkim, West Bengal. Endemic.

Habit. Epiphytic herb.

Agrostophyllum planicaule (Wall. ex Lindl.) Rchb. f., Ann. Bot. Syst. (Walpers) 6: 909. 1864. *Eria planicaulis* Wall. ex Lindl., Edwards's Bot. Reg. 26(Misc.): 8. 1840. TYPE: icon. *cult. Lemon s.n. (leg. Wallich) (1839)* (holotype, K-LINDL).

Agrostophyllum khasianum Griff., Calcutta J. Nat. Hist. 4: 376, t. 19. 1844 (as "*khasiyanum*"). TYPE: India, Khasi Hills, *cult. Calcutta (leg. Gibson) s.n.* (holotype, CAL).

Distribution. Andaman & Nicobar Islands, Arunachal Pradesh, Assam, Manipur, Meghalaya, Nagaland, Sikkim, West Bengal [Bhutan, Myanmar, Nepal].

Habit. Epiphytic herb.

———

Ania Lindl., Gen. Sp. Orchid.: 129. 1831.

TYPE: *Ania angustifolia* Lindl.

Ania angustifolia Lindl., Gen. Sp. Orch. Pl: 129. 1831. *Mitopetalum angustifolium* (Lindl.) Blume, Mus. Bot. 2: 185. 1856. *Tainia angustifolia* (Lindl) Benth. & Hook. f., Gen. Pl. 3: 515. 1883. *Ascotainia angustifolia* (Lindl.) Schltr., Repert. Spec. Nov. Regni Veg. Beih. 4: 246. 1919. TYPE: Myanmar, Tavoy [Dawei], Oct. 1827, *Wallich (leg. Gomez) 3740* (holotype, K-WALL; isotype, BM, E).

Distribution. Assam, Manipur [China, Myanmar].

Habit. Terrestrial herb.

Ania penangiana (Hook. f.) Summerh., Bot. Mag. 161: t. 9553. 1939. *Tainia penangiana* Hook. f., Fl. Brit. India [J. D. Hooker] 5(16): 820. 1890. *Ascotainia penangiana* (Hook. f.) Ridl., Mat. Fl. Malay. Penins. 1: 116. 1907. TYPE: Malaysia, Penang, *Maingay 1642* (holotype, K; isotype, CAL, L).

Tainia hookeriana King & Pantl., J. Asiat. Soc. Bengal, Pt. 2, Nat. Hist. 64: 336. 1895 [1896]. *Ascotainia hookeriana* (King & Pantl.) Ridl., Mat. Fl. Malay. Penins. 1: 116. 1907. *Ania hookeriana* (King & Pantl.) Tang & F. T. Wang ex Summerh., Bot. Mag. 161: t. 9553. 1939. TYPE: India, Sikkim Himalaya, valley of the Teesta, 1000 ft, Feb.–Mar. 1892, *R. Pantling s.n. [204]* (lectotype, BM, designated by Turner, 1992; isotype, CAL, FI, K, LE, P, W).

Ascotainia siamensis Rolfe ex Downie, Bull. Misc. Inform. Kew 1925: 378. 1925. TYPE: Thailand, Doi Sutep, 2200 ft, 17 Jan. 1911, *Kerr 214* (holotype, K; isotype, K).

Distribution. Arunachal Pradesh, Assam, Odisha, West Bengal [China, Nepal].

Habit. Terrestrial herb.

Ania viridifusca (Hook.) Tang & F. T. Wang ex Summerh., Bot. Mag. 161: t. 9553. 1939. *Calanthe viridifusca* Hook., Bot. Mag. 78: t. 4669. 1852. *Tainia viridifusca* (Hook.) Benth. ex Hook. f., Fl. Brit. India [J. D. Hooker] 5(16): 820. 1890. *Eria ania* Rchb. f., Ann. Bot. Syst. (Walpers) 6: 270. 1861. *Ascotainia viridifusca* (Hook.) Schltr., Orchideen (Schlechter) 317. 1914. TYPE: India, Assam, *cult. Kew (leg. Simon) s.n.* (not found).

Distribution. Assam, Manipur, Meghalaya, Nagaland, Uttar Pradesh [China, Myanmar].

Habit. Terrestrial herb.

———

Anoectochilus Blume, Bijdr. Fl. Ned. Ind. 8: 411. 1825 ("Anecochilus"), nom. et orth. cons.

TYPE: *Anoectochilus setaceus* Blume.

Anoectochilus brevilabris Lindl., Gen. Sp. Orchid. Pl.: 499. 1840 (as "*Anaectochilus*"). TYPE: India, Assam, *Mack s.n.* (holotype, K-LINDL; isotype, K).

Anoectochilus griffithii Hook. f., Fl. Brit. India [J. D. Hooker] 6(17): 96. 1890. SYNTYPES: Eastern Himalaya, *Griffith s.n.* (syntype, K); Assam, Naga Hills, July 1886, *Prain 10* (syntypes, CAL, K).

Anoectochilus sikkimensis King & Pantl., J. Asiat. Soc. Bengal, Pt. 2, Nat. Hist. 65: 124. 1896. TYPE: India, Sikkim, 3000–5000 ft, Aug. 1895, *Pantling 285* (syntypes, BM, BR, CAL, G, K [2×], L, P, W).

Distribution. Arunachal Pradesh, Assam, Meghalaya, Mizoram, Nagaland, Sikkim, West Bengal [Bhutan].

Habit. Terrestrial herb.

Remarks. Swami (2016, 2017) illustrated at least two additional species of *Anoectochilus* from Arunachal Pradesh, identified as *A. regalis* Blume, *A. setaceus*, and *A. sikkimensis*. They appear to represent different species not currently known from India. Members of this genus can be difficult to identify from photographs, as diagnostic characters reside in details of the column morphology that cannot be assessed without dissecting a flower.

Anoectochilus elatus Lindl., J. Proc. Linn. Soc., Bot. 1: 178. 1857. TYPE: India, Tamil Nadu, Otacamund, Wallaghaut, Jan.–May, *McIvor 59* (holotype, K-LINDL; isotype, K).

Distribution.   Karnataka, Kerala, Tamil Nadu [Sri Lanka].

Habit.   Terrestrial herb.

Anoectochilus narasimhanii Sumathi, Jayanthi, Karthig. & Sreek., Blumea 48: 285. 2003. TYPE: India, North Andamans, Saddle Peak National Park, 650 m, 20 Sep. 2001, *Sumathi, Jayanthi & Karthikeyan 17368* (holotype, CAL; isotype, PBL).

Distribution.   Andaman & Nicobar Islands. Endemic.

Habit.   Terrestrial herb.

Anoectochilus nicobaricus N. P. Balakr. & Chakr., Bull. Bot. Surv. India 20(1–4): 80. 1978 [1979]. TYPE: India, Great Nicobar Island, 6 km on east-west road, 15 m, 1 Dec. 1975, *P. Chakraborty 3226A* (holotype, CAL; isotype, PBL).

Distribution.   Andaman & Nicobar Islands. Endemic.

Habit.   Terrestrial herb.

Anoectochilus papillosus Aver., Taiwania 52: 287. 2007. TYPE: Vietnam, Hoa Binh Prov., Mai Chau Distr., Pa Co Municipality, Xa Linh village, Rung gia locality (old forests), around point 20°44′30″N, 104°56′26″E, 968 m, 23 July 2006, *Phan Ke Loc, Le Dong Tan, Nguyen Sinh Khang & Nguyen Tien Vinh, HAL 8978* (holotype, HN; isotype, LE).

Distribution.   Meghalaya [also in Vietnam].

Habit.   Terrestrial herb.

Remarks.   Singh et al. (2019) consider the Indian record of *Anoectochilus papillosus* by Singh (2015) to be referable to *A. roxburghii*. However, judging from Singh's (2015) photographs we agree with his identification.

Anoectochilus roxburghii (Wall.) Lindl., Ill. Bot. Himal. Mts. [Royle] 1: 368. 1839. *Chrysobaphus roxburghii* Wall., Tent. Fl. Napal. 37, t. 27. 1826. SYNTYPES: India, Sylhet, Nov. 1807, *Wallich 1220* (syntype, BM); Mt. Cachar, *H. R. Smith s.n.* (not found); Nepal, Sumbbunath & Pusputnath, *Anonymous s.n.* (not found); excl. Ceylon [Sri Lanka], *T. Hardwick s.n.* (not found); Amboina [Indonesia, Ambon], *Rumphius s.n.* (Herb. Amb. 6: 93, t. 41, fig. 3) [see *Remarks* below].

Distribution.   Arunachal Pradesh, Assam, Kerala, Manipur, Meghalaya, Mizoram, Nagaland, Sikkim, Uttarakhand, West Bengal [Bhutan, China, Myanmar, Nepal].

Habit.   Terrestrial herb.

Remarks.   The lectotypification by Bhattacharjee and Chowdhery (2014) of the name *Chrysobaphus roxburghii* must be rejected, as they selected an undated specimen at WU that cannot be considered original material with certainty. On the other hand, they correctly pointed out that one of the often-quoted "syntypes," *Wallich (leg. W. Gomez) 7387* from "Mt. Silhet" (Sylhet), was collected in 1828 by William Gomez. These specimens therefore postdate the publication of *C. roxburghii* in 1826 and cannot be type material. Unfortunately, Bhattacharjee and Chowdhery overlooked a specimen from Sylhet, *Wallich 1220* in BM, dated 11/1807, which clearly predates the publication of *C. roxburghii* and therefore would have made an acceptable choice of lectotype. Wallich's records in the protologue from Sri Lanka and Ambon are undoubtedly based on misidentifications and must be discarded. More detailed studies of all the available material are needed to carry out a proper lectotypification of this well-known orchid, which was clearly illustrated in Wallich's protologue.

Anoectochilus tetrapterus Hook. f., Fl. Brit. India [J. D. Hooker] 6(17): 96. 1890. *Odontochilus tetrapterus* (Hook. f.) Av. Bhattacharjee & H. J. Chowdhery, Novon 21: 20. 2011. TYPE: India, Manipur, Eerung, *C. B. Clarke 42191* (holotype, K).

Distribution.   ?Assam, Manipur. Endemic.

Habit.   Terrestrial herb.

———

Anthogonium Wall. ex Lindl., Gen. Sp. Orchid. Pl.: 425. 1840.

TYPE: *Anthogonium gracile* Wall. ex Lindl.

Anthogonium gracile Wall. ex Lindl., Gen. Sp. Orchid. Pl.: 426. 1840. TYPE: Nepal, Sheopore [Shivapuri], Sep. 1821, *Wallich 7398* (holotype, K-LINDL; isotype, K-WALL).

Distribution.   Arunachal Pradesh, Assam, Manipur, Meghalaya, Mizoram, Nagaland, Sikkim, Tripura, West Bengal [Bangladesh, Bhutan, China, Myanmar, Nepal, Sri Lanka].

Habit.   Terrestrial herb.

———

Aphyllorchis Blume, Tab. Pl. Jav. Orchid. ad fig. 77. 1825.

TYPE: *Aphyllorchis pallida* Blume.

Aphyllorchis alpina King & Pantl., Ann. Roy. Bot. Gard. (Calcutta) 8: 261, t. 347. 1898. TYPE: India, Sikkim, at Sin-ga-le-la & Mon Lepcha, alt. 14,000 ft, *Pantling 462* (lectotype, P-00345168, designated by Chakraborty et al., 2021; isotype, BM, CAL, G, K, W).

Aphyllorchis gollanii Duthie, J. Asiat. Soc. Bengal, Pt. 2, Nat. Hist. 71: 42. 1902 (as "*gollani*"). TYPE: Western Himalaya, Tehri Garwhal, Nag Tiba, 9000–10,000 ft, Aug. 1899, *Mackinnon's collector (Ramsukh) 23000* (lectotype, DD, designated by Chakraborty et al., 2021; isotype K).

Distribution. Arunachal Pradesh, Sikkim, Uttarakhand [China, Nepal].

Habit. Holomycotrophic terrestrial herb.

Remarks. The synonymy of *Aphyllorchis alpina* follows Chakraborty et al. (2021).

Aphyllorchis montana Rchb. f., Linnaea 41: 57. 1877 [1876]. SYNTYPES: Sri Lanka, Ambagumowa District, 1853, *Thwaites 3189* (syntypes, BM, K, PDA); Sri Lanka, *Mrs. Walker s.n.*

Aphyllorchis prainii Hook. f., Fl. Brit. India [J. D. Hooker] 6(17): 117. 1890. TYPE: India, Naga Hills, Aug. 1886, *Prain s.n. [68]* (holotype, K).

Distribution. Andhra Pradesh, Arunachal Pradesh, Assam, Karnataka, Kerala, Meghalaya, Mizoram, Nagaland, Tamil Nadu, West Bengal [China, Sri Lanka].

Habit. Holomycotrophic terrestrial herb.

Apostasia Blume, Bijdr. Fl. Ned. Ind. 8: 423. 1825.

TYPE: *Apostasia odorata* Blume.

Apostasia nuda R. Br., Pl. Asiat. Rar. (Wallich) 1: 76, t. 85. 1830. *Adactylus nudus* (R. Br.) Rolfe, Orchid Rev. 4: 329. 1896. TYPE: Malaysia, Penang, *Wallich (leg. Porter) 4449* (lectotype, K-LINDL, designated by de Vogel, 1969; isotype, K-WALL K001039034).

Distribution. Assam, ?Meghalaya [Bangladesh, Myanmar].

Habit. Terrestrial herb.

Apostasia odorata Blume, Bijdr. Fl. Ned. Ind. 8: 423. 1825. *Mesodactylis odorata* (Blume) Endl., Gen. Pl. [Endlicher] 221. 1837. TYPE: Indonesia, Java, Mt. Salak, *Blume s.n.* (lectotype, L, designated by de Vogel, 1969; isotype, L).

Distribution. Arunachal Pradesh, Assam, Meghalaya [China].

Habit. Terrestrial herb.

Apostasia wallichii R. Br., Pl. Asiat. Rar. (Wallich) 1: 75, t. 84. 1830. *Mesodactylis wallichii* (R. Br.) Endl., Gen. Pl. [Endlicher] 221. 1837. TYPE: Nepal, Noakote, Apr. 1821, *Wallich 4448* (lectotype, K-WALL K001039032, designated by de Vogel, 1969; isotype, E, K, K-WALL).

Distribution. Andaman & Nicobar Islands, Arunachal Pradesh, Assam, Meghalaya, Tripura [Bangladesh, Myanmar, Nepal, Sri Lanka].

Habit. Terrestrial herb.

Appendicula Blume, Bijdr. Fl. Ned. Ind. 7: 297. 1825.

TYPE: *Appendicula alba* Blume.

Appendicula cornuta Blume, Bijdr. Fl. Ned. Ind. 7: 302. 1825. *Podochilus cornutus* (Blume) Schltr., Mem. Herb. Boissier 21: 34. 1900. SYNTYPES: Indonesia, Java, Mt. Seribu & Mt. Panjar, *Blume s.n.* Probable type: *Blume s.n.* (no locality) (syntype, L).

Appendicula bifaria Lindl. ex Benth., Hooker's J. Bot. Kew Gard. Misc. 7: 35. 1855. TYPE: *Champion s.n.* (holotype, K-LINDL).

Distribution. Assam, Nagaland [China, Myanmar].

Habit. Epiphytic herb.

Appendicula nicobarica Jayanthi, Sumathi & Karthig., Edinburgh J. Bot. 68: 321. 2011. TYPE: India, Andaman & Nicobar Islands, Great Nicobar Island, Campbell Bay National Park, Mount Thullier, 30 May 2002, *Jayanthi, Sumathi & Karthigeyan 19343* (holotype, CAL; isotype, PBL).

Distribution. Andaman & Nicobar Islands. Endemic.

Habit. Epiphytic herb.

Appendicula reflexa Blume, Bijdr. Fl. Ned. Ind. 7: 301. 1825. *Podochilus reflexus* (Blume) Schltr., Mem. Herb. Boissier 21: 31. 1900. SYNTYPES: Indonesia, Java, Tjapus River, *Blume s.n.* (not found); Mt. Pantjar, *Blume s.n.* (syntypes, L, P).

Distribution. Andaman & Nicobar Islands [also in Indonesia].

Habit.   Epiphytic herb.

———

Arachnis Blume, Bijdr. Fl. Ned. Ind. 8: 365. 1825.

TYPE: *Arachnis moschifera* Blume, nom. superfl. (*Aerides arachnites* Sw. = *Arachnis flos-aeris* (L.) Rchb. f.).

Arachnis cathcartii (Lindl.) J. J. Sm., Natuurk. Tijd-schr. Ned.-Indië 72: 75. 1912. *Vanda cathcartii* Lindl., Fol. Orchid. 4(*Vanda*): 8. 1853 (as "*cath-carti*"). *Esmeralda cathcartii* (Lindl.) Rchb. f., Xenia Orchid. 2: 39. 1862 (as "*cathcarti*"). *Arachnanthe cathcartii* (Lindl.) Benth. & Hook. f., Gen. Pl. 3: 573. 1883. SYNTYPES: India, Sikkim, alt. 3000 ft, *J. D. Hooker s.n.* (syntype, CAL); alt. 2000–4000 ft, *Cathcart s.n.* (syntype, K-LINDL).

Distribution.   Arunachal Pradesh, Meghalaya, Na-galand, Sikkim, West Bengal [Bhutan, Nepal].

Habit.   Epiphytic herb.

Arachnis clarkei (Rchb. f.) J. J. Sm., Natuurk. Tijd-schr. Ned.-Indië 72: 76. 1912. *Esmeralda clarkei* Rchb. f., Gard. Chron., n.s., 26: 552. 1886. *Arachnanthe clarkei* (Rchb. f.) Rolfe, Gard. Chron. ser. 3, 4: 567. 1888. *Vanda clarkei* (Rchb. f.) N. E. Br., Bull. Misc. Inform. Kew 1888: 122. 1888. TYPE: India, Sikkim, Yoksun, alt. 8000 ft, 17 Oct. 1875, *C. B. Clarke s.n.* (holotype, W).

Distribution.   Arunachal Pradesh, Assam, Manipur, Meghalaya, Mizoram, Nagaland, Sikkim, West Bengal [Bhutan, China, Myanmar, Nepal].

Habit.   Epiphytic herb.

Remarks.   A specimen at Kew, *Clarke 25232*, was collected at 6000 ft on 11 Oct. 1875 and cannot be considered an isotype.

Arachnis labrosa (Lindl. ex Paxton) Rchb. f., Bot. Cen-tralbl. 28: 343. 1886. *Arrhynchium labrosum* Lindl. ex Paxton, Paxton's Fl. Gard. 1: 142. 1850. *Armo-dorum labrosum* (Lindl. ex Paxton) Schltr., Repert. Spec. Nov. Regni Veg. 10: 197. 1911. TYPE: Trop-ical Asia, *cult. Egerton s.n.* (holotype, K-LINDL).

Renanthera bilinguis Rchb. f., Xenia Orchid. 1: 7, t. 4. 1854. *Arachnanthe bilinguis* (Rchb. f.) Benth., J. Linn. Soc., Bot. 18: 332. 1881. TYPE: China, *cult. Booth s.n.* (holo-type, W).
Arachnis zhaoi Z. J. Liu, S. C. Chen & S. P. Lei, Acta Bot. Yunnan. 30: 529. 2008. *Arachnis labrosa* (Lindl. ex Pax-ton) Rchb. f. var. *zhaoi* (Z. J. Liu, S. C. Chen & S. P. Lei) S. C. Chen & J. J. Wood, Fl. China 25: 466. 2009. TYPE: China, Hainan, Changjiang, Qicha, Qichaling

Mountain, 600 m, 8 Sep. 2005, *Z. J. Liu 3055* (holotype, NOCC).

Distribution.   Arunachal Pradesh, Assam, Manipur, Meghalaya, Nagaland, West Bengal [Bhutan, China, Myanmar].

Habit.   Epiphytic herb.

Remarks.   *Arachnis labrosa* var. *zhaoi* seems noth-ing more than a form of *A. labrosa* that lacks brown pigment in the flower. This color form has been re-corded from Arunachal Pradesh (Jakha et al., 2015).

Arachnis senapatiana (Phukan & A. A. Mao) Kocyan & Schuit., Phytotaxa 161: 62. 2014. *Armodorum senapatianum* Phukan & A. A. Mao, Orchid Rev. 110: 299. 2002. TYPE: India, Assam (Manipur), Senapati Hills, *A. A. Mao 60271* (holotype, CAL).

Distribution.   Assam, Manipur, Nagaland. Endemic.

Habit.   Epiphytic herb.

———

Arundina Blume, Bijdr. Fl. Ned. Ind. 8: 401. 1825.

TYPE: *Arundina speciosa* Blume (designated by Garay & Sweet, 1974; = *Arundina graminifolia* (D. Don) Hochr.).

Arundina graminifolia (D. Don) Hochr., Bull. New York Bot. Gard. 6: 270. 1910. *Bletia graminifolia* D. Don, Prodr. Fl. Nepal.: 29. 1825. TYPE: Nepal, Suembu, 15 July 1802, *Buchanan-Hamilton s.n.* (holotype, BM).

Cymbidium bambusifolium Roxb., Fl. Ind. ed. 1832, 3: 460. 1832. *Arundina bambusifolia* (Roxb.) Lindl., Gen. Sp. Orchid. Pl.: 125. 1831. TYPE: Bangladesh, Chittagong, icon. *Cymbidium bambusifolium*, *Roxburgh 2336* (lecto-type, CAL, designated by Pearce & Cribb, 2002; iso-type, K).
Arundina chinensis Blume, Bijdr. Fl. Ned. Ind. 8: 402. 1825. TYPE: China, *Blume s.n.* (holotype, L).
Arundina affinis Griff., Not. Pl. Asiat. 3: 330. 1851. TYPE: Khasi Hills, Churra, 16 Oct. 1835, *Griffith in Assam Herb. 150* (holotype, K-LINDL).

Distribution.   Arunachal Pradesh, Assam, ?Him-achal Pradesh, Karnataka, Kerala, Manipur, Megha-laya, Mizoram, Nagaland, Sikkim, Tamil Nadu, Uttara-khand, West Bengal [Bangladesh, Bhutan, China, Myanmar, Nepal, Sri Lanka].

Habit.   Terrestrial herb.

———

Bambuseria Schuit., Y. P. Ng & H. A. Pedersen, Bot. J. Linn. Soc. 186: 193. 2018.

TYPE: *Bambuseria bambusifolia* (Lindl.) Schuit., Y. P. Ng & H. A. Pedersen (*Eria bambusifolia* Lindl.).

Bambuseria bambusifolia (Lindl.) Schuit., Y. P. Ng & H. A. Pedersen, Bot. J. Linn. Soc. 186: 193. 2018. *Eria bambusifolia* Lindl., J. Proc. Linn. Soc., Bot. 3: 61. 1859 [1858]. *Pinalia bambusifolia* (Lindl.) Kuntze, Revis. Gen. Pl. 2: 679. 1891. *Cylindrolobus bambusifolius* (Lindl.) Brieger, Orchideen (Schlechter) 1(11–12): 664. 1981. *Callostylis bambusifolia* (Lindl.) S. C. Chen & J. J. Wood, Fl. China 25: 359. 2009. SYNTYPES: India, Khasi Hills, alt. 2000 ft, *Griffith s.n.* (syntype, K-LINDL); *Simons 64 in herb. Hooker* (syntype, K-LINDL).

Distribution. Andhra Pradesh, Arunachal Pradesh, Assam, Manipur, Meghalaya, Mizoram, Nagaland, Odisha, Sikkim, West Bengal [China, Myanmar].

Habit. Epiphytic herb.

Bambuseria crassicaulis (Hook. f.) Schuit., Y. P. Ng & H. A. Pedersen, Bot. J. Linn. Soc. 186: 193. 2018. *Eria crassicaulis* Hook. f., Fl. Brit. India [J. D. Hooker] 5(16): 805. 1890. *Pinalia crassicaulis* (Hook. f.) Kuntze, Revis. Gen. Pl. 2: 679. 1891. *Trichotosia crassicaulis* (Hook. f.) Kraenzl., Bot. Jahrb. Syst. 101: 22. 1910. *Cylindrolobus crassicaulis* (Hook. f.) Brieger, Orchideen (Schlechter) ed. 3. 1(11–12): 664. 1981. TYPE: India, Khasi Hills, Pomrang, Hort Cave, alt. 4000–5000 ft, 13 Nov. 1850, *J. D. Hooker & J. Thomson s.n.* (holotype, K; isotype, K, K-LINDL).

Distribution. Assam, Meghalaya. Endemic.

Habit. Epiphytic herb.

———

Biermannia King & Pantl., J. Asiat. Soc. Bengal, Pt. 2, Nat. Hist. 66: 591. 1898 [1897].

TYPE: *Biermannia quinquecallosa* King & Pantl.

Biermannia arunachalensis A. N. Rao, Rheedea 16(1): 29. 2006. TYPE: India, Arunachal Pradesh, West Kameng District, Tipi, *A. N. Rao 34444* (holotype, Orchid Herbarium, Tipi).

Distribution. Arunachal Pradesh. Endemic.

Habit. Epiphytic herb.

Biermannia bimaculata (King & Pantl.) King & Pantl., Ann. Roy. Bot. Gard. (Calcutta) 8: 200, t. 267. 1898. *Sarcochilus bimaculatus* King & Pantl., J.

Asiat. Soc. Bengal, Pt. 2, Nat. Hist. 64: 340. 1895 [1896]. TYPE: India, Sikkim, Teesta, May 1892, *Pantling s.n.* [*209*] (holotype, CAL; isotype, K).

Distribution. Arunachal Pradesh, Meghalaya, Sikkim, West Bengal. Endemic.

Habit. Epiphytic herb.

Biermannia jainiana S. N. Hegde & A. N. Rao, Bull. Bot. Surv. India 26(1–2): 97, fig. 14. 1984 [1985]. TYPE: India, Arunachal Pradesh, West Kameng District, Sessa, alt. 3500 ft, 5 May 1982; *Hegde 4192* (holotype, Orchid Herbarium, Tipi).

Distribution. Arunachal Pradesh. Endemic.

Habit. Epiphytic herb.

Biermannia quinquecallosa King & Pantl., J. Asiat. Soc. Bengal, Pt. 2, Nat. Hist. 66: 591. 1898 [1897]. TYPE: India, Jaintia Hills, Jowai, 4000 ft, July, *Pantling 631* (not found).

Distribution. Assam, Meghalaya. Endemic.

Habit. Epiphytic herb.

———

Brachycorythis Lindl., Gen. Sp. Orchid. Pl.: 363. 1838.

TYPE: *Brachycorythis ovata* Lindl.

Brachycorythis acuta (Rchb. f.) Summerh., Kew Bull. 10(2): 238. 1955. *Gymnadenia acuta* Rchb. f., Otia Bot. Hamburg. 32. 1878. *Platanthera acuta* (Rchb. f.) Kraenzl., Orchid. Gen. Sp. 1: 611. 1899. *Phyllomphax acuta* (Rchb. f.) Schltr., Repert. Spec. Nov. Regni Veg. Beih. 4: 119. 1919. TYPE: Cambodia, Pursat Province, Prum Bat, Slap tien Koom, June 1875, *Godefroy-Leboeuf s.n.* (holotype, K; isotype, P).

Brachycorythis obovalis Summerh., Kew Bull. 10(2): 237. 1955. *Phyllomphax obovalis* (Summerh.) Szlach., Richardiana 6(2): 78. 2006. TYPE: Myanmar, Upper Chindwin District, Paungbyin Reserve, 600 ft, 21 Aug. 1908, *J. H. Lace 4222* (holotype, K; isotype, CAL).

Distribution. Mizoram [Myanmar].

Habit. Terrestrial herb.

Brachycorythis galeandra (Rchb. f.) Summerh., Kew Bull. 10(2): 241. 1955. *Platanthera galeandra* Rchb. f., Linnaea 25: 226. 1852. *Habenaria galeandra* (Rchb. f.) Benth., Fl. Hongk. 363. 1861. *Gymnadenia galeandra* (Rchb. f.) Rchb. f., Otia

Bot. Hamburg. 33. 1878. *Phyllomphax galeandra* (Rchb. f.) Schltr., Repert. Spec. Nov. Regni Veg. 16: 286. 1919. TYPE: China, Hongkong, *Fortune 78* (holotype, W; isotype, K, K-LINDL).

Distribution.   Manipur, Meghalaya, Mizoram, Tamil Nadu [China, Myanmar, Nepal].

Habit.   Terrestrial herb.

Brachycorythis helferi (Rchb. f.) Summerh., Kew Bull. 10(2): 235. 1955. *Gymnadenia helferi* Rchb. f., Flora 55: 276. 1872. *Habenaria helferi* (Rchb. f.) Hook. f., Fl. Brit. India [J. D. Hooker] 6(17): 164. 1890. *Platanthera helferi* (Rchb. f.) Kraenzl., Orchid. Gen. Sp. 1: 611. 1899. *Phyllomphax helferi* (Rchb. f.) Schltr., Repert. Spec. Nov. Regni Veg. Beih. 4: 119. 1919. TYPE: Myanmar, Terasserim, 1837, *Helfer s.n.* (holotype, W; possible isotype, CAL, K).

Distribution.   Assam, Meghalaya, Mizoram [Myanmar, Nepal].

Habit.   Terrestrial herb.

Brachycorythis iantha (Wight) Summerh., Kew Bull. 10(2): 238. 1955. *Platanthera iantha* Wight, Icon. Pl. Ind. Orient. [Wight] 5(1): 11, t. 1692. 1851. *Habenaria iantha* (Wight) Hook. f., Fl. Brit. India [J. D. Hooker] 6(17): 164. 1890 (as "*jantha*"). TYPE: India, Nilgiri Hills, Aug.–Sep., *Wight s.n.* (holotype, K; isotype, BM, CAL, K-LINDL).

Platanthera affinis Wight, Icon. Pl. Ind. Orient. [Wight] 5(1): 12, t. 1693. 1851. TYPE: India, Pulney Hills, Sep., *Wight s.n.* (holotype, K-LINDL).
Habenaria galeandra (Rchb. f.) Benth. var. *nilagirica* Hook. f., Fl. Brit. India [J. D. Hooker] 6(17): 164. 1890, p.p. SYNTYPES: India, Travancore, Pulney Hills, *Heyne s.n.*; *Wight s.n.*

Distribution.   Arunachal Pradesh, Assam, Kerala, Meghalaya, Tamil Nadu. Endemic.

Habit.   Terrestrial herb.

Brachycorythis obcordata (Lindl.) Summerh., Kew Bull. 10(2): 243. 1955. *Platanthera obcordata* Lindl., Gen. Sp. Orchid. Pl.: 290. 1835. *Orchis obcordata* Buch.- Ham. ex D. Don, Prodr. Fl. Nepal.: 23. 1825, nom. illeg., non P. Willemet, 1796. *Gymnadenia obcordata* (Lindl.) Rchb. f., Otia Bot. Hamburg. 32. 1878. *Habenaria obcordata* (Lindl.) Fyson, Fl. Nilgiri & Pulney Hill-Tops 1: 405. 1915. *Phyllomphax obcordata* (Lindl.) Schltr., Repert. Spec. Nov. Regni Veg. Beih. 4: 119. 1919. TYPE: Nepal, May, July & Aug. 1821, *Wallich 7050A* (holotype, K; isotype, K-WALL).

Habenaria galeandra (Rchb. f.) Benth. var. *major* Hook. f., Fl. Brit. India [J. D. Hooker] 6(17): 164. 1890. TYPE: India, Khasi Hills, Myrung, *Hooker & Thomson s.n.* (not found).

Distribution.   Arunachal Pradesh, Himachal Pradesh, Kerala, ?Meghalaya, Nagaland, Sikkim, Tamil Nadu, Uttarakhand [Bhutan, ?Myanmar, Nepal].

Habit.   Terrestrial herb.

Brachycorythis splendida Summerh., Kew Bull. 10(2): 240. 1955. *Phyllomphax splendida* (Summerh.) Szlach., Richardiana 6(2): 78. 2006. TYPE: India, Pulney Hills, Sep. 1826, *Wight s.n.* (holotype, K).

Distribution.   Kerala, Tamil Nadu. Endemic.

Habit.   Terrestrial herb.

Brachycorythis wightii Summerh., Kew Bull. 10(2): 242. 1955. *Phyllomphax wightii* (Summerh.) Szlach., Richardiana 6(2): 78. 2006. TYPE: India, Travancore, *Wight 1031* (holotype, K; isotype, K-LINDL).

Distribution.   Kerala, Tamil Nadu. Endemic.

Habit.   Terrestrial herb.

————

Brachypeza Garay, Bot. Mus. Leafl. 23: 163. 1972.

TYPE: *Brachypeza archytas* (Ridl.) Garay.

Brachypeza unguiculata (Lindl.) Kocyan & Schuit., Phytotaxa 161: 64. 2014. *Sarcochilus unguiculatus* Lindl., Edwards's Bot. Reg. 26(Misc.): 67. 1840. *Pteroceras unguiculatum* (Lindl.) H. A. Pedersen, Nordic J. Bot. 12(4): 388. 1992. *Thrixspermum unguiculatum* (Lindl.) Rchb. f., Xeni Orchid. 2: 122. 1868. TYPE: Philippines, Luzon, Manila; *Cuming s.n.* (holotype, K-LINDL).

Distribution.   Andaman & Nicobar Islands [also in Indonesia].

Habit.   Epiphytic herb.

————

Bryobium Lindl., Intr. Nat. Syst. Bot., ed. 2: 446. 1836.

TYPE: *Bryobium pubescens* Lindl.

Bryobium pudicum (Ridl.) Y. P. Ng & P. J. Cribb, Orchid Rev. 113: 272. 2005. *Eria pudica* Ridl.,

J. Linn. Soc., Bot. 32: 294. 1896. SYNTYPES: Singapore, Changi; *H. N. Ridley s.n.* (syntypes, K, SING); Peninsular Malaysia, Johore, Batu Pahat; Kuala Kahang; *H. N. Ridley s.n.* (syntype, SING).

Eria hindei Summerh., Bull. Misc. Inform. Kew 1932: 321. 1932. TYPE: India, Assam, *cult. Kew (leg. Hinde) s.n.* (holotype, K).

Distribution. Arunachal Pradesh, Assam, Manipur, Meghalaya, Sikkim, West Bengal [China].

Habit. Epiphytic herb.

————

Bulbophyllum Thouars, Hist. Orchid. tabl. esp. 3 sub u. 1822, nom. cons.

TYPE: *Bulbophyllum nutans* Thouars (typ. cons.).

Bulbophyllum aberrans Schltr., Repert. Spec. Nov. Regni Veg. 10: 177. 1911. *Bulbophyllum dischidiifolium* J. J. Sm. subsp. *aberrans* (Schltr.) J. J. Verm. & P. O'Byrne, Bulbophyllum Sulawesi: 216. 2011. *Diphyes aberrans* (Schltr.) Szlach. & Rutk., Acta Soc. Bot. Poloniae 77: 313. 2008. TYPE: Indonesia, Sulawesi, Toli-Toli, *Schlechter 20666* (holotype, B, lost).

Distribution. Andaman & Nicobar Islands [also in Indonesia, Malaysia].

Habit. Epiphytic herb.

Bulbophyllum acutiflorum A. Rich., Ann. Sci. Nat., Bot. sér. 2, 15: 18, t. 7. 1841. *Cirrhopetalum acutiflorum* (A. Rich.) Hook. f., Fl. Brit. India [J. D. Hooker] 5(16): 779. 1890. *Phyllorkis acutiflora* (A. Rich.) Kuntze, Revis. Gen. Pl. 2: 678. 1891 (as "*Phyllorchis*"). TYPE: India, Nilgiri Hills, Otacamund, Sep., *Perrottet s.n.* [*862 & 1134*] (syntypes, G [2×], P, W).

Cirrhopetalum albidum Wight, Icon. Pl. Ind. Orient. [Wight] 5(1): 7, t. 1653. 1851. *Bulbophyllum albidum* (Wight) Hook. f., Fl. Brit. India [J. D. Hooker] 5(16): 757. 1890. *Phyllorkis albida* (Wight) Kuntze, Revis. Gen. Pl. 2: 677. 1891 (as "*Phyllorchis*"). TYPE: India, Nilgiri Hills, near Kotergherry, St. Catherine's Falls, Aug.–Sep., *Wight s.n.* (holotype, K).

Distribution. ?Arunachal Pradesh, Karnataka, Kerala, Tamil Nadu. Endemic.

Habit. Epiphytic herb.

Bulbophyllum affine Lindl., Gen. Sp. Orchid. Pl.: 48. 1830. *Sarcopodium affine* (Wall. ex Lindl.) Lindl. & Paxton, Paxton's Fl. Gard. 1: 155. 1850. *Phyllorkis affinis* (Wall. ex Lindl.) Kuntze, Revis. Gen. Pl. 2: 677. 1891 (as "*Phyllorchis*"). TYPE: Nepal, Sheopore [Shivapuri], June 1821, *Wallich 1982* (holotype, K-LINDL; isotype, BM, E, G, K, K-WALL).

Distribution. Arunachal Pradesh, Assam, Meghalaya, Nagaland, Sikkim, Tripura, Uttarakhand, West Bengal [Bhutan, China, Myanmar, Nepal].

Habit. Epiphytic herb.

Bulbophyllum ambrosia (Hance) Schltr., Repert. Spec. Nov. Regni Veg. Beih. 4: 247. 1919 subsp. **ambrosia.** *Eria ambrosia* Hance, J. Bot. 21: 232. 1883. TYPE: Hong Kong, Victoria Peak, Mar. 1875, *Hance 22156 (leg. Ford)* (holotype, K; isotype, BM).

Distribution. Assam [China, Nepal].

Habit. Epiphytic herb.

Bulbophyllum ambrosia (Hance) Schltr. subsp. **nepalense** J. J. Wood, Kew Bull. 41(4): 820. 1986 (as "*nepalensis*"). TYPE: Nepal, Langtang Khola, near Sharpubes E of Trisuli River, Mar. 1981, *M. I. Williams 39* (holotype, K).

Distribution. ?Meghalaya [Nepal].

Habit. Epiphytic herb.

Bulbophyllum amplifolium (Rolfe) N. P. Balakr. & Sud. Chowdhury, Bull. Bot. Surv. India 9(1–4): 89. 1967 [1968]. *Cirrhopetalum amplifolium* Rolfe, Notes Roy. Bot. Gard. Edinburgh 8(36): 21, t. 10. 1913. SYNTYPES: China, Yunnan, banks of the Salwin, Salwin-Irrawadi divide, alt. 3000 ft, Nov. 1905, *G. Forrest 958* (syntype, E); *970*; *1117* (syntype, E).

Distribution. Arunachal Pradesh, Manipur [Bhutan, China, Myanmar].

Habit. Epiphytic herb.

Bulbophyllum andersonii (Hook. f.) J. J. Sm., Bull. Jard. Bot. Buitenzorg, sér. 2, 8: 22. 1912. *Cirrhopetalum andersonii* Hook. f., Fl. Brit. India [J. D. Hooker] 5(16): 777. 1890. *Phyllorkis andersonii* (Hook. f.) Kuntze, Revis. Gen. Pl. 2: 677. 1891 (as "*Phyllorchis*"). TYPE: India, Sikkim Himalaya, near Darjeeling, alt. 7000 ft, icon. *T. Anderson s.n.* (holotype, CAL; copy K).

Distribution. Arunachal Pradesh, Nagaland, Sikkim, West Bengal [China, ?Myanmar].

Habit. Epiphytic herb.

Bulbophyllum apodum Hook. f., Fl. Brit. India [J. D. Hooker] 5(16): 766. 1890. *Phyllorkis apoda* (Hook. f.) Kuntze, Revis. Gen. Pl. 2: 677. 1891 (as *"Phyllorchis"*). SYNTYPES: Malaysia, Malacca, 9 May 1867 or 1868, *Maingay s.n.* [3244] (syntype, K); Perak, top of Batu Kurau, *Scortechini s.n.* (syntypes, CAL, K).

Bulbophyllum ebulbum King & Pantl., J. Asiat. Soc. Bengal, Pt. 2, Nat. Hist. 64: 334. 1895 [1896]. TYPE: India, Sikkim, Sivoke, alt. 1000 ft, July, *King & Pantling s.n.* [323] (holotype, CAL; isotype, K).

Distribution. Andaman & Nicobar Islands, Arunachal Pradesh, Assam, Sikkim [Myanmar].

Habit. Epiphytic herb.

Bulbophyllum appendiculatum (Rolfe) J. J. Sm., Bull. Jard. Bot. Buitenzorg, sér. 2. 8: 22. 1912. *Cirrhopetalum appendiculatum* Rolfe, Bull. Misc. Inform. Kew 1901: 148. 1901. *Mastigion appendiculatum* (Rolfe) Garay, Hamer & Siegerist, Nordic J. Bot. 14(6): 637. 1994. TYPE: India, Sikkim, 1000 ft, Oct. 1892, *Pantling 197* (holotype, CAL; isotype, BM, K).

Cirrhopetalum ornatissimum sensu King & Pantl., Ann. Roy. Bot. Gard. (Calcutta) 8: 96, t. 133. 1898, non Rchb. f., 1882.

Distribution. Bihar, Manipur, Odisha, Sikkim [Bhutan].

Habit. Epiphytic herb.

Bulbophyllum arunachalense (A. N. Rao) J. J. Verm., Schuit. & de Vogel, Phytotaxa 166: 104. 2014. *Ione arunachalense* A. N. Rao, Rheedea 7: 48. 1997. *Sunipia arunachalensis* (A. N. Rao) J. M. H. Shaw, Orchid Rev. 122(1308): 78. 2014. TYPE: India, Arunachal Pradesh, Lower Subansiri District, Tale Valley, alt. 3840 ft, *Ingalhalli & Bennedict 20350* (holotype, Orchid Herbarium, Tipi).

Distribution. Arunachal Pradesh. Endemic.

Habit. Epiphytic herb.

Bulbophyllum aureoflavum Karupp. & V. Ravich., Taprobanica 5: 120. 2013. TYPE: India, Kerala, Munnar, on the road to Poopara near Gap Road, 12 May 2011, *S. Karuppusamy 1246* (holotype, MH; isotype, Sre Ganesan Herbarium at the Madura College).

Distribution. Kerala, Tamil Nadu. Endemic.

Habit. Epiphytic herb.

Remarks. Singh et al. (2019) consider *Bulbophyllum aureoflavum* to be a synonym of *B. fischeri*. This may be correct, but according to the protologue *B. aureoflavum* differs in having glabrous (vs. pubescent) sepals and acuminate (vs. obtuse) petals.

Bulbophyllum aureum (Hook. f.) J. J. Sm., Bull. Jard. Bot. Buitenzorg, sér. 2, 8: 22. 1912. *Cirrhopetalum aureum* Hook. f., Fl. Brit. India [J. D. Hooker] 5(16): 777. 1890. *Phyllorkis aurea* (Hook. f.) Kuntze, Revis. Gen. Pl. 2: 677. 1891 (as *"Phyllorchis"*). TYPE: India, Malabar, Wynaad, icon. *Jerdon s.n.* (holotype, K).

Distribution. Kerala, Tamil Nadu. Endemic.

Habit. Epiphytic herb.

Bulbophyllum balaeniceps Rchb. f., Hamburger Garten-Blumenzeitung 19: 280. 1863. *Rhytionanthos balaeniceps* (Rchb. f.) C. S. Kumar & Garay, Proc. 20th World Orchid Conf.: 114. 2013. TYPE: Origin unknown; *cult. Schiller s.n., 1863* (holotype, W).

Distribution. Kerala. Endemic.

Habit. Epiphytic herb.

Remarks. In view of their great similarity, it would be desirable to compare fresh material of this species with that of the recently described *Bulbophyllum indicum* (C. S. Kumar & Garay) Sushil K. Singh, Agrawala & Jalal. Records of *B. balaeniceps* from South America are undoubtedly erroneous.

Bulbophyllum bisetum Lindl., Ann. Mag. Nat. Hist. 10: 186. 1842. *Bolbophyllaria biseta* (Lindl.) Rchb. f., Ann. Bot. Syst. (Walpers) 6: 242. 1861. *Phyllorkis biseta* (Lindl.) Kuntze, Revis. Gen. Pl. 2: 677. 1891 (as *"Phyllorchis"*). TYPE: India, Khasi Hills, Chune, *Griffith s.n.* (holotype, K-LINDL; isotype, CAL, K [25 Oct. 1835]).

Distribution. Arunachal Pradesh, Meghalaya, Mizoram, West Bengal [also in Thailand].

Habit. Epiphytic herb.

Bulbophyllum blepharistes Rchb. f., Flora 55: 278. 1872. *Cirrhopetalum blepharistes* (Rchb. f.) Hook. f., Fl. Brit. India [J. D. Hooker] 5(16): 779. 1890. *Phyllorkis blepharistes* (Rchb. f.) Kuntze, Revis. Gen. Pl. 2: 677. 1891 (as *"Phyllorchis"*). *Tripudianthes blepharistes* (Rchb. f.) Szlach. & Kras, Richardiana 7: 95. 2007. TYPE: Myanmar, Tenasserim, Moulmein [Mawlamyine] *Parish s.n.* [99] (lectotype, K, designated by Clayton, 2017; isotype, W).

Cirrhopetalum longiscapum Teijsm. & Binn., Natuurk. Tijd-schr. Ned.-Indië 24: 310 1862 (as "*longescapum*"). *Phyllorkis longiscapa* (Teijsm. & Binn.) Kuntze, Revis. Gen. Pl. 2: 677. 1891 (as "*Phyllorchis*"). TYPE: Malaysia, Penang, *T. Lobb s.n.*

Distribution. Assam, Meghalaya [Myanmar].

Habit. Epiphytic herb.

Bulbophyllum bonaccordense (C. S. Kumar) J. J. Verm., Schuit. & de Vogel, Phytotaxa 166: 110. 2014. *Trias bonaccordensis* C. S. Kumar, Blumea 34(1): 105. 1989. TYPE: India, Kerala State, Trivandrum District, Bonaccord, 3500 ft, *C. S. Kumar 3668* (holotype, TBGT).

Distribution. Karnataka, Kerala, Tamil Nadu. Endemic.

Habit. Epiphytic herb.

Bulbophyllum candidum (Lindl.) Hook. f., Fl. Brit. India [J. D. Hooker] 5(16): 770. 1890. *Ione candida* Lindl., Fol. Orchid. 2(*Ione*): 3. 1853. *Sunipia candida* (Lindl.) P. F. Hunt, Kew Bull. 26(1): 183. 1971. *Phyllorkis candida* (Lindl.) Kuntze, Revis. Gen. Pl. 2: 677. 1891 (as "*Phyllorchis*"). TYPE: India, Khasi Hills, alt. 5000–6000 ft, *Hooker & Thomson s.n.* [*70*] (holotype, K-LINDL; isotype, K).

Distribution. Arunachal Pradesh, Manipur, Meghalaya, Nagaland, Sikkim, West Bengal [Bhutan, China, Myanmar].

Habit. Epiphytic herb.

Bulbophyllum capillipes C. S. P. Parish & Rchb. f., Trans. Linn. Soc. London 30: 150, t. 32a. 1874. *Phyllorkis capillipes* (C. S. P. Parish & Rchb. f.) Kuntze, Revis. Gen. Pl. 2: 677. 1891 (as "*Phyllorchis*"). TYPE: Myanmar, Moulmein [Mawlamyine], May 1872, *C. S. P. Parish s.n.* [*301*] (holotype, K; isotype, W).

Distribution. Arunachal Pradesh, Assam [Bhutan, Myanmar].

Habit. Epiphytic herb.

Bulbophyllum careyanum (Hook.) Spreng., Syst. Veg. (ed. 16) [Sprengel] 3: 732. 1826. *Anisopetalon careyanum* Hook., Exot. Fl. 2: t. 149. 1825. TYPE: Nepal, *Carey s.n.* (lost). NEOTYPE: Nepal, Toka, Oct. 1821, *Wallich 1990.1* (neotype, K-LINDL; isoneotype, G, K-WALL).

Bulbophyllum manipurense C. S. Kumar & P. C. S. Kumar, Rheedea 15(1): 12. 2005. TYPE: India, Manipur, Kasom Khullen, *C. Sathish Kumar 28751* (holotype and isotype, TBGT).

Distribution. Arunachal Pradesh, Assam, Kerala, Manipur, Meghalaya, Mizoram, Nagaland, Sikkim, Uttarakhand, West Bengal[Bhutan, Myanmar, Nepal].

Habit. Epiphytic herb.

Remarks. *Bulbophyllum manipurense* was found to fall within the range of variation of *B. careyanum* by Chowlu (2014).

Bulbophyllum cariniflorum Rchb. f., Ann. Bot. Syst. (Walpers) 6: 253. 1861. TYPE: India, Khasi Hills, *T. Lobb s.n.* (holotype, K-LINDL).

Bulbophyllum densiflorum Rolfe, Bull. Misc. Inform. Kew 1892: 139. 1892. TYPE: Eastern Himalaya?, Sep. 1891, *cult. N. Campany s.n.* (holotype, K).
Bulbophyllum flavidum Lucksom, J. Bombay Nat. Hist. Soc. 90: 71. 1993 (as "*flavida*"), nom. illeg., non Lindl., 1840. *Bulbophyllum pantlingii* Lucksom, J. Bombay Nat. Hist. Soc. 90: 551. 1993 [1994]. TYPE: India, Sikkim, Lachung Valley, Phyangla R. F., *Lucksom 207* (holotype, CAL; isotype, Gangtok Forest Dept. Herb.).

Distribution. Andhra Pradesh, Arunachal Pradesh, Assam, ?Meghalaya, Nagaland, Odisha, Sikkim, Uttarakhand, West Bengal [Bhutan, Myanmar, Nepal].

Habit. Epiphytic herb.

Bulbophyllum caudatum Lindl., Gen. Sp. Orchid. Pl.: 56. 1830. *Cirrhopetalum caudatum* (Lindl.) King & Pantl., Ann. Roy. Bot. Gard. (Calcutta) 8(1–2): 93. 1898, nom. illeg., non Wight, 1852. *Phyllorkis caudata* (Lindl.) Kuntze, Revis. Gen. Pl. 2: 677. 1891. TYPE: Nepal, Sankoo, June 1821, *Wallich 1983* (holotype, K-LINDL; isotype, G, K, K-WALL).

Bulbophyllum berenicis Rchb. f., Gard. Chron., n.s., 14: 588. 1880. TYPE: Origin unknown, *cult. Strickland s.n.* (holotype, W).

Distribution. Arunachal Pradesh, Meghalaya, Nagaland, Sikkim, West Bengal [China, Nepal].

Habit. Epiphytic herb.

Bulbophyllum cauliflorum Hook. f., Fl. Brit. India [J. D. Hooker] 5(16): 758. 1890. SYNTYPES: India, Sikkim, *Griffith's collectors s.n.* [*5165*] (syntype, CAL); Chungthang, 6000 ft, 9 Aug. 1849, *J. D. Hooker s.n.* [*36*] (syntype, K); Khasi Hills, *Griffith s.n.* (syntype, K).

Bulbophyllum collettii King & Pantl., J. Asiat. Soc. Bengal, Pt. 2, Nat. Hist. 66: 585. 1898 [1897]. SYNTYPES: India, Assam, 1 May 1896, *G. E. Rita s.n.* (syntype, CAL); *H. Collett s.n.* (not found).

Bulbophyllum cauliflorum Hook. f. var. *sikkimense* N. Pearce & P. J. Cribb, Edinburgh J. Bot. 58(1): 108. 2001. TYPE: India, Sikkim, Chungthang, 6000 ft, 9 Aug. 1849, *J. D. Hooker 36B* (holotype, K-LINDL; isotype, K).

Distribution. Arunachal Pradesh, Meghalaya, Sikkim, Nagaland, West Bengal [Myanmar].

Habit. Epiphytic herb.

Bulbophyllum clandestinum Lindl., Edwards's Bot. Reg. 27(Misc.): 77. 1841. TYPE: Singapore, *cult. Loddiges s.n.* (not found).

Epidendrum sessile J. Koenig, Observ. Bot. (Retzius) 6: 60. 1791, nom. illeg., non Sw. 1788. *Phyllorkis sessilis* Kuntze, Revis. Gen. Pl. 2: 676. 1891. *Bulbophyllum sessile* J. J. Sm., Orch. Java 6: 448. 1905. TYPE: Without precise locality, *J. Koenig s.n.* (holotype, K).

Distribution. Andaman & Nicobar Islands [Myanmar].

Habit. Epiphytic herb.

Bulbophyllum cornu-cervi King & Pantl., J. Asiat. Soc. Bengal, Pt. 2, Nat. Hist. 64: 332. 1895 [1896]. TYPE: India, Sikkim, Engo ridge, ca. 2500 ft, July, *Pantling s.n.* [*?264 from Engo, Apr. 1893*] (holotype, CAL; isotype, BM, G, K).

Distribution. Arunachal Pradesh, Sikkim [Myanmar].

Habit. Epiphytic herb.

Bulbophyllum crabro (C. S. P. Parish & Rchb. f.) J. J. Verm., Schuit. & de Vogel, Phytotaxa 166: 106. 2014. *Monomeria crabro* C. S. P. Parish & Rchb. f., Trans. Linn. Soc. London 30: 143. 1874. TYPE: Myanmar, near Taok; *Parish s.n.* [*312*] (Feb. 1871) (lectotype, W, designated by Clayton, 2017; isotype, K).

Monomeria barbata Lindl., Gen. Sp. Orchid. Pl.: 61. 1830. *Epicranthes barbata* (Lindl.) Rchb. f., Ann. Bot. Syst. (Walpers) 6: 265. 1861. TYPE: Nepal, Thoka, Oct. 1821, *Wallich 1978* (holotype, K-LINDL; isotype, BM, G, K).

Distribution. Arunachal Pradesh, ?Meghalaya, Nagaland, Sikkim, West Bengal [China, Myanmar, Nepal].

Habit. Epiphytic herb.

Bulbophyllum crassipes Hook. f., Fl. Brit. India [J. D. Hooker] 5(16): 760. 1890. *Phyllorkis crassipes* (Hook. f.) Kuntze, Revis. Gen. Pl. 2: 677. 1891 (as "*Phyllorchis*"). *Bulbophyllum careyanum* Spreng. var. *crassipes* (Hook. f.) Pradhan, Indian Orchids: Guide Identif. & Cult. 2: 713. 1979. SYNTYPES: India, Sikkim, Dulkaghar, Terai, 500 ft, 16 Oct. 1884, *C. B. Clarke 36904* (syntype, K);

Martaban, 28 Jan. 1827, *Wallich 1990.2* (syntypes, K, K-WALL); Myanmar, Arracau, without collector (syntype, CAL); Malaysia, Penang, without collector (syntype, K-LINDL).

Distribution. Andaman & Nicobar Islands, Arunachal Pradesh, Assam, Bihar, Jharkhand, ?Meghalaya, Odisha, Sikkim, Tripura, West Bengal [Bangladesh, Bhutan, Myanmar].

Habit. Epiphytic herb.

Bulbophyllum cupreum Lindl., Edwards's Bot. Reg. 24(Misc.): 95. 1838. *Bulbophyllum careyanum* (Hook.) Spreng. var. *ochracea* Hook. f., Fl. Brit. India [J. D. Hooker] 5(16): 760. 1890. *Phyllorkis cuprea* (Lindl.) Kuntze, Revis. Gen. Pl. 2: 677. 1891 (as "*Phyllorchis*"). TYPE: "Philippines" [almost certainly not the actual origin], *Cuming s.n.* (not found).

Distribution. Arunachal Pradesh [Myanmar].

Habit. Epiphytic herb.

Bulbophyllum cylindraceum Lindl., Gen. Sp. Orchid. Pl.: 53. 1830. *Phyllorkis cylindracea* (Lindl.) Kuntze, Revis. Gen. Pl. 2: 677. 1891 (as "*Phyllorchis*"). TYPE: India, icon. *Wallich s.n.* (not found).

Bulbophyllum imbricatum Griff., Not. Pl. Asiat. 3: 289. 1851, nom. illeg., non Lindl., 1841. SYNTYPES: India, Assam and Khasi Hills, Myrung, 12 Nov. 1835, *Griffith in Assam Herb. 138*; *234* (not found); *Griffith s.n.* [*151*] (probable syntypes, K, LE).

Distribution. Arunachal Pradesh, Assam, Manipur, Meghalaya, Mizoram, Nagaland, Sikkim, West Bengal [Bhutan, Myanmar, Nepal].

Habit. Epiphytic herb.

Bulbophyllum cylindricum King & Pantl., J. Asiat. Soc. Bengal, Pt. 2, Nat. Hist. 64(2): 333. 1895 [1896]. TYPE: Sikkim, Mungpoo, 3000 ft, June, *Pantling s.n.* (not found).

Distribution. Sikkim, West Bengal. Endemic.

Habit. Epiphytic herb.

Bulbophyllum delitescens Hance, J. Bot. 14: 44. 1876. *Cirrhopetalum delitescens* (Hance) Rolfe, Gard. Chron., n.s., 18: 461. 1882. TYPE: China, Hongkong, Mt. Victoria, *Ford 19111* (holotype, BM).

Bulbophyllum mannii Rchb. f., Flora 55: 275. 1872, nom. illeg., non Hook. f., 1864. *Bulbophyllum reichenbachianum* Kraenzl., Bot. Jahrb. Syst. 17: 49. 1893. *Cirrhopetalum mannii* Mukerjee, Notes Roy. Bot. Gard. Edinburgh

21: 151. 1953, nom. superfl. TYPE: India, Assam, *Mann s.n.* (holotype, W).

Cirrhopetalum setiferum Rolfe, Bull. Misc. Inform. Kew 1895: 35. 1895. *Bulbophyllum setiferum* (Rolfe) J. J. Sm., Bull. Jard. Bot. Buitenzorg, II, 8: 28. 1912. SYNTYPES: *cult. O'Brien s.n.* (Aug. 1891) (syntype, K); *cult. Lawrence s.n.* (Aug. 1892) (syntype, K).

Distribution. Arunachal Pradesh, Assam, Manipur, ?Meghalaya [China].

Habit. Epiphytic herb.

Bulbophyllum depressum King & Pantl., J. Asiat. Soc. Bengal, Pt. 2, Nat. Hist. 66: 585 1898 [1897]. TYPE: India, Khasi Hills, betw. Jowai & Jhorain, Jaintia Hills, 3000 ft, 19 June 1897, *Pantling 627* (holotype, CAL).

Bulbophyllum hastatum Tang & F. T. Wang, Acta Phytotax. Sin. 12(1): 44. 1974. TYPE: China, Hainan, Chang-Chiang Hsien, *S. K. Lau 1871* (holotype, PE).

Distribution. Arunachal Pradesh, Meghalaya [China].

Habit. Epiphytic herb.

Bulbophyllum dickasonii Seidenf., Dansk Bot. Ark. 33: 192. 1979. *Tripudianthes dickasonii* (Seidenf.) Szlach. & Kras, Richardiana 7: 95. 2007. TYPE: Myanmar, Chin Hill, Kanpetlet, 7500 ft, Apr. 1939, *Dickason 8444* (holotype, AMES).

Distribution. Assam [Myanmar].

Habit. Epiphytic herb.

Bulbophyllum disciflorum Rolfe, Bull. Misc. Inform. Kew 1895: 7. 1895. *Trias disciflora* (Rolfe) Rolfe, Hand-List Orch. Cult. Roy. Gard. Kew, ed. 2: 215. 1896. TYPE: Thailand or Laos, Oct. 1894, *Linden cult s.n.* (holotype, K).

Distribution. Andaman & Nicobar Islands, Arunachal Pradesh [also in Thailand, etc.].

Habit. Epiphytic herb.

Bulbophyllum elassonotum Summerh., Bot. Mag. 158: t. 9408. 1935. TYPE: India, Assam, Kanrup, *cult. Kew s.n. (leg. Hinde)* (holotype, K).

Distribution. Assam [also in Thailand].

Habit. Epiphytic herb.

Bulbophyllum elatum (Hook. f.) J. J. Sm., Bull. Jard. Bot. Buitenzorg, sér. 2, 8: 23. 1912. *Cirrhopetalum elatum* Hook. f., Fl. Brit. India [J. D. Hooker] 5(16): 775. 1890. *Phyllorkis elata* (Hook. f.) Kuntze, Revis. Gen. Pl. 2: 677. 1891 (as "*Phyllor-*

chis"). TYPE: India, Sikkim Himalaya, Dugul, alt. 5000 ft, Sep. 1882, *Gamble s.n.* [*10560*] (syntype, K); Sikkim, Rungbee, 5000 ft, 14 Aug. 1870, *C. B. Clarke s.n.* [*12353*] (syntype, K).

Distribution. Arunachal Pradesh, Manipur, Meghalaya, Nagaland, Sikkim, West Bengal [Bhutan, China, Myanmar, Nepal].

Habit. Epiphytic herb.

Bulbophyllum elegantulum (Rolfe) J. J. Sm., Bull. Jard. Bot. Buitenzorg, sér. 2, 8: 23. 1912. *Cirrhopetalum elegantulum* Rolfe, Gard. Chron., ser. 3, 9: 552. 1891. TYPE: India, Coorg, *cult. O'Brien s.n. (31 Mar. 1891)* (holotype, K).

Distribution. Karnataka, Kerala, Tamil Nadu [Myanmar].

Habit. Epiphytic herb.

Bulbophyllum emarginatum (Finet) J. J. Sm., Bull. Jard. Bot. Buitenzorg, sér. 2, 8: 24. 1912. *Cirrhopetalum emarginatum* Finet, Bull. Soc. Bot. France 44: 269. 1897. TYPE: China, *Prince Henri d'Orleans s.n.* (holotype, P).

Bulbophyllum brachypodum A. S. Rao & N. P. Balakr. var. *geei* A. S. Rao & N. P. Balakr., Bull. Bot. Surv. India 10(3–4): 350. 1968 [1969]. *Bulbophyllum yoksunense* J. J. Sm var. *geei* (A. S. Rao & N. P. Balakr.) Bennet, J. Econ. Taxon. Bot. 4(2): 592. 1983. TYPE: India, Kameng District, Dirang Dzong, *E. P. Gee 43067* (holotype, ASSAM).

Distribution. Arunachal Pradesh, Assam [Bhutan, China, Myanmar].

Habit. Epiphytic herb.

Bulbophyllum eublepharum Rchb. f., Ann. Bot. Syst. (Walpers) 6: 252. 1861. *Phyllorkis eublephara* (Rchb. f.) Kuntze, Revis. Gen. Pl. 2: 677. 1891 (as "*Phyllorchis*"). TYPE: India, Darjeeling, 1844, *Griffith s.n. (East India Company Herbarium 5296)* (holotype, K-LINDL; isotype, CAL, K).

Distribution. Arunachal Pradesh, Assam, Nagaland, Sikkim, West Bengal [Bhutan, Myanmar, China].

Habit. Epiphytic herb.

Bulbophyllum fimbriatum (Lindl.) Rchb. f., Ann. Bot. Syst. (Walpers) 6: 260. 1861. *Cirrhopetalum fimbriatum* Lindl., Edwards's Bot. Reg. 25(Misc.): 72. 1839. *Phyllorkis fimbriata* (Lindl.) Kuntze, Revis. Gen. Pl. 2: 677. 1891 (as "*Phyllorchis*"). TYPE: India, Bombay, *cult. Loddiges s.n.* (holotype, K-LINDL).

Distribution. Karnataka, Kerala, Maharashtra, Tamil Nadu. Endemic.

Habit. Epiphytic herb.

Bulbophyllum fischeri Seidenf., Dansk Bot. Ark. 29(1): 202. 1973 [1974]. *Cirrhopetalum gamblei* Hook. f., Fl. Brit. India [J. D. Hooker] 5(16): 778. 1890. *Phyllorkis gamblei* (Hook. f.) Kuntze, Revis. Gen. Pl. 2: 677. 1891 (as *"Phyllorchis"*). *Bulbophyllum gamblei* (Hook. f.) J. J. Sm., Bull. Jard. Bot. Buitenzorg, sér. 2, 8: 24. 1912, nom. illeg., non (Hook. f.) Hook. f., 1890. TYPE: India, Nilgiri Hills, Conoor, alt. 6000 ft, Nov. 1883, *J. S. Gamble s.n.* (holotype, K; isotype, CAL).

Cirrhopetalum thomsonii Hook. f., Fl. Brit. India [J. D. Hooker] 5(16): 778. 1890 (as *"thomsoni"*). *Bulbophyllum thomsonii* (Hook. f.) J. J. Sm., Bull. Jard. Bot. Buitenzorg, sér, 2, 8: 28. 1912, nom. illeg., non Hook. f., 1894. SYNTYPES: India, Nilgiri Hills, *Wight s.n.* (syntype, K); Nilgiri Hills, *G. Thomson s.n.* [*20*] (syntypes, CAL, K [2×]); Nilgiri Hills, Neddivuttum, alt. 7000 ft, 30 Mar. 1870, *C. B. Clarke s.n.* [*11384*] (syntype, K).

Distribution. Karnataka, Kerala, Tamil Nadu [Sri Lanka].

Habit. Epiphytic herb.

Bulbophyllum forrestii Seidenf., Dansk Bot. Ark. 29(1): 120. 1973 [1974]. *Cirrhopetalum aemulum* W. W. Sm., Notes Roy. Bot. Gard. Edinburgh 13: 195. 1921. *Rhytionanthos acmulus* (W. W. Sm.) Garay, Hamer & Siegerist, Nordic J. Bot. 14(6): 637. 1994 (as *"aemulum"*). SYNTYPES: China, Yunnan, Shweli Valley, 5000–6000 ft, June 1912, *Forrest 8130* (syntype, E, K); ibid., Aug. 1913, *Forrest 11901* (syntype, E), China, Yunnan, Shweli-Salwin Divide, 7000 ft, June 1919, *Forrest 17970* (syntype, E, K); Myanmar, Hpimaw, 17 May 1914, *Ward 1554* (syntype, E).

Distribution. Assam, Manipur, Meghalaya, Nagaland [China, Myanmar].

Habit. Epiphytic herb.

Bulbophyllum fuscopurpureum Wight, Icon. Pl. Ind. Orient. [Wight] 5(1): 6, t. 1651. 1851. *Phyllorkis fuscopurpurea* (Wight) Kuntze, Revis. Gen. Pl. 2: 677. 1891 (as *"Phyllorchis"*). SYNTYPES: India, Nilgiri Hills, on the banks of Kartairy river below Kaiti, Feb. 1850, also below Neddawuttim on the north-eastern slopes, *Wight s.n.* (syntypes, K ["Neelgherry Hill"], also a sheet with specimens without data and one labeled "Tellyengry Hill, April 1850, *Wight 2959*").

Distribution. Karnataka, Kerala, Tamil Nadu. Endemic.

Habit. Epiphytic herb.

Bulbophyllum gamblei (Hook. f.) Hook. f., Hooker's Icon. Pl. 21: t. 2039b. 1890. *Bulbophyllum leptanthum* Hook. f. var. *gamblei* Hook. f., Fl. Brit. India [J. D. Hooker] 5(16): 759. 1890. TYPE: India, Sikkim, Gumpahar, alt. 7000 ft, 6 July 1876, *Gamble s.n.* (holotype, K).

Distribution. Arunachal Pradesh, ?Karnataka, Manipur, Nagaland, Sikkim, West Bengal [Bangladesh, Bhutan].

Habit. Epiphytic herb.

Bulbophyllum gracilipes King & Pantl., J. Asiat. Soc. Bengal, Pt. 2, Nat. Hist. 65: 119. 1896. TYPE: India, Sikkim, 1500 ft, Sep.–Oct. 1892, *Pantling 242* (holotype, CAL; isotype, AMES, BM, G, K, W).

Distribution. Assam, Sikkim, West Bengal. Endemic.

Habit. Epiphytic herb.

Bulbophyllum griffithii (Lindl.) Rchb. f., Ann. Bot. Syst. (Walpers) 6: 247. 1861. *Sarcopodium griffithii* Lindl., Fol. Orchid. 2(*Sarcopodium*): 6. 1853. SYNTYPES: India, Khasi Hills, *T. Lobb s.n.* (syntype, K-LINDL); Khasi Hills, 5000 ft, *Hooker & Thomson 53* (syntype, K-LINDL); Myrung, *Griffith* (syntype, K not found).

Distribution. Arunachal Pradesh, Assam, Meghalaya, Sikkim, West Bengal [Bhutan, China, Nepal].

Habit. Epiphytic herb.

Bulbophyllum guttulatum (Hook. f.) N. P. Balakr., J. Bombay Nat. Hist. Soc. 67: 66. 1970. *Cirrhopetalum guttulatum* Hook. f., Fl. Brit. India [J. D. Hooker] 5(16): 776. 1890. *Phyllorkis guttulata* (Hook. f.) Kuntze, Revis. Gen. Pl. 2: 677. 1891 (as *"Phyllorchis"*). SYNTYPES: Nepal, *Wallich s.n.*; India, Sikkim, alt. 3000–6000 ft, *J. D. Hooker s.n.* [*32*] (syntype, K-LINDL).

Bulbophyllum chyrmangense D. Verma, Lavania & Sushil K. Singh, Phytotaxa 195: 94. 2015 (as *"chyrmangensis"*). TYPE: India, Assam, West Jaintia Hills District, Chyrmang, *D. Verma 53* (holotype, ASSAM; isotype, LWU).

Distribution. Arunachal Pradesh, Assam, Manipur, Meghalaya, Nagaland, Odisha, Sikkim, Uttarakhand, West Bengal [Bhutan, Myanmar, Nepal].

Habit. Epiphytic herb.

Remarks. Kumar et al. (2018) argued that *Bulbophyllum chyrmangense* falls within the range of variation of *B. guttulatum.*

Bulbophyllum gymnopus Hook. f., Fl. Brit. India [J. D. Hooker] 5: 764. 1890. *Phyllorkis gymnopus* (Hook. f.) Kuntze, Revis. Gen. Pl. 2: 677. 1891 (as "*Phyllorchis*"). *Drymoda gymnopus* (Hook. f.) Garay, Hamer & Siegerist, Nordic J. Bot. 14(6): 641. 1994. *Monomeria gymnopus* (Hook. f.) Aver., Opred. Orkhid. V'etnama 285. 1994. TYPE: Bhutan Himalaya, alt. 2000 ft, *Griffith s.n. (East India Company Herbarium 5133)* (lectotype, K, designated by Seidenfaden, 1979).

Distribution. Arunachal Pradesh, Manipur, Meghalaya, Mizoram, Nagaland, Sikkim [Bhutan, China].

Habit. Epiphytic herb.

Remarks. Indian specimens of *Bulbophyllum gymnopus* were mistakenly identified as *B. longipes* Rchb. f. by Pearce et al. (2001). The latter is not known to occur in India.

Bulbophyllum gyrochilum Seidenf., Dansk Bot. Ark. 33(3): 80. 1979. TYPE: Thailand, Doi Sutep, *GT 3484* (holotype, C).

Distribution. ?Meghalaya [China].

Habit. Epiphytic herb.

Remarks. We have not seen any evidence that this species occurs in India, although it is listed by Singh et al. (2019).

Bulbophyllum helenae (Kuntze) J. J. Sm., Bull. Jard. Bot. Buitenzorg, sér. 2, 8: 24. 1912. *Cirrhopetalum cornutum* Lindl., Edwards's Bot. Reg. 24 (Misc.): 75. 1838. *Phyllorkis helenae* Kuntze, Revis. Gen. Pl. 2: 676. 1891 (as "*Phyllorchis*"). *Rhytionanthos cornutus* (Lindl.) Garay, Hamer & Siegerist, Nordic J. Bot. 14: 637. 1994 (as "*cornutum*"). TYPE: India, Khasi Hills, Nungclow, *J. Gibson s.n.* (holotype, K-LINDL).

Distribution. Arunachal Pradesh, Assam, Manipur, Meghalaya, Mizoram, Nagaland, Sikkim, Uttarakhand, West Bengal [Bhutan, China, Myanmar, Nepal].

Habit. Epiphytic herb.

Bulbophyllum hirtum (Sm.) Lindl., Gen. Sp. Orchid. Pl.: 51. 1830. *Stelis hirta* Sm., Cycl. (Rees) 34: Stelis no. 11. 1816. *Tribrachia hirta* (Sm.) Lindl., Coll. Bot. (Lindley), t. 41. 1826. *Phyllorkis hirta* (Sm.) Kuntze, Revis. Gen. Pl. 2: 677. 1891 (as

"*Phyllorchis*"). TYPE: Nepal, Feb. 1821, *Wallich 1989* (holotype, K-LINDL; isotype, G, K, K-WALL).

Bulbophyllum suave Griff., Not. Pl. Asiat. 3: 292. 1851. TYPE: India, Darjeeling, *Campbell s.n.* (not found).

Distribution. Arunachal Pradesh, Meghalaya, Mizoram, Nagaland, Sikkim, Uttarakhand, West Bengal [China, ?Myanmar, Nepal].

Habit. Epiphytic herb.

Bulbophyllum hymenanthum Hook. f., Fl. Brit. India [J. D. Hooker] 5(16): 767. 1890. *Phyllorkis hymenantha* (Hook. f.) Kuntze, Revis. Gen. Pl. 2: 677. 1891 (as "*Phyllorchis*"). TYPE: India, Khasi Hills, Myrung, alt. 5000 ft, 7 July 1850, *Hooker & Thomson s.n.* [*1525*] (holotype, K-LINDL; isotype, K).

Distribution. Arunachal Pradesh, Meghalaya, Nagaland, Sikkim, West Bengal [Bhutan].

Habit. Epiphytic herb.

Bulbophyllum indicum (C. S. Kumar & Garay) Sushil K. Singh, Agrawala & Jalal, Orchids India: 95. 2019. *Rhytionanthos indicus* C. S. Kumar & Garay, Proc. 20th World Orchid Conf.: 114. 2013 (as "*indicum*"). *Bulbophyllum indicum* (C. S. Kumar & Garay) Kottaim., Int. J. Curr. Res. Biosci. Pl. Biol. 6(10): 40. 2019, isonym. TYPE: India, Kerala, Trivandrum District, Agastyamala, *Satish Kumar 1426* (holotype, TBGT).

Distribution. Kerala. Endemic.

Habit. Epiphytic herb.

Bulbophyllum iners Rchb. f., Gard. Chron., n.s., 13: 776. 1880. *Phyllorkis iners* (Rchb. f.) Kuntze, Revis. Gen. Pl. 2: 677. 1891 (as "*Phyllorchis*"). TYPE: ?India, ?Assam, *cult. Bull s.n.* (holotype, W).

Distribution. ?Assam. ?Endemic.

Habit. Epiphytic herb.

Remarks. We have not been able to study authentic material of this species. Reichenbach (1880: 776) wrote in the protologue that "[t]here can be little doubt of its being of Assamese origin." This is in fact very much in doubt, as no *Bulbophyllum* answering his description has been found since in Assam.

Bulbophyllum interpositum J. J. Verm., Schuit. & de Vogel, Phytotaxa 197: 59. 2015. *Ione intermedia* King & Pantl., J. Asiat. Soc. Bengal, Pt. 2, Nat. Hist. 65: 120. 1896. *Sunipia intermedia* (King & Pantl.) P. F. Hunt, Kew Bull. 26(1): 184. 1971.

TYPE: India, Sikkim, Tendong, alt. 6000 ft, June 1891, *Pantling 161* (syntypes, CAL, K [2×], P [2×]).

Distribution. Arunachal Pradesh, West Bengal [China].

Habit. Epiphytic herb.

Bulbophyllum jainii (Hynn. & Malhotra) J. J. Verm., Schuit. & de Vogel, Phytotaxa 166: 104. 2014. *Sunipia jainii* Hynn. & Malhotra, J. Indian Bot. Soc. 57: 31. 1978. *Ione jainii* (Hynn. & Malhotra) Seidenf., Opera Bot. 124: 53. 1995. TYPE: India, Nagaland, Tseminyu, alt. 4000–5000 ft, 14 Nov. 1973, *T. M. Hynniewta 56060A* (holotype, CAL).

Distribution. Arunachal Pradesh, Assam, Nagaland. Endemic.

Habit. Epiphytic herb.

Bulbophyllum kaitiense Rchb. f., Ann. Bot. Syst. (Walpers) 6: 262. 1861. *Cirrhopetalum neilgherrense* Wight, Icon. Pl. Ind. Orient. [Wight] 5(1): 7, t. 1654. 1851. *Phyllorkis kaitiensis* (Rchb. f.) Kuntze, Revis. Gen. Pl. 2: 677. 1891 (as "*Phyllorchis*"). TYPE: India, Kartairy [River] below Kaitie, *Wight s.n.* (holotype, K).

Distribution. Andhra Pradesh, Karnataka, Kerala, Tamil Nadu. Endemic.

Habit. Epiphytic herb.

Bulbophyllum keralense M. Kumar & Sequiera, J. Bombay Nat. Hist. Soc. 98(1): 87. 2001 (as "*keralensis*"). TYPE: India, Kerala, Palghat District, Silent Valley National Park, Sispara, 1800 m, *Stephen 007857* (holotype, KFRI).

Distribution. Kerala. Endemic.

Habit. Epiphytic herb.

Bulbophyllum khasyanum Griff., Not. Pl. Asiat. 3: 284. 1851. *Bulbophyllum cylindraceum* Lindl. var. *khasyanum* (Griff.) Hook. f., Fl. Brit. India [J. D. Hooker] 5(16): 765. 1890 (as "*khasiana*"). TYPE: India, Khasi Hills, Sinureem, *Griffith in Assam Herb. 180* (holotype, K-LINDL).

Bulbophyllum conchiferum Rchb. f., Ann. Bot. Syst. (Walpers) 6: 253. 1861. *Phyllorkis conchifera* (Rchb. f.) Kuntze, Revis. Gen. Pl. 2: 677. 1891 (as "*Phyllorchis*"). TYPE: India, Khasi Hills, *cult. T. Lobb s.n.* (holotype, K-LINDL).

Distribution. Arunachal Pradesh, Assam, Madhya Pradesh, Manipur, Meghalaya, Nagaland [China, Myanmar].

Habit. Epiphytic herb.

Bulbophyllum kingii Hook. f., Fl. Brit. India [J. D. Hooker] 5(16): 760. 1890. *Phyllorkis kingii* (Hook. f.) Kuntze, Revis. Gen. Pl. 2: 677. 1891 (as "*Phyllorchis*"). TYPE: India, Sikkim, alt. 6000 ft, Oct. 1877, *King s.n.* (holotype, K).

Acrochaene punctata Lindl., Fol. Orchid. 2(*Acrochaene*): 1. 1853. *Monomeria punctata* (Lindl.) Schltr., Orchideen (Schlechter) 338. 1914. TYPE: India, Sikkim, alt. 4000 ft, *J. D. Hooker s.n. [40]* (holotype, K-LINDL; isotype, CAL, K).

Distribution. Arunachal Pradesh, Assam, Manipur, ?Meghalaya, Mizoram, Nagaland, Sikkim, Tamil Nadu, West Bengal [Bhutan, Myanmar].

Habit. Epiphytic herb.

Bulbophyllum kipgenii (Kishor, Chowlu & Vij) J. J. Verm., Schuit. & de Vogel, Phytotaxa 166: 104. 2014. *Ione kipgenii* Kishor, Chowlu & Vij, Kew Bull. 67: 517. 2012. *Sunipia kipgenii* (Kishor, Chowlu & Vij) J. M. H. Shaw, Orchid Rev. 122: 78. 2014. TYPE: India, Manipur, Chandel District, 490 m, *COGCEHR/Herb/ 00001* (holotype, CAL).

Distribution. Manipur. Endemic.

Habit. Epiphytic herb.

Bulbophyllum leopardinum (Wall.) Lindl. ex Wall., Numer. List [Wallich] n. 1981. 1829. *Dendrobium leopardinum* Wall., Tent. Fl. Napal. 39, t. 28. 1826. *Sarcopodium leopardinum* (Wall.) Lindl. & Paxton, Paxton's Fl. Gard. 1: 155. 1851. *Phyllorkis leopardina* (Wall.) Kuntze, Revis. Gen. Pl. 2: 677. 1891 (as "*Phyllorchis*"). TYPE: Nepal, Mt. Chaudraghiry & Gosainthan [Shishapangma], June & Aug. 1821, *Wallich 1981* (syntypes, K, K-LINDL, K-WALL).

Bulbophyllum schmidtianum Rchb. f., Hamburger Garten-Blumenzeitung 21: 357. 1865. *Phyllorkis schmidtiana* (Rchb. f.) Kuntze, Revis. Gen. Pl. 2: 678. 1891 (as "*Phyllorchis*"). TYPE: India, *cult. Schiller s.n. (leg. Calcutta)* (holotype, W).

Distribution. Arunachal Pradesh, Assam, Manipur, Meghalaya, Mizoram, Nagaland, Sikkim, Uttarakhand, West Bengal [Bhutan, Myanmar, Nepal].

Habit. Epiphytic herb.

Bulbophyllum lepidum (Blume) J. J. Sm., Orch. Java 6: 471, fig. 361. 1905. *Ephippium lepidum* Blume, Bijdr. Fl. Ned. Ind. 7: 310. 1825. TYPE: Indonesia, Java, Mt. Pantjar, *Blume s.n.* [*1934*] (holotype, L; isotype, BO).

?*Epidendrum flabellum-veneris* J. Koenig, Observ. Bot. (Retzius) 6: 57. 1791. ?*Cirrhopetalum flabellum-veneris* (J. Koenig) Seidenf. & Ormerod in G. Seidenfaden, Descr. Epidendrorum J. G. König. 39. 1995 (as "*flabelloveneris*"). ?*Bulbophyllum flabellum-veneris* (J. Koenig) Aver., Updated Checkl. Orchids Vietnam 73. 2003 (as "*flabelloveneris*"). TYPE: Not designated.
Cirrhopetalum gamosepalum Griff., Not. Pl. Asiat. 3: 296. 1851. TYPE: Myanmar, Mergui [Myeik], *Griffith in Mergue* [Mergui] *Herb. 520* (not found).

Distribution. Andaman & Nicobar Islands [Myanmar].

Habit. Epiphytic herb.

Remarks. It is by no means certain that the well-known *Bulbophyllum lepidum* is the same species as *Epidendrum flabellum-veneris*, as some diagnostic details are missing in the latter's description, and we do not think *B. lepidum* should be relegated to synonymy (as proposed by Seidenfaden & Ormerod in Seidenfaden, 1995) without the firm evidence of an actual specimen of *E. flabellum-veneris*. In the case of a common species such as *B. lepidum* we believe it would be prudent to refrain from name changes if this can easily be avoided.

Bulbophyllum leptanthum Hook. f., Fl. Brit. India [J. D. Hooker] 5(16): 759. 1890. *Phyllorkis leptantha* (Hook. f.) Kuntze, Revis. Gen. Pl. 2: 677. 1891 (as "*Phyllorchis*"). SYNTYPES: India, Khasi Hills, alt. 4000–5000 ft, *Griffith s.n.* (not found); Nunklow, 13 July 1850, *J. D. Hooker & J. J. Thomson 1690* (syntype, K); Churra, 17 June 1850, *J. D. Hooker & J. J. Thomson 1014* (syntype, K [2×]) .

Distribution. Assam, Meghalaya, Nagaland, Sikkim, West Bengal. Endemic.

Habit. Epiphytic herb.

Bulbophyllum lilacinum Ridl., J. Linn. Soc., Bot. 32: 276. 1896. SYNTYPES: Malay Peninsula, Kedah, Kedah Peak, *Ridley s.n.* (syntype, BM); Southern Thailand, *C. Curtis s.n.* (syntype, BM).

Distribution. Andaman & Nicobar Islands [Bangladesh].

Habit. Epiphytic herb.

Bulbophyllum lobbii Lindl., Edwards's Bot. Reg. 33: t. 29. 1847. *Sarcopodium lobbii* (Lindl.) Lindl. & Paxton, Paxton's Fl. Gard. 1: 155. 1851. *Phyllorkis lobbii* (Lindl.) Kuntze, Revis. Gen. Pl. 2:

677. 1891 (as "*Phyllorchis*"). TYPE: Indonesia, Java, *T. Lobb s.n.* (not found).

Distribution. Arunachal Pradesh, Assam, Manipur, ?Meghalaya, Mizoram, Nagaland, Tripura [Myanmar].

Habit. Epiphytic herb.

Bulbophyllum longerepens Ridl., J. Straits Branch Roy. Asiat. Soc. 49: 28. 1908. TYPE: Malaysia, Sarawak, Santubong, *Hewitt s.n.* (holotype, SING).

Odontostylis multiflora Breda, Gen. Sp. Orchid. Asclep. 1: t. 4. 1827 [1828]. *Odontostylis minor* Breda, Gen. Sp. Orchid. Asclep. 1: t. 4. 1827 [1828] (sphalm.). *Bulbophyllum multiflorum* (Breda) Kraenzl., Gard. Chron., ser. 3, 19: 294. 1896, nom. illeg., non Ridl., 1885. *Bulbophyllum bakhuizenii* Steenis in A. Hamzah & M. Toha, Mountain Fl. Java: t. 36. 1972. TYPE: Indonesia, Java, *Kuhl & van Hasselt s.n.* (not found).

Distribution. Andaman & Nicobar Islands [also in Indonesia, Malaysia].

Habit. Epiphytic herb.

Bulbophyllum macraei (Lindl.) Rchb. f., Ann. Bot. Syst. (Walpers) 6: 263. 1861. *Cirrhopetalum macraei* Lindl., Gen. Sp. Orchid. Pl.: 59. 1830. *Phyllorkis macraei* (Lindl.) Kuntze, Revis. Gen. Pl. 2: 677. 1891 (as "*Phyllorchis*"). TYPE: Sri Lanka, *Macrae s.n.* (holotype, K-LINDL).

Distribution. Arunachal Pradesh, Kerala, Odisha, Tamil Nadu [Sri Lanka].

Habit. Epiphytic herb.

Bulbophyllum macranthum Lindl., Edwards's Bot. Reg. 30: t. 13. 1844. *Sarcopodium macranthum* (Lindl.) Lindl. & Paxton, Paxton's Fl. Gard. 1: 155. 1851. *Phyllorkis macrantha* (Lindl.) Kuntze, Revis. Gen. Pl. 2: 677. 1891 (as "*Phyllorchis*"). *Carparomorchis macrantha* (Lindl.) M. A. Clem. & D. L. Jones, Orchadian 13(11): 499. 2002. TYPE: Singapore, *cult. Loddiges s.n.* (not found; a single flower without annotation in K-LINDL mounted with the above-mentioned illustration is probably the type).

Distribution. Andaman & Nicobar Islands, Assam [Myanmar].

Habit. Epiphytic herb.

Bulbophyllum macrocoleum Seidenf., Dansk Bot. Ark. 33(3): 125, fig. 83. 1979. TYPE: Thailand, Umphang Rd., *Seidenfaden & Smitinand 7810* (holotype, C).

Bulbophyllum longibracteatum Seidenf., Dansk Bot. Ark. 33(3): 140. 1979. TYPE: Laos, *Seidenfaden & Smitinand 2968* (holotype, C).

Distribution.   Andaman & Nicobar Islands [also in Thailand].

Habit.   Epiphytic herb.

Bulbophyllum maskeliyense Livera, Ann. Roy. Bot. Gard. (Peradeniya) 10: 142. 1926. TYPE: Sri Lanka, Kandy District, Maskeliya, *S. B. Stedman s.n.* (holotype, PDA).

Distribution.   Kerala, Tamil Nadu [Sri Lanka].

Habit.   Epiphytic herb.

Bulbophyllum maxillare (Lindl.) Rchb. f., in Walp., Ann. Bot. Syst. 6: 248. 1861. *Cirrhopetalum maxillare* Lindl., Edwards's Bot. Reg. 29: sub t. 49. 1843. TYPE: Philippines, *Cuming s.n.* (holotype, K-LINDL).

Distribution.   Andaman & Nicobar Islands [also in Indonesia, Malaysia, Philippines].

Habit.   Epiphytic herb.

Bulbophyllum medioximum J. J. Verm., Schuit. & de Vogel, Phytotaxa 166: 104. 2014. *Ione annamensis* Ridl., J. Nat. Hist. Soc. Siam 4: 115. 1921. *Sunipia annamensis* (Ridl.) P. F. Hunt, Kew Bull. 26: 183. 1971. TYPE: Vietnam, Langbian Province, South Annam, Langbian Peaks, Apr. 1918, *C. Boden Kloss s.n.* (holotype, BM).

Distribution.   ?Arunachal Pradesh [China].

Habit.   Epiphytic herb.

Remarks.   We have not seen evidence that this species occurs in Arunachal Pradesh, as listed by Singh et al. (2019).

Bulbophyllum moniliforme C. S. P. Parish & Rchb. f., Trans. Linn. Soc. London 30: 151. 1874. *Phyllorkis moniliformis* (C. S. P. Parish & Rchb. f.) Kuntze, Revis. Gen. Pl. 2: 677. 1891 (as "*Phyllorchis*"). TYPE: Myanmar, Mergui [Myeik], 1858, *C. S. P. Parish s.n.* [*96*] (lectotype, K, designated by Seidenfaden, 1979; isotype, W).

Bulbophyllum paramjitii Agrawala, Sharief & B. K. Singh, Phytotaxa 273: 72. 2016. TYPE: India, Sikkim, East District, near Bhusuk, 1550 m (flowered in cultivation at Gangtok, Aug. 2015), *Sharief & Singh 37938* (holotype, BSHC).
Trias pusilla J. Joseph & H. Deka, J. Indian Bot. Soc. 51: 378, fig. 1. 1972 [1973]. *Jejosephia pusilla* (J. Joseph & H. Deka) A. N. Rao & Mani, J. Econ. Taxon. Bot. 7(1):

217, fig. 1. 1985. *Bulbophyllum jejosephii* J. J. Verm., Schuit. & de Vogel, Phytotaxa 166: 106. 2014. TYPE: India, Assam, Khasi & Jaintia Hills, 8 Sep. 1970, *Hareswar Deka 37255A* (holotype, ASSAM; isotype, CAL).

Distribution.   Assam, Bihar, ?Goa, Madhya Pradesh, Meghalaya, Mizoram, Nagaland, Sikkim, West Bengal [Myanmar].

Habit.   Epiphytic herb.

Remarks.   *Bulbophyllum paramjitii* and *B. jejosephii* were found to be conspecific with *B. moniliforme* by Kumar et al. (2018).

Bulbophyllum muscicola Rchb. f., Flora 55: 275. 1872. TYPE: West Himalaya, alt. 9000 ft, *Mann s.n.* (holotype, W).

Cirrhopetalum wallichii Lindl., Pl. Asiat. Rar. (Wallich) 1: 53. 1830. *Phyllorkis wallichii* (Lindl.) Kuntze, Revis. Gen. Pl. 2: 676. 1891 (as "*Phyllorchis*"). *Bulbophyllum wallichii* (Lindl.) Merr. & F. P. Metcalf, Lingnan Sci. J. 21: 7. 1945, nom. illeg., non Rchb. f., 1861. TYPE: Nepal, *Wallich s.n.* (holotype, BM; isotype, K-LINDL).
Cirrhopetalum hookeri Duthie, J. Asiat. Soc. Bengal, Pt. 2, Nat. Hist. 71: 38. 1902. *Bulbophyllum hookeri* (Duthie) J. J. Sm., Bull. Jard. Bot. Buitenzorg, sér. 2, 8: 25. 1912. TYPE: India, Western Himalaya, Tehri Garwhal, Tehri, 5000–6000 ft, Oct. 1901, *Mackinnon's collector s.n.* [*25402*] (holotype, K; isotype, K).

Distribution.   Arunachal Pradesh, Assam, Nagaland, Sikkim, Uttarakhand, West Bengal [Bhutan, Nepal].

Habit.   Epiphytic herb.

Bulbophyllum mysorense (Rolfe) J. J. Sm., Bull. Jard. Bot. Buitenzorg, sér. 2, 8: 26. 1912. *Cirrhopetalum mysorense* Rolfe, Bull. Misc. Inform. Kew 1895: 34. 1895. TYPE: India, Karnataka, Mysore, cult. *J. O'Brien s.n. (Sep. 1891)* (syntype, K); ibid., cult. *Lawrence s.n. (Dec. 1892)* (syntype, K).

Distribution.   Karnataka, Kerala, Tamil Nadu [Myanmar].

Habit.   Epiphytic herb.

Bulbophyllum nasutum Rchb. f., Gard. Chron. 1871: 1482. 1871. *Trias nasuta* (Rchb. f.) Stapf, Bot. Mag. 152: t. 9150. 1928. *Phyllorkis nasuta* (Rchb. f.) Kuntze, Revis. Gen. Pl. 2: 677. 1891 (as "*Phyllorchis*"). TYPE: Myanmar, cult. *Saunders s.n.* (holotype, W).

Trias vitrina Rolfe, Bull. Misc. Inform. Kew 1895: 282. 1895. TYPE: Thailand, Panga (?Punga), cult. *Kew s.n. (Oct. 1895, leg. Curtis s.n.)* (holotype, K).

Distribution.   Arunachal Pradesh [Myanmar].

Habit.   Epiphytic herb.

Bulbophyllum nodosum (Rolfe) J. J. Sm., Bull. Jard. Bot. Buitenzorg, sér. 2, 8: 26. 1912. *Cirrhopetalum nodosum* Rolfe, Bull. Misc. Inform. Kew 1895: 35. 1895. *Rhytionanthos nodosus* (Rolfe) Garay, Hamer & Siegerist, Nordic J. Bot. 14(6): 639. 1994 (as "*nodosum*"). TYPE: India, Nilgiri Hills, *cult. J. O'Brien s.n. (Aug. 1893)* (holotype, K).

Distribution. Arunachal Pradesh, Tamil Nadu. Endemic.

Habit. Epiphytic herb.

Bulbophyllum oblongum (Lindl.) Rchb. f., Ann. Bot. Syst. (Walpers) 6: 249. 1861. *Trias oblonga* Lindl., Gen. Sp. Orchid. Pl.: 60. 1830. TYPE: Myanmar, Moulmein [Mawlamyine], *Wallich 1977* (syntypes, K-LINDL, K-WALL).

Trias ovata Lindl., Gen. Sp. Orchid. Pl.: 60. 1830. TYPE: East Indies, icon. *Wallich s.n.* (holotype, K).
Dendrobium tripterum Wall. ex Hook. f., Fl. Brit. India [J. D. Hooker] 5(16): 790. 1890, nom. inval.
Bulbophyllum manabendrae D. K. Roy, Barbhuiya & Talukdar, Phytotaxa 164: 291. 2014. *Trias manabendrae* (D. K. Roy, Barbhuiya & Talukdar) S. Misra, Orchids India: 484. 2019. TYPE: India, Meghalaya, South Garo Hills, Balphakram National Park, Khundol Gup, 182 m, 12 Mar. 2013, *Roy 129694* (holotype, ASSAM).

Distribution. Andaman & Nicobar Islands, Assam, Meghalaya, West Bengal [Bangladesh, Myanmar].

Habit. Epiphytic herb.

Remarks. Kumar and Gale (2020) pointed out that Seidenfaden's (1976) illustration of *Trias oblonga* does not agree with the type in certain critical details, which were later believed to separate *T. oblonga* from *Bulbophyllum manabendrae* by Roy et al. (2014).

Bulbophyllum obrienianum Rolfe, Gard. Chron., ser. 3, 12: 332. 1892. TYPE: Himalaya Region, *cult. J. O'Brien s.n.* (holotype, K).

Bulbophyllum leopardinum (Wall.) Lindl. ex Wall. var. *tuberculatum* N. P. Balakr. & Sud. Chowdhury, Bull. Bot. Surv. India 9: 90. 1967. TYPE: Bhutan, Nyoth, betw. Shali & Tashiyangtsi, 1800 m, 13 Mar. 1965, flowered in cultivation at Shillong 3 May 1965, *Balakrishnan 41971* (holotype, CAL; isotype ASSAM [2×]).

Distribution. Assam, Arunachal Pradesh [Bhutan].

Habit. Epiphytic herb.

Bulbophyllum odoratissimum (Sm.) Lindl., Gen. Sp. Orchid. Pl.: 55. 1830. *Stelis odoratissima* Sm., Cycl. (Rees) 34: Stelis no. 12. 1816. *Tribrachia odoratissima* (Sm.) Lindl., Coll. Bot. (Lindley), t. 41. 1826. *Phyllorkis odoratissima* (Sm.) Kuntze,

Revis. Gen. Pl. 2: 677. 1891 (as "*Phyllorchis*"). TYPE: Nepal, icon. *Buchanan-Hamilton s.n.* (holotype, LINN).

Bulbophyllum congestum Rolfe, Bull. Misc. Inform. Kew 1912: 131. 1912. SYNTYPES: China, Yunnan, mountains S of Szemao, 4000 ft, *Henry 12291* (syntype, K); Myanmar, Kachin Hills, Aug. 1898, *Mokum s.n.* (syntype, K).
Bulbophyllum odoratissimum (Sm.) Lindl. var. *racemosum* N. P. Balakr., J. Bombay. Nat. Hist. Soc. 75(1): 157. 1978. *Bulbophyllum trichocephalum* (Schltr.) Tang & F. T. Wang var. *racemosum* (N. P. Balakr.) Lucksom, Orchids Sikkim N. E. Himalaya: 695. 2007. TYPE: Bhutan, Tashiyangtsi, 1800 m, 24 Oct. 1965, *Balakrishnan 43070* (holotype, CAL; isotype, ASSAM).

Distribution. Andaman & Nicobar Islands, Arunachal Pradesh, Assam, Manipur, Meghalaya, Mizoram, Nagaland, Sikkim, West Bengal [Bhutan, China, Myanmar, Nepal].

Habit. Epiphytic herb.

Bulbophyllum orezii C. S. Kumar, Orchid Memories: 170. 2004. *Bulbophyllum josephi* M. Kumar & Sequiera, J. Bombay Nat. Hist. Soc. 98(1): 89. 2001, nom. illeg., non (Kuntze) Summerh, 1945. TYPE: India, Kerala, Palghat District, Silent Valley National Park, Punnamala, 850 m, *Stephen 007521* (holotype, KFRI).

Distribution. Kerala. Endemic.

Habit. Epiphytic herb.

Bulbophyllum ornatissimum (Rchb. f.) J. J. Sm., Bull. Jard. Bot. Buitenzorg, sér. 2, 8: 26. 1912. *Cirrhopetalum ornatissimum* Rchb. f., Gard. Chron., n.s., 18: 424. 1882. *Phyllorkis ornatissima* (Rchb. f.) Kuntze, Revis. Gen. Pl. 2: 677. 1891 (as "*Phyllorchis*"). *Mastigion ornatissimum* (Rchb. f.) Garay, Hamer & Siegerist, Nordic J. Bot. 14(6): 637. 1994. TYPE: Unknown provenance, *W. Bull 407* (holotype, W; isotype, K).

Distribution. Assam [Myanmar].

Habit. Epiphytic herb.

Remarks. Due to confusion caused by a misidentified illustration by King and Pantling (1898) labeled *Bulbophyllum ornatissimum*, it would appear that most or all Indian records outside Assam may refer to the more common *B. appendiculatum*; they have here been discarded.

Bulbophyllum paleaceum (Lindl.) Benth. ex Hemsl., Gard. Chron., n.s., 18: 104. 1882. *Ione paleacea* Lindl., Fol. Orchid. 2(*Ione*): 2. 1853. *Sunipia*

paleacea (Lindl.) P. F. Hunt, Kew Bull. 26(1): 184. 1971. TYPE: India, Darjeeling, *Griffith s.n.* (syntypes, K-LINDL, P).

Ione cirrhata Lindl., Fol. Orchid. 2(*Ione*): 1. 1853. *Bulbophyllum cirrhatum* (Lindl.) Hook. f., Fl. Brit. India [J. D. Hooker] 5(16): 769. 1890. *Phyllorkis cirrhata* (Hook. f.) Kuntze, Revis. Gen. Pl. 2: 677. 1891 (as "*Phyllorchis*"). *Sunipia cirrhata* (Lindl.) P. F. Hunt, Kew Bull. 26(1): 184. 1971. TYPE: India, Sikkim Himalaya, alt. 4000 ft, icon. *Cathcart* (holotype, K-LINDL).
Ione fuscopurpurea Lindl., Fol. Orchid. 2(*Ione*): 2. 1853 (as "fusco-purpurea"). *Sunipia fuscopurpurea* (Lindl.) P. F. Hunt, Kew Bull. 26(1): 184. 1971. *Bulbophyllum purpureofuscum* J. J. Verm., Schuit. & de Vogel, Phytotaxa 166: 105. 2014. TYPE: Himalaya: Mishmee Mountains, Thumathya, on trees, *Griffith s.n.* (holotype, K-LINDL).
Ione virens Lindl., Fol. Orchid. 2(*Ione*): 1. 1853. *Bulbophyllum virens* (Lindl.) Hook. f., Fl. Brit. India [J. D. Hooker] 5(16): 770. 1890. *Phyllorkis virens* (Lindl.) Kuntze, Revis. Gen. Pl. 2: 678. 1891 (as "*Phyllorchis*"). *Sunipia virens* (Lindl.) P. F. Hunt, Kew Bull. 26(1): 185. 1971. TYPE: India, Assam, Mishmee Hills, *Griffith s.n.* (holotype, K-LINDL).
Bulbophyllum mishmeense Hook. f., Fl. Brit. India [J. D. Hooker] 5(16): 769. 1890. *Phyllorkis mishmeensis* (Hook. f.) Kuntze, Revis. Gen. Pl. 2: 677. 1891 (as "*Phyllorchis mischmeensis*"). TYPE: India, Upper Assam, Mishmee Mountains, Thumathya, *Griffith s.n.* (holotype, K-LINDL).

Distribution.   Arunachal Pradesh, Assam, Sikkim, West Bengal [Bhutan, China, Myanmar, Nepal].

Habit.   Epiphytic herb.

Bulbophyllum parviflorum C. S. P. Parish & Rchb. f., Trans. Linn. Soc. London 30: 152. 1874. *Phyllorkis parviflora* (C. S. P. Parish & Rchb. f.) Kuntze, Revis. Gen. Pl. 2: 677. 1891 (as "*Phyllorchis*"). TYPE: Myanmar, Moulmein [Mawlamyine], 1870, *C. S. P. Parish s.n.* [*305*] (holotype, K; isotype, C, W).

Bulbophyllum thomsonii Hook. f., Fl. Brit. India [J. D. Hooker] 5(16): 764. 1890. *Phyllorkis thomsonii* (Hook. f.) Kuntze, Revis. Gen. Pl. 2: 677. 1891 (as "*Phyllorchis*"). TYPE: India, Sikkim Himalaya, 1850, *T. Thomson s.n.* (holotype, K; isotype, CAL).

Distribution.   Arunachal Pradesh, Sikkim, West Bengal [Bhutan, Myanmar].

Habit.   Epiphytic herb.

Bulbophyllum pectinatum Finet, Bull. Soc. Bot. France 44: 268, pl. 7. 1897. TYPE: China, Yunnan, *Prince Henri d'Orleans s.n.* (holotype, P).

Bulbophyllum spectabile Rolfe, Bull. Misc. Inform. Kew 1898: 193. 1898. TYPE: India, Assam, May 1896, *cult. Glasnevin s.n.* (holotype, K).

Distribution.   Assam, Mizoram, Nagaland [China, Myanmar].

Habit.   Epiphytic herb.

Bulbophyllum penicillium C. S. P. Parish & Rchb. f., Trans. Linn. Soc. London 30: 151. 1874. *Phyllorkis penicillium* (C. S. P. Parish & Rchb. f.) Kuntze, Revis. Gen. Pl. 2: 677. 1891 (as "*Phyllorchis*"). TYPE: Myanmar, Moulmein [Mawlamyine], 1870, *C. S. P. Parish s.n.* [*303*] (lectotype, K, designated by Clayton, 2017; isotype W).

Distribution.   Arunachal Pradesh, Manipur, Meghalaya, Nagaland, West Bengal [Bhutan, Myanmar].

Habit.   Epiphytic herb.

Bulbophyllum picturatum (Lodd.) Rchb. f., Ann. Bot. Syst. (Walpers) 6: 262. 1861. *Cirrhopetalum picturatum* Lodd., Edwards's Bot. Reg. 26(Misc.): 49. 1840. *Phyllorkis picturata* (Lodd.) Kuntze, Revis. Gen. Pl. 2: 675. 1891 (as "*Phyllorchis*"). TYPE: India, *cult. Loddiges s.n.* (holotype, K-LINDL).

Distribution.   Assam, Mizoram [Myanmar].

Habit.   Epiphytic herb.

Bulbophyllum piluliferum King & Pantl., Ann. Roy. Bot. Gard. (Calcutta) 8: 76, t. 104. 1898. TYPE: India, Sikkim, Teesta Valley, 1000 ft, May 1891, *Pantling 141* (holotype, CAL; isotype, K).

Distribution.   Arunachal Pradesh, ?Meghalaya, Nagaland, Sikkim, West Bengal. Endemic.

Habit.   Epiphytic herb.

Bulbophyllum polyrhizum Lindl., Gen. Sp. Orchid. Pl.: 53. 1830 (as "*polyrrhizum*"). *Phyllorkis polyrhiza* (Lindl.) Kuntze, Revis. Gen. Pl. 2: 677. 1891 (as "*Phyllorchis*"). TYPE: India, icon. *Wallich s.n.* (holotype, K).

Distribution.   Arunachal Pradesh, Assam, Mizoram, Nagaland, Odisha, Sikkim, Uttarakhand, West Bengal [Myanmar, Nepal].

Habit.   Epiphytic herb.

Bulbophyllum propinquum Kraenzl., Orchis 2: 62. 1908. TYPE: Thailand, *cult. Fuerstenberg (leg. Hosseus) s.n.*

Distribution.   Manipur [also in Thailand].

Habit.   Epiphytic herb.

Bulbophyllum protractum Hook. f., Fl. Brit. India [J. D. Hooker] 5(16): 758. 1890. *Phyllorkis protracta* (Hook. f.) Kuntze, Revis. Gen. Pl. 2: 677.

1891 (as "*Phyllorchis*"). TYPE: India, Andaman Islands or Myanmar, Tenasserim, *Helfer s.n.* [*244*] (holotype, K; isotype, CAL, P).

Distribution. Andaman & Nicobar Islands, Arunachal Pradesh, Assam, ?Meghalaya, Sikkim, West Bengal [Myanmar].

Habit. Epiphytic herb.

Bulbophyllum proudlockii (King & Pantl.) J. J. Sm., Bull. Jard. Bot. Buitenzorg, sér. 2, 8: 27. 1912. *Cirrhopetalum proudlockii* King & Pantl., J. Asiat. Soc. Bengal, Pt. 2, Nat. Hist. 66: 588. 1898 [1897]. *Tripudianthes proudlockii* (King & Pantl.) Szlach. & Kras, Richardiana 7(2): 95. 2007, nom. inval. TYPE: India, Tamil Nadu, Nilgiri Hills, Otacamund, Apr. 1897, *R. L. Proudlock s.n.* (holotype, CAL; isotype, K, L, W).

Distribution. Karnataka, Tamil Nadu. Endemic.

Habit. Epiphytic herb.

Bulbophyllum pteroglossum Schltr., Repert. Spec. Nov. Regni Veg. Beih. 4: 71. 1919. TYPE: China, Yunnan, south of Szemao, 4000 ft, *Henry 12959* (holotype, K).

Bulbophyllum uniflorum Griff., Not. Pl. Asiat. 3: 293. 1851, nom. illeg., non (Blume) Hassk., 1844. *Sarcopodium uniflorum* Lindl., Fol. Orchid. 2: 6. 1853. *Phyllorkis monantha* Kuntze, Revis. Gen. Pl. 2: 676. 1891 (as "*Phyllorchis*"). *Bulbophyllum monanthum* (Kuntze) J. J. Sm., Bull. Jard. Bot. Buitenzorg, sér. 2 8: 26. 1912, nom. illeg., non *B. monanthos* Ridl., 1896. *Bulbophyllum devangiriense* N. P. Balakr., J. Bombay Nat. Hist. Soc. 67: 66. 1970 (as "*devangiriensis*"). *Bulbophyllum tiagii* A. S. Chauhan, J. Econ. Taxon. Bot. 5(3): 995. 1984, nom. superfl. TYPE: Bhutan, *Griffith in Bootan Herb. 138* (not found). NEOTYPE: Bhutan, Mishmee Hills, *Griffith s.n.* (neotype, K-LINDL, designated by Seidenfaden, 1994).

Distribution. Arunachal Pradesh, Assam, ?Meghalaya, Nagaland [Bhutan].

Habit. Epiphytic herb.

Bulbophyllum repens Griff., Not. Pl. Asiat. 3: 293. 1851. *Phyllorkis repens* (Griff.) Kuntze, Revis. Gen. Pl. 2: 677. 1891 (as "*Phyllorchis*"). TYPE: India, Khasi Hills, *Griffith in Khasyah Herb. 1021* (holotype, K-LINDL).

Distribution. Arunachal Pradesh, Meghalaya, Mizoram, Nagaland [Myanmar].

Habit. Epiphytic herb.

Bulbophyllum reptans (Lindl.) Lindl., Wallich Cat. 1988. 1829. *Tribrachia reptans* Lindl., Coll. Bot.

(Lindley), t. 41A. 1826. *Phyllorkis reptans* (Lindl.) Kuntze, Revis. Gen. Pl. 2: 677. 1891 (as "*Phyllorchis*"). TYPE: Nepal, *E. Rudge s.n.* (holotype, K-LINDL).

Stelis racemosa Sm., Cycl. (Rees) 34: Stelis no. 10. 1816. *Sunipia racemosa* (Sm.) Tang & F. T. Wang, Acta Phytotax. Sin. 1(1): 90. 1951. *Ione racemosa* (Sm.) Seidenf., Bot. Tidsskr. 64: 227. 1969. TYPE: Nepal, Bagmati Zone, Kathmandu, on trees, *Buchanan-Hamilton s.n.* (holotype, BM)
Bulbophyllum grandiflorum Griff., Not. Pl. Asiat. 3: 293. 1851, nom. illeg., non Blume, 1849. TYPE: Bhutan, *Griffith in Bootan Herb. 705* (holotype, K-LINDL).
Bulbophyllum clarkei Rchb. f., Flora 71: 155. 1888. SYNTYPES: India, Kohima, 6000 ft, 7 Nov. 1885, *Clarke 41813 (leg. P. Badgeley)* (syntype, K); Bhutan, Mishmee Hills, *Griffith s.n.* (syntype, K).
Bulbophyllum reptans (Lindl.) Lindl. var. *subracemosum* Hook. f., Fl. Brit. India [J. D. Hooker] 5(16): 769. 1890 (as "*subracemosa*"). SYNTYPES: India, Khasi Hills, *Griffith s.n. (East India Company Herbarium 5130)* (syntypes, CAL, K, W); Bhutan, Passuling, *Griffith s.n.* [*31*] (syntype, K-LINDL).
Bulbophyllum raui Arora, Bull. Bot. Surv. India 11(3–4): 440. 1969 [1972]. TYPE: India, Eastern Kumaon, Pithoragarh District, Shandev, on Shandev-Thal Road, *C. M. Arora 37802A* (holotype, BSD).
Bulbophyllum reptans (Lindl.) Lindl. var. *acuta* Malhotra & Balodi, Bull. Bot. Surv. India 26: 110. 1984 [1985]. TYPE: India, Kumaon Uttranchal, Pithoragarh, Ghorpatta, 10 Mar. 1965, *Rau 35340* (holotype, CAL).

Distribution. Arunachal Pradesh, Assam, Manipur, Meghalaya, Mizoram, Nagaland, Sikkim, Uttarakhand, West Bengal [Bhutan, China, Myanmar, Nepal].

Habit. Epiphytic herb.

Bulbophyllum restrepia (Ridl.) Ridl., Mat. Fl. Malay Penins.(1): 78. 1907. *Cirrhopetalum restrepia* Ridl., Trans. Linn. Soc. London, Bot. 3: 365. 1893. TYPE: Malaysia, Pahang, Pramau near Pekan, *Ridley s.n.* (holotype, SING).

Distribution. Andaman & Nicobar Islands [also in Indonesia, Malaysia].

Habit. Epiphytic herb.

Bulbophyllum retusiusculum Rchb. f., Gard. Chron. 1869: 1182. 1869. *Cirrhopetalum retusiusculum* (Rchb. f.) Hook. f., Fl. Brit. India [J. D. Hooker] 5(16): 776. 1890. *Phyllorkis retusiuscula* (Rchb. f.) Kuntze, Revis. Gen. Pl. 2: 677. 1891 (as "*Phyllorchis*"). TYPE: Myanmar, Moulmein [Mawlamyine], *Benson (Veitch 93) s.n.* (syntypes, K, W).

Distribution. Arunachal Pradesh, ?Meghalaya, Nagaland, West Bengal [Bhutan, China, Myanmar, Nepal].

Habit. Epiphytic herb.

Bulbophyllum rheedei Manilal & C. S. Kumar, Rheedea 1: 55. 1991. *Rhytionanthos rheedei* (Manilal & C. S. Kumar) Garay, Hamer & Siegerist, Nordic J. Bot. 14(6): 639. 1994. TYPE: India, Kerala, Trivandrum District, Palode, *C. Sathish Kumar 4643* (holotype, CAL).

Distribution. Karnataka, Kerala. Endemic.

Habit. Epiphytic herb.

Bulbophyllum rigidum King & Pantl., Ann. Roy. Bot. Gard. (Calcutta) 8: 69, t. 94. 1898. SYNTYPES: India, Darjeeling, *Griffith s.n. (East India Company Herbarium 5291)* (syntypes, CAL, K); Sikkim, 6000 ft, May–June [Aug.] 1892, *Pantling 42* (syntypes, BM, CAL, G, K, P, W).

Distribution. Arunachal Pradesh, Nagaland, Sikkim, West Bengal [Nepal].

Habit. Epiphytic herb.

Bulbophyllum rolfei (Kuntze) Seidenf., Dansk Bot. Ark. 33(3): 149. 1979. *Cirrhopetalum parvulum* Hook. f., Fl. Brit. India [J. D. Hooker] 5(16): 778. 1890. *Phyllorkis rolfei* Kuntze, Revis. Gen. Pl. 2: 676. 1891 (as "*Phyllorchis*"). *Bulbophyllum parvulum* (Hook. f.) J. J. Sm., Bull. Jard. Bot. Buitenzorg, sér. 2, 8: 27. 1912, nom. illeg., non Lindl., 1830. TYPE: India, Darjeeling, *Griffith's collector s.n. (East India Company Herbarium 5174)* (syntypes, K, K-LINDL).

Cirrhopetalum dyerianum King & Pantl., J. Asiat. Soc. Bengal, Pt. 2, Nat. Hist. 64: 335. 1895 [1896]. *Bulbophyllum dyerianum* (King & Pantl.) Seidenf., Dansk Bot. Ark. 29(1): 175, t. 88. 1973 [1974]. TYPE: India, Sikkim, Tendong, 7000 ft, Aug., *Pantling s.n.* (not found).

Distribution. Arunachal Pradesh, Assam, Nagaland, Sikkim, West Bengal [Bhutan, China, Myanmar, Nepal].

Habit. Epiphytic herb.

Bulbophyllum rosemarianum C. S. Kumar, P. C. S. Kumar & Saleem, Rheedea 11(2): 97. 2001. TYPE: India, Kerala, Idukki District, Thalakode, *Saleem 28211* (holotype, TBGT).

Distribution. Kerala. Endemic.

Habit. Epiphytic herb.

Remarks. This species should be critically compared with *Bulbophyllum sterile* (Lam.) Surish, of which it may just be a color form.

Bulbophyllum roseopictum J. J. Verm., Schuit. & de Vogel, Phytotaxa 166: 105. 2014. *Sunipia bicolor* Lindl., Gen. Sp. Orchid. Pl.: 179. 1833. *Ione bicolor* (Lindl.) Lindl., Fol. Orchid. 2(*Ione*): 3. 1853. *Bulbophyllum bicolor* (Lindl.) Hook. f., Fl. Brit. India [J. D. Hooker] 5(16): 770. 1890, non Lindl., 1830. TYPE: Nepal, icon. *Wallich s.n.* (holotype, BM).

Distribution. Arunachal Pradesh, Assam, Manipur, Meghalaya, Nagaland, Sikkim, Uttarakhand, West Bengal [Bhutan, China, ?Myanmar, Nepal].

Habit. Epiphytic herb.

Bulbophyllum rothschildianum (O'Brien) J. J. Sm., Bull. Jard. Bot. Buitenzorg, sér. 2, 8: 27. 1912. *Cirrhopetalum rothschildianum* O'Brien, Gard. Chron., ser. 3, 18: 608. 1895. *Mastigion rothschildianum* (O'Brien) Lucksom, Orchids Sikkim N. E. Himalaya: 682. 2007. TYPE: India, West Bengal, Darjeeling, *cult. Rothschild s.n.* (holotype, K).

Distribution. Arunachal Pradesh, Manipur, Nagaland, West Bengal [China].

Habit. Epiphytic herb.

Bulbophyllum roxburghii (Lindl.) Rchb. f., Ann. Bot. Syst. (Walpers) 6: 263. 1861. *Cirrhopetalum roxburghii* Lindl., Gen. Sp. Orchid. Pl.: 58. 1830. *Phyllorkis roxburghii* (Lindl.) Kuntze, Revis. Gen. Pl. 2: 677. 1891 (as "*Phyllorchis*"). TYPE: India, at the mouth of the Ganges, icon. *Aerides radiatum*, *Roxburgh 2351* (holotype, K; isotype, CAL).

Aerides radiata Roxb., Fl. Ind. ed. 1832, 3: 472. 1832 (as "*radiatum*"). TYPE: India, at the mouth of the Ganges, icon. *Aerides radiatum*, *Roxburgh 2351* (holotype, CAL; isotype, K).

Cirrhopetalum sikkimense King & Pantl., Ann. Roy. Bot. Gard. (Calcutta) 8: 90, t. 125. 1898. *Bulbophyllum sikkimense* (King & Pantl.) J. J. Sm., Bull. Jard. Bot. Buitenzorg, sér. 2, 8: 28. 1912. TYPE: India, Sikkim, June 1893, *Pantling 148* (holotype, CAL; isotype, AMES, K, L, P).

Distribution. Arunachal Pradesh, Assam, Nagaland, Sikkim, West Bengal [Bangladesh].

Habit. Epiphytic herb.

Bulbophyllum rufinum Rchb. f., Xenia Orchid. 3: 44. 1881. *Phyllorkis rufina* (Rchb. f.) Kuntze, Revis. Gen. Pl. 2: 677. 1891 (as "*Phyllorchis*"). TYPE: India, *cult. Hincks von Breckenborough (leg. Stevens) s.n.* (holotype, W).

Distribution. Andaman & Nicobar Islands [Bangladesh, Myanmar, China].

Habit. Epiphytic herb.

Bulbophyllum sarcophylloides Garay, Hamer & Siegerist, Nordic J. Bot. 14(6): 620. 1994. *Cirrhopetalum sarcophyllum* King & Pantl. var. *minor* King & Pantl., Ann. Roy. Bot. Gard. (Calcutta) 8: 91. 1898. TYPE: Sikkim, Siooke, alt. 1000 ft, 3 June 1891, *Pantling s.n.* [*95A*] (holotype, K; isotype, BM, CAL, W).

Distribution. Arunachal Pradesh, Assam, Sikkim, West Bengal. Endemic.

Habit. Epiphytic herb.

Bulbophyllum sarcophyllum (King & Pantl.) J. J. Sm., Bull. Jard. Bot. Buitenzorg, sér. 2, 8: 27. 1912. *Cirrhopetalum sarcophyllum* King & Pantl., J. Asiat. Soc. Bengal, Pt. 2, Nat. Hist. 64: 335. 1895 [1896]. TYPE: India, Sikkim, Rishap, alt. 2500 ft, Sep., *Pantling s.n.* [*95*] (holotype, CAL).

Bulbophyllum panigrahianum S. Misra, Nordic J. Bot. 6(1): 25, fig. 1. 1986. *Cirrhopetalum panigrahianum* (S. Misra) S. Misra, J. Orchid Soc. India 11(1–2): 54. 1997. TYPE: India, Odisha, Keonjhar District, Rebana reserve forests, Rimbeda, 500 m, 11 July 1982, *Sarat Misra 698* (holotype, CAL; isotype, K).

Bulbophyllum cherrapunjeense Barbhuiya & D. Verma, Phytotaxa 156: 298. 2014 (as "*cherrapunjeensis*"). TYPE: India, Meghalaya, East Khasi Hills District, Cherrapunji, 1460 m, 7 July 2013, *Barbhuiya & Verma 112212* (holotype, ASSAM).

Distribution. Andaman & Nicobar Islands, Arunachal Pradesh, Assam, Meghalaya, Odisha, Sikkim, West Bengal [Bhutan, ?Myanmar, Nepal].

Habit. Epiphytic herb.

Remarks. *Bulbophyllum cherrapunjeense* was synonymized with *B. sarcophyllum* by Kumar et al. (2018).

Bulbophyllum sasakii (Hayata) J. J. Verm., Schuit. & de Vogel, Phytotaxa 166: 105. 2014. *Ione sasakii* Hayata, Icon. Pl. Formosan. 2: 139. 1912. TYPE: Taiwan, Mt. Arisan, Jan. 1912, *B. Hayata & H. Sasaki s.n.*

Ione andersonii King & Pantl., Ann. Roy. Bot. Gard. (Calcutta) 8: 159, t. 217. 1898 (as "*andersoni*"). *Sunipia andersonii* (King & Pantl.) P. F. Hunt, Kew Bull. 26(1): 183. 1971. SYNTYPES: Bhutan: near Buxa, alt. 6000 ft, *Anderson s.n.* (syntype, CAL); *Pantling's drawing 473* (syntype, CAL).

Distribution. Arunachal Pradesh, Sikkim, West Bengal [Bhutan, China, Myanmar].

Habit. Epiphytic herb.

Remarks. This is not *Bulbophyllum andersonii* (Hook. f.) J. J. Sm.

Bulbophyllum scabratum Rchb. f., Ann. Bot. Syst. (Walpers) 6: 259. 1861. *Cirrhopetalum caespitosum* Wall. ex Lindl., Edwards's Bot. Reg. 24 (Misc.): 35. 1838. TYPE: Without exact locality (East Indies), *cult. Duke of Devonshire s.n.* (holotype, K-LINDL).

Bulbophyllum confertum Hook. f., Fl. Brit. India [J. D. Hooker] 5(16): 757. 1890. *Phyllorkis conferta* (Hook. f.) Kuntze, Revis. Gen. Pl. 2: 677. 1891 (as "*Phyllorchis*"). TYPE: India, Khasi Hills, *Griffith s.n.* (holotype, K; isotype, K-LINDL).

Bulbophyllum psychoon Rchb. f., Gard. Chron., n.s., 10: 170. 1878. *Phyllorkis psychoon* (Rchb. f.) Kuntze, Revis. Gen. Pl. 2: 677. 1891 (as "*Phyllorchis*"). TYPE: India, Assam, *Freeman s.n.* (holotype, W).

Distribution. Arunachal Pradesh, Assam, Meghalaya, Nagaland, Sikkim, West Bengal [Bhutan, China, Myanmar, Nepal].

Habit. Epiphytic herb.

Bulbophyllum secundum Hook. f., Fl. Brit. India [J. D. Hooker] 5(16): 764. 1890. *Phyllorkis secunda* (Hook. f.) Kuntze, Revis. Gen. Pl. 2: 678. 1891 (as "*Phyllorchis*"). TYPE: India, Upper Assam, Naga Hills, Kohima, June 1886, *Prain s.n.* [*41*] (holotype, K; isotype, CAL).

Distribution. Arunachal Pradesh, Assam, Madhya Pradesh, Meghalaya, Mizoram, Nagaland, ?Odisha, Sikkim, Uttarakhand, West Bengal [Bhutan, China, Myanmar, Nepal].

Habit. Epiphytic herb.

Bulbophyllum serratotruncatum Seidenf., Dansk Bot. Ark. 29: 50. 1973 [1974]. *Cirrhopetalum ochraceum* Ridl., J. Bot. 36: 212. 1898. *Bulbophyllum ochraceum* (Ridl.) Ridl., Mat. Fl. Malay. Penins. 1: 84. 1907, nom. illeg., non (Barb. Rodr.) Cogn., 1902. TYPE: Malaysia, Mt. Pahang, *Ridley s.n. (III-1897).*

Distribution. Andaman & Nicobar Islands [also in Malaysia].

Habit. Epiphytic herb.

Bulbophyllum silentvalliense M. P. Sharma & S. K. Srivast., J. Jap. Bot. 68(4): 209. 1993 (as "*silentvalliensis*"). TYPE: India, Kerala, Palghat, Silent Valley National Park, 1500 m, 21 Apr. 1987, *M. P. Sharma 16911* (holotype, CAL; isotype, CDRI).

Distribution. Kerala. Endemic.

Habit. Epiphytic herb.

Bulbophyllum spathulatum (Rolfe ex E. W. Cooper) Seidenf., Bot. Tidsskr. 65(4): 347. 1970. *Cirrhopetalum spathulatum* Rolfe ex E. W. Cooper, Orchid Rev. 37: 106. 1929. *Rhytionanthos spathulatus* (Rolfe ex E. W. Cooper) Garay, Hamer & Siegerist, Nordic J. Bot. 14(6): 639. 1994 (as "*spathulatum*"). TYPE: Thailand, Bangkok, *Roebelen 292-12* (holotype, K).

Distribution. Arunachal Pradesh, Manipur, Meghalaya, Mizoram, Nagaland, ?Odisha, Sikkim, West Bengal [Bhutan, China, Myanmar, Nepal].

Habit. Epiphytic herb.

Bulbophyllum stenobulbon C. S. P. Parish & Rchb. f., Trans. Linn. Soc. London 30: 153. 1874. *Phyllorkis stenobulbon* (C. S. P. Parish & Rchb. f.) Kuntze, Revis. Gen. Pl. 2: 678. 1891 (as "*Phyllorchis*"). TYPE: Myanmar, Moulmein (Mawlamyine), *C. S. P. Parish s.n.* [*319*] (holotype, K; isotype, W).

Bulbophyllum clarkeanum King & Pantl., J. Asiat. Soc. Bengal, Pt. 2, Nat. Hist. 64: 333. 1895 [1896]. TYPE: Bhutan, Kumai Forest, Jaldacca River, 1500 ft, June, *Pantling s.n.* [*319*] (holotype, CAL; isotype, K).

Distribution. Arunachal Pradesh, Manipur [Bhutan, Myanmar].

Habit. Epiphytic herb.

Bulbophyllum sterile (Lam.) Suresh in D. H. Nicolson, C. R. Suresh & K. S. Manilal, Interpret. Van Rheede's Hort. Malab.: 298. 1988. *Epidendrum sterile* Lam., Encycl. (Lamarck) 1(1): 189. 1783. TYPE: icon. Rheede, Hort. Malab. 12: t. 22. 1692. EPITYPE: India, Karnataka, Beltangadi [Belthangady], 25 Nov. 1900, *C. A. Barber 2532* (epitype, K, here designated).

Bulbophyllum neilgherrense Wight, Icon. Pl. Ind. Orient. [Wight] 5(1): 6, t. 1650. 1851. *Phyllorkis neilgherrensis* (Wight) Kuntze, Revis. Gen. Pl. 2: 677. 1891 (as "*Phyllorchis nilgherensis*"). TYPE: India, Nilgiri Hills & Malabar, *Wight s.n.* (holotype, K K000829149).

Distribution. Arunachal Pradesh, Assam, Goa, Karnataka, Kerala, Maharashtra, Meghalaya, Nagaland, Tamil Nadu [Bangladesh, Nepal].

Habit. Epiphytic herb.

Remarks. *Bulbophyllum sterile* was only known from an illustration of a non-flowering specimen in Rheede's *Hortus Malabaricus* (1692), where it is called "Thekamaravara." As there appears to be only one species of *Bulbophyllum* in the Malabar area that agrees with this illustration it seems justified to accept this name and to fix its identity by selecting an epitype, as above.

Bulbophyllum stocksii (Benth. ex Hook. f.) J. J. Verm., Schuit. & de Vogel, Phytotaxa 166: 111. 2014. *Trias stocksii* Benth. ex Hook. f., Fl. Brit. India [J. D. Hooker] 5(16): 781. 1890. SYNTYPES: India, Deccan Peninsula, Canara, *J. E. Stocks s.n.*; Concan, *Law s.n.* (syntypes, K, K-LINDL, P [2×]).

Distribution. Karnataka, Kerala, Maharashtra, Tamil Nadu. Endemic.

Habit. Epiphytic herb.

Bulbophyllum striatum (Griff.) Rchb. f., Ann. Bot. Syst. (Walpers) 6: 257. 1861. *Dendrobium striatum* Griff., Not. Pl. Asiat. 3: 318. 1851. *Sarcopodium striatum* (Griff.) Lindl., Fol. Orchid. 2(*Sarcopodium*): 5. 1853. *Phyllorkis striata* (Griff.) Kuntze, Revis. Gen. Pl. 2: 678. 1891 (as "*Phyllorchis*"). TYPE: Assam, Membree, 11 Nov. 1835, *Griffith in Assam Herb. 236* (holotype, K-LINDL).

Distribution. Arunachal Pradesh, Meghalaya, Nagaland, Sikkim, West Bengal [Bhutan, China, Myanmar, Nepal].

Habit. Epiphytic herb.

Bulbophyllum sunipia J. J. Verm., Schuit. & de Vogel, Phytotaxa 166: 105. 2014. *Sunipia scariosa* Lindl., Gen. Sp. Orchid. Pl.: 179. 1833. *Ione scariosa* (Lindl.) King & Pantl., Ann. Roy. Bot. Gard. (Calcutta) 8: 161, t. 219. 1898. TYPE: Nepal, *Wallich 7373* (holotype, K-LINDL; isotype, BM, K, K-WALL).

Distribution. Manipur, Meghalaya, Mizoram, Nagaland, Sikkim, West Bengal [China, Myanmar, Nepal].

Habit. Epiphytic herb.

Bulbophyllum tenuifolium (Blume) Lindl., Gen. Sp. Orchid. Pl.: 50. 1830. *Diphyes tenuifolia* Blume, Bijdr. Fl. Ned. Ind.: 316. 1825. *Phyllorkis tenuifolia* (Blume) Kuntze, Revis. Gen. Pl. 2: 678. 1891 (as "*Phyllorchis tenuiflora*"). TYPE: Indonesia, Java, Mt. Salak, *Blume 639* (holotype, L).

Distribution. Andaman & Nicobar Islands [also in Indonesia, Thailand].

Habit. Epiphytic herb.

Bulbophyllum tortuosum (Blume) Lindl., Gen. Sp. Orchid. Pl.: 50. 1830. *Diphyes tortuosa* Blume, Bijdr. Fl. Ned. Ind. 7: 310, t. 311. 1825. *Phyllorkis tortuosa* (Blume) Kuntze, Revis. Gen. Pl. 2: 678. 1891 (as "*Phyllorchis*"). TYPE: Indonesia, Java, Mt. Salak, *Blume s.n.* (holotype, L).

Bulbophyllum listeri King & Pantl., J. Asiat. Soc. Bengal, Pt. 2, Nat. Hist. 64: 334. 1895 [1896]. TYPE: Bhutan, Rumpti Lake, 1000 ft, Mar., *Lister s.n.* (holotype, CAL; isotype, AMES, BM, K, W).

Distribution. Arunachal Pradesh, Assam, West Bengal [Myanmar].

Habit. Epiphytic herb.

Bulbophyllum tremulum Wight, Icon. Pl. Ind. Orient. [Wight] 5(1): 20, t. 1749. 1851. *Phyllorkis tremula* (Wight) Kuntze, Revis. Gen. Pl. 2: 678. 1891 (as "*Phyllorchis*"). TYPE: India, Wynaud, *Jerdon & Cotton s.n.*

Distribution. Arunachal Pradesh, Karnataka, Kerala, Tamil Nadu. Endemic.

Habit. Epiphytic herb.

Bulbophyllum trichocephalum (Schltr.) Tang & F. T. Wang, Acta Phytotax. Sin. 1: 90. 1951. *Cirrhopetalum trichocephalum* Schltr., Repert. Spec. Nov. Regni Veg. Beih. 4: 72. 1919. TYPE: China, Yunnan, Szemao, *A. Henry 12086* (holotype, B, lost).

Bulbophyllum trichocephalum (Schltr.) Tang & F. T. Wang var. *wallongense* Agrawala, Sabap. & H. J. Chowdhery, Indian J. Forest. 27: 305. 2004. TYPE: India, Arunachal Pradesh, Lohit District, Wallong, *D. K. Agrawala & C. M. Sabapathy 32586* (holotype, CAL).
Bulbophyllum trichocephalum (Schltr.) Tang & F. T. Wang var. *capitatum* Lucksom, Orchids Sikkim N. E. Himalaya: 695. 2007. TYPE: India, Sikkim, Dikchu, 80–300 m, *Lucksom 348* (holotype, CAL; isotype, CAL).

Distribution. Arunachal Pradesh, Assam, ?Meghalaya, Sikkim [China].

Habit. Epiphytic herb.

Bulbophyllum tridentatum Kraenzl., Bot. Tidsskr. 24: 8. 1901. TYPE: Thailand, Koh Chang, *Schmidt 465* (holotype, C).

Distribution. Arunachal Pradesh [also in Thailand].

Habit. Epiphytic herb.

Bulbophyllum triste Rchb. f., Ann. Bot. Syst. (Walpers) 6: 253. 1861. TYPE: India, Khasi Hills, *cult. T. Lobb s.n.* (holotype, K-LINDL; isotype, W).

Bulbophyllum alopecurum Rchb. f., Gard. Chron., n.s., 14: 70. 1880. TYPE: Myanmar, *cult. S. Low (leg. R. Curnow) s.n.* (holotype, W).
Bulbophyllum micranthum Hook. f., Fl. Brit. India [J. D. Hooker] 5(16): 768. 1890, nom. illeg., non Barb. Rodr., 1877. TYPE: India, Tenasserim, Teongu, 1 Feb. 1868, *C. S. P. Parish s.n.* [207] (holotype, K).

Distribution. Arunachal Pradesh, Assam, Bihar, Madhya Pradesh, Manipur, Meghalaya, Mizoram, Nagaland, Odisha, Sikkim, Uttarakhand, West Bengal [Myanmar, Nepal].

Habit. Epiphytic herb.

Bulbophyllum umbellatum Lindl., Gen. Sp. Orchid. Pl.: 56. 1830. *Phyllorkis umbellata* (Lindl.) Kuntze, Revis. Gen. Pl. 2: 675. 1891 (as "*Phyllorchis*"). TYPE: Nepal, Sankoo, Apr. 1821, *Wallich 1984* (holotype, K-WALL; isotype, BM, K-LINDL, icon.).

Cirrhopetalum maculosum Lindl., Edwards's Bot. Reg. 27 (Misc.): 81. 1841. *Phyllorkis maculosa* (Lindl.) Kuntze, Revis. Gen. Pl. 2: 677 (1891) (as "*Phyllorchis*"). *Bulbophyllum maculosum* (Lindl.) Garay, Hamer & Siegerist, Nordic J. Bot. 14: 631 (1994), nom. illeg., non Ames, 1915. TYPE: India, *Wallich s.n.* (holotype, K-LINDL).
Cirrhopetalum bootanense Griff., Not. Pl. Asiat. 3: 296. 1851 (as "*bootanensis*"). *Bulbophyllum bootanense* (Griff.) C. S. P. Parish & Rchb. f., Trans. Linn. Soc. London 30: 153. 1874 (excl. descr. & pl. = *B. spathulatum*). *Phyllorkis bootanensis* (Griff.) Kuntze, Revis. Gen. Pl. 2: 677 (1891) (as "*Phyllorchis*"). *Rhytionanthos bootanensis* (Griff.) Garay, Hamer & Siegerist, Nordic J. Bot. 14: 639. 1994 (as "*bootanense*"). TYPE: Not designated.
Cirrhopetalum maculosum Lindl. var. *fuscescens* Hook. f., Fl. Brit India 5: 776. 1890. *Bulbophyllum umbellatum* Lindl. var. *fuscescens* (Hook. f.) P. K. Sarkar, J. Econ. Taxon. Bot. 5(5): 1007. 1984. SYNTYPES: Nepal and India, Sikkim, not designated.

Distribution. Assam, Manipur, Meghalaya, Mizoram, Nagaland, Odisha, Sikkim, Tripura, Uttarakhand, West Bengal [Bhutan, China, Myanmar, Nepal].

Habit. Epiphytic herb.

Bulbophyllum viridiflorum (Hook. f.) Schltr., Orchis 4: 108. 1910. *Cirrhopetalum viridiflorum* Hook. f., Fl. Brit. India [J. D. Hooker] 5(16): 779. 1890. *Phyllorkis viridiflora* (Hook. f.) Kuntze, Revis. Gen. Pl. 2: 677. 1891 (as "*Phyllorchis*"). *Tripudianthes viridiflora* (Hook. f.) Szlach. & Kras, Richardiana 7(2): 96. 2007. SYNTYPES: India, Sikkim Himalaya, alt. 6000–7000 ft, 21 Oct. 1874, *Treutler s.n.* [947 p.p.] (syntype, K); Khasi Hills, Shillong, alt. 5500 ft, 15 Oct. 1867, *C. B. Clarke s.n.* [5737] (syntype, K).

Distribution. Arunachal Pradesh, Meghalaya, Mizoram, Nagaland, Sikkim, West Bengal [Nepal].

Habit. Epiphytic herb.

Bulbophyllum wallichii Rchb. f., Ann. Bot. Syst. (Walpers) 6: 259. 1861. *Cirrhopetalum wallichii* Lindl., Edwards's Bot. Reg. 25(Misc.): 72. 1839, nom. illeg., non Lindl., 1830. *Bulbophyllum re-*

fractoides Seidenf., Bot. Tidsskr. 65: 342. 1970, nom. superfl. *Tripudianthes wallichii* (Rchb. f.) Szlach. & Kras, Richardiana 7(2): 96. 2007. TYPE: Nepal, *Wallich 1980* (holotype, K-WALL).

Distribution. Arunachal Pradesh, Assam, Meghalaya, Mizoram, Nagaland, Sikkim, Uttarakhand, West Bengal [Bhutan, Myanmar, Nepal].

Habit. Epiphytic herb.

Bulbophyllum xylophyllum C. S. P. Parish & Rchb. f., Trans. Linn. Soc. London 30: 151. 1874. *Phyllorkis xylophylla* (C. S. P. Parish & Rchb. f.) Kuntze, Revis. Gen. Pl. 2: 678. 1891 (as *"Phyllorchis"*). TYPE: Myanmar, Moulmein [Mawlamyine], *C. S. P. Parish s.n. [82B]* (holotype, K).

Bulbophyllum agastyamalayanum Gopalan & A. N. Henry, J. Bombay Nat. Hist. Soc. 90: 78. 1993. TYPE: India, Tamil Nadu, Tirunelveli, Kattabomman District, Agastyamalai, *Gopalan 96220* (holotype, CAL; isotype, MH).

Distribution. Arunachal Pradesh, Assam, Mizoram, Tamil Nadu [Myanmar].

Habit. Epiphytic herb.

Bulbophyllum yoksunense J. J. Sm., Bull. Jard. Bot. Buitenzorg, sér. 2, 8: 29. 1912. *Cirrhopetalum brevipes* Hook. f., Fl. Brit. India [J. D. Hooker] 5(16): 777. 1890. *Phyllorkis brevipes* (Hook. f.) Kuntze, Revis. Gen. Pl. 2: 677. 1891 (as *"Phyllorchis"*). *Bulbophyllum brachypodum* A. S. Rao & N. P. Balakr., Bull. Bot. Surv. India 10(3–4): 350. 1968 [1969], nom. superfl. TYPE: India, Sikkim Himalaya, betw. Yoksun & Jongri, alt. 8000 ft, *T. Anderson s.n. [1253]* (holotype, CAL).

Bulbophyllum brachypodum A. S. Rao & N. P. Balakr. var. *parviflorum* A. S. Rao & N. P. Balakr., Bull. Bot. Surv. India 10(3–4): 350. 1968 [1969], nom. superfl. *Bulbophyllum yoksunense* J. J. Sm. var. *parviflorum* (A. S. Rao & N. P. Balakr.) Bennet, J. Econ. Taxon. Bot. 4(2): 592. 1983. TYPE: Bhutan, Gumdrithang, *N. P. Balakrishnan 43004A* (holotype, CAL).

Distribution. Arunachal Pradesh, Nagaland, Sikkim, West Bengal [Bhutan, Myanmar, Nepal].

Habit. Epiphytic herb.

———

Calanthe R. Br., Bot. Reg. 7: ad t. 573 (as "578"). 1821, nom. cons.

TYPE: *Calanthe veratrifolia* R. Br. ex Ker Gawl. (Bot. Reg. 9: ad t. 720. 1823), nom. superfl. (*Limodorum veratrifolium* Willd., nom. superfl. = *Calanthe triplicata* (P. Willemet) Ames) (*Orchis triplicata* P. Willemet).

Calanthe alismifolia Lindl., Fol. Orchid. 6/7(*Calanthe*): 8. 1855 (as *"alismaefolia"*). *Alismorkis alismifolia* (Lindl.) Kuntze, Revis. Gen. Pl. 2: 650. 1891 (as *"Alismorchis"*). SYNTYPES: India, Sikkim, in hot valleys, alt. 2000 ft, *J. D. Hooker s.n.* (syntype, CAL); Khasi Hills, alt. 4000 ft, *Hooker & Thomson s.n. [239]* (syntype, K-LINDL).

Distribution. Arunachal Pradesh, Assam, Meghalaya, Nagaland, Sikkim, Uttarakhand, West Bengal [Bhutan, China, Myanmar].

Habit. Terrestrial herb.

Calanthe alpina Hook. f. ex Lindl., Fol. Orchid. 6/7(*Calanthe*): 4. 1855. *Alismorkis alpina* (Hook. f. ex Lindl.) Kuntze, Revis. Gen. Pl. 2: 650. 1891 (as *"Alismorchis"*). TYPE: India, Sikkim, Lachen, 10,000 ft, *J. D. Hooker s.n. [245]* (holotype, K-LINDL; isotype, AMES, K, P).

Distribution. Arunachal Pradesh, Nagaland, Sikkim, Uttarakhand [Bhutan, China, Myanmar, Nepal].

Habit. Terrestrial herb.

Calanthe anthropophora Ridl., J. Fed. Malay States Mus. 5: 167. 1914. TYPE: Thailand, Koh Samui, hills of Itoh, 7 May 1913, *H. C. Robinson 5701* (holotype, K).

Distribution. Assam, Meghalaya [also in Thailand].

Habit. Terrestrial herb.

Calanthe biloba Lindl., Fol. Orchid. 6/7(*Calanthe*): 3. 1855. *Alismorkis biloba* (Lindl.) Kuntze, Revis. Gen. Pl. 2: 650. 1891 (as *"Alismorchis"*). TYPE: India, Sikkim, alt. 4000 ft *J. D. Hooker s.n. [246]* (holotype, K-LINDL; isotype, AMES, CAL, K, P).

Calanthe biloba Lindl. var. *diptera* Hook. f., Fl. Brit. India [J. D. Hooker] 5(16): 848. 1890. TYPE: India, Naga Hills, Oct. 1886, *Prain 72* (holotype, K).
Calanthe biloba Lindl. var. *treutleri* Hook. f., Fl. Brit. India [J. D. Hooker] 5(16): 848. 1890. TYPE: India, Sikkim, 6000 ft, 1 Dec. 1874, *Treutler 1155* (holotype, K).

Distribution. Arunachal Pradesh, Assam, Manipur, Meghalaya, Nagaland, Sikkim, West Bengal [Bhutan, China, Myanmar, Nepal].

Habit. Terrestrial herb.

Calanthe brevicornu Lindl., Gen. Sp. Orchid. Pl.: 251. 1833. *Alismorkis brevicornu* (Lindl.) Kuntze, Revis. Gen. Pl. 2: 650. 1891 (as *"Alismorchis"*). TYPE: Nepal, *Wallich 7338* (syntypes, K-LINDL, K-WALL).

Calanthe brevicornu Lindl. var. *wattii* Hook. f., Fl. Brit. India [J. D. Hooker] 5(16): 848. 1890. TYPE: India, Manipur, Jopu, 7000 ft, 15 May 1882, *Watt 6898* (holotype, K; isotype, CAL).

Distribution. Arunachal Pradesh, Manipur, Mizoram, Nagaland, Sikkim, Uttarakhand, West Bengal [Bhutan, China, Myanmar, Nepal].

Habit. Terrestrial herb.

Calanthe ceciliae Low ex Rchb. f., Gard. Chron., n.s., 19: 432. 1883. TYPE: Malaysia Peninsula, Perak, *cult. Sir Hugh Low s.n.* (holotype, W).

Calanthe burmanica Rolfe, Bull. Misc. Inform. Kew 1907: 129. 1907. TYPE: Myanmar, Shan States, *cult. Royal Botanic Garden, Glasnevin s.n.* Sep. 1896 (holotype, K).
Calanthe wrayi Hook. f., Fl. Brit. India [J. D. Hooker] 5(16): 850. 1890. TYPE: Malaysia, Perak, upper part of Batang Padang Valley, alt. 2000 ft, *Wray s.n.* [*1451*] (holotype, K).

Distribution. ?Assam, Nagaland [Myanmar].

Habit. Terrestrial herb.

Calanthe chloroleuca Lindl., Fol. Orchid. 6/7(*Calanthe*): 10. 1855. *Alismorkis chloroleuca* (Lindl.) Kuntze, Revis. Gen. Pl. 2: 650. 1891 (as "*Alismorchis*"). TYPE: India, Sikkim, hot Valleys, alt. 2000 ft, *J. D. Hooker s.n.* [*244*] (holotype, K-LINDL).

Calanthe galeata Lindl., Fol. Orchid. 6/7 (*Calanthe*): 5. 1855. TYPE: India, Sikkim, near Darjeeling, *J. D. Hooker s.n.* (holotype, K-LINDL).

Distribution. Arunachal Pradesh, Nagaland, Sikkim, West Bengal [Bhutan].

Habit. Terrestrial herb.

Remarks. Singh et al. (2019) include *Calanthe chloroleuca* in *C. griffithii* Lindl. as a new synonym. We believe the two to be distinct species. *Calanthe griffithii* may be recognized by the short spur, which is a little shorter than the ovary (excluding the much longer pedicel), and by the distinct lamellate tooth on the mid-lobe of the uniformly yellow lip, whereas *C. chloroleuca* has a much longer spur, almost as long as the pedicel with ovary, and lacks a tooth on the mid-lobe, which is white with a yellow patch at the base. The species illustrated as *C. chloroleuca* by King and Pantling (1898) is *C. griffithii*.

Calanthe davidii Franch., Nouv. Arch. Mus. Hist. Nat., sér. 2, 10: 85. 1888. SYNTYPES: China, Sichuan, Moupin, *David s.n.* (syntype, P); Sichuan, *Legendre 46* (syntype, P).

Calanthe pachystalix Rchb. f. ex Hook. f., Fl. Brit. India [J. D. Hooker] 5(16): 850. 1890. *Alismorkis pachystalix* (Rchb. f.) Kuntze, Revis. Gen. Pl. 2: 650. 1891 (as "*Alismorchis pachystalyx*"). TYPE: Western Himalaya, *Falconer s.n.* [*1054*] (holotype, K-LINDL).

Distribution. Arunachal Pradesh, Himachal Pradesh, Uttarakhand [China, Nepal].

Habit. Terrestrial herb.

Remarks. This species is illustrated as *Calanthe whiteana* King & Pantl. in Swami (2016, 2017).

Calanthe densiflora Lindl., Gen. Sp. Orchid. Pl.: 250. 1833. *Alismorkis densiflora* (Lindl.) Kuntze, Revis. Gen. Pl. 2: 650. 1891 (as "*Alismorchis*"). *Styloglossum densiflorum* (Lindl.) T. Yukawa & P. J. Cribb, Bull. Natl. Mus. Nat. Sci., Tokyo, B. 40: 148. 2014. TYPE: India, Sylhet, Pundua Hills, *Wallich 7344 (leg. De Silva s.n.)* (syntypes, K-LINDL, K-WALL).

Calanthe clavata Lindl., Gen. Sp. Orchid. Pl.: 251. 1833. *Alismorkis clavata* (Lindl.) Kuntze, Revis. Gen. Pl. 2: 650. 1891 (as "*Alismorchis*"). *Styloglossum clavatum* (Lindl.) T. Yukawa & P. J. Cribb, Bull. Natl. Mus. Nat. Sci., Tokyo, B. 40: 148. 2014. TYPE: Bangladesh or India, Sylhet, *Wallich 7343 (leg. De Silva s.n.)* (holotype, K-WALL).
Phaius epiphyticus Seidenf., Nordic J. Bot. 5(2): 159, fig. 2. pl. Ia. 1985. TYPE: Thailand: Phuluang, Loei, alt. 4480–4800 ft, *GT 8687* (holotype, C).

Distribution. Arunachal Pradesh, Assam, Manipur, Meghalaya, Mizoram, Nagaland, Sikkim, West Bengal [Bhutan, China, Bangladesh, Myanmar, Nepal].

Habit. Terrestrial or sometimes epiphytic herb.

Remarks. Typical forms of *Calanthe clavata* and *C. densiflora* look rather different; the former having elongate inflorescences and distinctly clavate spurs, the latter having condensed inflorescences and narrowly cylindric spurs. However, intermediate specimens exist with condensed inflorescences and clavate spurs, while the position of the lamellae on the lip varies as well. Therefore, we agree with King and Pantling (1898), who, unlike some more recent authors, considered them conspecific.

Calanthe griffithii Lindl., Paxton's Fl. Gard. 3: 37 1852. *Alismorkis griffithii* (Lindl.) Kuntze, Revis. Gen. Pl. 2: 650. 1891 (as "*Alismorchis*"). SYNTYPES: Bhutan, Chuku, wet banks, alt. 6000 ft, *Griffith s.n.* [*33*] (syntype, K-LINDL); above Telagong, *Griffith s.n.* (syntype, K-LINDL).

Calanthe anjanae Lucksom, Indian J. Forest. 16: 386. 1993 [1994] (as "*anjanii*"). SYNTYPES: India, Sikkim, Fimphu, *Lucksom 206A, B, C & D* (syntypes, BSHC, CAL).

Distribution. Arunachal Pradesh, Assam, Nagaland, Sikkim, West Bengal [Bhutan, China, Nepal].

Habit. Terrestrial herb.

Calanthe hancockii Rolfe., Bull. Misc. Inform. Kew 1896: 197. 1896. TYPE: China, Yunnan, Mengtse, alt. 6600 ft, Apr. 1893, *Hancock 78* (holotype, K).

Distribution. Mizoram, Nagaland [China, Myanmar].

Habit. Terrestrial herb.

Calanthe herbacea Lindl., Fol. Orchid. 6/7(*Calanthe*): 10. 1855. *Alismorkis herbacea* (Lindl.) Kuntze, Revis. Gen. Pl. 2: 650. 1891 (as "*Alismorchis*"). TYPE: India, Sikkim, icon. *Cathcart s.n.* (holotype, K; partial copy K-LINDL).

Calanthe elytroglossa Rchb. f. ex Hook. f., Fl. Brit. India [J. D. Hooker] 5(16): 853. 1890. *Alismorkis elytroglossa* (Rchb. f. ex Hook. f.) Kuntze, Revis. Gen. Pl. 2: 650. 1891 (as "*Alismorchis*"). SYNTYPES: India, Sikkim, alt. 6000–8000 ft, *Treutler s.n.* [*537*] (syntype, K); *C. B. Clarke s.n.* [*12320*] (syntype, K); Darjeeling, *C. B. Clarke s.n.* [*8621*] (syntype, K).

Distribution. Arunachal Pradesh, Assam, Meghalaya, Nagaland, Sikkim, West Bengal [China, Myanmar, Nepal].

Habit. Terrestrial herb.

Calanthe keshabii Lucksom, Indian J. Forest. 15: 136. 1992. *Calanthe alpina* Hook. f. ex Lindl. var. *keshabii* (Lucksom) R. C. Srivast., Novon 8(2): 203. 1998. TYPE: India, Sikkim-Chungthang Valley, Shibgyar, *Lucksom 205a* (holotype, CAL; isotype, Gangtok, Forest Dept. Herb.).

Distribution. Arunachal Pradesh, Sikkim [Bhutan].

Habit. Terrestrial herb.

Calanthe longipes Hook. f., Fl. Brit. India [J. D. Hooker] 6(17): 195. 1890. *Alismorkis longipes* (Hook. f.) Kuntze, Revis. Gen. Pl. 2: 650. 1891 (as "*Alismorchis*"). *Phaius longipes* (Hook. f.) Holttum, Gard. Bull. Singapore 11: 286. 1947. *Cephalantheropsis longipes* (Hook. f.) Ormerod, Orchid Digest 62(4): 156. 1998. TYPE: India, Sikkim Himalaya, *King s.n.* (holotype, BM).

Distribution. Arunachal Pradesh, Assam, Manipur, Meghalaya, Mizoram, Nagaland, Sikkim [Bangladesh, China, Myanmar].

Habit. Terrestrial herb.

Calanthe lyroglossa Rchb. f., Otia Bot. Hamburg. 53. 1878. *Alismorkis lyroglossa* (Rchb. f.) Kuntze, Revis. Gen. Pl. 2: 650. 1891 (as "*Alismorchis*"). *Styloglossum lyroglossum* (Rchb. f.) T. Yukawa & P. J. Cribb, Bull. Natl. Mus. Nat. Sci., Tokyo, B. 40: 149. 2014. TYPE: Philippines, Mt. Mahahai, *Wilkes s.n.* (holotype, W; isotype, ?AMES, ?GH, HBG).

Calanthe foerstermannii Rchb. f., Gard. Chron., n.s., 19: 814. 1883. *Alismorkis foerstermannii* (Rchb. f.) Kuntze, Revis. Gen. Pl. 2: 650. 1891 (as "*Alismorchis*"). TYPE: Myanmar, *Foerstermann s.n.* (holotype, W).

Distribution. Nagaland [China, Myanmar].

Habit. Terrestrial herb.

Calanthe mannii Hook. f., Fl. Brit. India [J. D. Hooker] 5(16): 850. 1890. *Alismorkis mannii* (Hook. f.) Kuntze, Revis. Gen. Pl. 2: 650. 1891 (as "*Alismorchis*"). SYNTYPES: India, Kumaon, below Ranikhet, 30 May 1886, *Duthie s.n.* [*5996*] (syntypes, CAL, K); Khasi Hills, 4000 ft, Apr. 1882, *Mann s.n.* [*62*] (syntype, K); Khasi Hills, Myrung, 5000 ft, May 1890, *Mann s.n.* (syntype, K); E Khasi Hills, 18 May 1886, *C. B. Clarke 49321* (syntype, K).

Distribution. Arunachal Pradesh, Assam, Manipur, Meghalaya, Mizoram, Nagaland, Uttarakhand, West Bengal [Bhutan, China, Myanmar, Nepal].

Habit. Terrestrial herb.

Calanthe masuca (D. Don) Lindl., Gen. Sp. Orchid. Pl.: 249. 1833. *Bletia masuca* D. Don, Prodr. Fl. Nepal.: 30. 1825. *Alismorkis masuca* (D. Don) Kuntze, Revis. Gen. Pl. 2: 650. 1891 (as "*Alismorchis*"). TYPE: Nepal, Bagmati Zone, Narainhetty, 21 Feb. 1803, *Buchanan-Hamilton s.n.* (holotype, BM; isotype, LINN).

Calanthe purpurea Lindl., Gen. Sp. Orchid. Pl.: 249. 1833. *Alismorkis purpurea* (Lindl.) Kuntze, Revis. Gen. Pl. 2: 650. 1891 (as "*Alismorchis*"). TYPE: Sri Lanka, *Macrae s.n.* (holotype, K-LINDL).
Calanthe fulgens Lindl., Fol. Orchid. 6/7(*Calanthe*): 10. 1855. *Calanthe masuca* (D. Don) Lindl. var. *fulgens* (Lindl.) Hook. f., Fl. Brit. India [J. D. Hooker] 5(16): 851. 1890. TYPE: India, Sikkim, *J. D. Hooker 248* (holotype, K-LINDL).
Calanthe wightii Rchb. f., Ann. Bot. Syst. (Walpers) 6: 932. 1864. TYPE: India, Courtallum, *Wight s.n.* [*2992*] (holotype, K; isotype, AMES).

Distribution. Arunachal Pradesh, Assam, Karnataka, Kerala, Manipur, Meghalaya, Mizoram, Nagaland, Sikkim, Tamil Nadu, West Bengal [Bhutan, Myanmar, Nepal, Sri Lanka].

Habit. Terrestrial herb.

Remarks. Indian specimens identified as *Calanthe sylvatica* (Thouars) Lindl. belong to *C. masuca*, which is considered a similar but distinct species by Clayton and Cribb (2013). The photograph labeled *C. sylvatica* in Singh et al. (2019) shows *C. perrottetii* A. Rich.; these authors incorrectly treat the latter as a synonym of *C. triplicata* (P. Willemet) Ames.

Calanthe mishmensis (Lindl. & Paxton) M. W. Chase, Christenh. & Schuit., Phytotaxa 472(2): 163. 2020. *Limatodis mishmensis* Lindl. & Paxton, Paxton's Fl. Gard. 3: 36. 1852. *Phaius mishmensis* (Lindl. & Paxton) Rchb. f., Bonplandia 5(3): 43. 1857. TYPE: India, Mishmee Hills, *Griffith s.n.* (holotype, K-LINDL).

Distribution. Arunachal Pradesh, Assam, Manipur, Meghalaya, Mizoram, Nagaland, Sikkim, West Bengal [Bangladesh, Bhutan, China, Myanmar].

Habit. Terrestrial herb.

Calanthe nana (Hook. f.) M. W. Chase, Christenh. & Schuit., Phytotaxa 472(2): 164. 2020. *Phaius nanus* Hook. f., Fl. Brit. India [J. D. Hooker] 6(17): 192. 1890. TYPE: India, Bengal, Buxa Res., W Duars, 1 Feb. 1879, *Gamble 6672B* (holotype, K).

Distribution. Assam, Meghalaya, Nagaland, West Bengal [Bangladesh, Myanmar].

Habit. Terrestrial herb.

Calanthe obcordata (Lindl.) M. W. Chase, Christenh. & Schuit., Phytotaxa 472(2): 164. 2020. *Bletia obcordata* Lindl., Gen. Sp. Orchid. Pl.: 123. 1831. *Cephalantheropsis obcordata* (Lindl.) Ormerod, Orchid Digest 62: 157. 1998. TYPE: India, Sylhet, icon. *Wallich s.n.* (not found).

Calanthe gracilis Lindl., Gen. Sp. Orchid. Pl.: 251. 1833. *Limatodis gracilis* (Lindl.) Lindl., Fol. Orchid. 6/7(*Calanthe*): 1. 1855. *Alismorkis gracilis* (Lindl.) Kuntze, Revis. Gen. Pl. 2: 650. 1891. *Paracalanthe gracilis* (Lindl.) Kudô, J. Soc. Trop. Agric. 2: 236. 1930. *Cephalantheropsis gracilis* (Lindl.) S. Y. Hu, Quart. J. Taiwan Mus. 25: 213. 1972. *Phaius gracilis* (Lindl.) S. S. Ying, Coloured Ill. Indig. Orchids Taiwan 1(2): 278. 1977, nom. illeg., non Hayata, 1911. *Gastrorchis gracilis* (Lindl.) Aver., Prelim. list Vietnam. Orchids: 203. 1988. TYPE: India, Sylhet, Pundua Hills, *Wallich 7341 (leg. De Silva)* (holotype, K-WALL; isotype, AAU, K-LINDL).

Distribution. Assam, Sikkim [Bangladesh, China, Myanmar].

Habit. Terrestrial herb.

Calanthe odora Griff., Not. Pl. Asiat. 3: 365. 1851. *Alismorkis odora* (Griff.) Kuntze, Revis. Gen. Pl. 2: 650. 1891 (as "*Alismorchis*"). TYPE: India, Upper Assam, Suddyah, 10 Apr. 1836, *Griffith in Assam Herb. 477* (?syntype, W).

Calanthe angusta Lindl., Fol. Orchid. 6/7(*Calanthe*): 7. 1855. *Alismorkis angusta* (Lindl.) Kuntze, Revis. Gen. Pl. 2: 650. 1891 (as "*Alismorchis*"). TYPE: India, Khasi Hills, *T. Lobb s.n.* (holotype, K-LINDL; isotype, K).
Calanthe vaginata Lindl., Fol. Orchid. 6/7(*Calanthe*): 7. 1855. *Alismorkis vaginata* (Lindl.) Kuntze, Revis. Gen. Pl. 2: 650. 1891 (as "*Alismorchis*"). TYPE: India, Assam, 1829, *Jenkins s.n.* (holotype, K).

Distribution. Arunachal Pradesh, Assam, Manipur, Meghalaya, Nagaland, West Bengal [Bhutan, China, ?Myanmar, Nepal].

Habit. Terrestrial herb.

Remarks. The photo labeled *Calanthe odora* in Singh et al. (2019) represents *C. triplicata* (P. Willemet) Ames.

Calanthe perrottetii A. Rich., Ann. Sci. Nat., Bot. sér. 2, 15: 68. 1841. TYPE: India, Nilgiri Hills, Avalanchy, *Perrottet s.n.* [*1108 in G*] (holotype, P; isotype, G).

Calanthe comosa Rchb. f., Linnaea 19: 374. 1847 [1846]. TYPE: India, Nilgiri Hills, *Delessert s.n.* (holotype, W).

Distribution. Kerala, Tamil Nadu [Sri Lanka].

Habit. Terrestrial herb.

Remarks. The photo labeled *Calanthe sylvatica* (Thouars) Lindl. in Singh et al. (2019) shows *C. perrottetii*.

Calanthe plantaginea Lindl., Gen. Sp. Orchid. Pl.: 250. 1833. *Alismorkis lindleyana* Kuntze, Revis. Gen. Pl. 2: 650. 1891 (as "*Alismorchis*"). SYNTYPES: Nepal, Feb. 1821, *Wallich 7346A* (syntypes, K-LINDL, K-WALL); India, Kumaon, *Wallich 7346B (leg. Blinkworth)* (syntypes, K-LINDL, K-WALL, P).

Distribution. Arunachal Pradesh, Himachal Pradesh, Nagaland, Uttarakhand [Bhutan, China, Myanmar, Nepal].

Habit. Terrestrial herb.

Calanthe puberula Lindl., Gen. Sp. Orchid. Pl.: 252. 1833. *Alismorkis puberula* (Lindl.) Kuntze, Revis. Gen. Pl. 2: 650. 1891 (as "*Alismorchis*"). *Paracalanthe reflexa* (Maxim.) Kudô var. *puberula* (Lindl.) Kudô, J. Soc. Trop. Agric. 2: 235. 1930. *Calanthe reflexa* Maxim. var. *puberula* (Lindl.) Kudô, J. Soc. Trop. Agric. 2: 236. 1930, nom. superfl. TYPE: India, Sylhet, Pundua Hills, *Wallich 7342 (leg. De Silva s.n.)* (holotype, K-WALL).

Distribution. Arunachal Pradesh, Assam, Himachal Pradesh, Meghalaya, Nagaland, Sikkim, Uttarakhand, West Bengal [Bangladesh, Bhutan, China, Myanmar, Nepal].

Habit. Terrestrial herb.

Calanthe simplex Seidenf., Dansk Bot. Ark. 29(2): 42. 1975. TYPE: Thailand, Chiang Mai Prov., Doi Chiang Dao, 21 Dec. 1931, *Put 4470* (holotype, K).

Distribution. Manipur [China, Myanmar].

Habit. Terrestrial herb.

Calanthe tankervilleae (Banks) M. W. Chase, Christenh. & Schuit., Phytotaxa 472(2): 165. 2020. *Limodorum tankervilleae* Banks, icon. *"Limodorum Tankervilliae."* 1788. *Limodorum incarvillei* Pers., Syn. Pl. 2: 520. 1807, nom. superfl. *Bletia tankervilleae* (Banks) R. Br., Bot. Mag. 44: t. 1924. 1817. *Phaius tankervilleae* (Banks) Blume, Mus. Bot. 2: 177. 1856 (as *"tankervillii"*). *Phaius incarvillei* (Pers.) Kuntze, Revis. Gen. Pl. 2: 675. 1891, nom. superfl. TYPE: Icon. *"Limodorum Tankervilliae"* by J. Sowerby (lectotype, BM, designated by Mabberley, 2011).

Phaius veratrifolius Lindl., Gen. Sp. Orchid. Pl.: 127. 1831. TYPE: India or Bangladesh, Sylhet, *Wallich 3746 (leg. De Silva s.n.)* (holotype, K-WALL).
Phaius blumei Lindl. var. *pulchra* King & Pantl., Ann. Roy. Bot. Gard. (Calcutta) 8: 108, t. 151. 1898. *Phaius tankervilleae* (Banks) Blume var. *pulchra* (King & Pantl.) Karthik. in S. Karthikeyan et al., Fl. Ind. ser. 4, 1 (Monocotyledon): 163. 1989. TYPE: India, Sikkim, tropical valleys, 1891, *Pantling 139* (holotype, K; isotype, BM, CAL, LE, W).

Distribution. Arunachal Pradesh, Assam, Jharkhand, Madhya Pradesh, Manipur, Meghalaya, Mizoram, Nagaland, Odisha, Sikkim, Tripura, Uttarakhand, West Bengal [Bhutan, China, Myanmar, Nepal].

Habit. Terrestrial herb.

Calanthe testacea M. W. Chase, Christenh. & Schuit., Phytotaxa 472(2): 165. 2020. *Phaius luridus* Thwaites, Enum. Pl. Zeyl. [Thwaites]: 300. 1861. TYPE: Sri Lanka, Saffragam District, Rakwne, *Thwaites s.n. [Ceylon Plants 613]* (holotype, K; isotype, CAL).

Distribution. Kerala, Tamil Nadu [Sri Lanka].

Habit. Terrestrial herb.

Remarks. This is not *Calanthe lurida* Decne.

Calanthe tricarinata Lindl., Gen. Sp. Orchid. Pl.: 252. 1833. *Alismorkis tricarinata* (Lindl.) Kuntze,

Revis. Gen. Pl. 2: 650. 1891 (as *"Alismorchis"*). *Paracalanthe tricarinata* (Lindl.) Kudô, J. Soc. Trop. Agric. 2: 236. 1930. TYPE: Nepal, *Wallich 7339* (syntypes, K, K-LINDL, K-WALL).

Calanthe pantlingii Schltr., Repert. Spec. Nov. Regni Veg. Beih. 4: 240. 1919. TYPE: *Ann. Roy. Bot. Gard. (Calcutta) 8: t. 233. 1898* (iconotype).
Calanthe occidentalis Lindl., Fol. Orchid. 6/7(*Calanthe*): 3. 1855. TYPE: India, Himalayas, *Hooker & Thomson 237* (holotype, K-LINDL; isotype, P).

Distribution. Arunachal Pradesh, Himachal Pradesh, Jammu & Kashmir, Manipur, Meghalaya, Mizoram, Nagaland, Sikkim, Uttarakhand, West Bengal [Bhutan, China, Myanmar, Nepal, Pakistan].

Habit. Terrestrial herb.

Calanthe triplicata (P. Willemet) Ames, Philipp. J. Sci., C. 2(4): 326. 1907 (as *"triplicatis"*). *Orchis triplicata* P. Willemet, Ann. Bot. (Usteri) 18: 52. 1796. *Limodorum veratrifolium* Willd., Sp. Pl., ed. 4. 4: 122. 1805, nom. superfl. *Calanthe veratrifolia* R. Br. ex Ker Gawl., Bot. Reg. 9: t. 720. 1823, nom. superfl. TYPE: Icon. Rumphius, Herb. Amb. *6: t. 52 fig. 2. 1750* (lectotype, designated by Kores, 1989).

Distribution. Andaman & Nicobar Islands, Arunachal Pradesh, Assam, Meghalaya, Mizoram, Nagaland, West Bengal [Bhutan, China, Myanmar].

Habit. Terrestrial herb.

Calanthe trulliformis King & Pantl., J. Asiat. Soc. Bengal, Pt. 2, Nat. Hist. 64: 337. 1895 [1896]. TYPE: India, Sikkim, Mahaldaram Peak, 6000 ft, July 1891, *Pantling s.n. [168]* (holotype, CAL; isotype, ?BM, CAL, ?K, ?P).

Distribution. Arunachal Pradesh, Sikkim, West Bengal [Bhutan, Nepal].

Habit. Terrestrial herb.

Remarks. The potential isotypes at BM, K, and P either do not agree with the published collecting locality or were collected after 1895; only specimens in CAL agree with the protologue. Pantling numbered his collections by species rather than by gathering number; specimens collected at different times at different localities often received the same number. This can make it difficult to identify type specimens, especially if—as we suspect may have happened—mistakes were made copying dates and localities to labels.

Calanthe uncata Lindl., Fol. Orchid. 6/7(*Calanthe*): 6. 1855. *Alismorkis uncata* (Lindl.) Kuntze, Revis.

Gen. Pl. 2: 650. 1891 (as "*Alismorchis*"). TYPE: India, Sikkim, Darjeeling, *Griffith s.n.* (holotype, K-LINDL).

Distribution. West Bengal. Endemic.

Habit. Terrestrial herb.

Remarks. This species is still only known from the type collection and may be extinct.

Calanthe vestita Wall. ex Lindl., Gen. Sp. Orchid. Pl.: 250. 1833. *Phaius vestitus* (Wall. ex Lindl.) Rchb. f., Gard. Chron. 1867: 264. 1867. *Alismorkis vestita* (Wall. ex Lindl.) Kuntze, Revis. Gen. Pl. 2: 650. 1891 (as "*Alismorchis*"). *Preptanthe vestita* (Wall. ex Lindl.) Rchb. f., Fl. des Serres Jard. Eur. 8: 245. 1853. TYPE: Myanmar, Tavoy [Dawei], *Wallich 7345 (leg. Gomez 1176)* (holotype, K-WALL).

Distribution. ?Kerala, ?Tamil Nadu, ?Nagaland [Myanmar].

Habit. Terrestrial, lithophytic or epiphytic herb.

Remarks. Singh et al. (2019) consider all records from India doubtful, and we agree with this assessment. This showy species is common in cultivation, and there are no wild-collected specimens of *Calanthe vestita* from India in Kew's Herbarium. One specimen attributed to the flora of South India, *Abraham 3168*, had been "grown in pots" in a garden in Trivandrum.

Calanthe wallichii (Lindl.) M. W. Chase, Christenh. & Schuit., Phytotaxa 472(2): 166. *Phaius wallichii* Lindl., Pl. Asiat. Rar. (Wallich) 2: 46, t. 158. 1831. TYPE: India, Sylhet, Pundua Hills, Apr. 1824, *Wallich 3747 (leg. De Silva 1389)* (holotype, K-LINDL; isotype, K, K-WALL).

Phaius roeblingii O'Brien, Gard. Chron., ser. 3, 17: 358. 1895. TYPE: India, Khasi Hills, *cult. Roebling s.n.* (not found).

Distribution. Assam, Sikkim, West Bengal [Bangladesh, China, ?Myanmar, Sri Lanka].

Habit. Terrestrial herb.

Calanthe whiteana King & Pantl., J. Asiat. Soc. Bengal, Pt. 2, Nat. Hist. 65: 121. 1896. TYPE: India, Sikkim, Chungthang, 6000 ft, June 1895, *Pantling 365* (holotype, CAL; isotype, BM, K, P).

Distribution. Arunachal Pradesh, Nagaland, Sikkim [Bhutan, Myanmar].

Habit. Terrestrial herb.

Calanthe woodfordii (Hook.) M. W. Chase, Christenh. & Schuit., Phytotaxa 472(2): 167. 2020. *Bletia*

woodfordii Hook., Bot. Mag. 54: t. 2719. 1827. *Phaius woodfordii* (Hook.) Merr., J. Arnold Arbor. 29: 211. 1948. TYPE: "Trinidad" (ex cult.), cult. *Kew s.n. (leg. R. Woodford s.n.)* (not found).

Limodorum flavum Blume, Bijdr. Fl. Ned. Ind. 8: 375. 1825. *Phaius flavus* (Blume) Lindl., Gen. Sp. Orchid. Pl.: 128. 1831. *Paraphaius flavus* (Blume) J. W. Zhai, Z. J. Liu & F. W. Xing, Molec. Phylogen. Evol. 77: 221. 2014. TYPE: Indonesia, Java, Mt. Gede, *Blume s.n.* (holotype, L).
Dendrobium veratrifolium Roxb., Fl. Ind. ed. 1832, 3: 482. 1832. TYPE: India, Meghalaya, Garrow [Garo] Country, Apr. & May, icon. *Dendrobium veratrifolium*, *Roxburgh 2352* (syntypes, CAL, K).

Distribution. Arunachal Pradesh, Assam, Manipur, Meghalaya, Mizoram, Nagaland, Sikkim, Tripura, West Bengal [Bhutan, China, Myanmar, Nepal].

Habit. Terrestrial herb.

Remarks. This is not *Calanthe flava* (Blume) C. Morren. Although *Dendrobium veratrifolium* Roxb. is usually considered to be conspecific with *Phaius veratrifolius* Lindl. (= *C. tankervilleae*) (Stone & Cribb, 2017), we agree with Paul Ormerod (pers. comm.) that Roxburgh's drawing, although stylized, looks much more like *P. flavus* (Blume) Lindl. (= *C. woodfordii*). The description, which refers to a "plaited and curved" mid-lobe is also in agreement with this identification. When Lindley described *P. veratrifolius*, he cited *D. veratrifolium* Roxb. in the synonymy, which at that time was still a nomen nudum. The validly published names *D. veratrifolium* and *P. veratrifolius* are based on different types.

Calanthe yuksomnensis Lucksom, J. Bombay Nat. Hist. Soc. 95(2): 319. 1998. TYPE: India, Sikkim, Yoksam, *Lucksom 311a* (holotype, CAL; isotype, Gangtok, Forest Dept. Herb.).

Distribution. Meghalaya, Sikkim, West Bengal. Endemic.

Habit. Terrestrial herb.

———

Callostylis Blume, Bijdr. Fl. Ned. Ind.: 340. 1825.

TYPE: *Callostylis rigida* Blume.

Callostylis rigida Blume, Bijdr. Fl. Ned. Ind.: 340. 1825. *Tylostylis rigida* (Blume) Blume, Fl. Javae, Praef.: vi. 1828. *Eria rigida* (Blume) Rchb. f., Bonplandia (Hannover) 5: 55. 1857, nom. illeg., non Blume, 1856. SYNTYPES: Indonesia, Java, Mt. Salak, Mt. Gede, Pantjar, and Mt. Burangrang, *Blume s.n.* (syntype, L).

Eria discolor Lindl., J. Proc. Linn. Soc., Bot. 3: 51. 1859 [1858]. *Tylostylis discolor* (Lindl.) Hook. f., Ann. Roy. Bot. Gard. (Calcutta) 5: 22, t. 32. 1895. *Callostylis rigida* Blume subsp. *discolor* (Lindl.) Brieger, Orchideen (Schlechter) 1(11–12): 749. 1981. TYPE: India, Sikkim, Glen Cathcart, alt. 3000–4000 ft, May 1850, *J. D. Hooker 168* (holotype, K).
Liparis bidentata Griff., Not. Pl. Asiat. 3: 277. 1851. TYPE: India, Assam, Naga, 12 Mar. 1836, *Griffith in Assam Herb. 428* (not found).

Distribution. Arunachal Pradesh, Assam, Manipur, Meghalaya, Nagaland, Sikkim, West Bengal [China, Myanmar, Nepal].

Habit. Epiphytic herb.

———

Cephalanthera Rich., Mém. Mus. Hist. Nat. 4: 51. 1818.

TYPE: *Cephalanthera damasonium* (Mill.) Druce (*Serapias damasonium* Mill.).

Cephalanthera damasonium (Mill.) Druce, Ann. Scott. Nat. Hist. 15(60): 225. 1906. *Serapias damasonium* Mill., Gard. Dict., ed. 8, *Serapias* no. 2. 1768. TYPE: United Kingdom, Oxfordshire, Bledlow, 19 May 1960, *P. A. Sims s.n.* (neotype BM, designated by Alarcón & Aedo, 2002).

Distribution. Arunachal Pradesh [Bhutan, China, Myanmar].

Habit. Terrestrial herb.

Cephalanthera longifolia (L.) Fritsch, Oesterr. Bot. Z. 38: 81. 1888. *Serapias helleborine* L. var. *longifolia* L., Sp. Pl. 2: 950. 1753. *Serapias longifolia* (L.) L., Sp. Pl. ed. 2: 1345. 1763. TYPE: *Herb. Linn. No. 1057. 4* (neotype LINN, designated by Renz, 1984).

Cephalanthera thomsonii Rchb. f., Linnaea 41: 54. 1877 [1876] (as "*thomsoni*"). TYPE: India, Sikkim, 1857, *Thomson s.n.* (holotype, W).

Distribution. Arunachal Pradesh, Assam, Himachal Pradesh, Jammu & Kashmir, Manipur, Meghalaya, Mizoram, Nagaland, Sikkim, Uttarakhand, West Bengal [Bhutan, China, Myanmar, Nepal, Pakistan].

Habit. Terrestrial herb.

———

Ceratostylis Blume, Bijdr. Fl. Ned. Ind. 7: 304. 1825.

TYPE: *Ceratostylis subulata* Blume.

Ceratostylis himalaica Hook. f., Fl. Brit. India [J. D. Hooker] 5(16): 826. 1890. *Ritaia himalaica* (Hook. f.) King & Pantl., Ann. Roy. Bot. Gard. (Calcutta) 8: 157, t. 214. 1898. SYNTYPES: Bhutan, *Griffith s.n.* [*1187*] and *Griffith s.n.* [*13, East India Company Herbarium 5214*] (syntypes, CAL, K (13 & 1187), K-LINDL); Nepal, Dec., *J. D. Hooker s.n.* [*359*] (syntypes, K, K-LINDL); India, Khasi Hills, *Gibson s.n.* (syntypes, CAL, K).

Distribution. Arunachal Pradesh, Assam, Manipur, Meghalaya, Mizoram, Nagaland, Sikkim, West Bengal [Bhutan, China Myanmar, Nepal].

Habit. Epiphytic herb.

Ceratostylis radiata J. J. Sm., Orchid. Java: 295. 1905. SYNTYPES: Indonesia, Java, Sukabumi Garut, and S. Preanger, *Raciborski s.n.* (syntype(s), BO).

Distribution. Assam [Myanmar].

Habit. Epiphytic herb.

Ceratostylis subulata Blume, Bijdr. Fl. Ned. Ind. 7: 306. 1825. TYPE: Java, Mt. Salak & Mt. Pantjar, *Blume s.n.* (syntype, L).

Appendicula teres Griff., Not. Pl. Asiat. 3: 359. 1851. *Ceratostylis teres* (Griff.) Rchb. f., Bonplandia 2(7): 89. 1854. TYPE: India, Upper Assam, towards Negrigam, 19 Jan. 1836, *Griffith s.n.* (holotype, K-LINDL).

Distribution. Andaman & Nicobar Islands, Arunachal Pradesh, Assam, Meghalaya, Nagaland [China, Myanmar].

Habit. Epiphytic herb.

———

Chamaegastrodia Makino & F. Maek., Bot. Mag. (Tokyo) 49(585): 596 -597. 1935.

TYPE: *Chamaegastrodia shikokiana* Makino & F. Maek.

Chamaegastrodia poilanei (Gagnep.) Seidenf. & A. N. Rao, Nordic J. Bot. 14: 297. 1994. *Evrardia poilanei* Gagnep., Bull. Mus. Natl. Hist. Nat., sér. 2, 4: 596. 1932. *Hetaeria poilanei* (Gagnep.) Tang & F. T. Wang, Acta Phytotax. Sin. 1: 71. 1951. *Evrardianthe poilanei* (Gagnep.) Rauschert, Feddes Repert. 94: 433. 1983. *Evrardiana poilanei* (Gagnep.) Aver., Bot. Zhurn. (Moscow & Leningrad) 73: 432. 1988. *Odontochilus poilanei* (Gagnep.) Ormerod, Lindleyana 17: 225. 2002. TYPE: Vietnam, Dalat, arboretum, 15 November 2914, *F. Poilane 1807* (holotype, P).

Evrardia asraoa J. Joseph & Abbar., Bull. Bot. Surv. India 25(1–4): 232, fig. 1. 1983 [1985]. *Hetaeria asraoa* (J. Joseph & Abbar.) Karthik. in S. Karthikeyan et al., Fl. Ind. ser. 4, 1 (Monocotyledon): 146. 1989. *Chamaegastrodia asraoa* (J. Joseph & Abbar.) Seidenf. & A. N. Rao, Nordic J. Bot. 14: 299. 1994. *Evrardianthe asraoa* (J. Joseph & Abbar.) C. S. Kumar in C. Sathish Kumar & K. S. Manilal, Cat. Indian Orchids 94. 1994. *Odontochilus asraoa* (J. Joseph & Abbar.) Ormerod, Taiwania 50(1): 7. 2005. TYPE: India, Khasi Hills, Pynursla, 26 Aug. 1980, *Joseph 73566A* (holotype, ASSAM).

Distribution. Arunachal Pradesh, Meghalaya [Bhutan, China, Myanmar, Nepal].

Habit. Holomycotrophic terrestrial herb.

Remarks. The species described as *Evrardia asraoa* differs from *Chamaegastrodia poilanei* only in the absence of slender processes on the lip epichile. Other differences mentioned (in the flanges of the lip mesochile and the rostellum) appear to be spurious. Photographs (Swami, 2017) show that plants with slender epichile processes also occur in India, and we consider this a variable character within the species. Hence, *E. asraoa* is here treated as synonymous with *C. poilanei*. Swami (2017) illustrates as *Odontochilus poilanei* a plant that has longer, more subulate fringes along the mesochile than the plant he illustrates as *O. asraoa*. Such long-fringed specimens have also been recorded from Thailand and Taiwan. It remains to be studied if such specimens still fall within the natural variation of *O. poilanei* or if more species should be recognized.

Chamaegastrodia shikokiana Makino & F. Maek., Bot. Mag. (Tokyo) 49: 596. 1935. *Hetaeria shikokiana* (Makino & F. Maek.) Tuyama, Fl. Jap. (Ohwi) 341. 1965 (as "*Sikokiana*"). *Gastrodia shikokiana* Makino, Bot. Mag. (Tokyo) 6: 48. 1892, nom. inval. TYPE: Japan, Hondo, Kawachi Prov., near Chihaya, 29 Aug. 1889, *S. Matsuda s.n.* (holotype, TI).

Distribution. Arunachal Pradesh, Assam [China].

Habit. Holomycotrophic terrestrial herb.

Chamaegastrodia vaginata (Hook. f.) Seidenf., Nordic J. Bot. 14(3): 294. 1994. *Aphyllorchis vaginata* Hook. f., Fl. Brit. India [J. D. Hooker] 6(17): 117. 1890. TYPE: India, Khasi Hills, Mamlu, alt. 5000 ft, 27 Aug. 1850, *Hooker & Thomson s.n.* [*365*] (holotype, K).

Distribution. Meghalaya [China].

Habit. Holomycotrophic terrestrial herb.

———

Cheirostylis Blume, Bijdr. Fl. Ned. Ind. 8: 413. 1825.

TYPE: *Cheirostylis montana* Blume.

Cheirostylis flabellata (A. Rich.) Wight, Icon. Pl. Ind. Orient. [Wight] 5(1): 16, t. 1727. 1851. *Goodyera flabellata* A. Rich., Ann. Sci. Nat., Bot. sér. 2, 15: 79, t. 12. 1841 (as "*flabellatum*" in tab.). *Monochilus flabellatus* (A. Rich.) Wight, Icon. Pl. Ind. Orient. [Wight] 5(1): t. 1727. 1851. TYPE: India, Nilgiri Hills, Kaity, May–June, *Perrottet s.n.* (holotype, P).

Distribution. Karnataka, Kerala, Maharashtra, Tamil Nadu [Sri Lanka].

Habit. Terrestrial herb.

Cheirostylis griffithii Lindl., J. Proc. Linn. Soc., Bot. 1: 188. 1857. SYNTYPES: India, Khasi Hills, Mamlu, *T. Lobb s.n.* (syntype, K-LINDL); Khasi Hills, *Griffith s.n.* (syntype, K-LINDL).

Distribution. Arunachal Pradesh, Assam, Meghalaya, Mizoram, Nagaland, Sikkim, Uttarakhand, West Bengal [Bangladesh, China, Myanmar, Nepal].

Habit. Terrestrial or epiphytic herb.

Cheirostylis gunnarii A. N. Rao, Nordic J. Bot. 18(1): 23. 1998. TYPE: India, Arunachal Pradesh, Lohit District, Kamlang Reserve Forests, *A. N. Rao 28003* (holotype, Orchid Herbarium, Tipi).

Distribution. Arunachal Pradesh. Endemic.

Habit. Epiphytic herb.

Cheirostylis moniliformis (Griff.) Seidenf., Dansk Bot. Ark. 32(2): 69. 1978. *Goodyera moniliformis* Griff., Itin. Pl. Khasyah Mts. 143. 1848. *Zeuxine moniliformis* (Griff.) Griff., Not. Pl. Asiat. 3: 397, t. 350. 1851. TYPE: Bhutan, Kooree nuddi, *Griffith 679* (holotype, K; isotype, OXF).

Cheirostylis bhotanensis Tang & F. T. Wang, Acta Phytotax. Sin. 1: 86. 1951. TYPE: Bhutan, *Griffith 679* (holotype, K).
Cheirostylis chinensis Rolfe var. *glabra* Bhaumik & M. K. Pathak, Bull. Bot. Surv. India 47(1–4): 183. 2005 [2006]. TYPE: India, Arunachal Pradesh, Dibang Valley District, Bejari, *Bhaumik & Tham 104752* (holotype, CAL).

Distribution. Andhra Pradesh, Arunachal Pradesh [Bhutan].

Habit. Terrestrial or epiphytic herb.

Cheirostylis parvifolia Lindl., Edwards's Bot. Reg. 25(Misc.): 19. 1839. TYPE: Sri Lanka, *cult. Loddiges s.n.* (holotype, K-LINDL).

Cheirostylis seidenfadeniana C. S. Kumar & F. N. Rasm., Nordic J. Bot. 7: 409. 1987. TYPE: India, Kerala, Trivandum District, Ponmudi, *C. Sathish Kumar CU 36960* (holotype, TBGT; isotype, C, CALI).

Distribution.   Kerala, Maharashtra, Odisha, Tamil Nadu [Sri Lanka].

Habit.   Terrestrial herb.

Cheirostylis pusilla Lindl., Gen. Sp. Orchid. Pl.: 489. 1840. *Hetaeria pusilla* Lindl., Numer. List [Wallich] n. 7382. 1832, nom. inval. (as "*Etaria*"). TYPE: India, Sylhet mountains, Jentya (Jaintia), Sep. 1818, *Wallich 7382 (leg. W. Gomez 208)* (holotype, K-LINDL; isotype, K-WALL).

Distribution.   Arunachal Pradesh, Meghalaya, Nagaland [China, Myanmar].

Habit.   Terrestrial or epiphytic herb.

Cheirostylis sessanica A. N. Rao, Nordic J. Bot. 8(4): 339. 1988. TYPE: India, Arunachal Pradesh, West Kameng District, Sessa, *A. N. Rao 24016* (holotype, Orchid Herbarium, Tipi).

Distribution.   Arunachal Pradesh. Endemic.

Habit.   Terrestrial herb.

Cheirostylis tippica A. N. Rao, Arunachal Forest News 9(2): 17. 1991. TYPE: India, Arunachal Pradesh, West Kameng District, Tipi, *A. N. Rao 25750* (holotype, Orchid Herbarium, Tipi).

Distribution.   Arunachal Pradesh. Endemic.

Habit.   Terrestrial herb.

Cheirostylis yunnanensis Rolfe, Bull. Misc. Inform. Kew 1896: 201. 1896. TYPE: China: Yunnan, Mengtse, Apr. 1893, *Hancock 25* (holotype, K).

Cheirostylis munnacampensis A. N. Rao, Nordic J. Bot. 8(4): 340. 1988. TYPE: India, Arunachal Pradesh State, West Kameng District, Mummacamp, *A. N. Rao 14567* (holotype, Orchid Herbarium, Tipi).
Cheirostylis pabongensis Lucksom, Indian J. Forest. 20: 305. 1997 (as "*pabongnensis*"). TYPE: India, Sikkim, Pabong, 2000 ft, 20 Mar. 1996, *Lucksom 210a* (holotype, CAL; isotype, Herb. Forest Dept., Gangtok).

Distribution.   Arunachal Pradesh, Sikkim, West Bengal [China, Myanmar].

Habit.   Terrestrial herb.

Remarks.   This species is illustrated as *Cheirostylis griffithii* Lindl. in Swami (2016).

Chiloschista Lindl., Edwards's Bot. Reg. 18: t. 1522. 1832.

TYPE: *Chiloschista usneoides* (D. Don) Lindl. (*Epidendrum usneoides* D. Don).

Chiloschista fasciata (F. Muell.) Seidenf. & Ormerod, Opera Bot. 124: 64. 1995. *Sarcochilus fasciatus* F. Muell., Fragm. (Mueller) 5: 202. 1866. TYPE: Icon. Pl. Ind. Orient. [Wight] 5(1): t. 1741. 1851 (right hand figure, iconotype).

Sarcochilus minimifolius Hook. f., Fl. Brit. India [J. D. Hooker] 6(17): 37. 1890. *Thrixspermum minimifolium* (Hook. f.) Kuntze, Revis. Gen. Pl. 2: 682. 1891. *Chiloschista minimifolia* (Hook. f.) N. P. Balakr., J. Bombay Nat. Hist. Soc. 67(1): 66. 1970. TYPE: Sri Lanka, Central Province, Wattegodde Hills, March 1820, *Thwaites s.n.* (holotype, K).
Sarcochilus wightii Hook. f., Fl. Brit. India [J. D. Hooker] 6(17): 37. 1890. SYNTYPES: Malabar, Cochin, *Johnson s.n.* (not found); Nilgiri Hills, on the western slope of the Wynaad, *Wight s.n.* (syntype, K); Ceylon, on trees in the Botanical Gardens, *Trimen s.n.* (syntype, K).

Distribution.   Kerala [Sri Lanka].

Habit.   Epiphytic herb.

Chiloschista glandulosa Blatt. & McCann, J. Bombay Nat. Hist. Soc. 35(3): 488. 1932. TYPE: India, North Kanara, Karwar, *T. R. Bell 4969*.

Distribution.   Karnataka, Kerala. Endemic.

Habit.   Epiphytic herb.

Chiloschista himalaica Tobgay, C. Gyeltshen & Dalström, Lankesteriana 20(3): 296. 2020. TYPE: Bhutan, Chhukha, along old road to Darla Hydropower Adits. Chenlakha, alt. 1480 m, 19 May 2015, *S. Dalström 4225 (with K. Tobgay & D. Wangdi)* (holotype, THIM, spirit mat.).

Distribution.   Arunachal Pradesh, Assam, Manipur, Meghalaya, Mizoram, Nagaland, Odisha, Sikkim, West Bengal [Bhutan].

Habit.   Epiphytic herb.

Remarks.   Gyeltshen et al. (2020) believe that *Chiloschista lunifera* (Rchb. f.) J. J. Sm. does not occur in the Himalaya region and that material from there identified as such (or under synonyms such as *Sarcochilus lunifer* (Rchb. f.) Benth. ex Hook. f.) belongs to their *C. himalaica*.

Chiloschista parishii Seidenf., Opera Bot. 95: 176. 1988. TYPE: Myanmar, Moulmein [Mawlamyine], *Parish 55* (holotype, W; isotype, [*with Parish's drawing*] K).

Distribution. Andaman & Nicobar Islands, Arunachal Pradesh, Assam, Manipur, Meghalaya, Mizoram, Nagaland, Odisha, Sikkim, West Bengal [Bhutan, China, Myanmar, Nepal].

Habit. Epiphytic herb.

Remarks. Records from the Himalaya region may refer to *Chiloschista himalaica* and need to be verified (Gyeltshen et al., 2020).

Chiloschista usneoides (D. Don) Lindl., Edwards's Bot. Reg. 18: t. 1522. 1832. *Epidendrum usneoides* D. Don., Prodr. Fl. Nepal.: 37. 1825. *Sarcochilus usneoides* (D. Don) Rchb. f., Ann. Bot. Syst. (Walpers) 6: 497. 1863. *Thrixspermum usneoides* (D. Don) Rchb. f., Xenia Orchid. 2: 120. 1867. TYPE: Nepal, *Buchanan-Hamilton s.n.* (lost?). NEOTYPE: Nepal, *Wallich 7330* (neotype, K-LINDL; isoneotype, K-WALL, designated by Seidenfaden, 1988).

Distribution. Meghalaya, Uttarakhand [Bhutan, Myanmar, Nepal].

Habit. Epiphytic herb.

Remarks. Singh et al. (2019) record this from three more states but not from Meghalaya. Considering the confused state of the taxonomy of *Chiloschista* in India we suggest that more research is needed before the distribution of the various species can be established.

────

Chrysoglossum Blume, Bijdr. Fl. Ned. Ind. 7: 337. 1825.

TYPE: *Chrysoglossum ornatum* Blume.

Chrysoglossum assamicum Hook. f., Fl. Brit. India [J. D. Hooker] 5(16): 784. 1890. *Collabium assamicum* (Hook. f.) Seidenf., Opera Bot. 72: 24. 1983 [1984]. *Collabiopsis assamica* (Hook. f.) S. S. Ying, Coloured Ill. Indig. Orchids Taiwan 2: 456. 1990. TYPE: India, Assam, *Griffith s.n.* [*1233/1322*] (holotype, K; isotype, W).

Distribution. Assam, Meghalaya [China].

Habit. Terrestrial herb.

Chrysoglossum ornatum Blume, Bijdr. Fl. Ned. Ind. 7: 338. 1825. TYPE: Indonesia, Java, Mt. Salak, *Blume 295* (holotype, L).

Chrysoglossum erraticum Hook. f., Fl. Brit. India [J. D. Hooker] 5(16): 784. 1890. TYPE: India, Sikkim, Rishep, *C. B. Clarke s.n.* [*12314*] (holotype, K, W).

Ania maculata Thwaites, Enum. Pl. Zeyl. [Thwaites]: 301. 1861. *Chrysoglossum maculatum* (Thwaites) Hook. f., Fl. Brit. India [J. D. Hooker] 5(16): 784. 1890. *Tainia maculata* (Thwaites) Hook. f., Fl. Brit. India [J. D. Hooker] 5(16): 821. 1890. TYPE: Sri Lanka, Hapootelle, *Thwaites s.n.* [*Ceylon Plants 3515*].

Chrysoglossum hallbergii Blatt., J. Bombay Nat. Hist. Soc. 32: 519. 1928. TYPE: India, Tamil Nadu, Madura Dist., High Wavy Mountains, May 1917, *Blatter 26488* (isotype, K).

Distribution. Arunachal Pradesh, Assam, Karnataka, Kerala, Sikkim, Tamil Nadu, West Bengal [China, Myanmar, Nepal, Sri Lanka].

Habit. Terrestrial herb.

────

Cleisocentron Brühl, Guide Orchids Sikkim 136. 1926.

TYPE: *Cleisocentron trichromum* (Rchb. f.) Brühl (*Saccolabium trichromum* Rchb. f.).

Cleisocentron pallens (Cathcart ex Lindl.) N. Pearce & P. J. Cribb, Edinburgh J. Bot. 58(1): 118, fig. 8. 2001. *Saccolabium pallens* Cathcart ex Lindl., J. Proc. Linn. Soc., Bot. 3: 35. 1859 [1858]. TYPE: India, Sikkim, icon. *Cathcart* (holotype, K).

Saccolabium trichromum Rchb. f., Hamburger Garten-Blumenzeitung 15: 51. 1859. *Gastrochilus trichromus* (Rchb. f.) Kuntze, Revis. Gen. Pl. 2: 661. 1891. *Cleisocentron trichromum* (Rchb. f.) Brühl, Guide Orchids Sikkim 137. 1926. TYPE: Himalayas, *cult. Stange s.n.* (holotype, W; isotype, W).

Distribution. Arunachal Pradesh, Assam, Meghalaya, Nagaland, Sikkim, West Bengal. Endemic.

Habit. Epiphytic herb.

────

Cleisomeria Lindl. ex G. Don in J. C. Loudon, Encycl. Pl., new ed., Suppl. 2: 1447. 1855.

TYPE: *Cleisomeria lanatum* (Lindl.) Lindl. ex G. Don. (*Cleisostoma lanatum* Lindl.)

Cleisomeria pilosulum (Gagnep.) Seidenf. & Garay, Bot. Tidsskr. 67: 120. 1972. *Cleisostoma pilosulum* Gagnep., Bull. Soc. Bot. France 79: 35. 1932. *Saccolabium pilosulum* (Gagnep.) Tang & F. T. Wang, Acta Phytotax. Sin. 1: 97. 1951. TYPE: Cambodia, Kampong Speu Prov., between Pum Ho Tet and Sra Nam Chrom, 22 June 1930, *Poilane 17742* (holotype, P).

Distribution. Manipur [Myanmar].

Habit. Epiphytic herb.

———

Cleisostoma Blume, Bijdr. Fl. Ned. Ind. 8: 362. 1825.

TYPE: *Cleisostoma sagittatum* Blume.

Cleisostoma appendiculatum (Lindl.) Benth. &
Hook. f. ex B. D. Jacks., Index Kew. 1: 555. 1893.
Aerides appendiculata Lindl., Gen. Sp. Orchid.
Pl.: 242. 1833 (as "*appendiculatum*"). *Sarcanthus
appendiculatus* (Lindl.) E. C. Parish in Mason,
Burmah, ed. 4. 2: 181. 1883 (as "*appendiculatum*"). TYPE: Myanmar, Tennaserim, Tavoy [Dawei],
1827, *Wallich 7315 (leg. W. Gomez 1177)* (holotype, K-LINDL; isotype, K-WALL, L).

Sarcanthus hincksianus Rchb. f., Gard. Chron., n.s., 9: 73.
1878. *Cleisostoma hincksianum* (Rchb. f.) Garay, Bot.
Mus. Leafl. 23: 171. 1972. TYPE: Origin unknown, *cult.
Hincks s.n.* (XII-1876) (holotype, W).

Distribution. Arunachal Pradesh, Assam, Karnataka,
Manipur, Meghalaya, Mizoram, Nagaland, Odisha, Sikkim, West Bengal [Bangladesh, Bhutan, Myanmar, Nepal].

Habit. Epiphytic herb.

Remarks. Records from India could be referable
to *Cleisostoma chantaburiense* Seidenf., *C. schneideri*
Choltco, or *C. simondii* (Gagnep.) Seidenf. and need to
be verified.

Cleisostoma arietinum (Rchb. f.) Garay, Bot. Mus.
Leafl. 23(4): 169. 1972. *Sarcanthus arietinus* Rchb.
f., Gard. Chron. 1869: 416. 1869. *Echioglossum
arietinum* (Rchb. f.) Szlach., Ann. Bot. Fenn. 40:
68. 2003. TYPE: India, Assam, *J. S. Day s.n.* (holotype, W).

Distribution. ?Assam [Myanmar].

Habit. Epiphytic herb.

Remarks. This species is listed as "excluded" in
Singh et al. (2019).

Cleisostoma armigerum King & Pantl., J. Asiat. Soc.
Bengal, Pt. 2, Nat. Hist. 65: 123. 1896 (as "*ar-
migera*"). *Pomatocalpa armigerum* (King & Pantl.)
Tang & F. T. Wang, Acta Phytotax. Sin. 1(1): 98.
1951. *Sarcanthus armiger* (King & Pantl.) J. J.
Sm., Natuurk. Tijdschr. Ned.-Indië 72: 83. 1912.
TYPE: India, Sikkim, tropical valleys, Sep. 1892,
Pantling 252 (holotype, CAL; isotype, BM, K, W).

Distribution. Arunachal Pradesh, Assam, Mizoram, Sikkim, West Bengal. Endemic.

Habit. Epiphytic herb.

Cleisostoma aspersum (Rchb. f.) Garay, Bot. Mus.
Leafl. 23(4): 169. 1972. *Sarcanthus aspersus* Rchb.
f., Hamburger Garten- Blumenzeitung 21: 297.
1865. TYPE: Myanmar, Moulmein [Mawlamyine],
cult. Veitch 89 (syntypes, W).

Cleisostoma bicuspidatum Hook. f., Fl. Brit. India [J. D.
Hooker] 6(17): 75. 1890. *Stereochilus bicuspidatus* (Hook.
f.) King & Pantl., Ann. Roy. Bot. Gard. (Calcutta) 8: 236.
1898. *Sarcanthus khasiaensis* Tang & F. T. Wang, Acta
Phytotax. Sin. 1(1): 98. 1951. SYNTYPES: India, Sikkim Himalaya, Darjeeling, *Anderson s.n.* (not found);
Assam, Khasi Hills, *Mann 34/1884* (syntype, K); Tenasserim, *C. S. P. Parish s.n.* (not found).

Distribution. Arunachal Pradesh, Assam, Meghalaya, Mizoram, Nagaland, Sikkim, Uttarakhand, West
Bengal [Bhutan, Myanmar, Nepal].

Habit. Epiphytic herb.

Cleisostoma bambusarum (King & Pantl.) King &
Pantl., Ann. Roy. Bot. Gard. (Calcutta) 8: 233,
t. 310. 1898. *Sarcanthus bambusarum* King &
Pantl., J. Asiat. Soc. Bengal, Pt. 2, Nat. Hist. 65:
123 1896. *Saccolabium bambusarum* (King &
Pantl.) Tang & F. T. Wang, Acta Phytotax. Sin. 1:
96. 1951. *Pomatocalpa bambusarum* (King &
Pantl.) Garay, Bot. Mus. Leafl. 23(4): 190. 1972.
Robiquetia bambusarum (King & Pantl.) R. Rice,
Photo Intro Vandoid Orchid Gen. Asia, Rev. Ed.:
163. 2018 (as "*bambusara*"). TYPE: India, Sikkim, Sembree, 1500 ft, May 1894, *Pantling 211*
(holotype, CAL; isotype, K).

Distribution. Sikkim, West Bengal [Bhutan].

Habit. Epiphytic herb.

Remarks. Although the characters of this species
are somewhat ambiguous and, in the lamellate back-
wall callus in the spur, suggestive of *Pomatocalpa*,
Watthana (2007) found it to belong in *Cleisostoma*.

Cleisostoma chantaburiense Seidenf., Dansk Bot.
Ark. 29: 70. 1975. TYPE: Thailand, Khao Kuap,
Krat, *Kerr 0882* (cultivated in Bangkok, fl. 18 Oct.
1930) (holotype, K).

Distribution. Sikkim [also in Thailand].

Habit. Epiphytic herb.

Remarks. Choltco (2009) pointed out that plate 318
in King and Pantling (1898), labeled *Sarcanthus appendiculatus*, is referable to *Cleisostoma chantaburiense*.

Cleisostoma discolor Lindl., Edwards's Bot. Reg.
31(Misc.): 59. 1845. *Sarcanthus discolor* (Lindl.)
J. J. Sm., Natuurk. Tijdschr. Ned.-Indië 72: 85.

1912. TYPE: India, *cult. Loddiges s.n.* (holotype, K-LINDL).

Sarcanthus macrodon Rchb. f., Gard. Chron. 1872: 1555. 1872. TYPE: India, Madras, *cult. Veitch (leg. Benson) s.n.* (holotype, W).

Saccolabium rostellatum Hook. f., Fl. Brit. India [J. D. Hooker] 6(17): 59. 1890. *Gastrochilus rostellatus* (Hook. f.) Kuntze, Revis. Gen. Pl. 2: 661. 1891. TYPE: India, Sikkim, Darjeeling, *Gamble s.n.* (holotype, CAL).

Sarcanthus auriculatus Rolfe, Bull. Misc. Inform. Kew 1895: 9. 1895. *Cleisostoma auriculatum* (Rolfe) Garay, Bot. Mus. Leafl. 23, 4: 169. 1972. TYPE: Origin not known, *cult. O'Brien s.n. (June 1890)* (holotype, K).

Distribution. Arunachal Pradesh, Manipur, Sikkim, West Bengal [also in Indonesia, Thailand].

Habit. Epiphytic herb.

Cleisostoma duplicilobum (J. J. Sm.) Garay, Bot. Mus. Leafl. 23: 171. 1972. *Sarcanthus duplicilobus* J. J. Sm., Bull. Dépt. Agric. Indes Néerl. 13: 64. 1907. *Garayanthus duplicilobus* (J. J. Sm.) Szlach., Fragm. Florist. Geobot., Suppl. 3: 136. 1995. SYNTYPES: Java, Dieng?, *Z. Kamerling s.n.* (syntype, BO); *cult. Bogor s.n.* (E. Java?) (syntype, BO).

Sarcanthus curinatus Rolfe ex Downie, Bull. Misc. Inform. Kew 1925: 408. 1925. *Cleisostoma carinatum* (Rolfe ex Downie) Garay, Bot. Mus. Leafl. 23: 170. 1972. *Garayanthus carinatus* (Rolfe ex Downie) Szlach., Fragm. Florist. Geobot., Suppl. 3: 136. 1995. TYPE: Thailand, Chieng Mai, Doi Sutep, 1500–2500 ft, 30 May 1909, *Kerr 182* (holotype, K; isotype, K).

Distribution. Assam, Nagaland [Myanmar].

Habit. Epiphytic herb.

Cleisostoma filiforme (Lindl.) Garay, Bot. Mus. Leafl. 23(4): 171. 1972. *Sarcanthus filiformis* Lindl., Edwards's Bot. Reg. 28(Misc.): 61. 1842. TYPE: India, *cult. Fielding (leg. Wallich) s.n.* (holotype, K-LINDL).

Distribution. Arunachal Pradesh, Assam, Manipur, Meghalaya, Mizoram, Nagaland, Sikkim, Tripura, West Bengal [Bhutan, China, Myanmar, Nepal].

Habit. Epiphytic herb.

Cleisostoma linearilobatum (Seidenf. & Smitinand) Garay, Bot. Mus. Leafl. 23(4): 172. 1972. *Sarcanthus linearilobatus* Seidenf. & Smitinand, Orch. Thail. (Prelim. List) 4, 2: 684. 1965 (as "*linearilobata*"). *Ormerodia linearilobata* (Seidenf. & Smitinand) Szlach., Ann. Bot. Fenn. 40(1): 68. 2003. TYPE: Thailand, Chieng Mai, Kawng He, 300 m, 2 May 1915, *Kerr 363* (holotype, K).

Sarcanthus sagittatus King & Pantl., J. Asiat. Soc. Bengal, Pt. 2, Nat. Hist. 66: 595. 1898 [1897]. *Cleisostoma sagittiforme* Garay, Bot. Mus. Leafl. 23(4): 174. 1972. *Ormerodia sagittata* (King & Pantl.) Szlach., Ann. Bot. Fenn. 40(1): 68. 2003. TYPE: India, Khasi Hills, probably at Teria Ghat, June, *Pantling 629* (holotype, CAL).

Cleisostoma sikkimense Lucksom, Indian J. Forest. 15: 27. 1992. SYNTYPES: India, Sikkim, Namchebong, Gangtok, 1500–6000 ft, 8 June 1990, *S. Z. Lucksom 204* (holotype, CAL).

Distribution. Arunachal Pradesh, Assam, ?Meghalaya, Sikkim, West Bengal [Bhutan, China, Myanmar].

Habit. Epiphytic herb.

Cleisostoma paniculatum (Ker Gawl.) Garay, Bot. Mus. Leafl. 23(4): 173. 1972. *Aerides paniculata* Ker Gawl., Bot. Reg. 3: t. 220. 1817 (as "*peniculatum*"). *Vanda paniculata* (Ker Gawl.) R. Br., Bot. Reg. 6: t. 506. 1820. *Sarcanthus paniculatus* (Ker Gawl.) Lindl., Ill. Orch. Pl. [Bauer & Lindley] t. 9. 1830. *Garayanthus paniculatus* (Ker Gawl.) Szlach., Fragm. Florist. Geobot., Suppl. 3: 136. 1995. TYPE: China, 1817, *Banks s.n.* (holotype, K).

Distribution. Arunachal Pradesh, Assam, Manipur, ?Meghalaya, Mizoram, Nagaland, Tripura [China].

Habit. Epiphytic herb.

Cleisostoma parishii (Hook. f.) Garay, Bot. Mus. Leafl. 23(4): 173. 1972. *Sarcanthus parishii* Hook. f., Bot. Mag. 86: t. 5217. 1860. TYPE: Myanmar, Moulmein [Mawlamyine], *cult. Low s.n. (leg. Parish)* (holotype, K).

Sarcoglyphis manipurensis A. N. Rao, Vik. Kumar & H. B. Sharma, Nordic J. Bot. 34(2): 191. 2016. TYPE: India, Manipur, Chandel district, Songpiyang Hills, 420 m, 24°27.045′N, 94°27.866′E, 25 May 2014, *H. B. Sharma 596* (holotype, CAL; isotype, COGCEHR, herbarium, Hengbung, Manipur).

Distribution. Arunachal Pradesh, Manipur [China, Myanmar].

Habit. Epiphytic herb.

Cleisostoma racemiferum (Lindl.) Garay, Bot. Mus. Leafl. 23(4): 173. 1972. *Saccolabium racemiferum* Lindl., Gen. Sp. Orchid. Pl.: 224. 1833. *Sarcanthus racemifer* (Lindl.) Rchb. f., Ann. Bot. Syst. (Walpers) 6: 891. 1864 (as "*racemiferus*"). *Aerides racemifera* Wall. ex Lindl., Gen. Sp. Orchid. Pl.: 224. 1833, in syn. *Gastrochilus racemifer* (Lindl.) Kuntze, Revis. Gen. Pl. 2: 661 (as "*racemiferus*"). 1891. TYPE: India, icon. *Wallich 655* (holotype, K).

Sarcanthus lorifolius C. S. P. Parish ex Hook. f., Fl. Brit. India [J. D. Hooker] 6(17): 69. 1890. TYPE: India, Tenasserim, Moulmein [Mawlamyine], *C. S. P. Parish s.n.*

Sarcanthus pallidus Lindl., Edwards's Bot. Reg. 26(Misc.): 78. 1840. TYPE: India, *J. Gibson s.n.* (holotype, K-LINDL).

Distribution. Arunachal Pradesh, Assam, Manipur, Meghalaya, Mizoram, Nagaland, Tripura [Bhutan, China, Myanmar, Nepal].

Habit. Epiphytic herb.

Cleisostoma rolfeanum (King & Pantl.) Garay, Bot. Mus. Leafl. 23(4): 174. 1972. *Sarcanthus rolfeanus* King & Pantl., J. Asiat. Soc. Bengal, Pt. 2, Nat. Hist. 66: 594. 1898 [1897]. TYPE: Myanmar, Moulmein [Mawlamyine], *C. Peché s.n.* (holotype, CAL).

Distribution. Manipur [China, Myanmar].

Habit. Epiphytic herb.

Cleisostoma schneideri Choltco, Orchid Rev. 117 (1285): 41. 2009. TYPE: NE India, *cult. Choltco, TC 1001* (holotype, PAC; isotype, MU, all spirit mat.).

Distribution. India (Northeast, without precise locality). Endemic.

Habit. Epiphytic herb.

Cleisostoma simondii (Gagnep.) Seidenf., Dansk Bot. Ark. 29(3): 66. 1975. *Vanda simondii* Gagnep., Bull. Mus. Natl. Hist. Nat. Ser. 2, 22: 628. 1950. *Echioglossum simondii* (Gagnep.) Szlach., Fragm. Florist. Geobot., Suppl. 3: 137. 1995. TYPE: Vietnam, Tonkin, icon. *Simond (leg. Rives), Pl. 56* (holotype, P).

Distribution. Arunachal Pradesh, Assam, ?Meghalaya, Nagaland, Tripura [China].

Habit. Epiphytic herb.

Remarks. Records from India could be referable to *Cleisostoma chantaburiense* or *C. schneideri* and need to be verified.

Cleisostoma striatum (Rchb. f.) Garay, Bot. Mus. Leafl. 23(4): 175. 1972. *Echioglossum striatum* Rchb. f., Gard. Chron., n.s., 12: 390. 1879. *Sarcanthus striatus* (Rchb. f.) J. J. Sm., Natuurk. Tijdschr. Ned.-Indië 72: 93. 1912. *Raciborskanthos striatus* (Rchb. f.) Szlach., Ann. Bot. Fenn. 40(1): 68. 2003. TYPE: India, Sikkim, Darjeeling, *cult. Mackay s.n.* (holotype, W).

Cleisostoma brevipes Hook. f., Fl. Brit. India [J. D. Hooker] 6(17): 73. 1890. *Sarcanthus brevipes* (Hook. f.) J. J. Sm., Natuurk. Tijdschr. Ned.-Indië 72: 84. 1912. SYNTYPES: India, Sikkim, Rangit, alt. 5000 ft, July 1882,

Gamble s.n. [*10452*] (syntype, K); Assam, *cult. N. Campany s.n. Aug. 1890* (syntype, K).

Distribution. Assam, Meghalaya, Nagaland, Sikkim, West Bengal [China].

Habit. Epiphytic herb.

Cleisostoma subulatum Blume, Bijdr. Fl. Ned. Ind. 8: 363. 1825 (as "*subulata*"). *Sarcanthus subulatus* (Blume) Rchb. f., Bonplandia 5(3): 41. 1857. TYPE: Indonesia, Java, Mt. Parang, *Blume 1308* (holotype, L; isotype, L).

Sarcanthus secundus Griff., Not. Pl. Asiat. 3: 362. 1851. TYPE: India, Assam, Suddyah, Aug. 1836, *Griffith s.n.* (holotype, K).

Distribution. Arunachal Pradesh, Assam, Meghalaya, Mizoram, Nagaland, Sikkim, West Bengal [Bangladesh, Bhutan, Myanmar].

Habit. Epiphytic herb.

Cleisostoma tenuifolium (L.) Garay, Bot. Mus. Leafl. 23(4): 175. 1972. *Epidendrum tenuifolium* L., Sp. Pl. 2: 952. 1753. *Cymbidium tenuifolium* (L.) Willd., Sp. Pl., ed. 4 [Willdenow] 4(1): 103. 1805. *Aerides tenuifolia* (L.) Moon, Cat. Pl. Ceylon 60. 1824. *Sarcochilus tenuifolius* (L.) Náves, Fl. Filip., ed. 3 (Blanco). 4(13A): 238. 1880. *Saccolabium tenuifolium* (L.) Alston, Ann. Roy. Bot. Gard. (Peradeniya) 11: 205. 1929. *Sarcanthus tenuifolius* (L.) Seidenf., Dansk Bot. Ark. 27(4): 37. 1971. TYPE: India, Malabar. Icon. "Tsjerou-mau-maravara" in Rheede, Hort. Malab. 12: 11, t. 5. 1692 (lectotype, designated by Majumdar & Bakshi, 1979).

Sarcanthus pauciflorus Wight, Icon. Pl. Ind. Orient. [Wight] 5(1): 20, t. 1747. 1851. *Cleisostoma pauciflorum* (Wight) Senghas, Orchideen (Schlechter) 1(22): 1340. 1989. TYPE: India, Malabar, *Jerdon s.n.* (holotype, K).
Sarcanthus peninsularis Dalzell, Hooker's J. Bot. Kew Gard. Misc. 3: 343. 1851. *Saccolabium peninsulare* (Dalzell) Alston in H. Trimen, Handb. Fl. Ceylon 6(Suppl.): 278. 1931. TYPE: India, Bombay, Virdee, *Dalzell s.n.* [*35*] (holotype, K).

Distribution. Andhra Pradesh, Goa, Karnataka, Kerala, Maharashtra, Tamil Nadu [Sri Lanka].

Habit. Epiphytic herb.

Cleisostoma tricallosum S. N. Hegde & A. N. Rao, Orchid Rev. 91: 54, fig. 53. 1983. TYPE: India, Arunachal Pradesh, Sessa, 1100 m, 11 July 1980, *Hegde 4127* (holotype, Orchid Herbarium, Tipi).

Distribution. Arunachal Pradesh. Endemic.

Habit. Epiphytic herb.

Cleisostoma uraiense (Hayata) Garay & H. R. Sweet, Orchids S. Ryukyu Is. 156. 1974. *Sarcanthus uraiensis* Hayata, Icon. Pl. Formosan. 8: 130, fig. 58. 1919. TYPE: Taiwan, Urai, *B. Hayata s.n.*

Distribution. Andaman & Nicobar Islands [also in Indonesia].

Habit. Epiphytic herb.

Cleisostoma williamsonii (Rchb. f.) Garay, Bot. Mus. Leafl. 23(4): 176. 1972 (as *"williamsoni"*). *Sarcanthus williamsonii* Rchb. f., Hamburger Garten-Blumenzeitung 21: 333. 1865 (as *"williamsoni"*). *Echioglossum williamsonii* (Rchb. f.) Szlach., Fragm. Florist. Geobot., Suppl. 3: 137. 1995. TYPE: India, Assam, icon. *Williamson s.n.* (holotype, W).

Cleisostoma elegans Seidenf., Dansk Bot. Ark. 29(3): 46, fig. 19. 1975. *Echioglossum elegans* (Seidenf.) Szlach., Ann. Bot. Fenn. 40(1): 68. 2003. TYPE: Myanmar, Moulmein [Mawlamyine], *C. S. P. Parish 272* (holotype, W; isotype, K).

Distribution. Andaman & Nicobar Islands, Arunachal Pradesh, Assam, Manipur, Nagaland, Sikkim, West Bengal [Bhutan, China, Myanmar].

Habit. Epiphytic herb.

———

Coelogyne Lindl., Coll. Bot. (Lindley) sub t. 33. 1821 (as *"Caelogyne"*); corr. Lindl., Coll. Bot. (Lindley) sub t. 37. 1825.

TYPE: *Coelogyne cristata* Lindl.

Coelogyne alba (Lindl.) Rchb. f., Ann. Bot. Syst. (Walpers) 6: 236. 1861. *Otochilus albus* Lindl., Gen. Sp. Orchid. Pl.: 35. 1830 (as *"alba"*). TYPE: Nepal, 1821, *Wallich 1967.1* (holotype, K-LINDL; isotype, K-WALL).

Distribution. Arunachal Pradesh, Assam, Manipur, Meghalaya, Mizoram, Nagaland, Sikkim, West Bengal [Bangladesh, China, Myanmar, Nepal].

Habit. Epiphytic herb.

Coelogyne albolutea Rolfe, Bull. Misc. Inform. Kew 1908: 414. 1908. TYPE: India, *cult. Royal Botanical Garden, Glasnevin s.n.* (holotype, K).

Distribution. India. Precise locality unknown.

Habit. Epiphytic herb.

Remarks. This obscure species is possibly conspecific with *Coelogyne flaccida* Lindl.

Coelogyne apiculata (Lindl.) Rchb. f., Ann. Bot. Syst. (Walpers) 6: 225. 1861. *Panisea apiculata* Lindl., Fol. Orchid. 5(*Panisea*): 2. 1854. *Chelonistele apiculata* (Lindl.) Pfitzer, Pflanzenr. (Engler) IV, 50 (Heft II B 7): 138. 1907. TYPE: Myanmar, Moulmein [Mawlamyine], 5000 ft, *T. Lobb s.n.* (holotype, K-LINDL).

Distribution. Assam, Nagaland [Myanmar].

Habit. Epiphytic herb.

Coelogyne articulata (Lindl.) Rchb. f., Ann. Bot. Syst. (Walpers) 6: 238. 1861. *Pholidota articulata* Lindl., Gen. Sp. Orchid. Pl.: 38. 1830. TYPE: India, Pundua Hills, July 1825, *Wallich 1992 (leg. De Silva 1554)* (lectotype, K-LINDL, designated by de Vogel, 1988; isotype, G, K, K-WALL).

Pholidota khasyana Rchb. f., Bonplandia (Hannover) 4: 329. 1856. *Coelogyne khasyana* (Rchb. f.) Rchb. f., Ann. Bot. Syst. (Walpers) 6: 238. 1861. TYPE: India, Khasi Hills, *T. Lobb 53* (holotype, W; isotype, K).
Pholidota griffithii Hook. f., Hooker's Icon. Pl. 19: t. 1881. 1889. *Pholidota articulata* Lindl. var. *griffithii* (Hook. f.) King & Pantl., Ann. Roy. Bot. Gard. (Calcutta) 8: 147. 1898. TYPE: India, Sikkim, *J. D. Hooker s.n.* (lectotype, K, designated by de Vogel, 1988).
Pholidota obovata Hook. f., Fl. Brit. India [J. D. Hooker] 5(16): 845. 1890. TYPE: Bhutan, *Griffith s.n. [5034]* (lectotype, K, designated by de Vogel, 1988; isotype, CAL).

Distribution. Arunachal Pradesh, Assam, Himachal Pradesh, Manipur, Meghalaya, Mizoram, Nagaland, Sikkim, Tamil Nadu, Tripura, Uttar Pradesh, Uttarakhand, West Bengal [Bhutan, China, Myanmar, Nepal].

Habit. Epiphytic herb.

Coelogyne barbata Lindl. ex Griff., Itin. Pl. Khasyah Mts. 72. 1848. *Pleione barbata* (Lindl. ex Griff.) Kuntze, Revis. Gen. Pl. 2: 680. 1891. TYPE: India, Khasi Hills, Mamlu, *Griffith 1120 [s.n.]* (holotype, K-LINDL).

Distribution. Arunachal Pradesh, Assam, Manipur, Meghalaya, Mizoram, Nagaland, Sikkim, West Bengal [Bhutan, Myanmar].

Habit. Epiphytic herb.

Coelogyne breviscapa Lindl., Fol. Orchid. 5(*Coelogyne*): 4. 1854. *Pleione breviscapa* (Lindl.) Kuntze, Revis. Gen. Pl. 2: 680. 1891. TYPE: Sri Lanka, *Walker s.n.* (holotype, K-LINDL).

Distribution. Karnataka, Kerala, Tamil Nadu [Sri Lanka].

Habit. Epiphytic herb.

Coelogyne brunnea Lindl., Gard. Chron. 1848: 71. 1848. *Coelogyne fuscescens* Lindl. var. *brunnea* (Lindl.) Lindl., Fol. Orchid. 5(*Coelogyne*): 11. 1854. TYPE: East Indies, 1844, *cult. Syon s.n.* (holotype, K-LINDL).

Distribution. Arunachal Pradesh, Manipur, Meghalaya, Mizoram [Bhutan, Myanmar, Nepal].

Habit. Epiphytic herb.

Coelogyne bulleyia R. Rice, Photo Intro Asian *Bulbophyllum, Coelogyne & Dendrobium* Orchids: 171. 2019. *Bulleyia yunnanensis* Schltr., Notes Roy. Bot. Gard. Edinburgh 5(24): 108. 1912. TYPE: China, Yunnan, Lichiang Range, *G. Forrest 4879* (holotype, E; isotype, BM, CAL).

Distribution. Arunachal Pradesh, Nagaland [Bhutan, China, Myanmar].

Habit. Epiphytic herb.

Coelogyne calcicola Kerr, J. Siam Soc., Nat. Hist. Suppl. 9: 233. 1933. TYPE: Laos, Chiengkwang (=Xiangkhouang), Muang Cha, 1500 m, 16 Apr. 1932, *Kerr 978* (holotype, K; isotype, K).

Distribution. Nagaland [China, Myanmar].

Habit. Epiphytic or lithophytic herb.

Coelogyne convallariae C. S. P. Parish & Rchb. f., Flora 55: 277. 1872. *Pholidota convallariae* (C. S. P. Parish & Rchb. f.) Hook. f., Hooker's Icon. Pl. 19: t. 1880. 1889. TYPE: Myanmar, *D. Oliver s.n.* (holotype, K; isotype, W).

Pholidota convallariae (C. S. P. Parish & Rchb. f.) Hook. f. var. *breviscapa* Deori & J. Joseph, Bull. Bot. Surv. India 20(1–4): 159. 1978 [1979]. *Pholidota katakiana* Phukan, Orchid Rev. 104: 238. 1996. *Coelogyne katakiana* (Phukan) R. Rice, Photo Intro Asian *Bulbophyllum, Coelogyne & Dendrobium* Orchids: 173. 2019. TYPE: India, Arunachal Pradesh, Kameng, camp 43, 5332 ft (1666 m), 16 Nov. 1971, *A. S. Rao 50902* (holotype, CAL [50902 A]; isotype, ASSAM [50902 B–G]).

Distribution. Arunachal Pradesh, Assam, Manipur, Meghalaya, Mizoram, Nagaland [China, Myanmar].

Habit. Epiphytic herb.

Coelogyne corymbosa Lindl., Fol. Orchid. 5(*Coelogyne*): 7. 1854. *Pleione corymbosa* (Lindl.) Kuntze, Revis. Gen. Pl. 2: 680. 1891. TYPE: India, Sikkim, 6000–9000 ft, *J. D. Hooker 136* (holotype, K-LINDL, p.p.; isotype, K).

Distribution. Arunachal Pradesh, Manipur, Meghalaya, Nagaland, Sikkim, West Bengal [Bhutan, China, Myanmar, Nepal].

Habit. Epiphytic herb.

Coelogyne cristata Lindl., Coll. Bot. (Lindley) t. 33. 1821. TYPE: Nepal, Kathmandu, Tokha, *Wallich 1958.1* (holotype, K-LINDL; isotype, K, K-WALL).

Distribution. Arunachal Pradesh, Assam, Himachal Pradesh, ?Karnataka, Manipur, Meghalaya, Nagaland, Sikkim, Uttarakhand, West Bengal [Bangladesh, Bhutan, China, Nepal].

Habit. Epiphytic herb.

Coelogyne demissa (D. Don) M. W. Chase & Schuit., Phytotaxa 510(2): 108. 2021. *Dendrobium demissum* D. Don, Prodr. Fl. Nepal.: 34. 1825. *Panisea demissa* (D. Don) Pfitzer, Pflanzenr. (Engler) IV, 50 (Heft 32): 141, t. 49. 1907. TYPE: Nepal, Bagmati Zone, Rashuwa District, Gosainthan [Shishapangma], *Wallich s.n.* (holotype, BM).

Coelogyne parviflora Lindl., Gen. Sp. Orchid. Pl.: 44. 1830. *Panisea parviflora* (Lindl.) Lindl., Fol. Orchid. 5(*Panisea*): 1. 1854. TYPE: Nepal, *Wallich s.n.* (holotype, K-LINDL).

Distribution. Arunachal Pradesh, Assam, Manipur, Meghalaya, Nagaland, Sikkim, West Bengal [Bhutan, China, ?Myanmar, Nepal].

Habit. Epiphytic herb.

Coelogyne fimbriata Lindl., Bot. Reg. 11: t. 868. 1825. *Pleione fimbriata* (Lindl.) Kuntze, Revis. Gen. Pl. 2: 680. 1891. TYPE: China, 1824, *J. D. Parks s.n.* (holotype, K-LINDL; isotype, C, P).

Distribution. Arunachal Pradesh, Assam, Manipur, Meghalaya, Mizoram, Nagaland, Sikkim, Tripura, West Bengal [Bhutan, China, Myanmar, Nepal].

Habit. Epiphytic herb.

Coelogyne flaccida Lindl., Gen. Sp. Orchid. Pl.: 39. 1830. *Pleione flaccida* (Lindl.) Kuntze, Revis. Gen. Pl. 2: 680. 1891. TYPE: Nepal, Nuwakot ["Noakote"], *Wallich 1961* (holotype, K-LINDL; isotype, E, K, K-WALL).

Distribution. Arunachal Pradesh, Assam, Manipur, Meghalaya, Nagaland, Sikkim, Uttarakhand, West Bengal [Bhutan, Myanmar, Nepal].

Habit. Epiphytic herb.

Coelogyne fusca (Lindl.) Rchb. f., Ann. Bot. Syst. (Walpers) 6: 236. 1861. *Otochilus fuscus* Lindl., Gen. Sp. Orchid. Pl.: 35. 1830 (as *"fusca"*). TYPE: Nepal, 1821, *Wallich 1969* (holotype, K-LINDL; isotype, K-WALL).

Otochilus lancifolius Griff., Not. Pl. Asiat. 3: 278. 1851. SYNTYPES: India, Assam, Sarureem, *Griffith in Assam Herb. 181* (not found); *in Khasyah Herb. 1074* (not found).

Distribution. Arunachal Pradesh, Assam, Manipur, Meghalaya, Mizoram, Nagaland, Sikkim, Tripura, West Bengal [Bangladesh, Bhutan, China, Myanmar, Nepal].

Habit. Epiphytic herb.

Coelogyne fuscescens Lindl., Gen. Sp. Orchid. Pl.: 41. 1830. *Pleione fuscescens* (Lindl.) Kuntze, Revis. Gen. Pl. 2: 680. 1891. TYPE: Nepal, Kathmandu, Toka, *Wallich 1962* (holotype, K-LINDL; isotype, K-WALL).

Coelogyne assamica Linden & Rchb. f., Allg. Gartenzeitung 25(51): 403. 1857. *Coelogyne fuscescens* Lindl. var. *assamica* (Linden & Rchb. f.) Pfitzer, Pflanzenr. (Engler) IV, 50 (Heft 32): 43. 1907. TYPE: India, Assam, Dec., *Linden s.n.* (holotype, B, lost). NEOTYPE: Xenia Orchid. 2: t. 134, fig. 7–9 (1874) (iconotype, selected by Pearce & Cribb, 2002).
Coelogyne fuscescens Lindl. var. *viridiflorum* Pradhan, Indian Orchids: Guide Identif. & Cult. 2: 268. 1979. TYPE: India, West Bengal, Kalimpong, *U. C. Pradhan's collector 56* (holotype, Herbarium U. C. Pradhan).

Distribution. Arunachal Pradesh, Meghalaya, Mizoram, Nagaland, Sikkim, West Bengal [Bhutan, Myanmar, Nepal].

Habit. Epiphytic herb.

Remarks. The name *Coelogyne assamica* has long been misapplied to *C. annamensis* Rolfe (Schuiteman et al., 2020).

Coelogyne gardneriana Lindl., Pl. Asiat. Rar. (Wallich) 1: 33, t. 38. 1830. *Neogyna gardneriana* (Lindl.) Rchb. f., Bot. Zeitung (Berlin) 10: 931. 1852. *Pleione gardneriana* (Lindl.) Kuntze, Revis. Gen. Pl. 2: 680. 1891. TYPE: Nepal, Noakote, 1817, *E. Gardner s.n.* (not found).

Distribution. Arunachal Pradesh, Assam, Manipur, Meghalaya, Mizoram, Nagaland [Bangladesh, Bhutan, China, Myanmar, Nepal].

Habit. Epiphytic herb.

Coelogyne ghatakii T. K. Paul, Soumen K. Basu & M. C. Biswas, J. Bombay Nat. Hist. Soc. 86(3): 425. 1989 [1990]. TYPE: India, Manipur, Imphal Valley, Apr., *Ghatak 2213a* (holotype, CAL).

Distribution. Manipur. Endemic.

Habit. Epiphytic herb.

Coelogyne griffithii Hook. f., Fl. Brit. India [J. D. Hooker] 5(16): 838. 1890. *Pleione griffithii* (Hook. f.) Kuntze, Revis. Gen. Pl. 2: 680. 1891. SYNTYPES: India, Upper Assam, summit of Patkoye Mountain, *Griffith s.n. (East India Company Herbarium 5091)* (syntype, K); Manipur, Khongui Valley, 4000–5000 ft, 27 Apr. 1882, *G. Watt 6780* (syntypes, CAL, K).

Distribution. Arunachal Pradesh, Assam, Manipur, Mizoram, Nagaland [China, Myanmar].

Habit. Epiphytic herb.

Coelogyne hajrae Phukan, Orchid Rev. 105: 94. 1997. TYPE: India, Arunachal Pradesh, Tirap, Namdapha, 850 m, *B. K. Shukla 87943* (holotype, CAL).

Distribution. Arunachal Pradesh. Endemic.

Habit. Epiphytic herb.

Coelogyne hitendrae Sandh. Das & S. K. Jain, Orchid Rev. 86: 195. 1978. TYPE: India, Nagaland, Pulebadje, Mar., *Kataki 60202A* (holotype, CAL).

Distribution. Nagaland. Endemic.

Habit. Epiphytic herb.

Remarks. Ormerod et al. (2021) consider it likely that *Coelogyne hitendrae* is a form of the variable *C. corymbosa*.

Coelogyne holochila P. F. Hunt & Summerh., Kew Bull. 20: 52. 1966. TYPE: Myanmar, Chin Hills, Mt. Victoria, *cult. Royal Botanical Garden, Glasnevin (leg. Mrs. Wheeler Cuffe) s.n., 16 June 1914* (holotype, K).

Distribution. Arunachal Pradesh, Manipur, Meghalaya, Mizoram [Myanmar].

Habit. Epiphytic herb.

Coelogyne imbricata (Hook.) Rchb. f., Ann. Bot. Syst. (Walpers) 6: 238. 1861. *Pholidota imbricata* Hook., Exot. Fl. 2: t. 138. 1825. TYPE: Icon. *Exotic Fl. t. 138. 1825* (lectotype, designated by Seidenfaden, 1986).

Ptilocnema bracteata D. Don, Prodr. Fl. Nepal.: 33. 1825. *Pholidota bracteata* (D. Don) Seidenf., Opera Bot. 89: 100, fig. 57. 1986. SYNTYPES: Nepal, *Buchanan-Hamilton s.n.* (syntype, BM); *Wallich s.n.* (syntype, BM).

Cymbidium imbricatum Roxb., Fl. Ind. ed. 1832, 3: 460. 1832. SYNTYPES: Bangladesh, Chittagong and Sylhet, icon. *Cymbidium imbricatum*, *Roxburgh 2339* [?origin] (syntypes, CAL, K).

Pholidota assamica Regel, Gartenflora 39: 607. 1890, nom. inval.

Pholidota pygmaea H. J. Chowdhery & G. D. Pal, Nordic J. Bot. 15(4): 411 1995 [1996]. TYPE: India, Arunachal Pradesh: Lower Subansiri District, Itanagar, alt. 2000 ft, 30 July 1993, *Chowdhery 1760* (holotype, CAL).

Distribution. Andaman & Nicobar Islands, Andhra Pradesh, Arunachal Pradesh, Assam, Bihar, Chhattisgarh, Goa, Jharkhand, Karnataka, Kerala, Madhya Pradesh, Maharashtra, Manipur, Meghalaya, Mizoram, Nagaland, Odisha, Sikkim, Tamil Nadu, Tripura, Uttarakhand, West Bengal [Bhutan, China, Myanmar, Nepal, Sri Lanka].

Habit. Epiphytic or lithophytic herb.

Coelogyne lancilabia (Seidenf.) R. Rice, Photo Intro Asian *Bulbophyllum, Coelogyne & Dendrobium* Orchids: 173. 2019. *Otochilus lancilabius* Seidenf., Opera Botanica 89: 94. 1986. *Otochilus albus* Lindl. var. *lancilabius* (Seidenf.) Pradhan, Indian Orchids: Guide Identif. & Cult. 2: 706. 1979. TYPE: India, Sikkim, s. loc., 6000 ft, Oct.–Dec. 1893, *Pantling 26* (holotype, K; isotype, BM, W).

Distribution. Arunachal Pradesh, ?Meghalaya, Nagaland, ?Sikkim, Uttarakhand [Bhutan, China, Nepal].

Habit. Epiphytic herb.

Coelogyne longipes Lindl., Fol. Orchid. 5(*Coelogyne*): 10. 1854. *Pleione longipes* (Lindl.) Kuntze, Revis. Gen. Pl. 2: 680. 1891. TYPE: India, Khasi Hills, *Hooker & Thomson s.n.* [*129*] (holotype, K-LINDL.; isotype, BM, CAL).

Distribution. Arunachal Pradesh, Manipur, Meghalaya, Nagaland, Sikkim, West Bengal [Bhutan, ?Myanmar].

Habit. Epiphytic herb.

Coelogyne micrantha Lindl., Gard. Chron. 1855: 173. 1855. *Pleione micrantha* (Lindl.) Kuntze, Revis. Gen. Pl. 2: 680. 1891. TYPE: East Indies, Dale Park, 11 Mar. 1855, *A. Dick s.n.* (holotype, K-LINDL).

Distribution. Arunachal Pradesh, Manipur, Meghalaya, Mizoram, Nagaland [China, Myanmar].

Habit. Epiphytic herb.

Coelogyne missionariorum (Gagnep.) R. Rice, Photo Intro Asian *Bulbophyllum, Coelogyne & Dendrobium* Orchids: 174. 2019 (as "*missionarium*").

Pholidota missionariorum Gagnep., Bull. Mus. Natl. Hist. Nat., Ser. 2. 3: 146. 1931. SYNTYPES: China, Gan-pin, *Martin 2576* (lectotype, P, designated by de Vogel, 1988; isotype, E).

Distribution. ?Arunachal Pradesh [Bhutan, China, Myanmar].

Habit. Epiphytic herb.

Remarks. This species is listed as "excluded" in Singh et al. (2019).

Coelogyne mossiae Rolfe, Bull. Misc. Inform. Kew 1894: 156. 1894. TYPE: India, Nilgiri Hills, *cult. J. S. Moss s.n., June 1890* (holotype, K).

Coelogyne glandulosa Lindl. var. *bournei* Sandh. Das & S. K. Jain, Bull. Bot. Surv. India 18: 244. 1976 [1979]. TYPE: India, Tamil Nadu, Pulney Hills, Poombarai, *Bourne 2941A* (holotype, CAL).

Coelogyne glandulosa Lindl. var. *sathyanarayanae* Sandh. Das & S. K. Jain, Bull. Bot. Surv. India 18(1–4): 242. 1976 [1979]. TYPE: India, Tamil Nadu, Kodaikanal, *Saldanha 5211* (holotype, CAL; isotype, BLAT).

Distribution. Kerala, Tamil Nadu. Endemic.

Habit. Lithophytic herb.

Coelogyne nervosa A. Rich., Ann. Sci. Nat., Bot. sér. 2, 15: 16. 1841. *Pleione nervosa* (A. Rich.) Kuntze, Revis. Gen. Pl. 2: 680. 1891. SYNTYPES: India, Nilgiri Hills, Neddoubetta, July 1840, *Perrottet s.n.* [*522 & 868*] (syntypes, P [2×]).

Coelogyne corrugata Wight, Icon. Pl. Ind. Orient. [Wight] 5(1): 5, t. 1639. 1851. TYPE: India, Courtallam, Pulney Hills, Aug.–Sep. 1835, *Wight s.n.* [*905*] (holotype, K; isotype, K, MH).

Coelogyne glandulosa Lindl., Fol. Orchid. 5(*Coelogyne*): 6. 1854. *Pleione glandulosa* (Lindl.) Kuntze, Revis. Gen. Pl. 2: 680. 1891. TYPE: India, Nilgiri Hills, Pycarah, *Wight s.n.* (holotype, K-LINDL).

Distribution. Karnataka, Kerala, Tamil Nadu. Endemic.

Habit. Epiphytic or lithophytic herb.

Coelogyne nitida (Wall. ex D. Don) Lindl., Gen. Sp. Orchid. Pl.: 40. 1830. *Cymbidium nitidum* Wall. ex D. Don, Prodr. Fl. Nepal.: 35. 1825. *Pleione nitida* (Lindl.) Kuntze, Revis. Gen. Pl. 2: 680. 1891. TYPE: Nepal, *Wallich 1954* (lectotype, K-LINDL; isotype, BM, C, CAL, K, K-WALL, L, P, WU).

Coelogyne ochracea Lindl., Edwards's Bot. Reg. 32: t. 69. 1846. *Pleione ochracea* (Lindl.) Kuntze, Revis. Gen. Pl. 2: 680. 1891. SYNTYPES: India, Darjeeling, 1844, *Griffith s.n.* (syntype, K-LINDL); India, Mishmee Hills, *Griffith 19* (syntype, K-LINDL); Bhutan, *Griffith 24* (syntype, K-LINDL [2×]); Bhutan, *Griffith 25* (syntype, K-LINDL); *Brockelhurst cult s.n., April 1845* (not found, used to produce the above-mentioned plate in Edwards's Botanical Register).

Coelogyne goweri Rchb. f., Gard. Chron. 1869: 443. 1869. TYPE: India, Assam, *Gower s.n.* (holotype, W).

Distribution. Arunachal Pradesh, Assam, Manipur, Meghalaya, Mizoram, Nagaland, Sikkim, Uttarakhand, West Bengal [Bangladesh, Bhutan, China, Myanmar, Nepal].

Habit. Epiphytic herb.

Remarks. The distribution of this species is uncertain due to confusion with *Coelogyne punctulata* Lindl., which has very similar flowers. In *C. punctulata* the inflorescence arises from a fully grown shoot; in *C. nitida* it arises from an immature shoot before the leaves are visible.

Coelogyne occultata Hook. f., Fl. Brit. India [J. D. Hooker] 5(16): 832. 1890. *Pleione occultata* (Hook. f.) Kuntze, Revis. Gen. Pl. 2: 680. 1891. TYPE: India, Darjeeling, *Griffith s.n.* (holotype, K; isotype, CAL).

Coelogyne occultata Hook. f. var. *uniflora* N. P. Balakr., J. Bombay Nat. Hist. Soc. 75: 159, fig. 3. 1978. TYPE: Bhutan, Nyoth Forest, *Balakrishnan 43041A* (holotype, CAL; isotype, ASSAM).
Coelogyne pantlingii Lucksom, Orchid Rev. 113(1262): 108. 2005. TYPE: India, Sikkim, Tendong, 15 June 2004, *Lucksom 456a* (holotype, CAL; isotype, K).

Distribution. Arunachal Pradesh, Meghalaya, Nagaland, Sikkim, West Bengal [Bhutan, Myanmar].

Habit. Epiphytic herb.

Coelogyne odoratissima Lindl., Gen. Sp. Orchid. Pl.: 41. 1830. *Pleione odoratissima* (Lindl.) Kuntze, Revis. Gen. Pl. 2: 680. 1891. TYPE: Sri Lanka, Nuera Ellia, *Wallich 1960 (leg. Macrae 14, 1829)* (holotype, K-LINDL; isotype, K-WALL).

Coelogyne angustifolia A. Rich., Ann. Sci. Nat., Bot. sér. 2. 15: 16. 1841. *Coelogyne odoratissima* Lindl. var. *angustifolia* (A. Rich.) Lindl., Fol. Orchid. 5(*Coelogyne*): 5. 1854. TYPE: India, Nilgiri Hills, Neddoubetta, July, *Perrottet s.n.*

Distribution. Karnataka, Tamil Nadu [Sri Lanka].

Habit. Epiphytic herb.

Coelogyne ovalis Lindl., Edwards's Bot. Reg. 24(Misc.): 91. 1838. TYPE: Nepal and India, Kamaon, 1821, *Wallich 1957 p.p.* (syntypes, K-LINDL, K-WALL).

Coelogyne fuliginosa Lodd. ex Hook., Bot. Mag. 75: t. 4440. 1849. *Pleione fuliginosa* (Lodd. ex Hook.) Kuntze, Revis. Gen. Pl. 2: 680. 1891. TYPE: India, 1838, *Loddiges's collector s.n.* (holotype, K-LINDL).
Coelogyne ovalis Lindl. var. *latifolia* Hook. f., Fl. Brit. India [J. D. Hooker] 5(16): 836. 1890. TYPE: India, Munnipore, Kohima, alt. 4500 ft, *C. B. Clarke s.n.* (not found)
Coelogyne arunachalensis H. J. Chowdhery & G. D. Pal, Nordic J. Bot. 17: 369. 1997. TYPE: India, Assam (Arunachal Pradesh), Lower Subansiri District near Doimukh, *G. D. Pal 1790* (holotype, CAL).
Coelogyne mishmensis K. Gogoi, Richardiana 16: 375. 2016. TYPE: India, Arunachal Pradesh, Lower Dibang Valley District, Mehao Wildlife Sanctuary, 800 m, 24 July 2016, *Gogoi 0765A* (holotype, CAL; isotype, DU, Herb. Orchid Soc. E. Himalaya).

Distribution. Arunachal Pradesh, Assam, Karnataka, Manipur, Meghalaya, Mizoram, Nagaland, Sikkim, Tamil Nadu, Tripura, Uttarakhand, West Bengal [Bhutan, China, Myanmar, Nepal].

Habit. Epiphytic herb.

Remarks. We agree with Kumar et al. (2018) that *Coelogyne mishmensis* is indistinguishable from *C. ovalis*.

Coelogyne pallida (Lindl.) Rchb. f., Ann. Bot. Syst. (Walpers) 6: 288. 1861. *Pholidota pallida* Lindl. Edwards's Bot. Reg. 21: t. 1777. 1835. TYPE: Nepal, Gosainthan [Shishapangma], 1819, *Wallich s.n.* (lectotype, K-LINDL, designated by Hallé, 1977).

Pholidota calceata Rchb. f., Bonplandia 4: 329. 1856. *Coelogyne calceata* (Rchb. f.) Rchb. f., Ann. Bot. Syst. (Walpers) 6: 238. 1861. TYPE: India, Khasi Hills, *Hooker & Thomson s.n.* (holotype, K-LINDL; isotype, C, G, K, L, OXF, NY, S, W).
Pholidota imbricata Hook. var. *sessilis* Hook. f., Fl. Brit. India [J. D. Hooker] 5(16): 846. 1890. *Pholidota pallida* Lindl. var. *sessilis* (Hook. f.) P. K. Sarkar, J. Econ. Taxon. Bot. 5(5): 1008. 1984. TYPE: India, Naga Hills, Kohima, alt. 3500–5000 ft, Apr.–May 1886, *Prain s.n.* [25] (syntypes, CAL, K).

Distribution. Andaman & Nicobar Islands, Andhra Pradesh, Assam, Jharkhand, Karnataka, Kerala, Manipur, Meghalaya, Mizoram, Nagaland, Sikkim, Tamil Nadu, West Bengal [Bhutan, China, Myanmar, Nepal].

Habit. Epiphytic herb.

Coelogyne panchaseensis (Subedi) M. W. Chase & Schuit., Phytotaxa 510(2): 122. 2021. *Panisea panchaseensis* Subedi, Nordic J. Bot. 29: 362. 2011. TYPE: Nepal, Kaski district, Panchase forest, *Subedi 1980* (holotype, KATH; isotype, TUCH).

Distribution. Assam, Nagaland [Nepal].

Habit. Epiphytic herb.

Coelogyne pempahisheyana H. J. Chowdhery, Indian J. Forest. 27(1): 121. 2004. TYPE: India, Kalimpong, Holumba Estate, 3900 ft, 27 Mar. 2004, *H. J. Chowdhery 32598* (holotype, CAL).

Distribution. West Bengal. ?Endemic.

Habit. Epiphytic herb.

Remarks. This seems little more than a color form of *Coelogyne flaccida* Lindl.

Coelogyne pendula Summerh. ex D. A. Clayton & J. J. Wood, Orchid Rev. 118: 229. 2010. TYPE: India, Mizoram, Lushai Hills, Burpui, 2000 ft, 9 July 19--, *N. E. Parry 241* (holotype, K).

Distribution. Mizoram. Endemic.

Habit. Epiphytic herb.

Coelogyne porrecta (Lindl.) Rchb. f., Ann. Bot. Syst. (Walpers) 6: 236. 1861. *Otochilus porrectus* Lindl., Gen. Sp. Orchid. Pl.: 36. 1830 (as "*porrecta*"). TYPE: India, Pundua, Dec. 1820, *Wallich 1968 (leg. De Silva 401)* (holotype, K-LINDL; isotype, K-WALL).

Otochilus latifolius Griff., Not. Pl. Asiat. 3: 279. 1851. TYPE: India, Assam, *Griffith in Khasyah Herb. 1130* (holotype, K).

Distribution. Arunachal Pradesh, Assam, Manipur, Meghalaya, Mizoram, Nagaland, Sikkim, Tripura, West Bengal [China, Myanmar, Nepal].

Habit. Epiphytic herb.

Coelogyne prolifera Lindl., Gen. Sp. Orchid. Pl.: 40. 1830. *Pleione prolifera* (Lindl.) Kuntze, Revis. Gen. Pl. 2: 680. 1891. SYNTYPES: Nepal, Nuwakot ["Noakote"] & Kathmandu, Tokha, *Wallich 1956* (syntypes, BM, CAL, K-LINDL, K-WALL, OXF, W).

Coelogyne flavida Hook. f. ex Lindl., Fol. Orchid. 5(*Coelogyne*): 10. 1854. *Pleione flavida* (Hook. f. ex Lindl.) Kuntze, Revis. Gen. Pl. 2: 680. 1891. SYNTYPES: India, Khasi Hills, *T. Lobb s.n.* (syntypes, K, K-LINDL); icon. *Cathcart s.n.* (syntype, K).

Distribution. Arunachal Pradesh, Assam, Manipur, Meghalaya, Mizoram, Nagaland, Sikkim, Tamil Nadu, West Bengal [Bhutan, China, Myanmar, Nepal].

Habit. Epiphytic herb.

Coelogyne protracta (Hook. f.) R. Rice, Photo Intro Asian *Bulbophyllum, Coelogyne & Dendrobium* Orchids: 175. 2019. *Pholidota protracta* Hook. f., Hooker's Icon. Pl. 19: t. 1877. 1889. TYPE: India,

Sikkim, Neetay, *Clarke 25229* (lectotype, K, designated by de Vogel, 1988).

Distribution. Arunachal Pradesh, Assam, Manipur, Meghalaya, Mizoram, Nagaland, Sikkim, West Bengal [Bhutan, China, Myanmar, Nepal].

Habit. Epiphytic herb.

Coelogyne punctulata Lindl., Coll. Bot. (Lindley) t. 33. 1821. TYPE: Nepal, *Wallich s.n.* (holotype, BM).

Cymbidium nitidum Roxb., Fl. Ind. ed. 1832, 3: 459. 1832, nom. illeg., non Wall. ex D. Don, 1825. TYPE: India, Meghalaya, Garrow [Garo] Hills, icon. *Cymbidium nitidum, Roxburgh 2337* (syntypes, CAL, K).
Coelogyne ocellata Lindl., Gen. Sp. Orchid. Pl.: 40. 1830. TYPE: India, Sylhet, Pundua Hills, *Wallich 1953.1 (leg. De Silva 677)* (syntypes, K-LINDL, K-WALL).
Coelogyne brevifolia Lindl., Fol. Orchid. 5(*Coelogyne*): 7. 1854. *Pleione brevifolia* (Lindl.) Kuntze, Revis. Gen. Pl. 2: 680. 1891. *Coelogyne punctulata* Lindl. f. *brevifolia* (Lindl.) Sandh. Das & S. K. Jain, Fasc. Fl. India 5: 25. 1980. TYPE: India, Khasi Hills, alt. 4000–5000 ft, *Hooker & Thomson s.n.* (holotype, K-LINDL).

Distribution. Arunachal Pradesh, Manipur, Meghalaya, Mizoram, Nagaland [Bangladesh, Bhutan, Myanmar, Nepal].

Habit. Epiphytic herb.

Remarks. The distribution of this species is uncertain due to confusion with *Coelogyne nitida* (Wall. ex D. Don) Lindl.

Coelogyne quadratiloba Gagnep., Bull. Mus. Natl. Hist. Nat. Ser. 2. 22: 507. 1950. TYPE: Vietnam, Feb. 1910, *cult. Hanoi (leg. Deschamps s.n.)* (holotype, P).

Coelogyne thailandica Seidenf., Dansk Bot. Ark. 29(4): 46, fig. 17. 1975. TYPE: Thailand, Ta Kanun, Kanburi, 400 m, 19 Jan. 1924, *Kerr 0260* (holotype, K K000078777; isotype K K000078778).

Distribution. Andaman & Nicobar Islands [also in Cambodia, Thailand, Vietnam].

Habit. Epiphytic herb.

Coelogyne raizadae S. K. Jain & Sandh. Das, Proc. Indian Acad. Sci. Pl. Sci. 87(5): 119, fig. 1. 1978. TYPE: India, Khasi Hills, Mawsmai, 30 Apr. 1974, *S. Das 55419* (holotype, CAL; isotype, ASSAM).

Distribution. Arunachal Pradesh, Meghalaya, Nagaland [Bhutan, Nepal].

Habit. Epiphytic herb.

Coelogyne recurva (Lindl.) Rchb. f., Ann. Bot. Syst. (Walpers) 6: 237. 1861. *Pholidota recurva* Lindl.,

Gen. Sp. Orchid. Pl.: 37. 1830. TYPE: Nepal, *Wallich s.n.* (holotype, K-LINDL; isotype, BM, C, K).

Distribution. Meghalaya, Sikkim, West Bengal [China, Myanmar, Nepal].

Habit. Epiphytic herb.

Coelogyne rigida C. S. P. Parish & Rchb. f., Trans. Linn. Soc. London 30: 146. 1874. *Pleione rigida* (C. S. P. Parish & Rchb. f.) Kuntze, Revis. Gen. Pl. 2: 680. 1891. TYPE: Myanmar, Moulmein [Mawlamyine], *C. S. P. Parish 42* (syntypes, K, W).

Distribution. Arunachal Pradesh, ?Meghalaya, Nagaland [Myanmar].

Habit. Epiphytic herb.

Coelogyne rubra (Lindl.) Rchb. f., Ann. Bot. Syst. (Walpers) 6: 238. 1861. *Pholidota rubra* Lindl., Gen. Sp. Orchid. Pl.: 37. 1830. TYPE: India, *Wallich s.n.* (holotype, K-LINDL).

Pholidota undulata Wall. ex Lindl., Gen. Sp. Orchid. Pl.: 37. 1830. *Coelogyne undulata* (Wall. ex Lindl.) Rchb. f., Ann. Bot. Syst. (Walpers) 6: 238. 1861. TYPE: India, 1828, *Wallich 141* (holotype, K-LINDL).

Distribution. Arunachal Pradesh, Assam, Meghalaya, Mizoram, Nagaland, Sikkim [Bhutan, Myanmar].

Habit. Epiphytic herb.

Coelogyne schultesii S. K. Jain & Sandh. Das, Proc. Indian Acad. Sci. Pl. Sci. 87(5): 121, fig. 1978. TYPE: India, Khasi Hills, Cherrapunji, *S. Das 60256A* (holotype, CAL; isotype, ASSAM).

Distribution. Arunachal Pradesh, Assam, Manipur, Meghalaya, Nagaland [Bhutan, Myanmar, Nepal].

Habit. Epiphytic herb.

Remarks. This species is illustrated as *Coelogyne prolifera* Lindl. in Swami (2017).

Coelogyne stricta (D. Don) Schltr., Repert. Spec. Nov. Regni Veg. Beih. 4: 184. 1919. *Cymbidium strictum* D. Don, Prodr. Fl. Nepal.: 35. 1825. TYPE: Nepal, *Wallich s.n.* (holotype, BM).

Coelogyne elata Lindl., Gen. Sp. Orchid. Pl.: 40. 1830. *Pleione elata* (Lindl.) Kuntze, Revis. Gen. Pl. 2: 680. 1891. SYNTYPES: Nepal, *Wallich s.n.*; India, Sylhet, *Wallich 1959* (syntypes, E, K, K-LINDL, K-WALL).

Distribution. Arunachal Pradesh, Assam, Manipur, Meghalaya, Mizoram, Nagaland, Sikkim, Uttarakhand, Tripura, West Bengal [Bhutan, Myanmar, Nepal].

Habit. Epiphytic herb.

Coelogyne suaveolens (Lindl.) Hook. f., Fl. Brit. India [J. D. Hooker] 5(16): 832. 1890. *Pholidota suaveolens* Lindl., Gard. Chron. 1856: 372. 1856. *Pleione suaveolens* (Lindl.) Kuntze, Revis. Gen. Pl. 2: 680. 1891. TYPE: Origin unknown, *cult. Bishop of Winchester s.n., 19 May 1856* (holotype, K-LINDL).

Distribution. Arunachal Pradesh, Assam, Manipur, Meghalaya, Nagaland, Tripura [Myanmar].

Habit. Epiphytic herb.

Coelogyne tricallosa (Rolfe) M. W. Chase & Schuit., Phytotaxa 510(2): 130. 2021. *Panisea tricallosa* Rolfe, Bull. Misc. Inform. Kew 1901: 148. 1901. *Sigmatogyne tricallosa* (Rolfe) Pfitzer, Pflanzenr. (Engler) IV, 50 (Heft 32): 133. 1907. TYPE: India, Assam, Apr. 1896, *cult. Royal Botanical Gardens, Glasnevin s.n.* (holotype, K).

Sigmatogyne pantlingii Pfitzer, Pflanzenr. (Engler) IV, 50 (Heft 32): 134. 1907. *Panisea pantlingii* (Pfitzer) Schltr., Orchideen (Schlechter) 155. 1915. TYPE: Assam, *Watt s.n.* (holotype, B).

Distribution. Arunachal Pradesh, Assam, Manipur, Mizoram, Nagaland, Sikkim, West Bengal [Bhutan, China, Myanmar, Nepal].

Habit. Epiphytic herb.

Coelogyne trinervis Lindl., Gen. Sp. Orchid. Pl.: 41. 1830. *Pleione trinervis* (Lindl.) Kuntze, Revis. Gen. Pl. 2: 680. 1891. TYPE: Myanmar, Tennaserim, Tavoy [Dawei], *Wallich 1955 (leg. W. Gomez)* (holotype, K-LINDL, isotype, K-WALL).

Coelogyne rossiana Rchb. f., Gard. Chron., n.s., 22: 808. 1884. *Pleione rossiana* (Rchb. f.) Kuntze, Revis. Gen. Pl. 2: 680. 1891. TYPE: Myanmar, 1884, *H. T. Ross s.n.* (holotype, W).

Distribution. Andaman & Nicobar Islands, Assam, Manipur [Myanmar].

Habit. Epiphytic herb.

Coelogyne uniflora Lindl., Gen. Sp. Orchid. Pl.: 42. 1830. *Panisea uniflora* (Lindl.) Lindl., Fol. Orchid. 5(*Panisea*): 2. 1854. *Pleione uniflora* (Lindl.) Kuntze, Revis. Gen. Pl. 2: 680. 1891. TYPE: Nepal, May 1821, *Wallich 1966* (holotype, K-LINDL; isotype, K-WALL).

Coelogyne thuniana Rchb. f., Allg. Gartenzeitung 23(19): 145. 1855. *Pleione thuniana* (Rchb. f.) Kuntze, Revis. Gen. Pl. 2: 680. 1891. TYPE: Nepal, *cult. Graf von Thun-Hohenstein s.n. (leg. Tofft)* (holotype, W).

Distribution. Assam, Mizoram, Nagaland, Sikkim, Tamil Nadu, West Bengal [Bhutan, China, Myanmar, Nepal].

Habit. Epiphytic herb.

Coelogyne vernicosa (L. O. Williams) M. W. Chase & Schuit. Phytotaxa 510(2): 131. 2021. *Dickasonia vernicosa* L. O. Williams, Bot. Mus. Leafl. 9(2): 38. 1941. TYPE: Myanmar, Kyauksit Chaung, Chin Hill District, 6500 ft, Apr. 1939, *Dickason 8576* (holotype, AMES).

Kalimpongia narajitii Pradhan, Orchid Digest 41(5): 172. 1977. TYPE: India, Manipur, Hengshi Village, 25 Mar. 1977, *R. K. Mohendrajit Singh s.n.* (Pradhan's private herbarium, Kalimpong, India).

Distribution. Assam, Manipur, ?Sikkim, ?West Bengal [Bhutan, Myanmar].

Habit. Epiphytic herb.

Coelogyne viscosa Rchb. f., Allg. Gartenzeitung 24(28): 218. 1856. *Pleione viscosa* (Rchb. f.) Kuntze, Revis. Gen. Pl. 2: 680. 1891. TYPE: East Indies, *cult. Bonsen (leg. Booth) s.n.* (holotype, W).

Coelogyne graminifolia C. S. P. Parish & Rchb. f., Trans. Linn. Soc. London 30: 146. 1874. *Pleione graminifolia* (C. S. P. Parish & Rchb. f.) Kuntze, Revis. Gen. Pl. 2: 680. 1891. TYPE: Myanmar, Moulmein [Mawlamyine], 1869, *C. S. P. Parish s.n.* [*252*] (lectotype, W, designated by Clayton, 2017; isotype, K).

Distribution. Arunachal Pradesh, Assam, Manipur, Meghalaya, Mizoram, Nagaland, Tamil Nadu, Tripura [Bangladesh, China, Myanmar, Nepal].

Habit. Epiphytic herb.

Coelogyne wattii (King & Pantl.) M. W. Chase & Schuit., Phytotaxa 510(2): 131. 2021. *Pholidota wattii* King & Pantl., J. Asiat. Soc. Bengal, Pt. 2, Nat. Hist. 66: 590. 1898 [1897]. TYPE: India, Assam, *G. Watt s.n.* [*623*] (holotype, CAL).

Distribution. Arunachal Pradesh, Assam, Nagaland. Endemic.

Habit. Epiphytic herb.

———

Collabium Blume, Bijdr. Fl. Ned. Ind. 8: 357. 1825.

TYPE: *Collabium nebulosum* Blume.

Collabium chinense (Rolfe) Tang & F. T. Wang, Fl. Hainan. 4: 217, fig. 1101. 1977. *Nephelaphyllum chinense* Rolfe, Bull. Misc. Inform. Kew 1896: 194.

1896. *Tainia chinensis* (Rolfe) Gagnep., Bull. Mus. Natl. Hist. Nat., II, 4: 706. 1932. *Collabiopsis chinensis* (Rolfe) S. S. Ying, Coloured Ill. Indig. Orchids Taiwan 2: 456. 1990. TYPE: China, Kwangtung, Tingushan, *Hance 17733* (holotype, BM).

Distribution. Arunachal Pradesh, Assam [China].

Habit. Terrestrial herb.

———

Corallorhiza Gagnebin, Acta Helv. Phys.-Math. 2: 61. 1755 (as "*Corallorrhiza*"), nom. et orth. cons.

TYPE: *Corallorhiza trifida* Châtel.

Corallorhiza trifida Châtel., Spec. Inaug. Corallorhiza 8. 1760. *Ophrys corallorhiza* L., Sp. Pl. 2: 945. 1753. *Epipactis corallorhiza* (L.) Crantz, Stirp. Austr. Fasc., ed. 2, 6: 464. 1769. *Helleborine corallorhiza* (L.) F. W. Schmidt, Fl. Boëm. Cent. 1: 79. 1793. *Cymbidium corallorhiza* (L.) Sw., Kongl. Vetensk. Acad. Nya Handl. 21: 238. 1800. *Epidendrum corallorhizon* (L.) Poir., Encycl. (Lamarck) Suppl. 1. 377. 1810. *Corallorhiza corallorhiza* (L.) H. Karst., Deut. Fl. (Karsten): 448. 1883, nom. inval. *Neottia corallorhiza* (L.) Kuntze, Revis. Gen. Pl. 2: 674. 1891. TYPE: *Herb. Linn. 1056. 5*, middle specimen (lectotype, LINN, designated by Baumann et al., 1989).

Corallorhiza jacquemontii Decne., Voy. Inde [Jacquemont] 4(Bot): 165. 1844. TYPE: India, Pyrpendjal, June, *Jacquemont s.n.* [*566*] (holotype, P; isotype, L).
Corallorhiza anandae Malhotra & Balodi, Bull. Bot. Surv. India 26(1–2): 108. 1984 [1985]. TYPE: India, Martoli Bugyal, *T. A. Rao 6851A* (holotype, CAL; isotype, BSD).

Distribution. Sikkim, Uttarakhand [China, Nepal, Pakistan].

Habit. Holomycotrophic terrestrial herb.

———

Corybas Salisb., Parad. Lond. ad t. 83. 1807.

TYPE: *Corybas aconitiflorus* Salisb.

Corybas himalaicus (King & Pantl.) Schltr., Repert. Spec. Nov. Regni Veg. 19: 19. 1923. *Corysanthes himalaica* King & Pantl., J. Asiat. Soc. Bengal, Pt. 2, Nat. Hist. 65: 128. 1896. *Calcearia himalaica* (King & Pantl.) M. A. Clem. & D. L. Jones, Orchadian 13(10): 444. 2002. TYPE: India, Sikkim, Lachen Valley, Lamteng, 9000[–12,000] ft, July 1895, *Pantling 385* (syntypes, AMES, BM, CAL, E, K, L, W).

?*Corybas purpureus* J. Joseph & Yogan., Indian Forester
93(12): 815. 1967. TYPE: India, Assam [Meghalaya],
Khasi Hills, Shillong, 1500 m, 27 June 1966, *Joseph
36892A* (holotype, CAL).

Distribution. Meghalaya, Sikkim [Bhutan, China].

Habit. Terrestrial herb.

———

Corymborkis Thouars, Nouv. Bull. Sci. Soc. Philom.
Paris 1: 318. 1809.

TYPE: *Corymborkis corymbis* Thouars.

Corymborkis veratrifolia (Reinw.) Blume, Coll. Orchid. 125, t. 42, fig. 2a. 1859. *Hysteria veratrifolia*
Reinw., Syll. Pl. Nov. 2: 5. 1828 [1825]. *Corymbis
veratrifolia* (Reinw.) Rchb. f., Flora 48: 184. 1865.
TYPE: Not designated. NEOTYPE: Indonesia,
Java, *T. Lobb 162* (neotype, K; isoneotype BM,
designated by Rasmussen, 1977).

Corymborkis assamica Blume, Coll. Orchid.: 126. 1859.
TYPE: India, Assam, *Jenkins s.n. (East India Company
Herbarium 5325)* (lectotype, L, designated by Rasmus-
sen, 1977; isotype, K).

Distribution. Andaman & Nicobar Islands, Aruna-
chal Pradesh, Assam, Manipur, Meghalaya, Mizoram,
Nagaland, Sikkim, Tamil Nadu, Tripura, West Bengal
[China, Sri Lanka].

Habit. Terrestrial herb.

———

Cottonia Wight, Icon. Pl. Ind. Orient. [Wight] 5(1):
21, t. 1755. 1851.

TYPE: *Cottonia macrostachys* Wight (= *Cottonia peduncularis*
(Lindl.) Rchb. f. ex Schiller).

Cottonia peduncularis (Lindl.) Rchb. f. ex Schiller,
Cat. Orch. Samml. Schiller, ed. 3. 22. 1857.
Vanda peduncularis Lindl., Gen. Sp. Orchid. Pl.:
216. 1833. TYPE: Sri Lanka, *Macrae s.n.* (holotype, K-LINDL; isotype, L).

Cottonia macrostachya Wight, Icon. Pl. Ind. Orient. [Wight]
5(1): 21, t. 1755. 1851 (as "*macrostachys*"). TYPE:
India, near Tellicherry, *Jerdon s.n.* (syntype, MH).

Distribution. Andhra Pradesh, Goa, Karnataka,
Kerala, Maharashtra, Odisha, Tamil Nadu [Sri Lanka].

Habit. Epiphytic herb.

———

Cremastra Lindl., Gen. Sp. Orchid. Pl.: 172. 1833.

TYPE: *Cremastra wallichiana* Lindl., nom. superfl. (= *Cremastra appendiculata* (D. Don) Makino; *Cymbidium
appendiculatum* D. Don).

Cremastra appendiculata (D. Don) Makino, Bot.
Mag. (Tokyo) 18: 24. 1904. *Cymbidium appendiculatum* D. Don, Prodr. Fl. Nepal.: 36. 1825. *Cremastra wallichiana* Lindl., Gen. Sp. Orchid. Pl.:
173. 1833, nom. superfl. TYPE: Nepal, Bagmati
Zone, Kathmandu District, Shivapuri, *Wallich 7349
p.p.* (lectotype, W, designated by Lund, 1988; isotype, BM, K-LINDL, K-WALL, L).

Cremastra appendiculata (D. Don) Makino var. *sonamii* Lucksom, Orchids Sikkim N. E. Himalaya: 326. 2007, nom.
inval. TYPE: Not designated.

Distribution. Arunachal Pradesh, Assam, Nagaland,
Sikkim, West Bengal [China, Myanmar, Nepal].

Habit. Terrestrial herb.

———

Crepidium Blume, Bijdr. Fl. Ned. Ind. 7: 387. 1825.

TYPE: *Crepidium rheedii* Blume.

Crepidium acuminatum (D. Don) Szlach., Fragm.
Florist. Geobot., Suppl. 3: 123. 1995. *Malaxis
acuminata* D. Don, Prodr. Fl. Nepal.: 29. 1825.
Corymborkis acuminata (D. Don) M. R. Almeida,
Fl. Maharashtra 5A: 29. 2009. TYPE: Nepal,
Gosainthan [Shishapangma], *Wallich s.n. [1940.1]*
(lectotype, K, designated by Margońska, 2012;
isotype, BM, C, FI, K, K-WALL, SING).

Microstylis biloba Lindl., Gen. Sp. Orchid. Pl.: 20. 1830. *Microstylis wallichii* Lindl. var. *biloba* (Lindl.) Hook. f., Fl.
Brit. India [J. D. Hooker] 5(15): 686. 1888. *Malaxis biloba* (Lindl.) Ames, Orchidaceae 2: 122. 1908. *Malaxis
acuminata* D. Don var. *biloba* (Lindl.) Ames, Enum.
Philipp. Apost.: 302. 1926. *Malaxis acuminata* D. Don
f. *biloba* (Lindl.) Tuyama in H. Hara, Fl. E. Himalaya 1:
443. 1966. *Crepidium bilobum* (Lindl.) Szlach. ex Lucksom, Orchids Sikkim N. E. Himalaya: 323. 2007. TYPE:
Nepal, *Wallich 1940.1* (lectotype, K-LINDL, designated
by Margońska, 2012; isotype, BM, C, FI, K, K-WALL,
SING).
Microstylis wallichii Lindl., Gen. Sp. Orchid. Pl.: 20. 1830.
Malaxis wallichii (Lindl.) Deb, Bull. Bot. Surv. India
3(2): 128. 1961 [1962]. TYPE: Nepal, Chaudraghiry,
June 1821, *Wallich 1938.1* (lectotype, K-LINDL, designated by Margońska, 2012; isotype, C, G, K, K-WALL).

Distribution. Andaman & Nicobar Islands, Andhra
Pradesh, Arunachal Pradesh, Assam, Himachal Pradesh, Karnataka, Madhya Pradesh, Manipur, Meghalaya, Mizoram, Nagaland, Sikkim, Tamil Nadu, Utta-

rakhand, West Bengal [Bangladesh, Bhutan, China, Myanmar, Nepal].

Habit.   Terrestrial herb.

Crepidium andamanicum (King & Pantl.) Marg. & Szlach., Polish Bot. J. 46(1): 43. 2001. *Microstylis andamanica* King & Pantl., J. Asiat. Soc. Bengal, Pt. 2, Nat. Hist. 66: 582. 1898 [1897]. *Malaxis andamanica* (King & Pantl.) N. P. Balakr. & Vasudeva Rao, Bull. Bot. Surv. India 21(1–4): 177. 1979 [1981]. TYPE: India, South Andaman Island, 3 June 1884, *King's collectors s.n.* [*306*] (lectotype, K, designated by Margońska, 2012; isotype, BM C, CAL, G, K, L, P, UPS).

Distribution.   Andaman & Nicobar Islands. Endemic.

Habit.   Terrestrial herb.

Crepidium aphyllum (King & Pantl.) A. N. Rao, J. Orchid Soc. India 14(1–2): 65. 2000. *Microstylis aphylla* King & Pantl., Ann. Roy. Bot. Gard. (Calcutta) 8: 18. 1898. *Malaxis aphylla* (King & Pantl.) Tang & F. T. Wang, Acta Phytotax. Sin. 1: 71. 1951. TYPE: India, Sikkim Himalaya, Valley of Testa, *Pantling 455* (lectotype, K, designated by Margońska, 2012; isotype, BM, CAL, G, W).

Distribution.   Arunachal Pradesh, West Bengal [Bhutan].

Habit.   Holomycotrophic terrestrial herb.

Crepidium biauritum (Lindl.) Szlach., Fragm. Florist. Geobot., Suppl. 3: 124. 1995. *Microstylis biaurita* Lindl., Gen. Sp. Orchid. Pl.: 20. 1830 (excl. Andaman specimens). *Malaxis biaurita* (Lindl.) Kuntze, Revis. Gen. Pl. 2: 673. 1891. TYPE: India, Pundua, Apr. 1823, *Wallich 1941 (leg. De Silva)* (lectotype, K-LINDL, designated by Margońska, 2012; isotype, G, K, K-WALL, W).

Distribution.   Andaman & Nicobar Islands, Assam, Meghalaya, Mizoram, Nagaland, Tripura, Uttarakhand [Bangladesh, China, Myanmar].

Habit.   Terrestrial herb.

Crepidium calophyllum (Rchb. f.) Szlach., Fragm. Florist. Geobot., Suppl. 3: 125. 1995. *Microstylis calophylla* Rchb. f., Gard. Chron., n.s., 12: 718. 1879. *Malaxis calophylla* (Rchb. f.) Kuntze, Revis. Gen. Pl. 2: 673. 1891. TYPE: Malaysia, *cult. Hamburg Bot. Gard. s.n.*, *May 1878* (holotype, W).

Microstylis wallichii Lindl. var. *brachycheila* Hook. f., Fl. Brit. India [J. D. Hooker] 5(15): 686. 1888. *Malaxis calo-*

phylla (Rchb. f.) Kuntze var. *brachycheila* (Hook. f.) Tang & F. T. Wang, Acta Phytotax. Sin. 1: 71. 1951. TYPE: Myanmar, Moulmein [Mawlamyine], *Parish 191* (lectotype, K, designated by Margońska, 2012; isotype, W).

Microstylis scottii Hook. f., Fl. Brit. India [J. D. Hooker] 5(16): 687. 1890. TYPE: Myanmar, Pegu, Rangoon, *Scott s.n.* (holotype, K).

Distribution.   Assam, Meghalaya, Mizoram, Sikkim, West Bengal [Bangladesh, China, Myanmar, Nepal].

Habit.   Terrestrial herb.

Crepidium crenulatum (Ridl.) Sushil K. Singh, Agrawala & Jalal, Orchids India: 172. 2019. *Microstylis crenulata* Ridl., J. Linn. Soc., Bot. 24: 346. 1888. *Malaxis crenulata* (Ridl.) Kuntze, Revis. Gen. Pl. 2: 673. 1891, nom. illeg., non Blume, 1825. *Seidenfia crenulata* (Ridl.) Szlach., Fragm. Florist. Geobot. Suppl. 3: 122. 1995. *Crepidium crenulatum* (Ridl.) Kottaim., Int. J. Curr. Res. Biosci. Pl. Biol. 6(10): 40. 2019, isonym. TYPE: India, West Nilgiri Hills, *Beddome s.n.* (lectotype, BM, designated by Margońska, 2012; isotype, C).

Distribution.   Kerala, Tamil Nadu. Endemic.

Habit.   Terrestrial herb.

Crepidium densiflorum (A. Rich.) Sushil K. Singh, Agrawala & Jalal, Orchids India: 172. 2019. *Liparis densiflora* A. Rich., Ann. Sci. Nat., Bot. sér. 2, 15: 18, t. 1B. 1841. *Malaxis densiflora* (A. Rich.) Kuntze, Revis. Gen. Pl. 2: 673. 1891. *Microstylis densiflora* (A. Rich.) Alston in H. Trimen, Handb. Fl. Ceylon 6(Suppl.): 272. 1931. *Seidenfia densiflora* (A. Rich.) Szlach., Fragm. Florist. Geobot. Supp. 3: 122. 1995. *Corymborkis densiflora* (A. Rich.) M. R. Almeida, Fl. Maharashtra 5A: 29. 2009. TYPE: India, Nilgiri Hills, near Otacamund and Dodabetta, *Perrottet s.n.* [*878*] (lectotype, P, designated by Margońska, 2012).

Microstylis luteola Wight, Icon. Pl. Ind. Orient. [Wight] 5(1): 4, t. 1632. 1851. *Microstylis versicolor* Lindl. var. *luteola* (Wight) Hook. f., Fl. Brit. India [J. D. Hooker] 5(16): 691. 1890. *Microstylis densiflora* (A. Rich.) Kuntze var. *luteola* (Wight) P. K. Sarkar, J. Econ. Taxon. Bot. 5(5): 1008. 1984. TYPE: India, Nilgiri Hills, Otacamund, Aug. 1848, *Wight s.n.* (holotype, K).

Distribution.   Karnataka, Kerala, Tamil Nadu [Sri Lanka].

Habit.   Terrestrial herb.

Crepidium intermedium (A. Rich.) Sushil K. Singh, Agrawala & Jalal, Orchids India: 172. 2019. *Liparis intermedia* A. Rich., Ann. Sci. Nat., Bot. sér.

2, 15: 17. 1841. *Malaxis intermedia* (A. Rich.) Seidenf., Bot. Tidsskr. 73(2): 99. 1978. *Seidenfia intermedia* (A. Rich.) Szlach., Fragm. Florist. Geobot., Suppl. 3: 122. 1995. *Corymborkis intermedia* (A. Rich.) M. R. Almeida, Fl. Maharashtra 5A: 29. 2009. TYPE: India, Nilgiri Hills, Waterfall ("Water-Fat") not far from Kaiti, *Perrottet s.n.* (lectotype, W, designated by Margońska, 2012; isotype, G).

Microstylis stocksii Hook. f., Hooker's Icon. Pl. 19: t. 1833. 1889. *Malaxis stocksii* (Hook. f.) Kuntze, Revis. Gen. Pl. 2: 673. 1891. *Seidenfia stocksii* (Hook. f.) Szlach., Fragm. Florist. Geobot., Suppl. 3: 122. 1995. TYPE: India, Deccan Peninsula, Canara, Bababoodan [Baba Budan] Hills, *J. E. Stocks s.n.* (lectotype, K, designated by Margońska, 2012; isotype, BM, C, FI, G, K, L, P, W).

Distribution. Karnataka, Kerala, Tamil Nadu. Endemic.

Habit. Terrestrial herb.

Crepidium josephianum (Rchb. f.) Marg., Ann. Bot. Fenn. 39(1): 65. 2002. *Microstylis josephiana* Rchb. f., Bot. Mag. 103: t. 6325. 1877. *Malaxis josephiana* (Rchb. f.) Kuntze, Revis. Gen. Pl. 2: 673. 1891. *Seidenfia josephiana* (Rchb. f.) A. N. Rao, Bull. Arunachal Forest Res. 26: 104. 2010. TYPE: *cult. Calcutta s.n. (leg. Anderson)* (not found). NEOTYPE: India, Sikkim, Teesta Valley, *Pantling 312* (neotype, W; isoneotype FI, designated by Margońska, 2012).

Distribution. Arunachal Pradesh, Manipur, Meghalaya, Mizoram, Nagaland, Sikkim, West Bengal. Endemic.

Habit. Terrestrial herb.

Crepidium khasianum (Hook. f.) Szlach., Fragm. Florist. Geobot., Suppl. 3: 127. 1995. *Microstylis khasiana* Hook. f., Fl. Brit. India [J. D. Hooker] 5(15): 686. 1888. *Malaxis khasiana* (Hook. f.) Kuntze, Revis. Gen. Pl. 2: 673. 1891. SYNTYPES: India, Khasi Hills, *T. Lobb s.n.* (syntype, K); Khasi Hills, 5000–7000 ft, 11 July, *Hooker & Thomson 1653* (syntype, K).

Distribution. Arunachal Pradesh, Assam, Manipur, Meghalaya, Mizoram, Nagaland, Sikkim, Uttarakhand, West Bengal [China, Myanmar, Nepal].

Habit. Terrestrial herb.

Crepidium mackinnonii (Duthie) Szlach., Fragm. Florist. Geobot., Suppl. 3: 128. 1995. *Microstylis mackinnonii* Duthie, J. Asiat. Soc. Bengal, Pt. 2, Nat. Hist. 71: 37. 1902. *Malaxis mackinnonii* (Duthie) Ames, Orchidaceae (Ames) 6: 289. 1920. *Seidenforchis mackinnonii* (Duthie) Marg., Acta Soc. Bot. Poloniae 75(4): 303. 2006. TYPE: India, Dehra Dun, *Mackinnon 25429* (lectotype, K, designated by Margońska, 2012; isotype, K, P).

Microstylis cardonii Prain, Bengal Pl. 2: 1004. 1903 (as "*Cardoni*"). TYPE: India, Benghal, Chota Nagpur, icon. *Prain s.n.* (holotype, CAL).

Distribution. Bihar, Gujarat, Jharkhand, Kerala, Madhya Pradesh, Odisha, Uttarakhand, West Bengal [Bangladesh, China, Myanmar, Nepal].

Habit. Terrestrial herb.

Crepidium malabaricum (Marg. & Szlach.) J. M. H. Shaw, Orchid Rev. 122(1306, Suppl.): 37. 2014. *Seidenfia malabarica* Marg. & Szlach., Polish Bot. J. 46: 59. 2001. TYPE: India, Concan, *Stocks & Law s.n.* (lectotype, W, designated by Margońska, 2012; isotype, C, FI, G, W).

Distribution. Karnataka. Endemic.

Habit. Terrestrial herb.

Crepidium maximowiczianum (King & Pantl.) Szlach., Fragm. Florist. Geobot., Suppl. 3: 129. 1995. *Microstylis maximowicziana* King & Pantl., J. Asiat. Soc. Bengal, Pt. 2, Nat. Hist. 64: 329. 1895 [1896]. *Malaxis maximowicziana* (King & Pantl.) Tang & F. T. Wang, Acta Phytotax. Sin. 1(1): 72. 1951. TYPE: India, Darjeeling, Mungpoo Cinchoa Plantation, alt. 2000–4000 ft, July, *Pantling 226* (lectotype, BM, designated by Margońska, 2012; isotype, CAL, K, P).

Distribution. ?Meghalaya, Sikkim, West Bengal. Endemic.

Habit. Terrestrial herb.

Crepidium ophrydis (J. Koenig) M. A. Clem. & D. L. Jones, Lasianthera 1(1): 38. 1996. *Epidendrum ophrydis* J. Koenig, Observ. Bot. (Retzius) 6: 46. 1791. *Malaxis ophrydis* (J. Koenig) Ormerod in G. Seidenfaden, Descr. Epidendrorum J. G. Koenig: 18. 1995. *Dienia ophrydis* (J. Koenig) Seidenf. in G. Seidenfaden, Contr. Orchid Fl. Thailand 13: 18. 1997. *Gastroglottis ophrydis* (J. Koenig) A. N. Rao, Bull. Arunachal Forest Res. 26: 103. 2010. TYPE: Thailand, *Koenig s.n.* (lectotype, K, designated by Ormerod, 1995).

Malaxis latifolia Sm., Cycl. (Rees) 22: Malaxis no. 3. 1812. *Microstylis latifolia* (Sm.) J. J. Sm., Orch. Java 6: 248, t. 185. 1905. *Gastroglottis latifolia* (Sm.) Szlach., Fragm. Florist. Geobot., Suppl. 3: 123. 1995. *Dienia latifolia*

(Sm.) M. A. Clem. & D. L. Jones, Lasianthera 1: 41. 1996. *Corymborkis latifolia* (Sm.) M. R. Almeida, Fl. Maharashtra 5A: 30. 2009. TYPE: Nepal, Bagmati Zone, Kathmandu, Narayanhetty, 12 Aug. 1802, *Buchanan-Hamilton s.n.* (holotype, LINN).

Dienia congesta Lindl., Bot. Reg. 10: t. 825. 1824. *Microstylis congesta* (Lindl.) Rchb. f., Ann. Bot. Syst. (Walpers) 6: 206. 1861. *Malaxis congesta* (Lindl.) Deb, Bull. Bot. Surv. India 3(2): 128. 1961 [1962]. TYPE: Nepal, Aug. 1821, *Wallich 1936* (lectotype, K-LINDL, designated by Margońska, 2012; isotype, E, G, K-WALL).

Distribution. Andaman & Nicobar Islands, Arunachal Pradesh, Assam, Bihar, Manipur, Meghalaya, Mizoram, Nagaland, Odisha, Sikkim, Tripura, Uttarakhand, West Bengal [Bangladesh, Bhutan, China, Myanmar, Nepal, Sri Lanka].

Habit. Terrestrial herb.

Remarks. The genus name *Dienia* has priority over *Crepidium*, which needs to be conserved in order to prevent numerous new combinations becoming necessary.

Crepidium parryae (Tang & F. T. Wang) Marg., Ann. Bot. Fenn. 39: 65. 2002. *Malaxis parryae* Tang & F. T. Wang, Acta Phytotax. Sin. 1: 74. 1951. TYPE: India, Assam, Lushai Hills, Aijal, Sairep, 5000 ft, July 1926, *A. D. Parry 19* (holotype, K).

Distribution. Assam, Mizoram. Endemic.

Habit. Terrestrial herb.

Crepidium purpureum (Lindl.) Szlach., Fragm. Florist. Geobot., Suppl. 3: 131. 1995. *Microstylis purpurea* Lindl., Gen. Sp. Orchid. Pl.: 20. 1830. *Malaxis purpurea* (Lindl.) Kuntze, Revis. Gen. Pl. 2: 673. 1891. TYPE: Sri Lanka, near Galle, *Macrae s.n.* (lectotype, AMES, designated by Margońska, 2012; isotype, K-LINDL).

Microstylis wallichii Lindl. var. *biloba* King & Pantl., Ann. Roy. Bot. Gard. (Calcutta) 8: 16, t. 19. 1898, nom. illeg., non (Lindl.) Hook. f., 1890. TYPE: India, Sikkim, tropical Himalayan valleys, *Pantling 37* (lectotype, CAL, designated by Margońska, 2012; isotype, BM, C, FI, G, K).

Distribution. Jharkhand, Kerala, Odisha, Sikkim, Tamil Nadu, Uttarakhand, West Bengal [Bhutan, China, Nepal, Sri Lanka].

Habit. Terrestrial herb.

Crepidium saprophytum (King & Pantl.) A. N. Rao, J. Orchid Soc. India 13(1–2): 54. 2000. *Microstylis saprophyta* King & Pantl., J. Asiat. Soc. Bengal, Pt. 2, Nat. Hist. 65: 118. 1896. *Malaxis saprophyta* (King & Pantl.) Tang & F. T. Wang, Acta Phytotax. Sin. 1(1): 75. 1951. TYPE: India, Sikkim Himalaya, Chungthang, 6000 ft, July 1895,

Pantling 394 (lectotype, K, designated by Margońska, 2012; isotype, BM, CAL, G, P, W).

Distribution. Sikkim. Endemic.

Habit. Holomycotrophic terrestrial herb.

Crepidium versicolor (Lindl.) Sushil K. Singh, Agrawala & Jalal, Orchids Maharashtra: 57. 2018. *Microstylis versicolor* Lindl., Gen. Sp. Orchid. Pl.: 21. 1830. *Malaxis versicolor* (Lindl.) Abeyw., Ceylon J. Sci., Biol. Sci. 2: 147. 1959. *Seidenfia versicolor* (Lindl.) Marg. & Szlach., Polish Bot. J. 46(1): 56. 2001. *Corymborkis versicolor* (Lindl.) M. R. Almeida, Fl. Maharashtra 5A: 30. 2009. *Crepidium versicolor* (Lindl.) Sushil K. Singh, Agrawala & Jalal, Orchids India: 175. 2019, isonym. TYPE: Sri Lanka, *Macrae s.n.* [2] (lectotype, K-LINDL, designated by Jayaweera, 1981; isotype, K, K-LINDL, LE, NY, SING).

Distribution. Andhra Pradesh, Goa, Karnataka, Kerala, Madhya Pradesh, Maharashtra, Odisha, Tamil Nadu [Sri Lanka].

Habit. Terrestrial herb.

Remarks. As Ormerod (2017) pointed out, when Swartz described *Malaxis rheedei* (as "rhedii") he cited *Epidendrum resupinatum* G. Forst. as a synonym. This makes *M. rheedei*, of which the description was based on the plant called "Basaala Poula Maravara" by Rheede (1692, t. 27), a superfluous name. It is far from clear which species was illustrated by Rheede, but it certainly was not the Australasian *Crepidium resupinatum* (G. Forst.) Szlach. Margońska (2012) suggests that what she called *Seidenfia rheedii* [sic] (Sw.) Szlach., believing it to be Rheede's plant, may be recognized by the central tooth on the lip margin being longer than the others. This statement is not supported by Rheede's drawing, which shows a large number of short teeth of equal length. The number of teeth seems to vary between the various flowers depicted and may not be accurate. The illustration is not detailed enough to rule out *C. versicolor* or another species, which may as yet lack a name.

———

Cryptochilus Wall., Tent. Fl. Napal. 36. 1824.

TYPE: *Cryptochilus sanguineus* Wall. ("sanguinea").

Cryptochilus acuminatus (Griff.) Schuit., Y. P. Ng & H. A. Pedersen, Bot. J. Linn. Soc. 186: 194. 2018. *Xiphosium acuminatum* Griff., Calcutta J. Nat. Hist. 5: 364. 1845. TYPE: India, Khasi Hills,

Cherrapunji, 4300 ft, *cult. Calcutta (Gibson) s.n.* (holotype, K-LINDL; isotype, K).

Eria carinata Gibson ex Lindl., J. Proc. Linn. Soc., Bot. 3: 50. 1859 [1858]. *Pinalia carinata* (Gibson ex Lindl.) Kuntze, Revis. Gen. Pl. 2: 679. 1891. *Cryptochilus carinatus* (Gibson ex Lindl.) H. Jiang, Wild Orchids Yunnan: 306. 2010. TYPE: India, Khasi Hills, Cherrapunji, *Griffith s.n. (East India Company Herbarium 5121)* (holotype, K-LINDL; isotype, K).

Distribution. Arunachal Pradesh, Assam, Meghalaya, Mizoram, Sikkim, West Bengal [Bhutan].

Habit. Epiphytic herb.

Cryptochilus luteus Lindl., J. Proc. Linn. Soc., Bot. 3: 21. 1859 [1858] (as *"lutea"*). SYNTYPES: India, Mishmee Hills, *Griffith s.n.* (syntypes, CAL, K-LINDL); Darjeeling, *Griffith s.n.* (syntype, K-LINDL); Sikkim, *Cathcart s.n.* (syntypes, K-LINDL, icon., L).

Distribution. Arunachal Pradesh, Manipur, ?Meghalaya, Nagaland, Sikkim, Uttarakhand, West Bengal [Bhutan, China, Myanmar, Nepal].

Habit. Epiphytic herb.

Cryptochilus sanguineus Wall., Tent. Fl. Napal. 36, t. 26. 1824 (as *"sanguinea"*). SYNTYPE: Nepal, Chaudraghiry, *Wallich s.n.* (not found).

Distribution. Arunachal Pradesh, Manipur, Meghalaya, Mizoram, Nagaland, Sikkim, West Bengal [Bhutan, Myanmar, Nepal].

Habit. Epiphytic herb.

Remarks. *Wallich 7350 (leg. W. Gomez)* from Cherrapunji was collected in June 1829 and therefore cannot be a type, as is sometimes stated.

Cryptochilus strictus (Lindl.) Schuit., Y. P. Ng & H. A. Pedersen, Bot. J. Linn. Soc. 186: 195. 2018. *Eria stricta* Lindl., Coll. Bot. (Lindley) t. 41B. 1826. *Mycaranthes stricta* (Lindl.) Lindl., Numer. List [Wallich] n. 1970. 1829. *Pinalia stricta* (Lindl.) Kuntze, Revis. Gen. Pl. 2: 679. 1891. TYPE: Nepal, Toka, Feb. 1821, *Wallich 1970* (holotype, K-LINDL; isotype, K, K-WALL).

Eria secundiflora Griff., Not. Pl. Asiat. 3: 302. 1851. TYPE: India, Naga Hills, 3100 ft, Feb. 1837, *Griffith s.n.*(syntype, K [specimens labeled *Griffith 5118 & Griffith 5119*]).

Distribution. Arunachal Pradesh, Assam, Manipur, Meghalaya, Nagaland, Sikkim, West Bengal [Bhutan, China, Myanmar, Nepal].

Habit. Epiphytic herb.

Cryptostylis R. Br., Prodr. Fl. Nov. Holland. 317. 1810.

TYPE: *Cryptostylis longifolia* R. Br., *nom. superfl.* (*Malaxis subulata* Labill. = *Cryptostylis subulata* (Labill.) Rchb. f.).

Cryptostylis arachnites (Blume) Hassk., Cat. Hort. Bot. Bogor. (Hasskarl) 48. 1844. *Zosterostylis arachnites* Blume, Bijdr. Fl. Ned. Ind. 8: 419, t. 32. 1825. SYNTYPES: Indonesia, Java, Mt. Salak & Mt. Seribu, *Blume s.n.* (probable syntypes, L, P, s. loc.).

Distribution. Andaman & Nicobar Islands, Assam, Meghalaya [China, Myanmar, Sri Lanka].

Habit. Terrestrial herb.

Cylindrolobus Blume, Fl. Javae: vi. 1828.

TYPE: *Cylindrolobus compressus* (Blume) Brieger (*Ceratium compressum* Blume).

Cylindrolobus arunachalensis (A. N. Rao) A. N. Rao, Bull. Arunachal Forest Res. 26: 83. 2010. *Eria arunachalensis* A. N. Rao, J. Econ. Taxon. Bot. 21(3): 711. 1997 [1998]. TYPE: India, Arunachal Pradesh, West Siang District, Kaying, alt. 480 ft, *A. N. Rao 30108A* (holotype, Orchid Herbarium, Tipi).

Distribution. Arunachal Pradesh, Meghalaya. Endemic.

Habit. Epiphytic herb.

Cylindrolobus biflorus (Griff.) Rauschert, Feddes Repert. 94: 445. 1983. *Eria biflora* Griff., Not. Pl. Asiat. 3: 302. 1851. *Pinalia biflora* (Griff.) Kuntze, Revis. Gen. Pl. 2: 679. 1891. TYPE: Myanmar, near Peenma, Dec. 1834, *Griffith in Mergue* [Mergui] *Herb. 830* (holotype, K-LINDL).

Distribution. Arunachal Pradesh, Assam, Manipur, Meghalaya, Nagaland, Sikkim, West Bengal [Myanmar].

Habit. Epiphytic herb.

Cylindrolobus clavicaulis (Wall. ex Lindl.) Rauschert, Feddes Repert. 94: 445. 1983. *Eria clavicaulis* Wall. ex Lindl., Edwards's Bot. Reg. 26: (Misc.): 90. 1840. *Pinalia clavicaulis* (Wall. ex Lindl.) Kuntze, Revis. Gen. Pl. 2: 679. 1891. TYPE: India, *Wallich s.n.* (holotype, K-LINDL).

Distribution. Arunachal Pradesh, Assam, Manipur, Meghalaya [Myanmar].

Habit. Epiphytic herb.

Remarks. See note under *Cylindrolobus khasianus.*

Cylindrolobus cristatus (Rolfe) S. C. Chen & J. J. Wood, Fl. China 25: 349. 2009. *Eria cristata* Rolfe, Bull. Misc. Inform. Kew 1892: 139. 1892. TYPE: Myanmar, Moulmein [Mawlamyine], *cult. Kew s.n. (leg. C. Peché s.n.), Mar. 1892* (holotype, K).

Distribution. Arunachal Pradesh [Myanmar].

Habit. Epiphytic herb.

Cylindrolobus glandulifer (Deori & Phukan) A. N. Rao, Bull. Arunachal Forest Res. 26: 103. 2010 (as "*glanduliferus*"). *Eria glandulifera* Deori & Phukan, J. Orchid Soc. India 2: 55. 1988. TYPE: India, Meghalaya, Khasi Hills, Mawsmai forest, 18 Apr. 1978, *N. C. Deori 71816A* (holotype, CAL).

Eria lohitensis A. N. Rao, Harid. & S. N. Hegde, J. Bombay Nat. Hist. Soc. 86(2): 229. 1989. *Cylindrolobus lohitensis* (A. N. Rao, Harid. & S. N. Hedge) A. N. Rao, Bull. Arunachal Forest Res. 26: 83. 2010. TYPE: India, Arunachal Pradesh, Lohit District, Mailang, Mithumna, alt. 5100 ft, 11 May 1985, *Haridasan 2185* (holotype, Arunachal Forest Herbarium).

Distribution. Arunachal Pradesh, Assam, Meghalaya. Endemic.

Habit. Epiphytic herb.

Cylindrolobus gloensis (Ormerod & Agrawala) Schuit., Y. P. Ng & H. A. Pedersen, Bot. J. Linn. Soc. 186: 195. 2018. *Eria gloensis* Ormerod & Agrawala, Taiwania 59(3): 206. 2014. TYPE: India, Arunachal Pradesh, Mishmi Hills, Kamlang Valley, Glo, 4000 ft, 25 Mar. 1949, *F. Kingdon Ward 18451* (holotype, AMES; isotype, NY).

Distribution. Arunachal Pradesh. Endemic.

Habit. Epiphytic herb.

Cylindrolobus hegdei (Agrawala & H. J. Chowdhery) A. N. Rao, Bull. Arunachal Forest Res. 26: 83. 2010. *Eria hegdei* Agrawala & H. J. Chowdhery, Phytotaxonomy 8: 8. 2008 [2009]. *Eria jengingensis* S. N. Hegde, J. Orchid Soc. India 7(1–2): 13. 1993, nom. inval. TYPE: Arunachal Pradesh, Lower Subansiri district, Hapoli-Pangi, *A. N. Rao 26117* (holotype, OHT).

Distribution. Arunachal Pradesh. Endemic.

Habit. Epiphytic herb.

Cylindrolobus khasianus (Lindl.) Ormerod & C. S. Kumar, Harvard Pap. Bot. 25(1): 125. 2020. *Eria khasiana* Lindl., J. Proc. Linn. Soc., Bot. 3: 59. 1858. TYPE: India, Khasi Hills, *Griffith s.n.* (holotype, K-LINDL).

Distribution. Meghalaya. Endemic.

Habit. Epiphytic herb.

Remarks. *Cylindrolobus khasianus* was long considered a synonym of *C. clavicaulis* (Wall. ex Lindl.) Rauschert, but Ormerod and Kumar (2020) have recently argued for its reinstatement. As a result, the distribution of these species should be reassessed.

Cylindrolobus pauciflorus (Wight) Schuit., Y. P. Ng & H. A. Pedersen, Bot. J. Linn. Soc. 186: 195. 2018. *Eria pauciflora* Wight, Icon. Pl. Ind. Orient. [Wight] 5(1): 4, t. 1636. 1851. *Pinalia nilgherensis* Kuntze, Revis. Gen. Pl. 2: 679. 1891. TYPE: India, Nilgiri Hills, Kaitie Falls, Aug.–Sep., *Wight s.n.* (holotype, K).

Distribution. Karnataka, Kerala, Tamil Nadu. Endemic.

Habit. Epiphytic herb.

Cylindrolobus pseudoclavicaulis (Blatt.) Schuit., Y. P. Ng & H. A. Pedersen, Bot. J. Linn. Soc. 186: 195. 2018. *Eria pseudoclavicaulis* Blatt., J. Bombay Nat. Hist. Soc. 32: 519. 1928. TYPE: India, Bombay, cultivated, Sep. 1917, *Blatter 554* (holotype, St. Xavier's College, Bombay, kept in formalin).

Distribution. Kerala, Tamil Nadu. Endemic.

Habit. Epiphytic herb.

———

Cymbidium Sw., Nova Acta Regiae Soc. Sci. Upsal. ser. 2. 6: 70. 1799.

TYPE: *Cymbidium aloifolium* (L.) Sw. (*Epidendrum aloifolium* L.).

Cymbidium aloifolium (L.) Sw., Nova Acta Regiae Soc. Sci. Upsal. 6: 73. 1799. *Epidendrum aloifolium* L., Sp. Pl. 2: 953. 1753. TYPE: India, Malabar, icon. "Kansjiram-maravara" in Rheede, Hort. Malab. 12: 17, t. 8. 1692 (lectotype, designated by Seth, 1982).

Epidendrum pendulum Roxb., Pl. Coromandel 1: 35, t. 44. 1795. *Cymbidium pendulum* (Roxb.) Sw., Nova Acta Regiae Soc. Sci. Upsal. 6: 73. 1799. TYPE: India, Coast of

Coromandel, icon. *Roxburgh, loc. cit.* (lectotype, designated by Seth, 1982).

Aerides borassii Buch.-Ham. ex Sm. in A. Rees, Cycl. 39(1): no. 8. 1818. TYPE: India, Mysore, *Buchanan-Hamilton s.n.* (holotype, BM).

Cymbidium erectum Wight, Icon. Pl. Ind. Orient. [Wight] 5(1): 21, t. 1753. 1851, nom. illeg., non Sw., 1799. TYPE: India, Tamil Nadu, Iyamally Hills, near Coimbatore, *Wight s.n.* (holotype, ?BM).

Cymbidium simulans Rolfe, Orchid Rev. 25: 175. 1917. TYPE: India, Sikkim, *Pantling 268* (lectotype, K, designated by Seth, 1982; isotype, P).

Cymbidium intermedium H. G. Jones, Reinwardtia 9(1): 71. 1974. TYPE: India, Bombay State, *Herb. Jones C/85.*

Distribution. Andaman & Nicobar Islands, Andhra Pradesh, Arunachal Pradesh, Assam, Bihar, Chhattisgarh, ?Goa, Jharkhand, Karnataka, Kerala, Madhya Pradesh, Maharashtra, Manipur, Meghalaya, Mizoram, Nagaland, Odisha, Sikkim, Tamil Nadu, Tripura, Uttar Pradesh, Uttarakhand, West Bengal [Bangladesh, China, Myanmar, Nepal, Sri Lanka].

Habit. Epiphytic herb.

Cymbidium bicolor Lindl., Gen. Sp. Orchid. Pl.: 164. 1833 subsp. **bicolor.** TYPE: Sri Lanka, *Macrae 54* (holotype, K-LINDL).

Distribution. Andaman & Nicobar Islands, Goa, Karnataka, Kerala, Maharashtra, Odisha, Tamil Nadu [Sri Lanka].

Habit. Epiphytic herb.

Cymbidium bicolor Lindl. subsp. **obtusum** Du Puy & P. J. Cribb, Genus Cymbidium 70. 1988. TYPE: Thailand, Uttaradit, *Menzies & Du Puy 120* (holotype, K).

Cymbidium mannii Rchb. f., Flora 55: 274. 1872. TYPE: India, Assam, *Mann s.n.* (holotype, W).

Distribution. Arunachal Pradesh, Assam, Manipur, Meghalaya, Odisha, Sikkim, Uttarakhand, West Bengal [Bhutan, China, Myanmar].

Habit. Epiphytic herb.

Cymbidium bicolor Lindl. subsp. **pubescens** (Lindl.) Du Puy & P. J. Cribb, Genus Cymbidium. 73. 1988. *Cymbidium pubescens* Lindl., Edwards's Bot. Reg. 26(Misc.): 75. 1840. *Cymbidium aloifolium* (L.) Sw. var. *pubescens* (Lindl.) Ridl., J. Straits Branch Roy. Asiat. Soc. 59: 196. 1911. TYPE: Singapore, *cult. Loddiges s.n. (leg. Cuming s.n.)* (holotype, K-LINDL).

Distribution. Andaman & Nicobar Islands [also in Indonesia].

Habit. Epiphytic herb.

Cymbidium cochleare Lindl., J. Proc. Linn. Soc., Bot. 3: 28. 1859 [1858]. *Cyperorchis cochlearis* (Lindl.) Benth., J. Linn. Soc., Bot. 18: 318. 1881. TYPE: India, Sikkim, hot valleys, *J. D. Hooker 235* (holotype, K; isotype, K [s.n.]).

Distribution. Arunachal Pradesh, Assam, Meghalaya, Mizoram, Nagaland, Sikkim, West Bengal [Bangladesh, Bhutan, China, ?Myanmar].

Habit. Epiphytic herb.

Cymbidium cyperifolium Lindl., Gen. Sp. Orchid. Pl.: 163. 1833. TYPE: Bangladesh or India, Sylhet, *Wallich 7353 (leg. De Silva & Bruce s.n.)* (holotype, K-LINDL; isotype, K, K-WALL).

Limodorum longifolium Roxb., Fl. Ind. ed. 1832, 3: 468. 1832. *Geodorum longifolium* (Roxb.) Voigt, Hort. Suburb. Calcutt.: 628. 1845. TYPE: India, valleys among the Kasai [Khasi] Hills, icon. *Limodorum longifolium, Roxburgh 2345* (syntypes, CAL, K).

Distribution. Arunachal Pradesh, Assam, Manipur, Meghalaya, Mizoram, Nagaland, Sikkim, Uttarakhand, West Bengal [Bhutan, China, Myanmar, Nepal].

Habit. Terrestrial herb.

Cymbidium dayanum Rchb. f., Gard. Chron. 1869: 710. 1869. *Cymbidium eburneum* Lindl. var. *dayanum* (Rchb. f.) Hook. f., Fl. Brit. India [J. D. Hooker] 6(17): 12. 1890 (as *"dayana"*). TYPE: India, Assam, 1868, *cult. J. Day s.n.* (holotype, W, Herb. no. 45964; isotype, K).

Cymbidium simonsianum King & Pantl., J. Asiat. Soc. Bengal, Pt. 2, Nat. Hist. 64: 338. 1895 [1896]. TYPE: India, "Sikkim," valley of the Teesta, 1000 ft, Aug. 1895, *Pantling s.n. [51]* (lectotype, K; isotype, BM, CAL, P).

Distribution. Arunachal Pradesh, Assam, Manipur, Meghalaya [Bhutan, China, ?Myanmar].

Habit. Epiphytic herb.

Cymbidium devonianum Paxton, Paxton's Mag. Bot. 10: 97, fig. s.n. 1843. TYPE: Icon. in *Paxton, loc. cit.*

Cymbidium sikkimense Hook. f., Fl. Brit. India [J. D. Hooker] 6(17): 9. 1890. TYPE: India, Sikkim, Lachen Valley, alt. 6000 ft, Aug. 1845, *J. D. Hooker s.n.* (holotype, K).

Distribution. Arunachal Pradesh, Manipur, Meghalaya, Mizoram, Nagaland, Sikkim, West Bengal [Bhutan, China, Myanmar, Nepal].

Habit. Lithophytic or epiphytic herb.

Cymbidium eburneum Lindl., Edwards's Bot. Reg. 33: t. 67. 1847. *Cyperorchis eburnea* (Lindl.) Schltr.,

Repert. Spec. Nov. Regni Veg. 20: 107. 1924.
TYPE: India, Khasi Hills, *cult. Loddiges s.n.* (holotype, K- LINDL).

Cymbidium syringodorum Griff., Not. Pl. Asiat. 3: 338. 1851. TYPE: India, Khasi Hills, Myrung, 10 Nov. 1835, *Griffith in Assam Herb. 228* (holotype, K).

Distribution. Arunachal Pradesh, Assam, Manipur, Meghalaya, Mizoram, Nagaland, Sikkim, Uttarakhand, West Bengal [Bhutan, China, Myanmar, Nepal].

Habit. Epiphytic herb.

Cymbidium elegans Lindl., Gen. Sp. Orchid. Pl.: 163. 1833. *Cyperorchis elegans* (Lindl.) Blume, Rumphia 4: 47. 1849. TYPE: Nepal, Oct. 1821, *Wallich 7354* (holotype, K-LINDL; isotype, K-WALL).

Cymbidium longifolium D. Don, Prodr. Fl. Nepal.: 36. 1825, nom. rej. *Cyperorchis longifolia* (D. Don) Schltr., Repert. Spec. Nov. Regni Veg. 20: 108. 1924. TYPE: Nepal, Bagmati Zone, Rashuwa District, Gosainthan [Shishapangma], 1819, *Wallich s.n.* (lectotype, BM).
Cymbidium densiflorum Griff., Not. Pl. Asiat. 3: 337. 1851. TYPE: India, Khasi Hills, Myrung, *Griffith in Assam Herb. 229* (holotype, K-LINDL).
Cymbidium elegans Lindl. var. *lutescens* Hook. f., Fl. Brit. India [J. D. Hooker] 6(17): 15. 1890. TYPE: icon. ex CAL (holotype, K).

Distribution. Arunachal Pradesh, Assam, Manipur, Meghalaya, Mizoram, Nagaland, Sikkim, Uttarakhand, West Bengal [Bhutan, China, Myanmar, Nepal].

Habit. Epiphytic or lithophytic herb.

Cymbidium ensifolium (L.) Sw., Nova Acta Regiae Soc. Sci. Upsal. 6: 77. 1799 subsp. **ensifolium.** *Epidendrum ensifolium* L., Sp. Pl. 2: 954. 1753. TYPE: China, Canton, *Osbeck s.n.* (lectotype, LINN, designated by Du Puy & Cribb, 1988).

Distribution. Arunachal Pradesh, Manipur, Meghalaya, Mizoram, Nagaland, Sikkim [China, Myanmar].

Habit. Terrestrial herb.

Remarks. At least some records appear to be referable to *Cymbidium ensifolium* subsp. *haematodes* (Kumar, 2015).

Cymbidium ensifolium (L.) Sw. subsp. **haematodes** (Lindl.) Du Puy & P. J. Cribb ex Govaerts, World Checklist Seed Pl. 3(1): 20. 1999. *Cymbidium haematodes* Lindl., Gen. Sp. Orchid. Pl.: 162. 1833. *Cymbidium ensifolium* (L.) Sw. var. *haematodes* (Lindl.) Trimen, Handb. Fl. Ceylon 4: 180. 1898. *Cymbidium sinense* Willd. var. *haematodes* (Lindl.) Z. J. Liu & S. C. Chen, Gen. Cymbidium

China: 172. 2006. TYPE: Sri Lanka, *Macrae s.n.* [*12*] (holotype, K-LINDL).

Limodorum laxiflorum Lam., Encycl. 3: 516. 1792. TYPE: "East Indies," *Sonnerat s.n.* (holotype, P).

Distribution. Chhattisgarh, Karnataka, Kerala, Manipur, Tamil Nadu [China, Sri Lanka].

Habit. Terrestrial herb.

Remarks. The extent of occurrence of this subspecies in northeast India is unclear.

Cymbidium erythraeum Lindl., J. Proc. Linn. Soc., Bot. 3: 30. 1859 [1858]. TYPE: India, Sikkim, *J. D. Hooker 229* (holotype, K-LINDL).

Distribution. Arunachal Pradesh, Assam, Manipur, Meghalaya, Nagaland, Sikkim, Uttarakhand, West Bengal [Bhutan, China, Myanmar, Nepal].

Habit. Epiphytic herb.

Remarks. This species is illustrated as *Cymbidium elegans* Lindl. in Swami (2017).

Cymbidium faberi Rolfe var. **szechuanicum** (Y. S. Wu & S. C. Chen) Y. S. Wu & S. C. Chen, Acta Phytotax. Sin. 18: 299. 1980. *Cymbidium szechuanicum* Y. S. Wu & S. C. Chen, Acta Phytotax. Sin. 11: 33. 1966. *Cymbidium cyperifolium* Wall. ex Lindl. var. *szechuanicum* (Y. S. Wu & S. C. Chen) S. C. Chen & Z. J. Liu, Acta Phytotax. Sin. 41(1): 83. 2003. SYNTYPES: China, Sichuan, Chion-lai-shan, *Y. S. Wu 2040* (syntype, PE); *WU & Fee 2055* (syntype, PE); *Fee 2061* (syntype, PE).

Distribution. Arunachal Pradesh, Uttarakhand [Bhutan, China, Nepal].

Habit. Terrestrial herb.

Cymbidium × gammieanum King & Pantl., J. Asiat. Soc. Bengal, Pt. 2, Nat. Hist. 64: 339. 1895 [1896]. *Cyperorchis × gammieana* (King & Pantl.) Schltr., Repert. Spec. Nov. Regni Veg. 20: 107. 1924. *×Cyperocymbidium gammieanum* (King & Pantl.) A. D. Hawkes, Orchid Rev. 72: 420. 1964. TYPE: India, Sikkim, alt. 5000–7000 ft, Sep.–Oct., *Pantling s.n.* [*299*] (syntypes, CAL, W).

Distribution. Arunachal Pradesh, Sikkim, West Bengal [Nepal].

Habit. Epiphytic herb.

Remarks. This taxon is a natural hybrid: *Cymbidium elegans* × *C. erythraeum.*

Cymbidium goeringii (Rchb. f.) Rchb. f., Ann. Bot. Syst. (Walpers) 3: 547. 1852. *Maxillaria goeringii* Rchb. f., Bot. Zeitung (Berlin) 3(20): 334. 1845. TYPE: Japan, *Goering 592* (holotype, W).

Cymbidium mackinnonii Duthie, J. Asiat. Soc. Bengal, Pt. 2, Nat. Hist. 71: 41. 1902. *Cymbidium goeringii* (Rchb. f.) Rchb. f. var. *mackinnonii* (Duthie) A. N. Rao, J. Econ. Taxon. Bot. 24: 214. 2000. TYPE: India, Mussoorie, alt. 5500 ft, 5 Feb. 1899, *P. M. Mackinnon s.n. [22709]* (syntypes, CAL, DD, K).

Distribution. Arunachal Pradesh [Bhutan, China].

Habit. Terrestrial herb.

Cymbidium hookerianum Rchb. f., Gard. Chron. 1866: 7. 1866. *Cymbidium giganteum* Lindl. var. *hookerianum* (Rchb. f.) Desbois, Orchid: 119. 1893. TYPE: *cult. Veitch s.n.* (holotype, W, Herb. no. 44940).

Cymbidium grandiflorum Griff., Not. Pl. Asiat. 3: 342. 1851, nom. illeg., non Sw., 1799. *Cyperorchis grandiflora* Schltr., Repert. Spec. Nov. Regni Veg. 20: 107. 1924. TYPE: Bhutan, [Peemee], *Griffith in Bootan Herb. 698* (holotype, CAL).

Distribution. Arunachal Pradesh, Assam, Manipur, ?Meghalaya, Mizoram, Sikkim, Uttarakhand, West Bengal [Bhutan, China, Myanmar, Nepal].

Habit. Epiphytic or lithophytic herb.

Cymbidium insigne Rolfe, Gard. Chron., ser. 3, 35: 387. 1904. *Cyperorchis insignis* (Rolfe) Schltr., Repert. Spec. Nov. Regni Veg. 20: 108. 1924. TYPE: Vietnam, 4000–5000 ft, *cult. C. Schneider s.n. (leg. Bronckart s.n. [43]), 28 Sep. 1901* (holotype, K).

Distribution. Manipur, Meghalaya, West Bengal [China].

Habit. Terrestrial herb.

Cymbidium iridioides D. Don, Prodr. Fl. Nepal.: 36. 1825. TYPE: Nepal, 1819, *Wallich s.n.* (holotype, BM).

Cymbidium giganteum Lindl., Gen. Sp. Orchid. Pl.: 163. 1833, nom. illeg., non (L. f.) Sw., 1799. *Iridorchis gigantea* Blume, Coll. Orchid.: 90. 1859. *Cyperorchis gigantea* (Blume) Schltr., Repert. Spec. Nov. Regni Veg. 20: 107. 1924. TYPE: Nepal, 1821, *Wallich 7355* (lectotype, K-LINDL; isotype, K-WALL [3×]).

Distribution. Arunachal Pradesh, ?Assam, Manipur, Meghalaya, Mizoram, Nagaland, Sikkim, Uttarakhand, West Bengal [Bhutan, China, Myanmar, Nepal].

Habit. Epiphytic herb.

Cymbidium lancifolium Hook., Exot. Fl. 1: t. 51. 1823. *Cymbidiopsis lancifolia* (Hook.) H. J. Chowdhery, Indian J. Forest. 32: 157. 2009. TYPE: Nepal, *cult. Shepherd s.n. (leg. Wallich s.n.)* (holotype, K).

Cymbidium gibsonii Lindl. & Paxton, Paxton's Fl. Gard. 3: 144. 1852. TYPE: India, Meghalaya, Khasi Hills, *cult. Paxton s.n.* (holotype, K-LINDL).
Cymbidium javanicum Blume var. *pantlingii* F. Maek., J. Jap. Bot. 33: 320. 1958. TYPE: India, Sikkim, near Sureil, *Pantling 75* (holotype, K).

Distribution. Arunachal Pradesh, Assam, Manipur, Meghalaya, Mizoram, Nagaland, Sikkim, West Bengal [Bhutan, China, Myanmar, Nepal, Sri Lanka].

Habit. Terrestrial herb.

Cymbidium lowianum (Rchb. f.) Rchb. f., Gard. Chron., n.s., 11: 332, 405. 1879. *Cymbidium giganteum* Lindl. var. *lowianum* Rchb. f., Gard. Chron., n.s., 7: 685. 1877. *Cyperorchis lowiana* (Rchb. f.) Schltr., Repert. Spec. Nov. Regni Veg. 20: 108. 1924. *Cymbidium hookerianum* Rchb. f. var. *lowianum* (Rchb. f.) Y. S. Wu & S. C. Chen, Acta Phytotax. Sin. 18: 303. 1980. TYPE: Myanmar, *cult. Low s.n. (leg. Boxall s.n.)* (holotype, W).

Distribution. Arunachal Pradesh, Manipur, Meghalaya, Nagaland, Uttarakhand [China, Myanmar].

Habit. Epiphytic herb.

Cymbidium macrorhizon Lindl., Gen. Sp. Orchid. Pl.: 162. 1833. *Pachyrhizanthe macrorhizon* (Lindl.) Nakai, Bot. Mag. (Tokyo) 45: 109. 1931. *Cymbidiopsis macrorhizon* (Lindl.) H. J. Chowdhery, Indian J. Forest. 32: 155. 2009 (as "*macrorhiza*"). TYPE: India, Kashmir, *Royle s.n.* (holotype, K).

Distribution. Arunachal Pradesh, Assam, Himachal Pradesh, Jammu & Kashmir, Jharkhand, Madhya Pradesh, Manipur, Meghalaya, Mizoram, Nagaland, Sikkim, Tamil Nadu, Uttarakhand, West Bengal [China, ?Myanmar, Pakistan].

Habit. Holomycotrophic terrestrial herb.

Cymbidium mastersii Griff. ex Lindl., Edwards's Bot. Reg. 31: t. 50. 1845. *Cyperorchis mastersii* (Griff. ex Lindl.) Benth., J. Linn. Soc., Bot. 18: 318. 1881. TYPE: *cult. Loddiges Cat. no. 1233* (holotype, K-LINDL).

Cymbidium affine Griff., Not. Pl. Asiat. 3: 336. 1851. TYPE: India, Meghalaya, Khasi Hills, Churra, *Griffith in Assam Herb. 167 [s.n.]* (holotype, K-LINDL).

Cymbidium micromeson Lindl., J. Proc. Linn. Soc., Bot. 3: 29. 1859 [1858]. TYPE: India, Meghalaya, Khasi Hills, Churra, *Griffith s.n.* (holotype, K-LINDL).

Distribution. Arunachal Pradesh, Assam, Manipur, Meghalaya, Mizoram, Nagaland, Sikkim, West Bengal [Bangladesh, Bhutan, China, Myanmar].

Habit. Epiphytic or lithophytic herb.

Cymbidium munronianum King & Pantl., J. Asiat. Soc. Bengal, Pt. 2, Nat. Hist. 64: 338. 1895 [1896]. *Cymbidium ensifolium* (L.) Sw. var. *munroanum* (King & Pantl.) Tang & F. T. Wang, Acta Phytotax. Sin. 1: 91. 1951. TYPE: India, Sikkim, Teesta Valley, alt. 1500 ft, May, *Pantling 256* (holotype, CAL; isotype, ?AMES [June 1894]).

Distribution. Arunachal Pradesh, Manipur, ?Meghalaya, Sikkim, West Bengal [Bhutan].

Habit. Terrestrial herb.

Cymbidium nanulum Y. S. Wu & S. C. Chen, Acta Phytotax. Sin. 29(6): 551. 1991. TYPE: China, Yunnan, near Liuku, alt. 800–1600 m, 15 June 1989, *cult. Beijing Bot. Gard. (leg. D. P. Yu 66)* (holotype, PE).

Distribution. Manipur, Meghalaya [China].

Habit. Terrestrial herb.

Cymbidium sinense (Andrews) Willd., Sp. Pl., ed. 4 [Willdenow] 4(1): 111. 1805. *Epidendrum sinense* Andrews, Bot. Repos. 3: t. 216. 1802. TYPE: China, icon. in Andrews, Bot. Repos. 3: t. 216. 1802.

Distribution. Arunachal Pradesh, ?Manipur, Meghalaya [China, Myanmar].

Habit. Terrestrial herb.

Cymbidium tigrinum C. S. P. Parish ex Hook., Bot. Mag. 90: t. 5457. 1864. *Cyperorchis tigrina* (C. S. P. Parish ex Hook.) Schltr., Repert. Spec. Nov. Regni Veg. 20: 108. 1924. TYPE: Myanmar, Moulmein [Mawlamyine], Mulayit, 1863, *Parish 144* (holotype, K).

Distribution. ?Assam, Manipur, Nagaland [China, Myanmar].

Habit. Epiphytic herb.

Cymbidium whiteae King & Pantl., Ann. Roy. Bot. Gard. (Calcutta) 8: 193, t. 258. 1898. *Cyperorchis whiteae* (King & Pantl.) Schltr., Repert. Spec. Nov. Regni Veg. 20: 107. 1924. TYPE: India, Sik-kim, Gantok, 5000 ft, Oct. 1896, *Pantling 425* (holotype, K; isotype, BR, CAL, P).

Distribution. Sikkim. Endemic.

Habit. Epiphytic herb.

———

Cymbilabia D. K. Liu & Ming H. Li, Molec. Phylogen. Evol. 145: art. 106729: 7. 2020.

TYPE: *Cymbilabia undulata* (Lindl.) D. K. Liu & Ming H. Li (*Vanda undulata* Lindl.).

Cymbilabia undulata (Lindl.) D. K. Liu & Ming H. Li, Molec. Phylogen. Evol. 145: art. 106729: 7. 2020. *Vanda undulata* Lindl., J. Proc. Linn. Soc., Bot. 3: 42. 1859 [1858]. *Stauropsis undulatus* (Lindl.) Benth. ex Hook. f., Fl. Brit. India [J. D. Hooker] 6(17): 27. 1890. *Vandopsis undulata* (Lindl.) J. J. Sm., Natuurk. Tijdschr. Ned.-Indië 72: 77. 1912. TYPE: India, Sikkim, icon. *Cathcart s.n.* (holotype, K-LINDL).

Vanda gowerae Gower, The Garden 42: 277. 1892. TYPE: India, Upper Assam, without collector, probably not preserved.

Distribution. Arunachal Pradesh, Assam, Manipur, Meghalaya, Nagaland, Sikkim, Uttarakhand, West Bengal [Bhutan, China, Myanmar, Nepal].

Habit. Terrestrial or epiphytic climber.

———

Cypripedium L., Sp. Pl. 2: 951. 1753.

TYPE: *Cypripedium calceolus* L.

Cypripedium cordigerum D. Don, Prodr. Fl. Nepal.: 37. 1825. TYPE: Nepal, Bagmati Zone, Rashuwa District, Gosainthan [Shishapangma], July 1818, *Wallich s.n.* (lectotype, BM).

Distribution. Himachal Pradesh, Jammu & Kashmir, Uttarakhand [Bhutan, China, Nepal, Pakistan].

Habit. Terrestrial herb.

Cypripedium elegans Rchb. f., Flora 69: 561. 1886. TYPE: Tibet, Kang Me, 3 Aug. 1879, *King 54* (holotype, K; isotype, W).

Distribution. Sikkim, Uttarakhand [Bhutan, China, Nepal].

Habit. Terrestrial herb.

Cypripedium himalaicum Rolfe, J. Linn. Soc., Bot. 29: 319. 1892. TYPE: India, Sikkim, Lachen, 11,500 ft, 21 July 1849, *Hooker & Thomson 317* (holotype, K).

Cypripedium macranthos Sw. var. *himalaicum* (Rolfe) Kraenzl., Orchid. Gen. Sp. 1: 26. 1897.

Distribution. Himachal Pradesh, ?Jammu & Kashmir, Sikkim, Uttarakhand [Bhutan, China, ?Myanmar, Nepal].

Habit. Terrestrial herb.

Cypripedium tibeticum King ex Rolfe, J. Linn. Soc., Bot. 29: 320. 1892. TYPE: China, Sichuan, Tachienlu, 9000–13,500 ft, *Pratt 301* (lectotype, K, designated by Cribb, 1997).

Cypripedium macranthos Sw. var. *tibeticum* (King ex Rolfe) Kraenzl., Orchid. Gen. Sp. 1: 26. 1897.

Distribution. Arunachal Pradesh, Uttarakhand [Bhutan, China].

Habit. Terrestrial herb.

———

Cyrtosia Blume, Bijdr. Fl. Ned. Ind. 8: 396. 1825.

TYPE: *Cyrtosia javanica* Blume.

Cyrtosia falconeri (Hook. f.) Aver., Turczaninowia 14(2): 38. 2011. *Galeola falconeri* Hook. f., Fl. Brit. India [J. D. Hooker] 6(17): 88. 1890. SYNTYPES: India, Subtropical Himalaya, Gharwal, *Falconer s.n.* (syntype, K); Sikkim, *Thomson s.n.* (syntype, K).

Distribution. Arunachal Pradesh, Assam, Manipur, Meghalaya, Mizoram, Nagaland, Sikkim, Uttarakhand, West Bengal [Bhutan, China, Nepal].

Habit. Holomycotrophic terrestrial herb.

Cyrtosia javanica Blume, Bijdr. Fl. Ned. Ind. 8: 396. 1825. *Galeola javanica* (Blume) Benth. & Hook. f., Gen. Pl. (Bentham & Hooker f.) 3: 590. 1883. SYNTYPES: Indonesia, Java, *Blume 329* (syntypes, BO, L); Java, *Blume s.n.* (syntype, L); Java, Jati Kalangan, *Waitz s.n.* (syntype, L).

Distribution. Arunachal Pradesh, Assam, Meghalaya [Sri Lanka].

Habit. Holomycotrophic terrestrial herb.

Cyrtosia lindleyana Hook. f. & Thomson, Ill. Himal. Pl. t. 22. 1855. *Galeola lindleyana* (Hook. f. &

Thomson) Rchb. f., Xenia Orchid. 2: 78. 1865. *Erythrorchis lindleyana* (Hook. f. & Thomson) Rchb. f., Bonplandia 5(3): 37. 1857. SYNTYPES: India, West Bengal and Khasi Hills, 5000–7000 ft, July, *J. D. Hooker s.n.* (syntypes, K, P).

Distribution. Arunachal Pradesh, Assam, Himachal Pradesh, Manipur, Meghalaya, Mizoram, Nagaland, Sikkim, Uttarakhand, West Bengal [Bhutan, China, Myanmar, Nepal].

Habit. Holomycotrophic terrestrial herb.

Cyrtosia nana (Rolfe ex Downie) Garay, Bot. Mus. Leafl. 30(4): 233. 1986. *Galeola nana* Rolfe ex Downie, Bull. Misc. Inform. Kew 1925: 409. 1925. TYPE: Thailand, Doi Suthep, 3000 ft, 4 Aug. 1912, *Kerr 313* (holotype, K).

Distribution. Manipur [also in Thailand].

Habit. Holomycotrophic terrestrial herb.

———

Dactylorhiza Neck. ex Nevski, Trudy Bot. Inst. Akad. Nauk S. S. S. R., Ser. 1, Fl. Sist. Vyssh. Rast. 4: 332. 1937, nom. cons.

TYPE: *Dactylorhiza umbrosa* (Kar. & Kir.) Nevski (*Orchis umbrosa* Kar. & Kir.).

Dactylorhiza hatagirea (D. Don) Soó, Nom. Nov. Gen. Dactylorhiza: 4. 1962. *Orchis hatagirea* D. Don, Prodr. Fl. Nepal.: 23. 1825. SYNTYPES: India, Sirinagur & Kamroop, *Buchanan-Hamilton s.n.* (syntypes, BM, lost?).

Orchis graggeriana Soó, J. Bot. 66: 15. 1928. *Dactylorhiza graggeriana* (Soó) Soó, Nom. Nov. Gen. Dactylorhiza: 4. 1962. TYPE: India, Kashmir, Siddar Valley, Baranginala, Baisaran, 13 Aug. 1901, *Inayat 25376* (holotype, K).

Distribution. Arunachal Pradesh, Himachal Pradesh, Jammu & Kashmir, Uttarakhand [China, Nepal, Pakistan].

Habit. Terrestrial herb.

Dactylorhiza kafiriana Renz, Fl. Iranica (Parsa) 126: 125. 1978. TYPE: Pakistan, Chitral, Brumboret (Bomboret) Valley near Karkal, *Renz 10273* (holotype, G).

Distribution. Jammu & Kashmir [Pakistan].

Habit. Terrestrial herb.

Dactylorhiza umbrosa (Kar. & Kir.) Nevski, Trudy Bot. Inst. Akad. Nauk S.S.S.R., Ser. 1, Fl. Sist.

Vyssh. Rast. 4: 332. 1937. *Orchis umbrosa* Kar. & Kir., Bull. Soc. Imp. Naturalistes Moscou 15: 504. 1842. *Dactylorchis umbrosa* (Kar. & Kir.) Wendelbo, Nytt Mag. Bot. 1: 24. 1952. TYPE: Kazakhstan, Lepsy River, 1840, *G. S. Karelin & I. P. Kirilow s.n.* (holotype, LE).

Distribution. Jammu & Kashmir [Pakistan].

Habit. Terrestrial herb.

Dactylorhiza viridis (L.) R. M. Bateman, Pridgeon & M. W. Chase, Lindleyana 12(3): 129. 1997. *Satyrium viride* L., Sp. Pl. 2: 944. 1753. *Orchis viridis* (L.) Crantz, Stirp. Austr. Fasc., ed. 2, 6: 491. 1769. *Habenaria viridis* (L.) R. Br., Hort. Kew., ed. 2 [W. T. Aiton] 5: 192. 1813. *Gymnadenia viridis* (L.) Rich., De Orchid. Eur. 35. 1817. *Sieberia viridis* (L.) Spreng., Anleit. Kenntn. Gew., ed. 2, 2(1): 282. 1817. *Coeloglossum viride* (L.) Hartm., Handb. Skand. Fl. 329. 1820. *Entaticus viridis* (L.) Gray, Nat. Arr. Brit. Pl. 2: 206. 1821 [1822]. *Chamorchis viridis* (L.) Dumort., Fl. Belg. (Dumortier) 133. 1827. *Platanthera viridis* (L.) Lindl., Syn. Brit. Fl. 261. 1829. *Himantoglossum viride* (L.) Rchb., Fl. Germ. Excurs. 120. 1830. *Peristylus viridis* (L.) Lindl., Syn. Brit. Fl., ed. 2. 261. 1835. TYPE: Europe, *Herb. Linn. No. 1055. 3* (holotype, LINN, lectotype, designated by Renz, 1984).

Coeloglossum kaschmirianum Schltr., Repert. Spec. Nov. Regni Veg. 16: 374. 1920. *Coeloglossum bracteatum* (Muhl. ex Willd.) Parl. var. *kaschmirianum* (Schltr.) Soó, Ann. Hist.-Nat. Mus. Natl. Hung. 26: 356. 1929 (as "*kaschmiricum*"). TYPE: India, Kashmir, Kamri Pass, Nai Gund, *Inayat Khan 25387* (holotype, B, lost).

Distribution. Himachal Pradesh, Jammu & Kashmir, Uttarakhand [Bhutan, China, Myanmar, Nepal, Pakistan].

Habit. Terrestrial herb.

————

Dendrobium Sw., Nova Acta Regiae Soc. Sci. Upsal. ser. 2. 6: 82. 1799, nom. cons.

TYPE: *Dendrobium moniliforme* (L.) Sw. (*Epidendrum moniliforme* L.) (typ. cons.).

Dendrobium aduncum Lindl., Edwards's Bot. Reg. 28(Misc.): 58. 1842 [sect. *Dendrobium*]. *Callista adunca* (Lindl.) Kuntze, Revis. Gen. Pl. 2: 654. 1891. TYPE: India, icon. *cult. Loddiges (leg. Wallich s.n.)* (holotype, AMES).

Distribution. Arunachal Pradesh, Assam, Sikkim, West Bengal [Bhutan, China, Myanmar].

Habit. Epiphytic herb.

Dendrobium amoenum Wall. ex Lindl., Gen. Sp. Orchid. Pl.: 78. 1830 [sect. *Dendrobium*]. *Callista amoena* (Wall. ex Lindl.) Kuntze, Revis. Gen. Pl. 2: 654. 1891. TYPE: Nepal, icon. *Wallich s.n.*

Dendrobium egertoniae Lindl., Edwards's Bot. Reg. 33: t. 36. 1847. TYPE: India, *cult. Egerton s.n.* (holotype, K-LINDL).
Dendrobium mesochlorum Lindl., Edwards's Bot. Reg. 33: t. 36. 1847. TYPE: India, icon. *cult. Veitch s.n.* (holotype, K-LINDL).

Distribution. Arunachal Pradesh, Himachal Pradesh, Sikkim, Uttarakhand, West Bengal [Bangladesh, Bhutan, Myanmar, Nepal].

Habit. Epiphytic herb.

Dendrobium amplum Lindl., Pl. Asiat. Rar. (Wallich) 1: 25, t. 29. 1830 [sect. *Sarcopodium*]. *Sarcopodium amplum* (Lindl.) Lindl., Fol. Orchid. 2(*Sarcopodium*): 1. 1853. *Bulbophyllum amplum* (Lindl.) Rchb. f., Ann. Bot. Syst. (Walpers) 6: 244. 1861. *Callista ampla* (Lindl.) Kuntze, Revis. Gen. Pl. 2: 654. 1891. *Katherinea ampla* (Lindl.) A. D. Hawkes, Lloydia 19(2): 95. 1956. *Epigeneium amplum* (Lindl.) Summerh., Kew Bull. 12: 260. 1957. TYPE: Nepal, Thoka, Sheopore [Shivapuri], Oct. 1821, *Wallich 2001.1* (holotype, K-LINDL; isotype, K-WALL).

Distribution. Arunachal Pradesh, Assam, Manipur, Meghalaya, Mizoram, Nagaland, Sikkim, West Bengal [Bhutan, China, Myanmar, Nepal].

Habit. Epiphytic herb.

Dendrobium anamalayanum Chandrab., V. Chandras. & N. C. Nair, J. Bombay Nat. Hist. Soc. 78: 575, fig. 10. 1981 [sect. *Stachyobium*]. TYPE: India, Tamil Nadu, Coimbatore District, Anamalai, Kavarakal, 1450 m, 22 July 1978, *Chandrabose 57259* (holotype, CAL; isotype, MH).

Distribution. Kerala, Tamil Nadu. Endemic.

Habit. Epiphytic herb.

Dendrobium anceps Sw., Kongl. Vetensk. Acad. Nya Handl. 21: 246. 1800 [sect. *Aporum*]. *Aporum anceps* (Sw.) Lindl., Gen. Sp. Orchid. Pl.: 71. 1830. *Ditulima anceps* (Sw.) Raf., Fl. Tellur. 4: 41. 1836 [1838]. *Callista anceps* (Sw.) Kuntze, Revis. Gen.

Pl. 2: 654. 1891. TYPE: India, *Swartz's collector s.n.* (holotype, UPS).

Distribution. Andaman & Nicobar Islands, Arunachal Pradesh, Assam, Manipur, Meghalaya, Mizoram, Nagaland, Sikkim, Tripura, West Bengal [Bangladesh, Bhutan, Myanmar, Nepal].

Habit. Epiphytic herb.

Dendrobium anilii P. M. Salim, J. Mathew & Szlach., Ann. Bot. Fennici 53: 342. 2016 [sect. *Stachyobium*]. TYPE: India, Kerala, Waynad Dist., Vellarimala tract, Chembra Hills, 1750 m, 20 July 2015, *P. M. Salim 4127* (holotype, MSSRF; isotype, MSSRF).

Distribution. Kerala, Tamil Nadu. Endemic.

Habit. Epiphytic herb.

Remarks. Singh et al. (2019) reduce this to the synonymy of *Dendrobium anamalayanum*, but *D. anilii* is quite distinct. It has a lip with short, subacute sidelobes (vs. long, acuminate sidelobes), a subtruncate (vs. subacute) mid-lobe, and a callus that is emarginate at the apex (vs. distinctly 3-dentate at the apex). Although the type of *D. anilii* was collected in 2015, there is a specimen at Kew collected in 1934 (*Barnes 893*, near Naduvattam, Tamil Nadu, fl. 22 Sep., elevation not stated). This shows that the inflorescence can be up to 6-flowered, whereas the type material of *D. anilii* had only 1- to 2-flowered inflorescences. The mid-lobe of *D. anilii* was wrongly described as ovate; it is in fact transversely oblong-elliptic. The collection from Tamil Nadu had the same distinctive lip color ("mid lobe yellow, side lobes crimson") as the type and was reportedly "very common on trees." It apparently co-occured there with *D. nanum* Hook. f., as one of the several specimens on the sheet belongs to this species, which is easily recognized by the fringed mid-lobe.

Dendrobium aphyllum (Roxb.) C. E. C. Fisch., Fl. Madras 3(6): 1416. 1928 [sect. *Dendrobium*]. *Limodorum aphyllum* Roxb., Pl. Coromandel 1: 34, t. 41. 1795. *Cymbidium aphyllum* (Roxb.) Sw., Nova Acta Regiae Soc. Sci. Upsal. 6: 73. 1799. *Epidendrum aphyllum* (Roxb.) Poir., Encycl. (Lamarck) Suppl. 1. 371. 1810. *Callista aphylla* (Roxb.) Kuntze, Revis. Gen. Pl. 2: 653. 1891. TYPE: India, Coromandel, icon. *Roxburgh, loc. cit.*

Dendrobium cucullatum R. Br., Bot. Reg. 7: t. 548. 1821. *Dendrobium pierardii* Roxb. ex R. Br. var. *cucullatum* (R. Br.) Hook. f., Fl. Brit. India [J. D. Hooker] 5(16): 738. 1890 (as "*cucullata*"), nom. superfl. *Dendrobium aphyllum* (Roxb.) C. E. C. Fisch. var. *cucullatum* (R. Br.) P. K.

Sarkar, J. Econ. Taxon. Bot. 5: 1007. 1984. TYPE: *cult. Springove s.n. (Banks)* (not found).

Dendrobium pierardii Roxb. ex R. Br., Bot. Reg. 7: t. 548. 1821 (as "*pierardi*"). *Dendrobium pierardii* Roxb. ex Hook., Exot. Fl. 1: t. 9. 1822 (as "*pierardi*"), nom. illeg., non Roxb. ex R. Br., 1821. *Dendrobium madrasense* A. D. Hawkes, Orquidea (Mexico City) 25: 102. 1963, nom. superfl. TYPE: Bangladesh, Chittagong, cult. Calcutta s.n. (leg. Pierard) [cf. Roxburgh, Fl. Ind. ed. 1932: 482, 1832 (as "*pieradi*")], icon. *Dendrobium pieradi* [sic], *Roxburgh 1646* (syntypes, CAL, K).

Dendrobium aphyllum (Roxb.) C. E. C. Fisch. var. *katakianum* I. Barua, Orchid Fl. Kamrup Distr. Assam: 170 (2001) (as "*katakinum*"). TYPE: India, Assam, Kamrup District, *Barua 2075* (holotype, CAL; isotype, ASSAM).

Distribution. Andaman & Nicobar Islands, Andhra Pradesh, Arunachal Pradesh, Assam, Jharkhand, Manipur, Meghalaya, Mizoram, Nagaland, Odisha, Sikkim, Tamil Nadu, Tripura, Uttarakhand, West Bengal [Bangladesh, Bhutan, China, Myanmar, Nepal].

Habit. Epiphytic herb.

Remarks. The name *Dendrobium aphyllum* has been misapplied to the species we now call *D. macrostachyum* Lindl. See Schuiteman (2011a).

Dendrobium aqueum Lindl., Edwards's Bot. Reg. 29(Misc.): 5. 1843 [sect. *Dendrobium*]. *Callista aquea* (Lindl.) Kuntze, Revis. Gen. Pl. 2: 654. 1891. TYPE: India, Bombay, *cult. Loddiges s.n.* (not found).

Dendrobium album Wight, Icon. Pl. Ind. Orient. [Wight] 5(1): 6, t. 1645. 1851, non Hook., 1825. TYPE: India, Tamil Nadu, Iyamally Hills, *Wight s.n.*

Distribution. Andhra Pradesh, Goa, Karnataka, Kerala, Maharashtra, Tamil Nadu. Endemic.

Habit. Epiphytic herb.

Dendrobium assamicum Sud. Chowdhury, Kew Bull. 43(4): 667. 1988 [sect. *Dendrobium*]. TYPE: India, Assam, Nowgong District, near Nellie village, 31 Mar. 1969, *S. Chowdhury 354* (holotype, CAL; isotype, K).

Distribution. Assam. Endemic.

Habit. Epiphytic herb.

Remarks. This is possibly an abnormal form of *Dendrobium aphyllum* with distorted flowers and a petaloid lip.

Dendrobium barbatulum Lindl., Gen. Sp. Orchid. Pl.: 84. 1830 [sect. *Fytchianthe*]. *Callista barbatula* (Lindl.) Kuntze, Revis. Gen. Pl. 2: 654. 1891. TYPE: India, 1 Feb. 1817, *Wallich 2013* *(leg. B. Heyne s.n.)* (holotype, K-LINDL; isotype, K-WALL).

Dendrobium kallarense J. Mathew, Kad. V. George, Yohannan & K. Madhus., Int. J. Advanced Res. 2(2): 799. 2014 (as *"kallarensis"*). TYPE: India, Kerala, Pathinamthitta District, *J. Mathew & K. V. George CMS 02748* (holotype, MH; isotype, CMS College, Kerala).

Distribution. Goa, Gujarat, Karnataka, Kerala, Maharashtra, Tamil Nadu. Endemic.

Habit. Epiphytic or lithophytic herb.

Dendrobium bellatulum Rolfe, J. Linn. Soc., Bot. 36: 10. 1903 [sect. *Formosae*]. TYPE: China: Yunnan, Mengtze, 5000 ft, *A. Henry 11109* (holotype, K; isotype, K).

Dendrobium bellatulum Rolfe var. *cleistogamia* Pradhan, Indian Orchids: Guide Identif. & Cult. 2: 331. 1979. TYPE: India, *U. C. Pradhan's collector, Nov. 1975* (Pradhan Herbarium).

Distribution. Assam, Manipur [China, Myanmar].

Habit. Epiphytic herb.

Dendrobium bensoniae Rchb. f., Bot. Zeitung (Berlin) 25(29): 230. 1867 (as *"bensonae"*) [sect. *Dendrobium*]. *Callista bensoniae* (Rchb. f.) Kuntze, Revis. Gen. Pl. 2: 654. 1891. TYPE: Myanmar, Thounggo, Pegu, *cult. Veitch s.n. (leg. Benson)* (holotype, W).

Distribution. Assam, Manipur, Mizoram, Nagaland [Myanmar].

Habit. Epiphytic herb.

Dendrobium bicameratum Lindl., Edwards's Bot. Reg. 25(Misc.): 59. 1839 [sect. *Dendrobium*]. *Callista bicamerata* (Lindl.) Kuntze, Revis. Gen. Pl. 2: 654. 1891. TYPE: India, icon. *J. Gibson s.n.* (holotype, K-LINDL).

Dendrobium bolboflorum Falc. ex Hook. f., Fl. Brit. India [J. D. Hooker] 5(16): 729. 1890. SYNTYPES: India, Sikkim, Rungbee, alt. 3000 ft, 29 July 1890, *C. B. Clarke s.n.* [*12268A*] (syntype, K); Darjeeling, alt. 7000 ft, July 1882, *Gamble s.n.* [*10510*] (syntype, K).

Distribution. Andhra Pradesh, Arunachal Pradesh, Assam, Bihar, Jharkhand, Manipur, Meghalaya, Mizoram, Nagaland, Odisha, Sikkim, Uttarakhand, West Bengal [Bhutan, Myanmar, Nepal].

Habit. Epiphytic herb.

Dendrobium brunneum Schuit. & Peter B. Adams, Muelleria 29: 65. 2011 [sect. *Sarcopodium*]. *Epigeneium chapaense* Gagnep., Bull. Mus. Natl. Hist. Nat., sér. 2, 4: 596. 1932. *Sarcopodium chapaense* (Gagnep.) Tang & F. T. Wang, Acta Phytotax. Sin. 1: 83. 1951. SYNTYPES: Vietnam, Tonkin, near Chapa, *Poilane 12605* (syntype, P); *Poilane 12607* (syntype, P).

Epigeneium arunachalense A. N. Rao, Bull. Arunachal Forest Res. 25: 3. 2009 [2010]. *Dendrobium subansiriense* D. Verma & Barbhuiya, Phytotaxa 167: 150. 2014. *Dendrobium deuteroarunachalense* J. M. H. Shaw, Orchid Rev. 122(1306, Suppl.): 38. 2014, nom. superfl. TYPE: India, Arunachal Pradesh, Lower Subansiri district, Hakhetari-Rizampak, ca. 1200 m, *Sastry 44804* (holotype, ASSAM).

Dendrobium nageswarayanum Chowlu, Natl. Acad. Sci. Lett. 43(1): 659. 2020. TYPE: India, Arunachal Pradesh, Kurung Kumey, Koloriang Hill, 823 m, 27°53′57.3″ N, 93°19′53.8″ E, 11 Nov. 2016, *Chowlu-40066 A* (holotype, CAL).

Distribution. Arunachal Pradesh [also in Vietnam].

Habit. Epiphytic herb.

Remarks. This species was recently photographed in Arunachal Pradesh by Swami (2017); his photographs and annotations show that *Epigeneium arunachalense* is just a small-flowered form of *Dendrobium brunneum*, Swami's material being intermediate in size. In our opinion, the recently described *D. nageswarayanum* also falls within the range of variation of this species.

This is not *Dendrobium chapaense* Aver.

Dendrobium brymerianum Rchb. f., Gard. Chron., n.s., 4: 323. 1875 [sect. *Dendrobium*]. *Callista brymeriana* (Rchb. f.) Kuntze, Revis. Gen. Pl. 2: 654. 1891. SYNTYPES: ?Myanmar, *cult. Brymer s.n.* (syntype, W); Myanmar, Bhamo, *cult. Low (leg. Rimann) s.n.* (syntype, W).

Distribution. Assam, Manipur [China, Myanmar].

Habit. Epiphytic herb.

Dendrobium calocephalum (Z. H. Tsi & S. C. Chen) Schuit. & Peter B. Adams, Muelleria 29: 66. 2011 [sect. *Crinifera*]. *Flickingeria calocephala* Z. H. Tsi & S. C. Chen, Acta Phytotax. Sin. 33: 203. 1995. TYPE: China, Yunnan, Jinghong, 8 June 1991, *Z. H. Tsi 91-720* (holotype, PE).

Flickingeria abhaycharanii Phukan & A. A. Mao, Orchid Rev. 113(1261): 22. 2005. *Dendrobium abhaycharanii* (Phukan & A. A. Mao) Schuit. & Peter B. Adams, Muelleria 29: 66. 2011. TYPE: India, Manipur, Senapati Hills, cultivated in National Orchidarium, Bot. Surv. India, Shillong, *S. Phukan 68274* (holotype, CAL).

Distribution. Manipur, ?Meghalaya, Nagaland [China].

Habit. Epiphytic herb.

Dendrobium capillipes Rchb. f., Gard. Chron. 1867: 997. 1867 [sect. *Dendrobium*]. *Callista capillipes* (Rchb. f.) Kuntze, Revis. Gen. Pl. 2: 654. 1891. TYPE: Myanmar, Moulmein [Mawlamyine], *cult. Low (leg. Parish 186) s.n.* (holotype, W).

Distribution. Assam, Mizoram [China, Myanmar].

Habit. Epiphytic herb.

Dendrobium cariniferum Rchb. f., Gard. Chron. 1869: 611. 1869 [sect. *Formosae*]. *Callista carinifera* (Rchb. f.) Kuntze, Revis. Gen. Pl. 2: 654. 1891. TYPE: India, *cult. Marshall s.n.* (holotype, W).

Distribution. Manipur, ?Meghalaya, Mizoram, Nagaland [Myanmar].

Habit. Epiphytic herb.

Dendrobium chrysanthum Wall ex Lindl., Edwards's Bot. Reg. 15: t. 1299. 1830 [sect. *Dendrobium*]. *Callista chrysantha* (Wall. ex Lindl.) Kuntze, Revis. Gen. Pl. 2: 654. 1891. TYPE: Nepal, Churopang [Cherapang], Dec. 1820, *Wallich 2012* (holotype, K-LINDL; isotype, K, K-WALL).

Distribution. Arunachal Pradesh, Assam, Manipur, Meghalaya, Mizoram, Nagaland, Sikkim, Tripura, Uttarakhand, West Bengal [Bhutan, China, Myanmar, Nepal].

Habit. Epiphytic herb.

Dendrobium chryseum Rolfe, Gard. Chron., ser. 3, 3: 233. 1888 [sect. *Dendrobium*]. TYPE: India, Assam, *cult. J. Veitch s.n., Jan. 1888* (holotype, K).

Dendrobium denneanum Kerr, J. Siam Soc., Nat. Hist. Suppl. 9: 229. 1933. TYPE: Laos, Xiangkhoang, Muang Cha, 1100 m, 19 Apr. 1932, *A. F. G. Kerr 965A* (holotype, K).
Dendrobium aurantiacum Rchb. f., Gard. Chron., ser. 3, 2: 98. 1887, nom. illeg., non (F. Muell.) F. Muell., 1865. SYN-TYPES: 1854, *cult. Kammerrath s.n.* (syntype, W); India, Assam, 1870, *Mann 7* (syntype, K, W); India, *cult. Bull s.n.* (syntype, W).
Dendrobium clavatum Wall. ex Lindl., Paxton's Fl. Gard. 2: 104, fig. 189. 1852, non Roxb., 1832. *Dendrobium rolfei* A. D. Hawkes & A. H. Heller, Lloydia 20: 123. 1957. TYPE: India, Assam, *cult. T. Denne s.n., May 1851* (holotype, K-LINDL).

Distribution. Assam, Manipur, Meghalaya, Nagaland, Sikkim, Uttarakhand, West Bengal [Bangladesh, Bhutan, China, Myanmar, Nepal].

Habit. Epiphytic herb.

Dendrobium chrysotoxum Lindl., Edwards's Bot. Reg. 33, sub t. 19. 1847 [sect. *Dendrobium*]. *Callista chrysotoxa* (Lindl.) Kuntze, Revis. Gen. Pl. 2: 654. 1891. TYPE: India, *cult. Henderson s.n.* (?holotype, K-LINDL [single flower without annotation]).

Distribution. Arunachal Pradesh, Assam, Manipur, Mizoram, Nagaland, Tripura, West Bengal [Bangladesh, Myanmar].

Habit. Epiphytic herb.

Dendrobium crepidatum Lindl. & Paxton, Paxton's Fl. Gard. 1: 63, fig. 45. 1850 [sect. *Dendrobium*]. *Callista crepidata* (Lindl. & Paxton) Kuntze, Revis. Gen. Pl. 2: 654. 1891. TYPE: India, cult. P. Basset, icon. *Basset* (holotype, K-LINDL).

Dendrochilum roseum Dalzell, Hooker's J. Bot. Kew Gard. Misc. 4: 291. 1852. *Dendrobium lawianum* Lindl., J. Proc. Linn. Soc., Bot. 3: 10. 1859 [1858] (as "*lawanum*"). *Callista lawiana* (Lindl.) Kuntze, Revis. Gen. Pl. 2: 655. 1891. TYPE: India, Syhadree Hills, *Stocks 31 (leg. Dalzell)*.
Dendrobium actinomorphum Blatt. & Hallb., J. Indian Bot. Soc. 2: 50. 1921. TYPE: India, Bombay Province, Castle Rock, Mar. 1919, *McCann s.n.* (holotype, Herb. S. X. C. 13768).
Dendrobium crepidatum Lindl. & Paxton var. *avita* Gammie, J. Bombay Nat. Hist. Soc. 17: 33. 1906. TYPE: India, Belgaum and Kanara Ghats, *Gammie s.n.*

Distribution. Arunachal Pradesh, Assam, Bihar, Chhattisgarh, Goa, Jharkhand, Karnataka, Kerala, Madhya Pradesh, Maharashtra, Manipur, Meghalaya, Mizoram, Nagaland, Odisha, Sikkim, Tamil Nadu, Uttarakhand, West Bengal [Bangladesh, Bhutan, China, Myanmar, Nepal].

Habit. Epiphytic herb.

Remarks. Jalal and Jayanthi (2018) and Jalal (2018, 2019) suggest that *Dendrobium lawianum* deserves recognition as a separate species, and they mention a number of diagnostic characters, including white flowers without a yellow patch on the lip and with an indistinct mentum. However, *D. lawianum* was described as having rose-tinted flowers. Judging from Jalal and Jayanthi's photograph, what they call *D. lawianum* looks like a semi-peloric form of *D. crepidatum*, but more research is needed.

Dendrobium crispum Dalzell, Hooker's J. Bot. Kew Gard. Misc. 4: 111. 1852 [sect. *Stachyobium*]. *Dendrobium humile* Wight, Icon. Pl. Ind. Orient. [Wight] 5(1): t. 1643. 1851, nom. illeg., non (Sm.) Sm., 1808. TYPE: India, Tamil Nadu, Iyamally Hills, July–Aug., *Wight s.n.* (holotype, K).

Dendrobium pygmaeum Lindl., Gen. Sp. Orchid. Pl.: 85. 1830, nom. illeg., non Sm., 1808. TYPE: Myanmar, Martaban, *Wallich s.n.* (holotype, BM).
Dendrobium peguanum Lindl., J. Proc. Linn. Soc., Bot. 3: 19. 1859 [1858]. SYNTYPES: Myanmar, Pegu, 5 Jan. 1854,

M. Lelland s.n. (syntype, K); "Borneo," *T. Lobb s.n.* (syntype, K-LINDL).

Distribution.   Chhattisgarh, Goa, Gujarat, Jharkhand, Karnataka, Kerala, Madhya Pradesh, Maharashtra, Mizoram, Odisha, Sikkim, West Bengal [Bhutan, Myanmar, Nepal].

Habit.   Epiphytic herb.

Remarks.   Ormerod and Kumar (2018) argued that *Dendrobium crispum* Dalzell (excluding the description, which is referable to *Dendrobium turbinatum* Ormerod & C. S. Kumar) must be regarded as a replacement name for *D. humile* Wight, and that both are earlier names for the well-known *D. peguanum* Lindl.

Dendrobium crumenatum Sw., J. Bot. (Schrader) 2(1): 237. 1799 [sect. *Aporum*]. *Onychium crumenatum* (Sw.) Blume, Bijdr. Fl. Ned. Ind. 326. 1825. *Callista crumenata* (Sw.) Kuntze, Revis. Gen. Pl. 2: 654. 1891. *Aporum crumenatum* (Sw.) Brieger, Orchideen (Schlechter) 1(11–12): 671. 1981. TYPE: Indonesia, Java, collector? (not found).

Distribution.   Andaman & Nicobar Islands, ?Kerala [China, Myanmar].

Habit.   Epiphytic herb.

Remarks.   Records from South India are possibly of introduced plants, as *Dendrobium crumenatum* has only recently been found there and the species is considered to be introduced in Sri Lanka (Fernando & Ormerod, 2008).

Dendrobium cumulatum Lindl., Gard. Chron. 1855: 756. 1855 [sect. *Pedilonum*]. *Callista cumulata* (Lindl.) Kuntze, Revis. Gen. Pl. 2: 654. 1891. *Eurycaulis cumulatus* (Lindl.) M. A. Clem., Telopea 10(1): 286. 2003. TYPE: Origin and collector not stated (holotype, K-LINDL).

Dendrobium cumulatum Lindl. var. *jenkinsii* Hook. f., Fl. Brit. India [J. D. Hooker] 5(16): 731. 1890. TYPE: Ic. s.n. (holotype, CAL).

Distribution.   Arunachal Pradesh, Assam, Manipur, Meghalaya, Sikkim, West Bengal [Bhutan, Myanmar, Nepal].

Habit.   Epiphytic herb.

Dendrobium darjeelingense Pradhan, Indian Orchids: Guide Identif. & Cult. 2: 336. 1979 (as "*darjeelingensis*") [sect. *Stachyobium*]. TYPE: West Bengal, Darjeeling, alt. 6400 ft, 19/10/1975, *U. C. Pradhan's collector s.n.* (holotype, Herbarium U. C. Pradhan, destroyed). NEOTYPE: India,

Assam: Cachar, near Loharbond Beat Office, 37 m, 6 Jan. 2012, Barbhuiya 846 (neotype, ASSAM, designated by Verma & Barbhuiya, 2014).

Distribution.   Mizoram, ?Sikkim, West Bengal. Endemic.

Habit.   Epiphytic herb.

Dendrobium delacourii Guillaumin, Bull. Mus. Natl. Hist. Nat. 30: 522. 1924 [sect. *Stachyobium*]. TYPE: Vietnam, Quang tri, 1924, *Delacourt s.n.* (holotype, P).

Dendrobium ciliatum C. S. P. Parish ex Hook. var. *breve* Rchb. f., Gard. Chron., n.s., 20: 328. 1883. SYNTYPES: Myanmar, *Veitch 17 (leg. Benson)*; unknown origin, *cult. Harding s.n.*; *cult. Veitch s.n.* (syntypes, W).

Distribution.   Manipur [Bangladesh, Myanmar].

Habit.   Epiphytic herb.

Dendrobium densiflorum Lindl., Pl. Asiat. Rar. (Wallich) 1: 34, t. 40. 1830 [sect. *Densiflora*]. *Callista densiflora* (Lindl.) Kuntze, Revis. Gen. Pl. 2: 654. 1891. TYPE: Nepal, 1821, *Wallich 2000* (holotype, K-LINDL; isotype, K-WALL).

Dendrobium clavatum Roxb., Fl. Ind. ed. 1832, 3: 481. 1832. *Endeisa flava* Raf., Fl. Tellur. 2: 51. 1836 [1837], nom. superfl. TYPE: India, Meghalaya, Garrow [Garo] Hills, icon. *Dendrobium clavatum*, *Roxburgh 2354* (syntypes, CAL, K).

Distribution.   Arunachal Pradesh, Assam, Manipur, Meghalaya, Mizoram, Nagaland, Sikkim, West Bengal [Bangladesh, Bhutan, China, Myanmar, Nepal].

Habit.   Epiphytic herb.

Dendrobium denudans D. Don, Prodr. Fl. Nepal.: 34. 1825 [sect. *Stachyobium*]. *Callista denudans* (D. Don) Kuntze, Revis. Gen. Pl. 2: 654. 1891. TYPE: Nepal, *Wallich s.n.* (holotype, BM).

Distribution.   Arunachal Pradesh, Manipur, ?Meghalaya, Mizoram, Nagaland, Sikkim, Uttarakhand, West Bengal [Bhutan, Nepal].

Habit.   Epiphytic herb.

Dendrobium devonianum Paxton, Paxton's Mag. Bot. 7: 169, fig. s.n. 1840 [sect. *Dendrobium*]. *Dendrobium pulchellum* Roxb. ex Lindl. var. *devonianum* (Paxton) Rchb. f., Ann. Bot. Syst. (Walpers) 6: 284. 1861. *Callista devoniana* (Paxton) Kuntze, Revis. Gen. Pl. 2: 654. 1891. TYPE: India, Khasi Hills, alt. 4500 ft, *leg. Gibson s.n.*, *icon. Paxton's Mag. Bot. 7: fig. s.n. 1840* (iconotype).

Distribution. Arunachal Pradesh, Assam, Manipur, Meghalaya, Mizoram, Nagaland [Bhutan, China, Myanmar].

Habit. Epiphytic herb.

Dendrobium dickasonii L. O. Williams, Bot. Mus. Leafl. 8: 107. 1940 [sect. *Dendrobium*]. TYPE: Myanmar, near Haka, 6000 ft, Apr. 1938, *Dickason 7779* (holotype, AMES).

Dendrobium arachnites Rchb. f., Gard. Chron., n.s., 2: 354. 1874, nom. illeg., non Thouars, 1822. *Callista arachnites* Kuntze, Revis. Gen. Pl. 2: 654. 1891. *Dendrobium seidenfadenii* Senghas & Bockemühl, Orchidee (Hamburg) 29(5): cppo 2. 1978. TYPE: Myanmar, *Boxall s.n.* (holotype, W).

Distribution. Assam, Manipur, Mizoram [Myanmar].

Habit. Epiphytic herb.

Dendrobium diodon Rchb. f. subsp. **kodayarensis** Gopalan & A. N. Henry, J. Econ. Taxon. Bot. 12(2): 487. 1988, publ. 1989 [sect. *Stachyobium*]. TYPE: India, Tamil Nadu, Muthukuzhivayal, Upper Kodayaar, 1300 m, 26 Mar. 1984, *R. Gopalan 81452* (holotype, CAL).

Distribution. Kerala, Tamil Nadu. Endemic.

Habit. Epiphytic herb.

Dendrobium draconis Rchb. f., Bot. Zeitung (Berlin) 20: 214. 1862 [sect. *Formosae*]. *Callista draconis* (Rchb. f.) Kuntze, Revis. Gen. Pl. 2: 654. 1891. TYPE: Myanmar, Moulmein [Mawlamyine], *cult. Low (leg. Parish) s.n.* (holotype, W).

Dendrobium eburneum H. Low, Proc. Roy. Hort. Soc. London 4: 8. 1864. TYPE: Myanmar, Moulmein [Mawlamyine], *cult. Low (leg. Parish) s.n.* (holotype, K).

Distribution. Assam, Manipur [Myanmar].

Habit. Epiphytic herb.

Dendrobium eriiflorum Griff., Not. Pl. Asiat. 3: 316. 1851 (as "*eriaeflorum*") [sect. *Stachyobium*]. *Callista eriiflora* (Griff.) Kuntze, Revis. Gen. Pl. 2: 654. 1891. TYPE: India, Khasi Hills, Myrung, 9 Nov. 1835, *Griffith in Assam Herb. 230* (holotype, K-LINDL; isotype, K).

Dendrobium eriiflorum Griff. var. *sikkimense* Lucksom, Orchids Sikkim N. E. Himalaya: 631. 2007, nom. inval. TYPE: Not designated.

Distribution. Andaman & Nicobar Islands, Arunachal Pradesh, Assam, Manipur, Meghalaya, Nagaland, Sikkim, West Bengal [Bhutan, Myanmar, Nepal].

Habit. Epiphytic herb.

Dendrobium falconeri Hook., Bot. Mag. 82: t. 4944. 1856 [sect. *Dendrobium*]. *Callista falconeri* (Hook.) Kuntze, Revis. Gen. Pl. 2: 654. 1891. TYPE: Bhutan, 4000 ft, *cult. G. Reid s.n., 1856* (holotype, K).

Dendrobium falconeri Hook. var. *senapatianum* C. Deori, Gogoi & A. A. Mao, Orchid Rev. 118: 21. 2010. TYPE: India, Senapati District, 1350 m, 15 May 2008, *A. A. Mao & R. Gogoi 115084* (holotype, ASSAM).

Distribution. Arunachal Pradesh, Assam, Manipur, ?Meghalaya, Mizoram, Nagaland, Sikkim, West Bengal [Bhutan, China, Myanmar].

Habit. Epiphytic herb.

Dendrobium fargesii Finet, Bull. Soc. Bot. France 50: 374, t. 12, fig. 11, 18. 1903 [sect. *Sarcopodium*]. *Desmotrichum fargesii* (Finet) Kraenzl., Pflanzenr. (Engler) IV, 50 (Heft 21 II B): 358. 1910. *Epigeneium fargesii* (Finet) Gagnep., Bull. Mus. Natl. Hist. Nat. Ser. 2, 4: 595. 1932. *Sarcopodium fargesii* (Finet) Tang & F. T. Wang, Acta Phytotax. Sin. 1: 83. 1951. TYPE: China, Sichuan, Chengkou District, Ta-han-Ky, *R. P. Farges 1506* (holotype, P; isotype, CAL, K, LE, NY).

Distribution. Arunachal Pradesh, Uttarakhand [Bhutan, China].

Habit. Epiphytic herb.

Dendrobium farmeri Paxton, Paxton's Mag. Bot. 15: 241, fig. s.n. 1849 (as "*farmerii*") [sect. *Densiflora*]. *Callista farmeri* (Paxton) Kuntze, Revis. Gen. Pl. 2: 654. 1891. TYPE: Icon. *Dendrobium farmerii* [sic], Paxton's Mag. Bot. 15: fig. s.n. (iconotype).

Dendrobium palpebrae Lindl., J. Hort. Soc. London 5: 33. 1850. *Callista palpebrae* (Lindl.) Kuntze, Revis. Gen. Pl. 2: 655. 1891. TYPE: Myanmar, Moulmein [Mawlamyine], *cult. Veitch s.n.* (holotype, K-LINDL).

Distribution. Arunachal Pradesh, Assam, Manipur, Mizoram, Nagaland, Sikkim, West Bengal [Bangladesh, Myanmar, Nepal].

Habit. Epiphytic herb.

Dendrobium fimbriatum Hook., Exot. Fl. 1: t. 71. 1823 [sect. *Dendrobium*]. *Callista fimbriata* (Hook.) Kuntze, Revis. Gen. Pl. 2: 653. 1891. TYPE: Nepal, 1821, *Wallich 2011* (holotype, K-LINDL; isotype, K, K-WALL).

Dendrobium normale Falc., Proc. Linn. Soc. London 1: 14. 1839. *Callista normalis* (Falc.) Kuntze, Revis. Gen. Pl. 2: 655. 1891. TYPE: India, *Falconer 1004* (holotype, W).

Dendrobium paxtonii Paxton, Paxton's Mag. Bot. 6: 169. 1839, fig. s.n., nom. illeg., non Lindl., 1839. TYPE: India, Khasi Hills, leg. Gibson, *icon., loc. cit.*

Dendrobium fimbriatum Hook. var. *oculatum* Hook., Bot. Mag. 71: t. 4160. 1845. *Callista oculata* (Hook.) Kuntze, Revis. Gen. Pl. 2: 653. 1891. TYPE: Nepal, *Wallich s.n.* (holotype, K).

Distribution. Arunachal Pradesh, Assam, Bihar, Jharkhand, Manipur, Meghalaya, Mizoram, Nagaland, Odisha, Sikkim, Tripura, Uttarakhand, West Bengal [Bangladesh, Bhutan, China, Myanmar, Nepal].

Habit. Epiphytic or lithophytic herb.

Dendrobium flexuosum Griff., Not. Pl. Asiat. 3: 317. 1851 [sect. *Formosae*]. TYPE: India, Assam, Khasi Hills, Churra, *Griffith in Assam Herb. 162* [*s.n.*] (holotype, K-LINDL).

Dendrobium bulleyi Rolfe, Notes Roy. Bot. Gard. Edinburgh 8: 20. 1913. TYPE: China, Yunnan, Teng-Yueh [Teng-chong] to Talifu [Dali] Road, Tien-Ching-pu, *Forrest 1091* (holotype, E).

Dendrobium chapaense Aver., Rheedea 16: 3. 2006. TYPE: Vietnam, Lao Cai Prov., *Averyanov & Loc HAL 8321* (holotype, HN; isotype, LE).

Distribution. Assam, Nagaland [China, Myanmar, Nepal].

Habit. Epiphytic herb.

Remarks. See remarks under *Dendrobium longicornu*.

Dendrobium formosum Roxb. ex Lindl., Pl. Asiat. Rar. (Wallich) 1: 34, t. 39. 1830 [sect. *Formosae*]. *Callista formosa* (Roxb. ex Lindl.) Kuntze, Revis. Gen. Pl. 2: 654. 1891. TYPE: India, Sylhet, 1813, *Roxburgh s.n.* (holotype, BM).

Distribution. Andaman & Nicobar Islands, Arunachal Pradesh, Assam, Bihar, Jharkhand, Manipur, Meghalaya, Mizoram, Nagaland, Odisha, Tripura, West Bengal [Myanmar, Nepal].

Habit. Epiphytic herb.

Dendrobium fugax Rchb. f., Gard. Chron. 1871: 1257. 1871 [sect. *Crinifera*]. *Callista fugax* (Rchb. f.) Kuntze, Revis. Gen. Pl. 2: 654. 1891. *Flickingeria fugax* (Rchb. f.) Seidenf., Dansk Bot. Ark. 34(1): 46. 1980. TYPE: Myanmar, Maymyo, *C. S. P. Parish s.n.* (holotype, W; isotype, W).

Dendrobium sordidum King & Pantl., J. Asiat. Soc. Bengal, Pt. 2, Nat. Hist. 66(3): 583. 1898 [1897]. TYPE: Myanmar, Moulmein [Mawlamyine], June, *cult. Calcutta s.n. (leg. C. Peché)* (holotype, CAL).

Distribution. Arunachal Pradesh, Assam, Manipur, Meghalaya, Mizoram, Nagaland, Tripura, Uttarakhand [Bhutan, Myanmar].

Habit. Epiphytic herb.

Dendrobium fuscescens Griff., Not. Pl. Asiat. 3: 308. 1851 [sect. *Sarcopodium*]. *Sarcopodium fuscescens* (Griff.) Lindl., Fol. Orchid. 2(*Sarcopodium*): 2. 1853. *Bulbophyllum fuscescens* (Griff.) Rchb. f., Ann. Bot. Syst. (Walpers) 6: 244. 1861. *Callista fuscescens* (Griff.) Kuntze, Revis. Gen. Pl. 2: 654. 1891. *Katherinea fuscescens* (Griff.) A. D. Hawkes, Lloydia 19(2): 95. 1956. *Epigeneium fuscescens* (Griff.) Summerh., Kew Bull. 12: 262. 1957. SYN-TYPES: India, Assam, Khasi Hills, Churra, *Griffith in Assam Herb. 16, 60, 149* (syntypes, K, K-LINDL, K-WALL).

Katherinea navicularis N. P. Balakr. & Sud. Chowdhury, Bull. Bot. Surv. India 8(3–4): 316. 1966 [1967]. *Epigeneium naviculare* (N. P. Balakr. & Sud. Chowdhury) Hynn. & Wadhwa, Orchids Nagaland 163. 2000. TYPE: Bhutan, near Narfong, *Balakrishnan 43014A* (holotype, CAL).

Distribution. Arunachal Pradesh, Assam, Manipur, Meghalaya, Mizoram, Nagaland, Sikkim, West Bengal [Bhutan, China, Myanmar, Nepal].

Habit. Epiphytic herb.

Remarks. A quite different species was illustrated by King and Pantling (1898: t. 88) as *Dendrobium fuscescens*. Their illustration looks closer to *D. treutleri* (Hook. f.) Schuit. & Peter B. Adams, although the shape of the lip mid-lobe does not match very well. The photo labeled *Epigeneium fuscescens* in Singh et al. (2019) undoubtedly shows *D. treutleri*. The true *D. fuscescens* is illustrated by these authors as *Epigeneium navicularis* [sic], which is a synonym of *D. fuscescens*, as a perusal of Griffith's illustration (1851: t. 309) will demonstrate.

Dendrobium gibsonii Lindl., Paxton's Mag. Bot. 5: 169. 1838 [sect. *Dendrobium*]. *Callista gibsonii* (Paxton) Kuntze, Revis. Gen. Pl. 2: 654. 1891. TYPE: India, Khasi Hills, 1837, *Gibson s.n.* (holotype, K-LINDL).

Dendrobium fuscatum Lindl., J. Proc. Linn. Soc., Bot. 3: 8. 1859 [1858]. SYNTYPES: India, Sikkim, icon. *Cathcart* (syntype, K-LINDL); India, Sikkim, *Hooker 12* (syntype, K-LINDL); India, Khasi Hills, *Hooker & Thomson 12* (syntype, K-LINDL).

Distribution. Arunachal Pradesh, Assam, Manipur, Meghalaya, Mizoram, Nagaland, Sikkim, West Bengal [Bhutan, China, Myanmar, Nepal].

Habit. Epiphytic herb.

Dendrobium grande Hook. f., Fl. Brit. India [J. D. Hooker] 5(16): 724. 1890 [sect. *Aporum*]. *Aporum grande* (Hook. f.) Rauschert, Feddes Repert. 94: 439. 1983. *Callista grandis* (Hook. f.) Kuntze, Revis. Gen. Pl. 2: 654. 1891. SYNTYPES: Malaysia, Perak, *Scortechini s.n.* (syntypes, CAL, K); India, South Andaman Islands, *Kurz s.n.* (syntype, K).

Distribution. Andaman & Nicobar Islands [?Myanmar].

Habit. Epiphytic herb.

Dendrobium gratiosissimum Rchb. f., Bot. Zeitung (Berlin) 23: 99. 1865 [sect. *Dendrobium*]. *Callista gratiosissima* (Rchb. f.) Kuntze, Revis. Gen. Pl. 2: 654. 1891. TYPE: Myanmar, Moulmein [Mawlamyine], *cult. Low (leg. Parish) s.n.* (holotype, W).

Distribution. Assam, Manipur [China, Myanmar].

Habit. Epiphytic herb.

Dendrobium griffithianum Lindl., Edwards's Bot. Reg. 21: t. 1756. 1835 [sect. *Densiflora*]. *Callista griffithiana* (Lindl.) Kuntze, Revis. Gen. Pl. 2: 654. 1891. TYPE: Myanmar, Zimjaik, 1834, *Griffith 351* (holotype, K-LINDL; isotype, K).

Distribution. Assam [Myanmar].

Habit. Epiphytic herb.

Dendrobium herbaceum Lindl., Edwards's Bot. Reg. 26(Misc.): 69. 1840 [sect. *Herbacea*]. *Callista herbacea* (Lindl.) Kuntze, Revis. Gen. Pl. 2: 654. 1891. TYPE: India, *cult. Loddiges s.n.* (holotype, K-LINDL).

Dendrobium ramosissimum Wight, Icon. Pl. Ind. Orient. [Wight] 5(1): t. 1648. 1851. TYPE: India, Karnataka, Coorg, Jan. 1847, *Jerdon s.n.* (holotype, K).
Dendrobium georgei J. Mathew, Telopea 16: 89. 2014. *Dendrobium herbaceum* Lindl. subsp. *georgei* (J. Mathew) S. Misra, Orchids India: 484. 2019. TYPE: India, Kerala, Kollam District, Kottavasal, ca. 1250 m, 25 Feb. 2011, *Mathew CMS2721* (holotype, MH; isotype, KFRI, CMSH, SESH).

Distribution. Andaman & Nicobar Islands, Andhra Pradesh, Bihar, Chhattisgarh, Goa, Jharkhand, Karnataka, Kerala, Madhya Pradesh, Maharashtra, Mizoram, Odisha, Tamil Nadu [Bangladesh].

Habit. Epiphytic herb.

Dendrobium hesperis (Seidenf.) Schuit. & Peter B. Adams, Muelleria 29: 67. 2011 [sect. *Crinifera*]. *Flickingeria hesperis* Seidenf., Nordic J. Bot. 2(1):

16. 1982. TYPE: India, Kumaon, Pithoragarh, Ogla road, 1500 m, *Arora 66130* (holotype, BSD).

Distribution. Manipur, Uttar Pradesh, Uttarakhand [also in Laos].

Habit. Epiphytic herb.

Dendrobium heterocarpum Wall. ex Lindl., Gen. Sp. Orchid. Pl.: 78. 1830 [sect. *Dendrobium*]. *Callista heterocarpa* (Wall. ex Lindl.) Kuntze, Revis. Gen. Pl. 2: 654. 1891. TYPE: Nepal, *Wallich s.n.* (holotype, K-LINDL).

Dendrobium aureum Lindl., Gen. Sp. Orchid. Pl.: 77. 1830. *Callista aurea* (Lindl.) Kuntze, Revis. Gen. Pl. 2: 653. 1891. TYPE: Sri Lanka, Nuera Ellia, *Macrae s.n.* (holotype, K-LINDL).

Distribution. Arunachal Pradesh, Assam, Karnataka, Kerala, Manipur, Meghalaya, Mizoram, Nagaland, Sikkim, Tamil Nadu, Uttarakhand, West Bengal [Bhutan, China, Myanmar, Nepal, Sri Lanka].

Habit. Epiphytic herb.

Dendrobium heyneanum Lindl., Gen. Sp. Orchid. Pl.: 90. 1830 [sect. *Stachyobium*]. *Callista heyneana* (Lindl.) Kuntze, Revis. Gen. Pl. 2: 654. 1891. TYPE: India, Aug. 1818, *Wallich 1995 (leg. Heyne s.n.)* (syntypes, E, K, K-WALL).

Distribution. Karnataka, Kerala, Tamil Nadu. Endemic.

Habit. Epiphytic herb.

Dendrobium hirsutum Griff., Not. Pl. Asiat. 3: 318. 1851 [sect. *Formosae*]. *Dendrobium longicornu* Lindl. var. *hirsutum* (Griff.) Hook. f., Fl. Brit. India [J. D. Hooker] 5(16): 720. 1890 (as "*hirsuta*"). TYPE: India, Khasi Hills, Churra Ponjee, *Griffith s.n.* (holotype, K-LINDL).

Dendrobium meghalayense C. Deori, S. K. Sarma, Hynn. & Phukan, Rheedea 16(1): 55. 2006, nom. illeg., non Y. Kumar & S. Chowdhury, 2003. *Dendrobium jaintianum* Sabap., Ind. J. For. 30(3): 371. 2007. TYPE: India, Meghalaya, Jaintia Mountains, *H. Deka 18385* (holotype, ASSAM; isotype, ASSAM).
Dendrobium tamenglongense Kishor, Y. N. Devi, H. B. Sharma, Tongbram & Vij, Nordic J. Bot. 32: 151. 2013. TYPE: India, Manipur, Tamenglong District, Kahulong, 1355 m, 1 Mar. 2011, *Nanda et al. 71* (holotype, ASSAM; isotype, ASSAM, CAL).

Distribution. Arunachal Pradesh, Assam, Manipur, Meghalaya, Nagaland [China, Myanmar].

Habit. Epiphytic herb.

Remarks. See remarks under *Dendrobium longicornu.*

Dendrobium hkinhumense Ormerod & C. S. Kumar, Rheedea 18(2): 75. 2008 [sect. *Dendrobium*]. TYPE: Myanmar, North Triangle (Hkinhum), 1370 m, 21 July 1953, *Kingdon Ward 21198* (holotype, BM).

Dendrobium tuensangense Odyuo & C. Deori, Phytotaxa 311(2): 185. 2017. TYPE: India, Nagaland, Tuensang district, Bakhong forest reserve, Chingmei village, 3 June 2016, *N. Odyuo 132814* (holotype, ASSAM; isotype, CAL).

Distribution. Nagaland [Myanmar].

Habit. Epiphytic herb.

Dendrobium hookerianum Lindl., J. Proc. Linn. Soc., Bot. 3: 8. 1859 [1858] [sect. *Dendrobium*]. *Callista hookeriana* (Lindl.) Kuntze, Revis. Gen. Pl. 2: 654. 1891. TYPE: India, Sikkim, *J. D. Hooker 8* (holotype, K-LINDL).

Dendrobium chrysotis Rchb. f., Gard. Chron. 1870: 1311. 1870. TYPE: India, Assam, *cult. Brooke s.n.* (holotype, K).

Distribution. Arunachal Pradesh, Assam, Manipur, Meghalaya, Mizoram, Nagaland, Sikkim, West Bengal [Bangladesh, Bhutan, Myanmar].

Habit. Epiphytic herb.

Dendrobium incurvum Lindl., J. Proc. Linn. Soc., Bot. 3: 18. 1859 [1858] [sect. *Stachyobium*]. *Callista incurva* (Lindl.) Kuntze, Revis. Gen. Pl. 2: 655. 1891. TYPE: Myanmar, Mergui [Myeik], 1844, *Griffith 808* (holotype, K-LINDL; isotype, K).

Dendrobium aclinia Lindl., J. Proc. Linn. Soc., Bot. 3: 9. 1859 [1858], nom. illeg., non Rchb. f., 1856. TYPE: Myanmar, Mergui [Myeik], 1844, *Griffith 809* (not found).

Distribution. Andaman & Nicobar Islands [Myanmar].

Habit. Epiphytic herb.

Dendrobium indragiriense Schltr., Repert. Spec. Nov. Regni Veg. 9: 164. 1911 [sect. *Grastidium*]. *Grastidium indragiriense* (Schltr.) Rauschert, Feddes Repert. 94: 450. 1983. TYPE: Indonesia, Sumatra, Sungei Lalah *R. Schlechter 13277* (holotype, B, lost; isotype, BO).

Distribution. Andaman & Nicobar Islands [also in Indonesia].

Habit. Epiphytic herb.

Dendrobium infundibulum Lindl., J. Proc. Linn. Soc., Bot. 3: 16. 1859 [1858] [sect. *Formosae*]. *Callista infundibulum* (Lindl.) Kuntze, Revis. Gen. Pl. 2: 655. 1891. TYPE: India, Moulmein [Maw-

lamyine], alt. 5000 ft on Thoung-gyun, *T. Lobb s.n.* (holotype, K-LINDL; isotype, K-LINDL).

Distribution. Assam, Manipur, Mizoram, Nagaland, Sikkim [Myanmar].

Habit. Epiphytic herb.

Dendrobium jenkinsii Wall. ex Lindl., Edwards's Bot. Reg. 25: t. 37. 1839 [unplaced]. *Dendrobium aggregatum* Roxb. var. *jenkinsii* (Wall. ex Lindl.) King & Pantl., Ann. Roy. Bot. Gard. (Calcutta) 8: 61. 1898. TYPE: India, Assam, Gualpara, Nov. 1836 *Jenkins s.n.* (holotype, K-LINDL).

Distribution. Arunachal Pradesh, Assam, Manipur, Meghalaya, Mizoram, Nagaland, Sikkim, West Bengal [Bhutan, China, Myanmar].

Habit. Epiphytic herb.

Dendrobium jerdonianum Wight, Icon. Pl. Ind. Orient. [Wight] 5(1): 6, t. 1644. 1851 [unplaced]. *Callista jerdoniana* (Wight) Kuntze, Revis. Gen. Pl. 2: 655. 1891. SYNTYPES: India, Karnataka, Coorg, Aug.–Sep., *Jerdon s.n.* (syntype, K); India, Tamil Nadu, Iyamally Hills, *Wight s.n.* (syntype, K).

Dendrobium villosulum Lindl., Paxton's Fl. Gard. 2: 82. 1851, nom. illeg., non Wall. ex Lindl., 1830. TYPE: India, Kerala, Kannur district, Tellicherry, *cult. Loddiges 355* (holotype, K-LINDL).

Distribution. Karnataka, Kerala. Endemic.

Habit. Epiphytic herb.

Dendrobium keithii Ridl., J. Linn. Soc., Bot. 32: 247. 1896 [sect. *Aporum*]. *Aporum keithii* (Ridl.) M. A. Clem., Telopea 10(1): 295. 2003. SYNTYPES: Thailand, Bangtaphan, *Keith Ridley s.n.* (syntype, BM); Pungah, *Curtis s.n.* (syntype, BM).

Distribution. Andaman & Nicobar Islands [also in Thailand].

Habit. Epiphytic herb.

Dendrobium kentrophyllum Hook. f., Fl. Brit. India [J. D. Hooker] 5(16): 725. 1890 [sect. *Aporum*]. *Callista kentrophylla* (Hook. f.) Kuntze, Revis. Gen. Pl. 2: 655. 1891. *Aporum kentrophyllum* (Hook. f.) Brieger, Orchideen (Schlechter) 1(11–12): 676. 1981. TYPE: Malaysia, Perak, G. Ijiak, Sep. 1854, *Scortechini s.n.* [*1247*] (holotype, K; isotype, CAL).

Distribution. ?Arunachal Pradesh, ?Assam, Manipur [also in Indonesia, Thailand].

Habit. Epiphytic herb.

Remarks. Singh et al. (2019) believe Indian records to have been misidentified, which is well possible.

Dendrobium khasianum Deori, J. Orchid Soc. India 2(1–2): 73. 1988 [sect. *Dendrobium*]. TYPE: India, Meghalaya, Khasi Hills, Pontong forest, *Deori 71823* (holotype, CAL; isotype, ASSAM).

Distribution. Assam, Meghalaya, Nagaland. Endemic.

Habit. Epiphytic herb.

Remarks. This looks intermediate between *Dendrobium fimbriatum* and *D. hookerianum*. If it is not one of these species it is possibly a natural hybrid between the two.

Dendrobium kratense Kerr, Bull. Misc. Inform. Kew 1927: 217. 1927 [sect. *Stachyobium*]. TYPE: Thailand, Khao Saming, Krat, 25 Nov. 1925, *Kerr 193* (holotype, K; isotype, BKF).

Distribution. Andaman & Nicobar Islands [also in Thailand].

Habit. Epiphytic herb.

Dendrobium lindleyi Steud., Nomencl. Bot., ed. 2 (Steudel) 1: 490. 1840 [unplaced]. *Dendrobium aggregatum* Roxb., Fl. Ind. ed. 1832, 3: 477. 1832, nom. illeg., non Kunth, 1815. *Callista aggregata* Kuntze, Revis. Gen. Pl. 2: 654. 1891. TYPE: Myanmar, Arracan, icon. *Dendrobium aggregatum, Roxburgh 1289 (leg. Pierard s.n.)* (syntypes, CAL, K).

Distribution. Arunachal Pradesh, Assam, Manipur, Meghalaya, Mizoram, Nagaland, Sikkim, Tripura, West Bengal [Bangladesh, China, Myanmar].

Habit. Epiphytic herb.

Dendrobium linguella Rchb. f., Gard. Chron. n.s., 18: 552. 1882 [sect. *Dendrobium*]. TYPE: Malay Archipelago, *cult. Veitch s.n.* (holotype, W).

Distribution. ?Manipur [also in Indonesia, Laos, Malaysia, Thailand, Vietnam].

Habit. Epiphytic herb.

Dendrobium lituiflorum Lindl., Gard. Chron. 1856: 372. 1856 [sect. *Dendrobium*]. *Callista lituiflora* (Lindl.) Kuntze, Revis. Gen. Pl. 2: 655. 1891. SYNTYPES: Unknown locality, *R. Hanbury s.n.*; *John Edwards s.n.* (syntype, K-LINDL).

Distribution. Arunachal Pradesh, Assam, Manipur, ?Meghalaya, Mizoram, Nagaland, Sikkim, West Bengal [Bangladesh, China, Myanmar].

Habit. Epiphytic herb.

Dendrobium longicornu Lindl., Edwards's Bot. Reg. 16: t. 1315. 1830 [sect. *Formosae*]. *Froscula hispida* Raf., Fl. Tellur. 4: 44. 1836 [1838]. *Callista longicornu* (Lindl.) Kuntze, Revis. Gen. Pl. 2: 655 (as "*longicornis*"). 1891. SYNTYPES: Nepal, Thoka, Oct. 1821, *Wallich 1997.1* (syntypes, C, K, K-LINDL, K-WALL, P); Bangladesh or India, Sylhet, Collangpaong, June 1823, *Wallich 1997.2 (leg. F. De Silva s.n.)* (syntypes, K, K-LINDL, K-WALL).

Dendrobium arunachalense C. Deori, S. K. Sarma, Phukan & A. A. Mao, J. Orchid Soc. India 20: 81. 2006 (as "*arunachalanse*"). TYPE: India, Arunachal Pradesh, Dibang Valley District, 1300 m, *Deori 101163* (holotype, CAL; isotype, ASSAM).

Distribution. Arunachal Pradesh, Manipur, Meghalaya, Mizoram, Nagaland, Sikkim, Uttarakhand, West Bengal [Bangladesh, Bhutan, China, Myanmar, Nepal].

Habit. Epiphytic herb.

Remarks. The taxonomy of *Dendrobium flexuosum*, *D. hirsutum*, and *D. longicornu* needs further study; these species are undoubtedly closely related. We follow here the unpublished thesis work of Sathapattayanon (2008).

Dendrobium macraei Lindl., Gen. Sp. Orchid. Pl.: 75. 1830 [sect. *Crinifera*]. *Callista macraei* (Lindl.) Kuntze, Revis. Gen. Pl. 2: 655. 1891. *Ephemerantha macraei* (Lindl.) P. F. Hunt & Summerh., Taxon 10: 105. 1961. *Flickingeria macraei* (Lindl.) Seidenf., Dansk Bot. Ark. 34(1): 39. 1980. TYPE: Sri Lanka, Peradenia, *Macrae 21* (holotype, K-LINDL; isotype, K).

Distribution. ?Goa, Karnataka, Odisha, Sikkim, West Bengal [Myanmar, Sri Lanka].

Habit. Epiphytic herb.

Dendrobium macrostachyum Lindl., Gen. Sp. Orchid. Pl.: 78. 1830 [sect. *Dendrobium*]. TYPE: Sri Lanka, *Macrae s.n.* [*17*] (holotype, K-LINDL).

Dendrobium gamblei King & Pantl., J. Asiat. Soc. Bengal, Pt. 2, Nat. Hist. 66: 584. 1898 [1897]. TYPE: India, Dehra Dun, July, *J. S. Gamble s.n.* (not found [a sheet at K is labeled "Dehra Dun, Paleo Hill, Malkot, 4000 ft, Feb. 1895, *J. S. Gamble 25575*"]).

Distribution. Andaman & Nicobar Islands, Goa, Jharkhand, Karnataka, Maharashtra, Odisha, Tamil Nadu, Uttarakhand [Myanmar, Sri Lanka].

Habit. Epiphytic herb.

Dendrobium mannii Ridl., J. Linn. Soc., Bot. 32: 246. 1896 [sect. *Aporum*]. *Aporum mannii* (Ridl.) Rauschert, Feddes Repert. 94(7–8): 440. 1983. SYNTYPES: India, Assam, *G. Mann s.n.* (syntype, BM); Malaysia, Mount Ophir, *Derry s.n.* (syntype, SING).

Distribution. Arunachal Pradesh, Assam, Manipur [also in Thailand].

Habit. Epiphytic herb.

Dendrobium metrium Kraenzl., Pflanzenr. (Engler) IV, 50 II B 21: 221. 1910 [sect. *Dendrobium*]. *Dendrobium modestum* Ridl., J. Bot. 36: 211. 1898, nom. illeg., non Rchb. f., 1855. TYPE: Malaysia, Penang Hill, Mar. 1896, *Ridley s.n. [7238]* (holotype, K).

Dendrobium sociale J. J. Sm., Bull. Jard. Bot. Buitenzorg, sér. 2, 3: 61. 1912. TYPE: Indonesia, Sumatra, Proeba Toea, *Hagen s.n.* (holotype, BO).

Distribution. Nagaland [also in Cambodia, Indonesia, Laos, Thailand, Vietnam].

Habit. Terrestrial or epiphytic herb.

Remarks. The specimens recorded from Nagaland by Deori et al. (2007, as *D. sociale*) differ from typical *D. metrium* in the greenish instead of white main color of the flowers and, judging from their illustrations, in that the mentum lacks the closed, spur-like extension that is present in *D. metrium*. They probably represent a different taxon, in which case *D. metrium* is not known to occur in India. This species is usually terrestrial but the specimens from Nagaland were said to be epiphytic. *Dendrobium metrium* does not belong in section *Pedilonum*, as often assumed, but in section *Dendrobium*, as confirmed by molecular data (Pérez Escobar, pers. comm.).

Dendrobium microbulbon A. Rich., Ann. Sci. Nat., Bot. sér. 2, 15: 19, t. 8. 1841 (as "*microbolbon*") [sect. *Stachyobium*]. *Callista microbulbon* (A. Rich.) Kuntze, Revis. Gen. Pl. 2: 655. 1891. TYPE: India, Nilgiri Hills, Otacamund, July–Sep., *Perrottet s.n.* (syntype, W).

Distribution. Goa, Gujarat, Karnataka, Kerala, Maharashtra, Tamil Nadu. Endemic.

Habit. Epiphytic herb.

Dendrobium miserum Rchb. f., Gard. Chron. 1869: 388. 1869 [sect. *Stachyobium*]. *Callista misera* (Rchb. f.) Kuntze, Revis. Gen. Pl. 2: 655. 1891. TYPE: India, Assam, *J. Day s.n.* (holotype, W).

Distribution. Assam, Meghalaya. Endemic.

Habit. Epiphytic herb.

Dendrobium moniliforme (L.) Sw., Nova Acta Regiae Soc. Sci. Upsal. 6: 85. 1799 [sect. *Dendrobium*]. *Epidendrum moniliforme* L., Sp. Pl. 2: 954. 1753. *Epidendrum monile* Thunb., Fl. Jap.: 30. 1784, nom. superfl. *Limodorum monile* Thunb., Trans. Linn. Soc. London 2: 327. 1794, nom. superfl. *Ormostema albiflora* Raf., Fl. Tellur. 4: 38. 1836 [1838], nom. superfl. *Callista moniliformis* (L.) Kuntze, Revis. Gen. Pl. 2: 655. 1891. *Dendrobium monile* (Thunb.) Kraenzl., Pflanzenr. (Engler) IV, 50 II B 21: 50. 1910, nom. superfl. TYPE: Japan, icon. "Fu Ran" in Kaempfer, Amoen. Exot. Fasc., 864, 865. 1712 (lectotype, designated by Merrill, 1938).

Dendrobium candidum Wall. ex Lindl., Edwards's Bot. Reg., Misc.: 36. 1838. *Callista candida* (Wall. ex Lindl.) Kuntze, Revis. Gen. Pl. 2: 654. 1891. TYPE: India, Meghalaya, Khasi Hills, Nungclow, *cult. Duke of Devonshire (leg. Gibson) s.n.* (holotype, K-LINDL).
Dendrobium spathaceum Lindl., J. Proc. Linn. Soc., Bot. 3: 15. 1859 [1858]. *Callista spathacea* (Lindl.) Kuntze, Revis. Gen. Pl. 2: 655. 1891. TYPE: India, Sikkim, Lachen, 6000–7000 ft, 26 May 1849, *J. D. Hooker 143* (holotype, K-LINDL; isotype, K).

Distribution. Arunachal Pradesh, Assam, Manipur, Meghalaya, Nagaland, Sikkim, Uttarakhand, West Bengal [Bhutan, China, Myanmar, Nepal].

Habit. Epiphytic herb.

Dendrobium monticola P. F. Hunt & Summerh., Taxon 10(4): 110. 1961 [sect. *Stachyobium*]. *Dendrobium alpestre* Royle, Ill. Bot. Himal. Mts. [Royle] 1: 370, t. 88, fig. 2. 1839, nom. illeg., non (Sw.) Sw., 1799. *Callista alpestris* Kuntze, Revis. Gen. Pl. 2: 654. 1891. *Dendrobium roylei* A. D. Hawkes & A. H. Heller, Orquídea (Rio de Janeiro) 24: 114. 1962, nom. superfl. TYPE: India, Uttarakhand, Gharwal, *Royle 1006* (holotype, K).

Distribution. Arunachal Pradesh, Uttarakhand [China, Nepal].

Habit. Epiphytic herb.

Dendrobium moschatum (Banks) Sw., Neues J. Bot. 1: 94. 1805 [sect. *Dendrobium*]. *Epidendrum moschatum* Banks, Account Embassy Kingd. Ava: 478. 1800. *Cymbidium moschatum* (Banks) Willd.,

Sp. Pl., ed. 4 [Willdenow] 4(1): 98. 1805. *Thicuania moschata* (Banks) Raf., Fl. Tellur. 4: 47. 1836 [1838]. *Callista moschata* (Banks) Kuntze, Revis. Gen. Pl. 2: 655. 1891. TYPE: Myanmar, Rangoon, *Buchanan-Hamilton s.n.* (holotype, BM).

Dendrobium calceolaria Carey ex Hook., Exot. Fl. 3: t. 184. 1825. TYPE: Without provenance, *cult. Shepherd (leg. Carey, 1820) s.n.* (holotype, K-LINDL).
Dendrobium moschatum (Banks) Sw. var. *unguipetalum* I. Barua, Orchid Fl. Kamrup Distr. Assam: 160. 2001. TYPE: India, Assam, Loharghat, Kamrup District, *Barua 2285* (holotype, CAL; isotype, ASSAM).

Distribution. Andhra Pradesh, Arunachal Pradesh, Assam, Bihar, Jharkhand, Manipur, Meghalaya, Mizoram, Nagaland, Odisha, Sikkim, Tripura, Uttarakhand, West Bengal [Bangladesh, Bhutan, China, Myanmar, Nepal].

Habit. Epiphytic herb.

Dendrobium nanum Hook. f., Hooker's Icon. Pl. 19: t. 1853. 1889 [sect. *Stachyobium*]. *Callista nana* (Hook. f.) Kuntze, Revis. Gen. Pl. 2: 655. 1891. TYPE: India, Malabar, Bababoodan [Baba Budan] Hills, *J. S. Law s.n.* (holotype, K; isotype, CAL).

?*Dendrobium mabeliae* Gammie, J. Bombay Nat. Hist. Soc. 16: 567. 1905 (as "*mabelae*"). TYPE: India, Karnataka, Belgaum Ghats, Oct., *Gammie s.n.*

Distribution. Goa, Karnataka, Kerala, Maharashtra, Tamil Nadu. Endemic.

Habit. Epiphytic herb.

Remarks. Dendrobium mabeliae may differ from *D. nanum* in the distinctly fimbriate-pectinate (vs. denticulate) lip mid-lobe margin and the thin-textured (vs. fleshy) mid-lobe, but more material must be examined.

Dendrobium nareshbahadurii H. B. Naithani, Indian Forester 112: 66. 1986 [sect. *Stachyobium*]. TYPE: India, Arunachal Pradesh, Kameng District, Jameri, *K. N. Bahadur & H. B. Naithani ser II 160* (holotype, DD).

Distribution. Arunachal Pradesh. Endemic.

Habit. Epiphytic herb.

Dendrobium nathanielis Rchb. f., Cat. Orch.-Samml. Schiller, ed. 3. 26. 1857 [sect. *Aporum*]. *Callista nathanielis* (Rchb. f.) Kuntze, Revis. Gen. Pl. 2: 655. 1891. *Aporum nathanielis* (Rchb. f.) M. A. Clem., Telopea 10(1): 296. 2003. TYPE: Myanmar, *cult. Schiller s.n.* (holotype, W).

Aporum cuspidatum Wall. ex Lindl., Edwards's Bot. Reg. 27(Misc.): 2. 1841. *Dendrobium cuspidatum* (Wall. ex Lindl.) Lindl., J. Proc. Linn. Soc., Bot. 3: 4. 1859 [1858], nom. illeg., non Lindl., 1830. TYPE: *cult. Loddiges (leg. Wallich) s.n.*

Distribution. Assam [Myanmar].

Habit. Epiphytic herb.

Dendrobium nobile Lindl., Gen. Sp. Orchid. Pl.: 79. 1830 [sect. *Dendrobium*]. *Callista nobilis* (Lindl.) Kuntze, Revis. Gen. Pl. 2: 655. 1891. TYPE: China, icon. *Reeves s.n.* (holotype, Royal Horticultural Society 30; isotype, BM 890).

Dendrobium coerulescens Wall. ex Lindl., Sert. Orchid. 3: t. 18. 1838. TYPE: India, Khasi Hills, Nungclow, *cult. Gibson s.n.* (holotype, K-LINDL).
Dendrobium lindleyanum Griff., Not. Pl. Asiat. 3: 309. 1851. TYPE: India, Assam, Suddhya, 14 Apr. 1836, *Griffith s.n.* (not found).

Distribution. Arunachal Pradesh, Assam, Manipur, Meghalaya, Mizoram, Nagaland, ?Odisha, Sikkim, West Bengal [Bhutan, China, Myanmar, Nepal].

Habit. Epiphytic herb.

Dendrobium nodosum Dalzell, Hooker's J. Bot. Kew Gard. Misc. 4: 292. 1852 [sect. *Crinifera*]. *Desmotrichum nodosum* (Dalzell) Tang & F. T. Wang, Acta Phytotax. Sin. 1: 83. 1951. *Flickingeria nodosa* (Dalzell) Seidenf., Dansk Bot. Ark. 34(1): 41. 1980. TYPE: India, Deccan, Ram Ghat, *J. E. Stocks 30* (holotype, K; isotype, K-LINDL).

Distribution. Goa, Gujarat, Karnataka, Kerala, Maharashtra, Tamil Nadu. Endemic.

Habit. Epiphytic herb.

Dendrobium numaldeorii C. Deori, Hynn. & Phukan, Orchid Rev. 112(1259): 279. 2004 [sect. *Stachyobium*]. TYPE: India, Arunachal Pradesh, Dibang Valley District, Mehao WLS, 1300 m, 10 Oct. 2003, *C. Deori 101122* (syntypes, ASSAM, CAL).

Distribution. Arunachal Pradesh, Nagaland [Myanmar].

Habit. Epiphytic herb.

Dendrobium nutantiflorum A. D. Hawkes & A. H. Heller, Lloydia 20(2): 122. 1957 [unplaced]. *Dendrobium nutans* Lindl., Gen. Sp. Orchid. Pl.: 90. 1830, nom. illeg., non C. Presl, 1827. *Callista nutans* (Lindl.) Kuntze, Revis. Gen. Pl. 2: 655. 1891. TYPE: Sri Lanka, Peradeniya, *Macrae s.n.* (holotype, K-LINDL).

Dendrobium nutans Lindl. var. *rubrilabre* Blatt. ex C. E. C. Fisch., Fl. Madras 3(8): 1416. 1928 (as "*rubrilaba*"). TYPE: India, Tamil Nadu, Madura Dist., High Wavy Mountain, May 1917, *Blatter & Hallberg s.n.* [*335*] (holotype, K).

Distribution. Kerala, Tamil Nadu [Sri Lanka].

Habit. Epiphytic herb.

Dendrobium ochreatum Lindl., Edwards's Bot. Reg. 21: t. 1756. 1835 [sect. *Dendrobium*]. *Callista ochreata* (Lindl.) Kuntze, Revis. Gen. Pl. 2: 655. 1891. TYPE: Bangladesh, Chittagong, Feb. 1828, *Wallich 7410 (leg. H. Bruce)* (holotype, K; isotype, K-WALL).

Dendrobium cambridgeanum Paxton, Paxton's Mag. Bot. 6: 265. 1839. TYPE: India, Meghalaya, Khasi Hills, *Gibson s.n.*

Distribution. Assam, Manipur, Meghalaya, Mizoram, Nagaland [Bangladesh, Myanmar, Nepal].

Habit. Epiphytic herb.

Dendrobium ovatum (L.) Kraenzl., Pflanzenr. (Engler) IV, 50 (Heft 45): 71. 1910 [sect. *Fytchianthe*]. *Epidendrum ovatum* L., Sp. Pl. 2: 952. 1753. *Cymbidium ovatum* (L.) Willd., Sp. Pl., ed. 4. 4: 101. 1805. *Callista ovata* (L.) Kuntze, Revis. Gen. Pl. 2: 653. 1891. TYPE: India, icon. "Anantalimaravara" in Rheede, Hort. Malab. 12: 15, t. 7 (lectotype, designated by Cribb, 1999). EPITYPE: India, Malabar, *Stocks & Law s.n.* (epitype, K, designated by Cribb in Cafferty & Jarvis (eds.), 1999).

Dendrobium chlorops Lindl., Edwards's Bot. Reg. 30(Misc.): 44. 1844. TYPE: India, Maharashtra, *cult. Loddiges s.n.* (holotype, K-LINDL).

Distribution. Andhra Pradesh, Dadra & Nagar Haveli, Goa, Gujarat, Karnataka, Kerala, Maharashtra, Tamil Nadu. Endemic.

Habit. Epiphytic or lithophytic herb.

Dendrobium pachyphyllum (Kuntze) Bakh. f., Blumea 12: 69. 1963 [sect. *Aporum*]. *Dendrobium carnosum* Teijsm. & Binn., Natuurk. Tijdschr. Ned.-Indië 5: 489. 1853, nom. illeg., non C. Presl, 1832. *Callista pachyphylla* Kuntze, Revis. Gen. Pl. 2: 654. 1891. TYPE: Indonesia, Java, Mt. Salak, *cult. Bogor s.n.* (not found).

Dendrobium pumilum Roxb., Fl. Ind. ed. 1832, 3: 479. 1832, nom. illeg., non Sw., 1805. *Callista pumila* Kuntze, Revis. Gen. Pl. 2: 655. 1891. *Polystachya pumila* (Kuntze) Kraenzl., Repert. Spec. Nov. Regni Veg. Beih. 39: 126. 1926. *Desmotrichum pumilum* (Kuntze) A. D. Hawkes,

Lloydia 20: 126. 1957. *Flickingeria pumila* (Kuntze) A. D. Hawkes, Orchid Weekly 2: 458. 1961. *Bolbodium pumilum* (Kuntze) Brieger, Orchideen (Schlechter) 1(11–12): 721. 1981 (as "*Bolbodium*"). TYPE: Bangladesh, Chittagong, cult. Calcutta s.n., icon. *Dendrobium pumilum*, *Roxburgh 2358* (syntypes, CAL, K).

Desmotrichum pusillum Blume, Bijdr. Fl. Ned. Ind. 7: 331. 1825. *Dendrobium pusillum* (Blume) Lindl., Gen. Sp. Orchid. Pl.: 77. 1830, nom. illeg., non Kunth, 1816, non D. Don, 1825. *Callista pusilla* (Blume) Kuntze, Revis. Gen. Pl. 2: 655. 1891. *Dendrobium perpusillum* N. P. Balakr., J. Bombay Nat. Hist. Soc. 67: 66. 1970. *Bolbodium pusillum* (Blume) Rauschert, Feddes Repert. 94: 443. 1983 (as "*Bolbidium*"). TYPE: Indonesia, Java, Nusa Kambangan, *Blume 1706* (holotype, L).

Distribution. Assam [Myanmar].

Habit. Epiphytic herb.

Dendrobium panduratum Lindl., J. Proc. Linn. Soc., Bot. 3: 19. 1859 [1858] [sect. *Stachyobium*]. TYPE: Sri Lanka, *Thwaites 2353* (lectotype, K, designated by Priyadarshana et al., 2020; isotype, CAL).

Dendrobium panduratum Lindl. subsp. *villosum* Gopalan & A. N. Henry, J. Bombay Nat. Hist. Soc. 87(1): 128. 1990. TYPE: India, Tamil Nadu, Triunelveli Kattabomman District, Kannikatty, Inchikkuzhi, *Gopalan 88699* (holotype, CAL).

Distribution. Kerala, Tamil Nadu [Sri Lanka].

Habit. Epiphytic herb.

Dendrobium parciflorum Rchb. f. ex Lindl., J. Proc. Linn. Soc., Bot. 3: 4. 1859 [1858] [sect. *Aporum*]. *Aporum jenkinsii* Griff., Calcutta J. Nat. Hist. 5: 367. 1845. *Callista jenkinsii* (Griff.) Kuntze, Revis. Gen. Pl. 2: 654. 1891. TYPE: India, Assam, *Jenkins s.n.* (holotype, K).

Dendrobium perula Rchb. f., Hamburger Garten-Blumenzeitung 21: 298. 1865. *Callista perula* (Rchb. f.) Kuntze, Revis. Gen. Pl. 2: 655. 1891. *Eurycaulis perula* (Rchb. f.) M. A. Clem., Telopea 10(1): 287. 2003. TYPE: India, Assam, *Day s.n.* (holotype, W).

Distribution. Arunachal Pradesh, Assam, ?Meghalaya, Sikkim, West Bengal [China].

Habit. Epiphytic herb.

Remarks. This is not *Dendrobium jenkinsii* Wall. ex Lindl.

Dendrobium parcum Rchb. f., Gard. Chron. 1866: 1042. 1866 [sect. *Herbacea*]. *Callista parca* (Rchb. f.) Kuntze, Revis. Gen. Pl. 2: 655. 1891. TYPE: Myanmar, *cult. Low (leg. Parish) s.n.* (holotype, W; isotype, ?K).

Dendrobium hexadesmia Rchb. f., Gard. Chron. 1869: 710. 1869. TYPE: *cult. Veitch s.n.* (holotype, W).

Dendrobium listeroglossum Kraenzl. in H. G. Reichenbach, Xenia Orchid. 3: 108. 1892. TYPE: Provenance unknown, *cult. Eisgrub (leg. Sander) s.n.*

Distribution.　Manipur, Mizoram [Myanmar].

Habit.　Epiphytic herb.

Dendrobium parishii Rchb. f., Bot. Zeitung (Berlin) 21(31): 237. 1863 [sect. *Dendrobium*]. *Callista parishii* (Rchb. f.) Kuntze, Revis. Gen. Pl. 2: 655. 1891. TYPE: Myanmar, Moulmein [Mawlamyine], *Parish s.n.* [*18*] (holotype, W; isotype, K).

Distribution.　Arunachal Pradesh, Assam, Manipur, Mizoram, Nagaland, Tripura [Bangladesh, China, Myanmar].

Habit.　Epiphytic herb.

Dendrobium pendulum Roxb., Fl. Ind. ed. 1832, 3: 484. 1832 [sect. *Dendrobium*]. *Callista pendula* (Roxb.) Kuntze, Revis. Gen. Pl. 2: 655. 1891. TYPE: Bangladesh, Chittagong, *cult. Calcutta s.n.* (not found).

Distribution.　Arunachal Pradesh, Assam, Manipur, Mizoram, Nagaland [Bangladesh, China, Myanmar].

Habit.　Epiphytic herb.

Dendrobium pensile Ridl., J. Linn. Soc., Bot. 32: 253. 1896 [sect. *Grastidium*]. *Grastidium pensile* (Ridl.) Rauschert, Feddes Repert. 94: 451. 1983. SYNTYPES: Singapore, Selitar, *Ridley s.n.* (syntype, SING); Malaysia, Johore, *Ridley's collector s.n.* (syntype, SING); Riau ("Rhio"), *Ridley's collector s.n.* (syntype, SING).

Distribution.　Andaman & Nicobar Islands [also in Indonesia].

Habit.　Epiphytic herb.

Dendrobium plicatile Lindl., Edwards's Bot. Reg. 26(Misc.): 10. 1840 [sect. *Crinifera*]. *Desmotrichum fimbriatum* Blume, Bijdr. Fl. Ned. Ind. 7: 329. 1825. *Dendrobium fimbriatum* (Blume) Lindl., Gen. Sp. Orchid. Pl.: 76. 1830, nom. illeg., non Hook., 1823. *Dendrobium flabellum* Rchb. f., Bonplandia (Hannover) 5: 56. 1857, nom. superfl. *Callista flabella* (Rchb. f.) Kuntze, Revis. Gen. Pl. 2: 654. 1891, nom. superfl. *Flickingeria fimbriata* (Blume) A. D. Hawkes, Orchid Weekly 2(46): 454. 1961. *Ephemerantha fimbriata* (Blume) P. F. Hunt & Summerh., Taxon 10: 103. 1961. SYNTYPES: Indonesia, Java, Mt. Panjar, *Blume s.n.*

(syntype, L); Batu Tulis, *Blume s.n.* (syntype, L); s. loc., *Blume s.n.* (syntype, L).

Dendrobium rabanii Lindl., J. Proc. Linn. Soc., Bot. 3: 7. 1859 [1858]. *Flickingeria rabanii* (Lindl.) Seidenf., Dansk Bot. Ark. 34, 1: 43. 1980. TYPE: India, Meghalaya, Khasi Hills, *cult. Raban (Hooker f. 25) s.n.* (holotype, K-LINDL; isotype, K).

Dendrobium kunstleri Hook. f., Fl. Brit. India [J. D. Hooker] 5(16): 714. 1890. *Callista kunstleri* (Hook. f.) Kuntze, Revis. Gen. Pl. 2: 655. 1891. *Desmotrichum kunstleri* (Hook. f.) Kraenzl., Pflanzenr. (Engler) IV, 50 II B 21: 357. 1910. *Flickingeria kunstleri* (Hook. f.) A. D. Hawkes, Orchid Weekly 2(46): 456. 1961. *Ephemerantha kunstleri* (Hook. f.) P. F. Hunt & Summerh., Taxon 10: 104. 1961. SYNTYPES: Malaysia, Perak, *Scortechini 253b* (syntypes, CAL, K, G); *King's collector 1877* (syntype, K); *King's collector 6897* (syntypes, K, L).

Distribution.　Andaman & Nicobar Islands, ? Assam [China, Nepal].

Habit.　Epiphytic herb.

Dendrobium polyanthum Wall. ex Lindl., Gen. Sp. Orchid. Pl.: 81. 1830 [sect. *Dendrobium*]. TYPE: Myanmar, Moulmein [Mawlamyine], 1827, *Wallich 2009* (holotype, K-WALL).

Dendrobium cretaceum Lindl., Edwards's Bot. Reg. 33: t. 62. 1847. SYNTYPES: Myanmar, Moulmein [Mawlamyine], *cult. Veitch (leg. T. Lobb) s.n.*; India, Meghalaya, Khasi Hills, *Griffith s.n.* (syntype, K-LINDL); Myanmar, Mergui [Myeik], *s. coll., s.n.*

Dendrobium primulinum Lindl., Gard. Chron. 1858: 400. 1858. *Callista primulina* (Lindl.) Kuntze, Revis. Gen. Pl. 2: 655. 1891. SYNTYPES: India, Sikkim, *J. D. Hooker s.n.*; *cult. Warner s.n.* (syntype, K-LINDL).

Dendrobium nobile Lindl. var. *pallidiflorum* Hook., Bot. Mag. 83: t. 5003. 1857. SYNTYPES: *cult. Parker s.n.* (not found); *cult. van Houtte s.n.* (not found).

Distribution.　Andaman & Nicobar Islands, Arunachal Pradesh, Assam, Manipur, Meghalaya, Mizoram, Nagaland, Sikkim, Uttar Pradesh, Uttarakhand [China, Myanmar, Nepal].

Habit.　Epiphytic herb.

Dendrobium porphyrochilum Lindl., J. Proc. Linn. Soc., Bot. 3: 18. 1859 [1858] [sect. *Stachyobium*]. *Callista porphyrochila* (Lindl.) Kuntze, Revis. Gen. Pl. 2: 655. 1891. SYNTYPES: India, Assam, *Griffith s.n.* (syntype, K-LINDL); India, Meghalaya, Khasi Hills, *T. Lobb s.n.* (syntype, K-LINDL); India, Meghalaya, Khasi Hills, *Hooker & Thomson 28* (syntype, K-LINDL).

Dendrobium caespitosum King & Pantl., J. Asiat. Soc. Bengal, Pt. 2, Nat. Hist. 64: 332. 1895 [1896] (as "*coespitosum*"). TYPE: India, Sikkim, Naru Valley, 6000 ft, June, *Pantling s.n.* [*320*] (holotype, K).

Distribution.   Arunachal Pradesh, Assam, Manipur, Meghalaya, Nagaland, Sikkim, West Bengal [China, Myanmar, Nepal].

Habit.   Epiphytic herb.

Dendrobium praecinctum Rchb. f., Gard. Chron., n.s., 7: 750. 1877 [sect. *Dendrobium*]. *Callista praecincta* (Rchb. f.) Kuntze, Revis. Gen. Pl. 2: 655. 1891. TYPE: Provenance unknown, *cult. Veitch s.n.* (holotype, W, Herb. No. 39641).

Dendrobium pauciflorum King & Pantl., J. Asiat. Soc. Bengal, Pt. 2, Nat. Hist. 64: 332. 1895 [1896]. *Dendrobium sikkimense* A. D. Hawkes & A. H. Heller, Lloydia 20: 124. 1957. TYPE: India, Sikkim, above Engo, alt. 4000 ft, June 1891, *Pantling s.n.* [*172*, ?also Namgah, 4000 ft, July 1895] (holotype, CAL; isotype, BM, K).

Distribution.   Arunachal Pradesh, Assam, ?Meghalaya, Sikkim, West Bengal [Bhutan, China, ?Myanmar].

Habit.   Epiphytic herb.

Dendrobium pulchellum Roxb. ex Lindl., Gen. Sp. Orchid. Pl.: 82. 1830 [sect. *Dendrobium*]. *Callista pulchella* (Roxb. ex Lindl.) Kuntze, Revis. Gen. Pl. 2: 655. 1891. TYPE: India, Sylhet, icon. *Dendrobium pulchellum*, *Roxburgh 2356* (syntypes, CAL, K).

Dendrobium dalhousieanum Wall. ex Paxton, Paxton's Mag. Bot. 11: 145. 1844. TYPE: India, *cult. Calcutta (leg. Gibson) s.n.*

Distribution.   Arunachal Pradesh, Assam, Manipur, ?Meghalaya, Mizoram, Nagaland, Sikkim, West Bengal [Bangladesh, China, Myanmar, Nepal].

Habit.   Epiphytic herb.

Dendrobium pycnostachyum Lindl., J. Proc. Linn. Soc., Bot. 3: 19. 1859 [1858] [sect. *Stachyobium*]. *Callista pycnostachya* (Lindl.) Kuntze, Revis. Gen. Pl. 2: 655. 1891. TYPE: Myanmar, Moulmein [Mawlamyine], *T. Lobb s.n.* (not found).

Distribution.   Assam, ?Mizoram [Myanmar].

Habit.   Epiphytic herb.

Dendrobium regium Prain, J. Asiat. Soc. Bengal, Pt. 2, Nat. Hist. 71: 80. 1902 [sect. *Dendrobium*]. TYPE: India, 1901, *cult. Calcutta s.n.* (not found).

Distribution.   Andhra Pradesh, Bihar, Chhattisgarh, Jharkhand, Madhya Pradesh, Nagaland, Odisha. Endemic.

Habit.   Epiphytic herb.

Dendrobium ritaeanum King & Pantl., J. Asiat. Soc. Bengal, Pt. 2, Nat. Hist. 66: 583. 1898 [1897] [sect. *Crinifera*]. *Desmotrichum ritaeanum* (King & Pantl.) Kraenzl., Pflanzenr. (Engler) IV, 50 (II B 21): 351. 1910. *Flickingeria ritaeana* (King & Pantl.) A. D. Hawkes, Orchid Weekly 2(46): 459. 1961. *Ephemerantha ritaeana* (King & Pantl.) P. F. Hunt & Summerh., Taxon 10(4): 106. 1961. TYPE: India, Meghalaya, Khasi Hills, alt. 3000–4000 ft, *Rita s.n.* (not found [two sheets at G are labeled as isotypes; however, these were collected in May 1899 at Jowai in the Jaintia Hills by Dr. Prain's collector]).

Distribution.   Arunachal Pradesh, Assam, Meghalaya [also in Vietnam].

Habit.   Epiphytic herb.

Dendrobium rotundatum (Lindl.) Hook. f., Fl. Brit. India [J. D. Hooker] 5(16): 712. 1890 [sect. *Sarcopodium*]. *Sarcopodium rotundatum* Lindl., Fol. Orchid. 2 (*Sarcopodium*): 2. 1853. *Bulbophyllum rotundatum* (Lindl.) Rchb. f., Ann. Bot. Syst. (Walpers) 6: 244. 1861. *Callista rotundata* (Lindl.) Kuntze, Revis. Gen. Pl. 2: 655. 1891. *Katherinea rotundata* (Lindl.) A. D. Hawkes, Lloydia 19(2): 97. 1956. *Epigeneium rotundatum* (Lindl.) Summerh., Kew Bull. 12: 264. 1957. TYPE: India, Sikkim, Libong, Apr. 1850, *J. D. Hooker s.n.* (holotype, K-LINDL; isotype, CAL, K [2×]).

Distribution.   Arunachal Pradesh, Assam, Meghalaya, Nagaland, Sikkim, West Bengal [Bhutan, China, Myanmar, Nepal].

Habit.   Epiphytic herb.

Dendrobium ruckeri Lindl., Edwards's Bot. Reg. 29(Misc.): 25. 1843 [sect. *Dendrobium*]. *Callista ruckeri* (Lindl.) Kuntze, Revis. Gen. Pl. 2: 655. 1891. TYPE: "Philippines, Manila," *cult. Rucker s.n. (leg. Cuming)* (holotype, K-LINDL) [This species does not occur in the Philippines].

Dendrobium ramosum Lindl., Gen. Sp. Orchid. Pl.: 82. 1830, nom. illeg., non (Ruiz & Pav.) Pers., 1807. TYPE: India, Pundua, Mar. 1824, *Wallich 2003 (leg. F. De Silva 1340)* (holotype, K-WALL; isotype, K).

Distribution.   Arunachal Pradesh, Assam, Bihar, Meghalaya, Mizoram, West Bengal [Bangladesh, Myanmar].

Habit.   Epiphytic herb.

Dendrobium salaccense (Blume) Lindl., Gen. Sp. Orchid. Pl.: 86. 1830 [sect. *Grastidium*]. *Grastidium salaccense* Blume, Bijdr. Fl. Ned. Ind. 7: 333.

1825. *Callista salaccensis* (Blume) Kuntze, Revis. Gen. Pl. 2: 655. 1891. TYPE: Indonesia, Java, Mt. Salak, *Blume s.n.* (holotype, L; isotype, AMES, L).

Dendrobium cathcartii Hook. f., Fl. Brit. India [J. D. Hooker] 5(16): 727. 1890. *Callista cathcartii* (Hook. f.) Kuntze, Revis. Gen. Pl. 2: 654. 1891. *Grastidium cathcartii* (Hook. f.) M. A. Clem. & D. L. Jones, Lasianthera 1(2): 67. 1997. TYPE: India, Sikkim, icon. *Cathcart* (holotype, CAL; isotype, K).
Dendrobium haemoglossum Thwaites, Enum. Pl. Zeyl. [Thwaites]: 429. 1864. *Callista haemoglossa* (Thwaites) Kuntze, Revis. Gen. Pl. 2: 654. 1891. *Grastidium haemoglossum* (Thwaites) M. A. Clem. & D. L. Jones, Lasianthera 1(2): 78. 1997. TYPE: Sri Lanka, Dolosbagey Dist., Matelle, *Gardner s.n.*(incl. icon., not found) [*C.V. 3842*] (syntypes, K [2×]).
Dendrobium bambusifolium C. S. P. Parish & Rchb. f., Trans. Linn. Soc. London 30: 149. 1874 (as *"bambusaefolium"*). TYPE: Myanmar, Moulmein [Mawlamyine], *C. S. P. Parish s.n.* [*188*] (holotype, W; isotype, K).

Distribution. Andaman & Nicobar Islands, Arunachal Pradesh, Assam, Karnataka, Kerala, Meghalaya, Mizoram, Odisha, Sikkim, Tamil Nadu, Tripura, West Bengal [Bhutan, China, Myanmar, Sri Lanka].

Habit. Epiphytic herb.

Dendrobium secundum (Blume) Lindl., Edwards's Bot. Reg. 15: t. 1291. 1829 [sect. *Pedilonum*]. *Pedilonum secundum* Blume, Bijdr. Fl. Ned. Ind. 7: 322. 1825. *Callista secunda* (Blume) Kuntze, Revis. Gen. Pl. 2: 653. 1891. TYPE: Indonesia, Java, Tjikao [Cikao], collector not named (syntype, L).

Distribution. Andaman & Nicobar Islands [China, Myanmar].

Habit. Epiphytic herb.

Dendrobium sessanicum Apang, J. Econ. Taxon. Bot. 36(2): 372. 2012 [sect. *Stachyobium*]. TYPE: India, Arunachal Pradesh, West Kameng district, Sessa Orchid Sanctuary, *Ona Apang 37576* (holotype, Orchid Herbarium, Tipi).

Distribution. Arunachal Pradesh. Endemic.

Habit. Epiphytic herb.

Dendrobium shompenii B. K. Sinha & P. S. N. Rao, Nordic J. Bot. 18(1): 27. 1998 [sect. *Aporum*]. *Aporum shompenii* (B. K. Sinha & P. S. N. Rao) M. A. Clem., Telopea 10(1): 297. 2003. TYPE: India, Andaman & Nicobar Islands, Great Nicobar Island, Galathea river, 5 Apr. 1995, *B. K. Sinha & V. Maina 16516* (holotype, CAL).

Distribution. Andaman & Nicobar Islands. Endemic.

Habit. Epiphytic herb.

Dendrobium simplicissimum (Lour.) Kraenzl., in Engl., Pflanzenr. IV. 50. II. B. 21: 235. 1910 [sect. *Aporum*]. *Ceraia simplicissima* Lour., Fl. Cochinch. 2: 518. 1790. *Dendrobium ceraia* Lindl., Gen. Sp. Orchid. Pl.: 89. 1830, nom. superfl. TYPE: *Loureiro s.n.* (holotype, BM).

Dendrobium angulatum Lindl., Gen. Sp. Orchid. Pl.: (1830) 88, nom. illeg., non (Blume) Lindl., 1830. *Dendrobium podagraria* Hook. f., Fl. Brit. India [J. D. Hooker] 5(16): 728. 1890. TYPE: Myanmar, Attran [Ataran] River, 27 Mar. 1817, *Wallich 2010* (syntypes, K-LINDL, K-WALL).

Distribution. Arunachal Pradesh, Assam, Manipur, ?Meghalaya, Mizoram, Tripura [Myanmar].

Habit. Epiphytic herb.

Dendrobium sinominutiflorum S. C. Chen, J. J. Wood & H. P. Wood, Fl. China 25: 394. 2009 [sect. *Stachyobium*]. *Dendrobium minutiflorum* S. C. Chen & Z. H. Tsi, Bull. Bot. Res., Harbin 9 (2): 27. 1989, nom. illeg., non Kraenzlin, 1914, non Gagnep., 1950. TYPE: China, Yunnan, Menghai Xian, *Tsi 23* (holotype, PE).

Distribution. Manipur [China, Myanmar].

Habit. Epiphytic herb.

Dendrobium spatella Rchb. f., Hamburger Garten-Blumenzeitung 21: 298. 1865 [sect. *Aporum*]. *Callista spatella* (Rchb. f.) Kuntze, Revis. Gen. Pl. 2: 655. 1891. *Aporum spatella* (Rchb. f.) M. A. Clem., Telopea 10(1): 297. 2003. TYPE: India, Assam, *cult. J. Day s.n.* (holotype, W).

Distribution. Andaman & Nicobar Islands, Arunachal Pradesh, Assam, Manipur, Meghalaya, Mizoram, Nagaland [Bhutan, China, Myanmar].

Habit. Epiphytic herb.

Remarks. This is often misidentified as *Dendrobium acinaciforme* Roxb., a species from eastern Indonesia (Moluccas).

Dendrobium stuposum Lindl., Edwards's Bot. Reg. 24 (Misc.): 52. 1838 [sect. *Dendrobium*]. *Callista stuposa* (Lindl.) Kuntze, Revis. Gen. Pl. 2: 653. 1891. TYPE: India, *cult. Loddiges s.n.* (holotype, K-LINDL).

Dendrobium sphegidoglossum Rchb. f., Bonplandia 2(7): 88. 1854 (as *"sphegidiglossum"*). TYPE: East Indies, *cult. Schiller s.n.* (holotype, W).

Distribution. Arunachal Pradesh, Assam, Manipur, Meghalaya, Nagaland, Sikkim, West Bengal [Bhutan, China, Myanmar].

Habit. Epiphytic herb.

Dendrobium sulcatum Lindl., Edwards's Bot. Reg. 24: t. 65. 1838 [sect. *Densiflora*]. *Callista sulcata* (Lindl.) Kuntze, Revis. Gen. Pl. 2: 655. 1891. TYPE: India, *J. Gibson s.n.*, *Edwards's Bot. Reg. 24: t. 65. 1838* (lectotype).

Dendrobium meghalayense Y. Kumar & S. Chowdhury, Orchidee (Hamburg) 54(4): 455. 2003. TYPE: India, Meghalaya, East Khasi Hills District, Pyrnursula, *Y. Kumar 7371* (holotype, CAL).

Distribution. Arunachal Pradesh, Assam, Manipur, Meghalaya, Mizoram, Sikkim, West Bengal [China, Myanmar].

Habit. Epiphytic herb.

Dendrobium tenuicaule Hook. f., Fl. Brit. India [J. D. Hooker] 6(17): 184. 1890 [sect. *Aporum*]. *Callista tenuicaulis* (Hook. f.) Kuntze, Revis. Gen. Pl. 2: 655. 1891. *Aporum tenuicaule* (Hook. f.) Brieger, Orchideen (Schlechter) 1(11–12): 674. 1981. *Ceraia tenuicaulis* (Hook. f.) M. A. Clem., Telopea 10(1): 294. 2003. TYPE: India, Andaman Islands, icon. in *Herb. Calcutta* (holotype, CAL).

Distribution. Andaman & Nicobar Islands. Endemic.

Habit. Epiphytic herb.

Dendrobium terminale C. S. P. Parish & Rchb. f., Trans. Linn. Soc. London 30: 149. 1874 [sect. *Aporum*]. *Aporum terminale* (C. S. P. Parish & Rchb. f.) M. A. Clem., Telopea 10(1): 297. 2003. *Callista terminalis* (C. S. P. Parish & Rchb. f.) Kuntze, Revis. Gen. Pl. 2: 655. 1891. TYPE: Myanmar, Moulmein [Mawlamyine], *Parish s.n.* [*33*] (holotype, W; isotype, K).

Distribution. Arunachal Pradesh, Assam, Meghalaya, Mizoram, Nagaland, West Bengal [Bhutan, Myanmar].

Habit. Epiphytic herb.

Dendrobium thyrsiflorum Rchb. f. ex André, Ill. Hort. 22: 88, t. 207. 1875 [sect. *Densiflora*]. *Callista thyrsiflora* (Rchb. f. ex André) M. A. Clem., Telopea 10(1): 289. 2003. TYPE: Myanmar, *Parish 190* (holotype, K).

Dendrobium densiflorum Lindl. ex Wall. var. *alboluteum* Hook. f., Bot. Mag. 95: t. 5780. 1869 (as "*albo-lutea*"). TYPE: Myanmar, Moulmein [Mawlamyine], *Parish s.n.*

Dendrobium galliceanum Linden, Lindenia 6: 5, t. 241. 1890. TYPE: Not designated.

Distribution. Assam, Manipur, Nagaland, Sikkim, West Bengal [China, Myanmar].

Habit. Epiphytic herb.

Dendrobium tortile Lindl., Gard. Chron. 1847: 797. 1847 [sect. *Dendrobium*]. TYPE: Myanmar. Mergui [Myeik], *cult. Veitch s.n.* (holotype, K-LINDL).

Distribution. ?Andaman & Nicobar Islands, Meghalaya, Tripura [Bangladesh, Myanmar].

Habit. Epiphytic herb.

Dendrobium transparens Lindl., Gen. Sp. Orchid. Pl.: 79. 1830 [sect. *Dendrobium*]. *Callista transparens* (Lindl.) Kuntze, Revis. Gen. Pl. 2: 655. 1891. TYPE: Nepal, May 1821, *Wallich 2008.1* (holotype, K-LINDL; isotype, E, K, K-WALL).

Distribution. Arunachal Pradesh, Assam, Bihar, Jharkhand, Manipur, Meghalaya, Mizoram, Nagaland, Odisha, Sikkim, Tripura, Uttarakhand, West Bengal [Bangladesh, Bhutan, Myanmar, Nepal].

Habit. Epiphytic herb.

Dendrobium treutleri (Hook. f.) Schuit. & Peter B. Adams, Muelleria 29: 66. 2011 [sect. *Sarcopodium*]. *Coelogyne treutleri* Hook. f., Fl. Brit. India [J. D. Hooker] 5(16): 837. 1890. *Pleione treutleri* (Hook. f.) Kuntze, Revis. Gen. Pl. 2: 680. 1891. *Epigeneium treutleri* (Hook. f.) Ormerod, Oasis 1(3): 3. 2000. TYPE: India, "Sikkim Himalaya," *Treutler s.n.* (holotype, K).

Distribution. Assam, Sikkim [China, Myanmar].

Habit. Epiphytic herb.

Dendrobium trinervium Ridl., J. Linn. Soc., Bot. 32: 242. 1896 [sect. *Stachyobium*]. TYPE: Thailand, Pungah [Phang Nga], *Curtis s.n.* (holotype, SING).

Dendrobium gunnarii P. S. N. Rao, Nordic J. Bot. 12(2): 227. 1992. TYPE: India, North Andaman Island, 0 m, 17 Dec. 1990, *P. S. N. Rao 15791* (holotype, CAL; isotype, PBR).

Distribution. Andaman & Nicobar Islands [also in Thailand, etc.].

Habit. Epiphytic herb.

Remarks. *Dendrobium gunnarii* was found to be synonymous with *D. trinervium* by Priyadarshana et al. (2020).

Dendrobium turbinatum Ormerod & C. S. Kumar, Harvard Pap. Bot. 23(2): 283. 2018 [sect. *Stachyobium*]. TYPE: West India, without locality and collector, *s.n.* (holotype, GH 4663).

Distribution. "Malabar." Endemic.

Habit. Epiphytic herb.

Remarks. According to Omerod and Kumar (2018), the description, but not the nomenclatural type, of *Dendrobium crispum* Dalzell is referable to this species.

Dendrobium versicolor Cogn., J. Orchidées 6: 153. 1895 [sect. *Pedilonum*]. TYPE: India, Assam, *cult. l'Horticulture International s.n.*

Distribution. Assam. Endemic.

Habit. Epiphytic herb.

Remarks. *Dendrobium versicolor* is a species of unclear identity that probably did not originate from Assam.

Dendrobium vexabile Rchb. f., Gard. Chron., n.s., 21: 27. 1884 [sect. *Dendrobium*]. TYPE: Myanmar, *cult. Low s.n.* (holotype, W).

Distribution. Arunachal Pradesh [Bhutan, China, Myanmar].

Habit. Epiphytic herb.

Remarks. *Dendrobium vexabile* is sometimes considered to be a natural hybrid of *D. ruckeri* and another species, and the name is then written as *Dendrobium* × *vexabile*. Its status as a species needs confirmation.

Dendrobium wardianum R. Warner, Select Orchid. Pl. Ser. 1, t. 19. 1863 [sect. *Dendrobium*]. *Callista wardiana* (R. Warner) Kuntze, Revis. Gen. Pl. 2: 655. 1891. TYPE: India, Assam, cult. *J. Day s.n.* (not found).

Distribution. Arunachal Pradesh, Assam, Manipur, Meghalaya, Nagaland [China, Myanmar].

Habit. Epiphytic herb.

Dendrobium wattii (Hook. f.) Rchb. f., Gard. Chron., ser. 3, 4: 725. 1888 [sect. *Formosae*]. *Dendrobium cariniferum* Rchb. f. var. *wattii* Hook. f., Bot. Mag. 109: t. 6715. 1883. *Callista wattii* (Hook. f.) Kuntze, Revis. Gen. Pl. 2: 655. 1891. TYPE: India, Manipur, *cult. Kew s.n. (leg. G. Watt s.n.), 26 Oct. 1882* (holotype, K).

Distribution. Assam, Manipur [Myanmar].

Habit. Epiphytic herb.

Remarks. Ormerod et al. (2021) consider *Dendrobium wattii* to be merely a small-flowered form of *D. infundibulum* Lindl.

Dendrobium wightii A. D. Hawkes & A. H. Heller, Orquidea (Rio de Janeiro) 24: 16. 1962 [sect. *Stachyobium*]. *Dendrobium graminifolium* Wight, Icon. Pl. Ind. Orient. [Wight] 5(1): 6, t. 1649. 1851, nom. illeg., non (L.) Willd., 1805. *Callista graminifolia* Kuntze, Revis. Gen. Pl. 2: 654. 1891. *Dendrobium wightii* N. P. Balakr., J. Bombay Nat. Hist. Soc. 67: 66. 1970, nom. illeg., non A. D. Hawkes & Heller, 1962. SYNTYPES: India, Courtallum, *Wight s.n. [904 & 2954]* (syntypes, CAL, K, K-LINDL).

Distribution. Karnataka, Kerala, Tamil Nadu. Endemic.

Habit. Epiphytic herb.

Remarks. Priyadarshana et al. (2020) consider earlier records from Sri Lanka to have been in error.

Dendrobium williamsonii J. Day & Rchb. f., Gard. Chron. 1869: 78. 1869 (as "*williamsoni*") [sect. *Formosae*]. *Callista williamsonii* (J. Day & Rchb. f.) Kuntze, Revis. Gen. Pl. 2: 655. 1891. TYPE: India, Assam, *W. J. Williamson s.n.* (holotype, W).

Distribution. Assam, Manipur, Meghalaya, Nagaland [China, Myanmar].

Habit. Epiphytic herb.

————

Dendrolirium Blume, Bijdr. Fl. Ned. Ind.: 343. 1825.

TYPE: *Dendrolirium ornatum* Blume.

Dendrolirium andamanicum (Hook. f.) Schuit., Y. P. Ng & H. A. Pedersen, Bot. J. Linn. Soc. 186: 195. 2018. *Eria andamanica* Hook. f., Fl. Brit. India [J. D. Hooker] 5(16): 801. 1890. *Pinalia andamanica* (Hook. f.) Kuntze, Revis. Gen. Pl. 2: 679. 1891. TYPE: India, South Andaman Islands, *Kurz s.n.* (not found).

Distribution. Andaman & Nicobar Islands. Endemic.

Habit. Epiphytic herb.

Dendrolirium ferrugineum (Lindl.) A. N. Rao, Bull. Arunachal Forest Res. 26: 103. 2010. *Eria ferruginea* Lindl., Edwards's Bot. Reg. 25: t. 35. 1839. *Pinalia ferruginea* (Lindl.) Kuntze, Revis. Gen.

Pl. 2: 679. 1891. *Trichotosia ferruginea* (Lindl.) Kraenzl., Bot. Jahrb. Syst. 101: 21. 1910. TYPE: India, Calcutta, *Messrs. Loddiges s.n.* (holotype, K-LINDL).

Eria ferruginea Lindl. var. *assamica* Gogoi, Das & R. Yonzone, McAllen Int. Orchid Soc. J. 15(2): 5. 2014. *Dendrolirium ferrugineum* (Lindl.) A. N. Rao var. *assamicum* (Gogoi, Das & R. Yonzone) Kottaim., Int. J. Curr. Res. Biosci. Pl. Biol. 6(10): 40. 2019. TYPE: India, Assam, Tinsukia district, Dowarmara Reserve Forest, 118 m, 4 July 2010, *Gogoi, Das and Yonzone 0046* (holotype, CAL; isotype, NBU, TOSEHIM).

Distribution. Arunachal Pradesh, Assam, Meghalaya, Mizoram [Bhutan].

Habit. Epiphytic herb.

Dendrolirium kamlangensis (A. N. Rao) A. N. Rao, Bull. Arunachal Forest Res. 26: 103. 2010. *Eria kamlangensis* A. N. Rao, J. Orchid Soc. India 16(1–2): 61. 2002. TYPE: India, Arunachal Pradesh, Lohit District, Kamlang wild life sanctuary, 1600 ft, *A. N. Rao 28186* (holotype, Orchid Herbarium, Tipi).

Eria nepalensis Bajrach. & K. K. Shreshta, J. Jap. Bot. 78: 158. 2003. TYPE: Nepal, Royal Chitwan National Park, Shawara, Narayani, 200 m, *D. M. Bajracharya CN 260* (holotype, KATH; isotype, TUCH).

Distribution. Arunachal Pradesh [Nepal].

Habit. Epiphytic herb.

Dendrolirium laniceps (Rchb. f.) Schuit., Y. P. Ng & H. A. Pedersen, Bot. J. Linn. Soc. 186: 195. 2018. *Eria laniceps* Rchb. f., Hamburger Garten-Blumenzeitung 19: 10. 1863. TYPE: East India, *cult. Schiller s.n.* (holotype, W).

Eria elongata Lindl., J. Proc. Linn. Soc., Bot. 3: 49. 1859 [1858], non Blume, 1825. *Pinalia lobbii* Kuntze, Revis. Gen. Pl. 2: 679. 1891. SYNTYPES: Myanmar, Moulmein [Mawlamyine], *Griffith 346* (syntype, K-LINDL); Zimjaik, *Griffith 347* (syntype, K-LINDL); Moulmein [Mawlamyine], *T. Lobb s.n.* (syntypes, K, K-LINDL).

Distribution. Assam, Meghalaya [Myanmar].

Habit. Epiphytic herb.

Remarks. Ormerod et al. (2021) consider this to be a synonym of *Dendrolirium lasiopetalum* (Willd.) S. C. Chen & J. J. Wood, which is probably correct.

Dendrolirium lasiopetalum (Willd.) S. C. Chen & J. J. Wood, Fl. China 25: 351. 2009. *Epidendrum flos-aeris* J. Koenig, Observ. Bot. (Retzius) 6: 64. 1791, nom. illeg., non L., 1753. *Aerides lasiopetala* Willd., Sp. Pl., ed. 4 [Willd.]. 4(1): 130. 1805. *Epi-*

dendrum lasiopetalum (Willd.) Poir., Encycl. (Lamarck) Suppl. 1. 384. 1810. *Eria lasiopetala* (Willd.) Ormerod, Opera Bot. 124: 22. 1995. TYPE: Not designated.

Dendrobium pubescens Hook., Exot. Fl. 2: t. 124. 1824. *Eria pubescens* (Hook.) Lindl. ex G. Don in J. C. Loudon, Hort. Brit.: 372. 1830. *Eria flava* Lindl., Gen. Sp. Orchid. Pl.: 65. 1830, nom. superfl. TYPE: *cult. Liverpool Garden (leg. Wallich, 1824)* (holotype, BM; isotype, K).

Distribution. Arunachal Pradesh, Assam, Madhya Pradesh, Manipur, Meghalaya, Mizoram, Odisha, Sikkim, Tamil Nadu, Uttar Pradesh, Uttarakhand, West Bengal [Bangladesh, Bhutan, China, Myanmar, Nepal].

Habit. Epiphytic or lithophytic herb.

Dendrolirium tomentosum (J. Koenig) S. C. Chen & J. J. Wood, Fl. China 25: 350. 2009. *Epidendrum tomentosum* J. Koenig, Observ. Bot. (Retzius) 6: 53. 1791. *Eria tomentosa* (J. Koenig) Hook. f., Fl. Brit. India [J. D. Hooker] 5(16): 803. 1890. *Pinalia tomentosa* (J. Koenig) Kuntze, Revis. Gen. Pl. 2: 679. 1891. TYPE: Not designated.

Distribution. Arunachal Pradesh, Assam, Manipur, Meghalaya, Mizoram, Tripura [Bangladesh, China, Myanmar].

Habit. Epiphytic herb.

———

Didymoplexis Griff., Calcutta J. Nat. Hist. 4: 383, t. 17. 1844.

TYPE: *Didymoplexis pallens* Griff.

Didymoplexis himalaica Schltr., Bull. Herb. Boissier ser. 2. 6(4) : 299. 1906. TYPE: Himalaya, Teesta Valley, June 1884, *local collector s.n.* (holotype, ?G).

Distribution. Assam, Sikkim. Endemic.

Habit. Holomycotrophic terrestrial herb.

Remarks. Ormerod (pers. comm.) suspects this to be a synonym of *Didymoplexis micradenia* (Rchb. f.) Hemsl.

Didymoplexis pallens Griff., Calcutta J. Nat. Hist. 4: 383, t. 17. 1844. *Epiphanes pallens* (Griff.) Rchb. f. in B. Seemann, Fl. Vit. 296. 1868. *Gastrodia pallens* (Griff.) F. Muell., Contr. Phytogr. New Hebrides 22. 1873. TYPE: India, Serampur, *local collector s.n.* (holotype, CAL).

Apetalon minutum Wight, Icon. Pl. Ind. Orient. [Wight] 5(1): 22, t. 1758. 1851. TYPE: India, Coorg, near Sultan's Battery, *Jerdon s.n.* (holotype, K).

Cheirostylis kanarensis Blatt. & McCann, J. Bombay Nat. Hist. Soc. 35(4): 732. 1932. TYPE: India, Kanara, Tatwal, *T. R. Bell s.n.* (holotype, BLAT).

Distribution. Andhra Pradesh, Arunachal Pradesh, Assam, Karnataka, Kerala, Meghalaya, Odisha, Sikkim, Tamil Nadu, West Bengal [Afghanistan, Bangladesh, Myanmar].

Habit. Holomycotrophic terrestrial herb.

Didymoplexis seidenfadenii C. S. Kumar & Ormerod, Orchid Memories: 182. 2004. TYPE: India, Kerala, Trivandrum District, Agastyamala, *P. C. Suresh Kumar 31322* (holotype, TBGT; isotype, TBGT).

Distribution. Kerala [Sri Lanka].

Habit. Holomycotrophic terrestrial herb.

Remarks. Singh et al. (2019) consider this a synonym of *Didymoplexis pallens*, which is to be confirmed.

———

Diglyphosa Blume, Bijdr. Fl. Ned. Ind. 7: 336. 1825.

TYPE: *Diglyphosa latifolia* Blume.

Diglyphosa latifolia Blume, Bijdr. Fl. Ned. Ind. 7: 336, t. 60. 1825. *Chrysoglossum latifolium* (Blume) Benth. ex Hemsl., Gard. Chron., n.s., 18: 428. 1882. TYPE: Indonesia, Java, Meggamedong, *Blume s.n.* (holotype, L).

Chrysoglossum macrophyllum King & Pantl., J. Asiat. Soc. Bengal, Pt. 2, Nat. Hist. 64: 335. 1895 [1896]. *Diglyphosa macrophylla* (King & Pantl.) King & Pantl., Ann. Roy. Bot. Gard. (Calcutta) 8: 98, t. 136. 1898. TYPE: India, Sikkim, Chel Valley, 4000 ft, May, *Pantling s.n.* [?315] (holotype, CAL; isotype, AMES, BM, K [all labeled May 1897, perhaps a mistake]).

Distribution. Arunachal Pradesh, Nagaland, Sikkim, West Bengal [China].

Habit. Terrestrial herb.

———

Diplocentrum Lindl., Edwards's Bot. Reg. 18: t. 1522. 1832.

TYPE: *Diplocentrum recurvum* Lindl.

Diplocentrum congestum Wight, Icon. Pl. Ind. Orient. [Wight] 5(1): 10, t. 1682. 1851. TYPE: India, Iyamally, July–Oct., *Wight s.n.* (holotype, K).

Distribution. Goa, Karnataka, Kerala, ?Odisha. Endemic.

Habit. Epiphytic herb.

Diplocentrum recurvum Lindl., Edwards's Bot. Reg. 18: t. 1522. 1832. TYPE: India, *Wallich 7331 (leg. Heyne)* (holotype, K-WALL).

Diplocentrum longifolium Wight, Icon. Pl. Ind. Orient. [Wight] 5(1): 10, t. 1681. 1851. SYNTYPES: India, Tamil Nadu, Nilgiri & Iyamally Hills, Orange Valley, June–July, *Wight s.n.* (syntype, K, Coimbatore, July 1847).

Distribution. Andhra Pradesh, Karnataka, Kerala, Tamil Nadu [Sri Lanka].

Habit. Epiphytic herb.

———

Diplomeris D. Don, Prodr. Fl. Nepal.: 26. 1825.

TYPE: *Diplomeris pulchella* D. Don.

Diplomeris hirsuta (Lindl.) Lindl., Gen. Sp. Orchid. Pl.: 330. 1835. *Diplochilus hirsutus* Lindl., Edwards's Bot. Reg. 18: t. 1499. 1832 (as "*Diplochilos hirsutum*"). TYPE: Nepal, Rasuwa District, Gosainthan [Shishapangma], Aug. 1821, *Wallich 7065* (holotype, K-LINDL; isotype, K-WALL).

Distribution. Arunachal Pradesh, Nagaland, Sikkim, Uttarakhand, West Bengal [China, Nepal].

Habit. Terrestrial or lithophytic herb.

Diplomeris josephii A. N. Rao & Swamin., Indian Orchid J. 2: 5. 1987. TYPE: India, Arunachal Pradesh, *Swaminathan 85101A* (holotype, ASSAM; isotype, BCAL, Orchid Herbarium, Tipi).

Distribution. Arunachal Pradesh. Endemic.

Habit. Terrestrial or lithophytic herb.

Diplomeris pulchella D. Don, Prodr. Fl. Nepal.: 26. 1825. *Paragnathis pulchella* (D. Don) Spreng., Syst. Veg. (ed. 16) [Sprengel] 3: 695. 1826. TYPE: Nepal [corrected by an unknown hand to "Silhet" (India, Sylhet) on the presumed type sheet], *Wallich s.n.* (?holotype, BM).

Orchis uniflora Roxb., Fl. Ind. ed. 1832, 3: 452. 1832. *Habenaria uniflora* (Roxb.) Griff., Icon. Pl. Asiat. 3: t. 338, f. 2. 1851, nom. illeg., non Buch.-Ham. ex D. Don, 1825. TYPE: India, Meghalaya, Garrow [Garo] Hills, icon. *Orchis uniflora*, *Roxburgh 2335* (syntypes, CAL, K).

Distribution. Arunachal Pradesh, Assam, Meghalaya, Nagaland [China, Myanmar, Nepal].

Habit. Terrestrial or lithophytic herb.

Diploprora Hook. f., Fl. Brit. India [J. D. Hooker] 6(17): 26. 1890.

TYPE: *Diploprora championii* (Lindl.) Hook. f. (*Cottonia championii* Lindl.) (as "*championi*").

Diploprora championii (Lindl.) Hook. f., Fl. Brit. India [J. D. Hooker] 6(17): 26. 1890 (as "*championi*"). *Cottonia championii* Lindl., Hooker's J. Bot. Kew Gard. Misc. 7: 35. 1855 (as "*championi*"). *Stauropsis championii* (Lindl.) Tang & F. T. Wang, Acta Phytotax. Sin. 1: 93. 1951. TYPE: China, Hongkong, *Champion s.n.* [*?277*] (holotype, K-LINDL; isotype, K).

Distribution.   Andaman & Nicobar Islands, Arunachal Pradesh, Assam, Karnataka, Kerala, Meghalaya, Nagaland, Odisha, Sikkim, West Bengal [China, Myanmar, Sri Lanka].

Habit.   Epiphytic herb.

Diploprora truncata Rolfe ex Downie, Bull. Misc. Inform. Kew 1925: 385. 1925. *Stauropsis truncata* (Rolfe ex Downie) Tang & F. T. Wang, Acta Phytotax. Sin. 1: 94. 1951. TYPE: Thailand, Doi Sutep, 5500 ft, 25 Feb. 1911, *A. F. G Kerr 270* (holotype, K).

Distribution.   Arunachal Pradesh, Sikkim, West Bengal [also in Thailand].

Habit.   Epiphytic herb.

Disperis Sw., Kongl. Vetensk. Acad. Nya Handl. 21: 218. 1800.

TYPE: *Disperis secunda* (Thunb.) Sw., nom. superfl. (*Arethusa secunda* Thunb., nom. superfl. = *Ophrys circumflexa* L.; *Disperis circumflexa* (L.) T. Durand & Schinz).

Disperis neilgherrensis Wight, Icon. Pl. Ind. Orient. [Wight] 5(1): 15, t. 1719. 1851. TYPE: India, Nilgiri Hills, July–Aug., *Wight s.n.* [*173*] (lectotype, K, designated by Kurzweil, 2005 (as *Wight 3018*); isotype, AMES, C, K-LINDL, S, W).

Disperis walkerae Rchb. f., Linnaea 41: 101. 1877 [1876]. TYPE: Sri Lanka, *Walker s.n.* [*180*] (syntypes, K).
Disperis zeylanica Trimen, J. Bot. 23: 245. 1885. SYNTYPES: Sri Lanka, near Rambodde, *Walker 16* (syntype, K, icon.); Sri Lanka, *Walker 180* (syntype, K).
Disperis monophylla Blatt. ex C. E. C. Fisch., Fl. Madras 3: 1478. 1928. TYPE: India, Tamil Nadu, Madura Dist., High Wavy Mountain, May 1917, *Blatter & Hallberg s.n.* (holotype, K).

Distribution.   Karnataka, Kerala, Tamil Nadu [China, Sri Lanka].

Habit.   Terrestrial herb.

Remarks.   It is possible that more than one species is found in the South India–Sri Lanka area.

Epipactis Zinn, Cat. Pl. Hort. Gott. 85. 1757, nom. cons.

TYPE: *Epipactis helleborine* (L.) Crantz (*Serapias helleborine* L.) (typ. cons.).

Epipactis helleborine (L.) Crantz, Stirp. Austr. Fasc., ed. 2. 2: 467. 1769. *Serapias helleborine* L., Sp. Pl. 2: 949. 1753. TYPE: *Herb. Burser X: 39* (lectotype, UPS, designated by Cribb & Wood in Cafferty & Jarvis (eds.), 1999).

Epipactis intrusa Lindl., J. Proc. Linn. Soc., Bot. 1: 175. 1857. *Epipactis latifolia* (L.) All. var. *intrusa* (Lindl.) Hook. f., Fl. Brit. India [J. D. Hooker] 6(17): 126. 1890. *Epipactis helleborine* (L.) Crantz var. *intrusa* (Lindl.) S. N. Mitra, Indian Forester 99: 101. 1973. *Epipactis helleborine* (L.) Crantz var. *intrusa* (Lindl.) Karthik., in S. Karthikeyan et al., Fl. Ind. ser. 4, 1 (Monocotyledon): 132. 1989, isonym. TYPE: India, Sikkim, 11,000 ft, *J. D. Hooker 323B* (holotype, K-LINDL).

Distribution.   Himachal Pradesh, Jammu & Kashmir, Sikkim, Uttarakhand [Bhutan, China, Myanmar, Nepal, Pakistan].

Habit.   Terrestrial herb.

Epipactis persica (Soó) Hausskn. ex Nannf., Bot. Not. 1946(1): 21. 1946. *Helleborine persica* Soó, Repert. Spec. Nov. Regni Veg. 24: 36. 1927. *Epipactis microphylla* (Ehrh.) Sw. subsp. *persica* (Soó) Hautz., Verh. Zool.-Bot. Ges. Wien 115: 42. 1976. *Epipactis helleborine* (L.) Crantz subsp. *persica* (Soó) H. Sund., Europ. Medit. Orchid., ed. 3: 41. 1980, nom. inval. SYNTYPES: Iran, Sultanabad, *Strauss s.n.*; Dagestan, Elbrus near Derbent, *Kotschy [1843] 921*; Askedessie, *Pollak s.n.*

Distribution.   Jammu & Kashmir [Afghanistan, Pakistan].

Habit.   Terrestrial herb.

Epipactis royleana Lindl., Gen. Sp. Orchid. Pl.: 461. 1840. *Cephalanthera royleana* (Lindl.) Regel, Trudy Imp. S.-Peterburgsk. Bot. Sada 6: 490. 1879. *Limodorum royleanum* (Lindl.) Kuntze, Revis. Gen. Pl. 2: 671. 1891. *Amesia royleana* (Lindl.) Hu, Rhodora 27: 106. 1925. *Helleborine royleana*

(Lindl.) Soó, Repert. Spec. Nov. Regni Veg. 24: 35. 1927. *Arthrochilium royleanum* (Lindl.) Szlach., Orchidee (Hamburg) 54: 588. 2003. TYPE: North India, Kunawur, Lippa, *Royle s.n.* (holotype, LIV).

Distribution. Himachal Pradesh, Jammu & Kashmir, Uttarakhand [Afghanistan, Bhutan, China, Nepal, Pakistan].

Habit. Terrestrial herb.

Epipactis veratrifolia Boiss. & Hohen., Diagn. Pl. Orient. ser. 1, 2(13): 11. 1854. *Arthrochilium veratrifolium* (Boiss. & Hohen.) Szlach., Orchidee (Hamburg) 54: 588. 2003. TYPE: Dagestan, Mt. Elbrus near Derbend, *Kotschy 401.632* (holotype, G; isotype, E, K-LINDL, LE, W).

Epipactis consimilis Wall. ex Hook. f., Fl. Brit. India [J. D. Hooker] 6(17): 126. 1890, non D. Don, 1825. SYN-TYPES: Nepal, Sivapur, 4 Feb. 1814, *Wallich 7403A (leg. Buchanan-Hamilton)* (syntype, K-WALL); Nepal, Bechiaco, Dec. 1820, *Wallich 7403B* (syntype, K-WALL); India, Dehradun, Sansedarra, Apr. 1825, *Wallich 7403C* (syntype, K-WALL).

Distribution. Bihar, Himachal Pradesh, Rajasthan, Sikkim, Tamil Nadu, Uttarakhand [Afghanistan, China, Myanmar, Nepal, Pakistan].

Habit. Terrestrial herb.

———

Epipogium J. G. Gmel. ex Borkh., Tent. Disp. Pl. German. 139. 1792.

TYPE: *Epipogium aphyllum* Sw. (*Satyrium epipogium* L.).

Epipogium aphyllum Sw., Summa Veg. Scand. (Swartz): 32. 1814. *Satyrium epipogium* L., Sp. Pl.: 945. 1753. *Orchis aphylla* F. W. Schmidt in Mayer, Samml. Phys. Aufs. 1: 240. 1791, nom. illeg., non Forssk., 1775. TYPE: icon. Gmelin, Fl. Sibir. 1: t. 2, fig. 2 (lectotype, designated by Renz & Taubenheim, 1984).

Distribution. Himachal Pradesh, Jammu & Kashmir, Sikkim, Uttarakhand [Bhutan, China, Myanmar, Nepal, Pakistan].

Habit. Holomycotrophic terrestrial herb.

Epipogium japonicum Makino, Bot. Mag. (Tokyo) 18: 131. 1904. SYNTYPES: Japan, Shimotsuke Prov., Nikko, Mt. Nyoho, *Aoki & Kurushima s.n.* (28 Oct. 1904); *Kurushima & Ioki s.n.* (16 Oct. 1904).

?*Epipogium tuberosum* Duthie, Ann. Roy. Bot. Gard. (Calcutta) 9: 151. 1906. TYPE: Not designated.

Distribution. ?Jammu & Kashmir, West Bengal [Bhutan, China, Nepal].

Habit. Holomycotrophic terrestrial herb.

Remarks. The taxonomy of *Epipogium roseum* and its allies, such as *E. japonicum*, is in need of revision.

Epipogium roseum (D. Don) Lindl., J. Proc. Linn. Soc., Bot. 1: 177. 1857. *Limodorum roseum* D. Don, Prodr. Fl. Nepal.: 30. 1825. *Ceratopsis rosea* (D. Don) Lindl., Gen. Sp. Orchid. Pl.: 384. 1840. *Galera rosea* (D. Don) Blume, Mus. Bot. 2: 188. 1856. TYPE: Nepal, 1818, *Wallich s.n.* (holotype, BM).

Epipogium nutans (Blume) Rchb. f., Bonplandia 5(3): 36. 1857. *Galera nutans* Blume, Bijdr. Fl. Ned. Ind. 8: 415. 1825. TYPE: Indonesia, Java, Mt. Salak, *Blume s.n.* (holotype, ?L [*Blume 723*]).
Podanthera pallida Wight, Icon. Pl. Ind. Orient. [Wight] 5(1): 22, t. 1759. 1851. TYPE: India, Wynaud, *Jerdon s.n.* (holotype, K-LINDL; isotype, K, MH).
Epipogium sessanum S. N. Hegde & A. N. Rao, J. Econ. Taxon. Bot. 3(2): 598. 1982. TYPE: India, Arunachal Pradesh, West Kameng District, Sessa Orchid Sanctuary, 11 June 1981, *S. N. Hegde 3945* (holotype, CAL).
Epipogium indicum H. J. Chowdhery, G. D. Pal & G. S. Giri, Nordic J. Bot. 13(4): 419. 1993. TYPE: India, Arunachal Pradesh, Lower Subansiri District, Sankie View, *H. J. Chowdhery 1746* (holotype, CAL).

Distribution. Arunachal Pradesh, Assam, Himachal Pradesh, Jammu & Kashmir, Karnataka, Kerala, Maharashtra, Manipur, Meghalaya, Mizoram, Nagaland, Sikkim, Tamil Nadu, Uttar Pradesh, Uttarakhand, West Bengal [China, Myanmar, Nepal, Pakistan, Sri Lanka].

Habit. Holomycotrophic terrestrial herb.

Remarks. Peloric forms lacking a spur are not uncommon in some parts of the huge distribution range of this species. We have little doubt that *Epipogium sessanum* was based on such a peloric specimen.

———

Eria Lindl., Bot. Reg. 11: t. 904. 1825, nom. cons.

TYPE: *Eria stellata* Lindl. (= *Eria javanica* (Thunb. ex Sw.) Blume).

Eria clausa King & Pantl., J. Asiat. Soc. Bengal, Pt. 2, Nat. Hist. 65: 121. 1896. *Eria corneri* Rchb. f. var. *clausa* (King & Pantl.) A. N. Rao, J. Econ. Taxon. Bot. 20: 708. 1996. TYPE: India, Sikkim, 3000–5000 ft, Feb.–Mar., *Pantling s.n.* [?559: Sittong Point, 6000 ft, Feb. 1891] (?syntypes, CAL, K).

Distribution. Arunachal Pradesh, Assam, Sikkim, West Bengal [Bhutan, China, Myanmar, Nepal].

Habit. Epiphytic herb.

Eria coronaria (Lindl.) Rchb. f., Ann. Bot. Syst. (Walpers) 6: 271. 1861. *Coelogyne coronaria* Lindl., Edwards's Bot. Reg. 27(Misc.): 83. 1841. *Trichosma suavis* Lindl., Edwards's Bot. Reg. 28: t. 21. 1842, nom. superfl. *Trichosma coronaria* (Lindl.) Kuntze, Revis. Gen. Pl. 2: 691. 1891. TYPE: India, Chirree District, Khasi Hills, *J. Gibson s.n.* (holotype, K-LINDL [presumably the specimen labeled "Chatsworth, Oct. 1841," as Gibson collected for the Duke of Devonshire at Chatsworth]).

Eria cylindripoda Griff., Not. Pl. Asiat. 3: 299. 1851. TYPE: India, Khasi Hills, Myrung, 9 Nov. 1835, *Griffith in Assam Herb. 231* (holotype, K).

Dendrobium crepidatum Griff., Not. Pl. Asiat. 3: 319. 1851, nom. illeg., non Lindl. & Paxton, 1850. TYPE: India, Khasi Hills, Churra, 23 Oct. 1835, *Griffith in Assam Herb. 165* (not found).

Distribution. Arunachal Pradesh, Manipur, Meghalaya, Mizoram, Nagaland, Sikkim, Uttarakhand, West Bengal [Bhutan, China, Nepal].

Habit. Epiphytic herb.

Eria javanica (Thunb. ex Sw.) Blume, Rumphia 2: 23. 1839. *Dendrobium javanicum* Thunb. ex Sw., Kongl. Vetensk. Acad. Nya Handl. 21: 247. 1800. TYPE: Indonesia, East Java, *Thunberg s.n.* (not found).

Eria fragrans Rchb. f., Bot. Zeitung (Berlin) 22(53): 415. 1864. TYPE: Myanmar, Moulmein [Mawlamyine], *Low s.n.* (holotype, W).

Distribution. Arunachal Pradesh, Manipur, Meghalaya, West Bengal [China, Myanmar].

Habit. Epiphytic herb.

Remarks. As Ormerod et al. (2021) pointed out, the description of *Dendrobium javanicum*, which mentions obtuse leaves, does not accord with what is generally called *Eria javanica* today. Unfortunately, the type material appears to be lost and the description is too vague to be of use; it could well refer to something like *Bulbophyllum gibbosum* (Blume) Lindl., but the flowers are not described.

Eria scabrilinguis Lindl., J. Proc. Linn. Soc., Bot. 3: 51. 1859 [1858]. *Pinalia scabrilinguis* (Lindl.) Kuntze, Revis. Gen. Pl. 2: 679. 1891. TYPE: India, Sikkim, icon. *Cathcart s.n.* (holotype, CAL; isotype, K).

Eria corneri Rchb. f., Gard. Chron., n.s., 10: 106. 1878. TYPE: Taiwan (Formosa), *A. Corner s.n.* (holotype, W).

Distribution. Arunachal Pradesh, Assam, Manipur, ?Sikkim [China, Nepal].

Habit. Epiphytic herb.

Remarks. This species was described from a painting after a plant said to have been collected in Sikkim. It has never been found there since, and the original record may have been in error. It is better known under its synonym *Eria corneri*, and the correct name has only recently been established (Schuiteman, 2011b).

Eria vittata Lindl., J. Proc. Linn. Soc., Bot. 3: 51. 1859 [1858]. *Pinalia vittata* (Lindl.) Kuntze, Revis. Gen. Pl. 2: 679. 1891. TYPE: India, Sikkim, icon. *Cathcart s.n.* (holotype, CAL; isotype, K).

Distribution. Arunachal Pradesh, Manipur, Mizoram, Nagaland, Sikkim, West Bengal [China, Myanmar].

Habit. Epiphytic herb.

———

Eriodes Rolfe, Orchid Rev. 23: 327. 1915.

TYPE: *Eriodes barbata* (Lindl.) Rolfe (*Tainia barbata* Lindl.).

Eriodes barbata (Lindl.) Rolfe, Orchid Rev. 23: 326. 1915. *Tainia barbata* Lindl., Gard. Chron. 1857: 68. 1857. *Eria barbata* (Lindl.) Rchb. f., Ann. Bot. Syst. (Walpers) 6: 270. 1861. *Pinalia barbata* (Lindl.) Kuntze, Revis. Gen. Pl. 2: 679. 1891. *Tainiopsis barbata* (Lindl.) Schltr., Orchis 9: 12. 1915. *Neotainiopsis barbata* (Lindl.) Bennett & Raizada, Indian Forester 107(7): 433. 1981. TYPE: India, Meghalaya, Khasi Hills, *Griffith s.n.* (lectotype, K-LINDL, designated by Seidenfaden, 1986, isotype, CAL, K).

Distribution. Arunachal Pradesh, Assam, Manipur, Meghalaya, Mizoram [Bhutan, China, Myanmar].

Habit. Epiphytic herb.

———

Erythrodes Blume, Bijdr. Fl. Ned. Ind. 8: 410. 1825.

TYPE: *Erythrodes latifolia* Blume.

Erythrodes blumei (Lindl.) Schltr., Fl. Schutzgeb. Südsee (Lauterbach) 87. 1905. *Physurus blumei* Lindl., Gen. Sp. Orchid. Pl.: 504. 1840. *Microchilus blumei* (Lindl.) D. Dietr., Syn. Pl. (D. Dietrich) 5: 166. 1852. TYPE: India, Sylhet, *Wallich 7397 (leg. De Silva)* (holotype, K-LINDL; isotype, K-WALL).

Erythrodes seshagiriana A. N. Rao, Indian Forester 123: 643. 1997. TYPE: India, Changlang District, Motijeel, Namdapha National Park, *Rao 28269* (holotype, CAL).

Distribution. Andaman & Nicobar Islands, Arunachal Pradesh, Manipur, Meghalaya [Bangladesh, China, Myanmar].

Habit. Terrestrial herb.

Remarks. We agree with Bhattacharjee and Chowdhery (2018) that *Erythrodes seshagiriana* is a synonym of *E. blumei* rather than *E. hirsuta* (Griff.) Ormerod. A bifid spur apex, as recorded in *E. seshagiriana*, can be found in *E. blumei* and cannot be considered a differentiating character for the former.

Erythrodes hirsuta (Griff.) Ormerod in G. Seidenfaden, Contr. Orchid Fl. Thailand 13: 12. 1997. *Goodyera hirsuta* Griff., Not. Pl. Asiat. 3: 393. 1851. *Physurus hirsutus* (Griff.) Lindl., J. Proc. Linn. Soc., Bot. 1: 180. 1857. TYPE: India, Assam, towards Nempean, 1837, icon. *Griffith, Icon. Pl. Asiat. 3: t. 347. 1851* (lectotype, designated by Bhattacharjee & Chowdhery, 2011); India, Sikkim, West District, near Khechiperi, 16 Mar. 2005, *Bhattacharjee 34804* (epitype, CAL, designated by Bhattacharjee & Chowdhery, 2011).

Physurus herpysmoides King & Pantl., J. Asiat. Soc. Bengal, Pt. 2, Nat. Hist. 65: 124. 1896. *Erythrodes herpysmoides* (King & Pantl.) Schltr., Bot. Jahrb. Syst. 45: 392. 1911. TYPE: India, West Bengal, above Engo, alt. 5000 ft [Kew specimen has 4000 ft], Apr. 1893, *Pantling 255* (holotype, K; isotype, CAL).

Distribution. Arunachal Pradesh, Assam, West Bengal [Bhutan, China, Myanmar].

Habit. Terrestrial herb.

———

Erythrorchis Blume, Rumphia 1: 200, t. 70. 1835 [1837].

TYPE: *Erythrorchis altissima* (Blume) Blume (*Cyrtosia altissima* Blume).

Erythrorchis altissima (Blume) Blume, Rumphia 1: 200, t. 70. 1835 [1837]. *Cyrtosia altissima* Blume, Bijdr. Fl. Ned. Ind. 8: 396. 1825. *Galeola altissima* (Blume) Rchb. f., Xenia Orchid. 2: 77. 1865. TYPE: Indonesia, Java, Mt. Seribu, *Blume s.n.* (holotype, L; isotype, BO).

Distribution. Andaman & Nicobar Islands, Arunachal Pradesh, Assam [Myanmar].

Habit. Holomycotrophic terrestrial climber.

———

Eulophia R. Br., Bot. Reg. 7: sub t. 573 (as "578"). 1821 (as "*Eulophus*"), nom. et orth. cons.

TYPE: *Eulophia guineensis* Ker Gawl. (typ. cons.).

Eulophia andamanensis Rchb. f., Flora 55: 276. 1872. *Graphorkis andamanensis* (Rchb. f.) Kuntze, Revis. Gen. Pl. 2: 662. 1891 (as "*Graphorchis*"). *Cyrtopera andamanensis* (Rchb. f.) Rolfe, Gard. Chron., ser. 3, 18: 581. 1895. TYPE: India, Andaman Island, *Mann s.n.* (holotype, W).

Distribution. Andaman & Nicobar Islands [Myanmar].

Habit. Terrestrial herb.

Eulophia bicallosa (D. Don) P. F. Hunt & Summerh., Kew Bull. 20(1): 60. 1966. *Bletia bicallosa* D. Don, Prodr. Fl. Nepal.: 30. 1825. *Limodorum bicallosum* (D. Don) Buch.-Ham. ex D. Don, Prodr. Fl. Nepal.: 30. 1825. *Graphorkis bicallosa* (D. Don) Kuntze, Revis. Gen. Pl. 2: 662. 1891 (as "*Graphorchis*"). *Liparis bicallosa* (D. Don) Schltr., Repert. Spec. Nov. Regni Veg. Beih. 4: 196. 1919. TYPE: Nepal, *Buchanan-Hamilton s.n.* (syntypes, BM, LINN).

Cyrtopera bicarinata Lindl., Gen. Sp. Orchid. Pl.: 190. 1833. *Eulophia bicarinata* (Lindl.) Hook. f., Fl. Brit. India [J. D. Hooker] 6(17): 6. 1890. TYPE: Bangladesh, Patgong, 23 Mar. 1809, *Wallich 7363 (leg. Buchanan-Hamilton s.n.)* (holotype, K-LINDL; isotype, K-WALL). *Cyrtopera candida* Lindl., J. Proc. Linn. Soc., Bot. 3: 31. 1859 [1858]. *Eulophia candida* (Lindl.) Hook. f., Fl. Brit. India [J. D. Hooker] 6(17): 6. 1890. SYNTYPES: India, Sikkim, *Cathcart s.n.* (not found); *J. D. Hooker 241* (syntype, K). *Eulophia bicarinata* (Lindl.) Hook. f. var. *major* King & Pantl., Ann. Roy. Bot. Gard. (Calcutta) 8: 181, t. 244A. 1898. *Eulophia bicallosa* (D. Don) P. F. Hunt & Summerh. var. *major* (King & Pantl.) Pradhan, Indian Orchids: Guide Identif. & Cult. 2: 461. 1979. TYPE: India, Sikkim, at the foot of the range, *Pantling 436* (holotype, CAL).

Distribution. Arunachal Pradesh, Assam, Manipur, Meghalaya, Mizoram, Nagaland, Sikkim, Tripura, Uttar Pradesh, Uttarakhand, West Bengal [Bangladesh, Myanmar, Nepal].

Habit. Terrestrial herb.

Eulophia bracteosa Lindl., Gen. Sp. Orchid. Pl.: 180. 1833. *Graphorkis bracteosa* (Lindl.) Kuntze, Revis. Gen. Pl. 2: 662. 1891 (as "*Graphorchis*"). TYPE: Bangladesh, Chittagong, *Wallich 7366 (leg. Bruce s.n.)* (holotype, K-LINDL; isotype, K-WALL).

Distribution. Meghalaya [Bangladesh, Myanmar].

Habit. Terrestrial herb.

Eulophia campbellii Prain, J. Asiat. Soc. Bengal,
Pt. 2, Nat. Hist. 73: 190. 1904. SYNTYPES:
India, Chora Nagpur, Manbhum, Pokhuria, *Campbell 7560* (syntype, CAL); Singbhum, Gidlung,
near Manharpur, *Burkill's collector 19734* (syntype, CAL).

Distribution.   Bihar, Jharkhand. Endemic.

Habit.   Terrestrial herb.

Remarks.   *Eulophia campbellii* is apparently an extremely rare endemic that has not been recollected for
over a century (Kumar, 2015). A conservation assessment is urgently required.

Eulophia dabia (D. Don) Hochr., Bull. New York Bot.
Gard. 6: 270. 1910. *Bletia dabia* D. Don, Prodr.
Fl. Nepal.: 30. 1825. *Limodorum dabia* (D. Don)
Buch.-Ham. ex D. Don, Prodr. Fl. Nepal.: 30.
1825. *Graphorkis dabia* (D. Don) Kuntze, Revis.
Gen. Pl. 2: 662. 1891 (as "*Graphorchis*"). TYPE:
Nepal, *Buchanan-Hamilton s.n.* (syntypes, BM,
LINN).

Limodorum ramentaceum Roxb., Fl. Ind. ed. 1832, 3: 467.
1832. *Eulophia ramentacea* (Roxb.) Lindl., Gen. Sp. Or-
chid. Pl.: 185. 1833. *Geodorum ramentaceum* (Roxb.)
Voigt, Hort. Suburb. Calcutt.: 628. 1845. TYPE: India,
Bengal, cult. Calcutta s.n. (leg. R. Smith, 1810), icon.
Limodorum ramentaceum, *Roxburgh 2344* (syntypes,
CAL, K).
Eulophia rupestris Wall. ex Lindl., Gen. Sp. Orchid. Pl.: 185.
1833. *Graphorkis rupestris* Kuntze, Revis. Gen. Pl. 2:
882. 1891 (as "*Graphorchis*"). SYNTYPES: Nepal, May
1821, *Wallich [7368]* (syntypes, K-LINDL, K-WALL);
Buchanan-Hamilton (syntype, BM).
Eulophia campestris Wall. ex Lindl., Gen. Sp. Orchid. Pl.:
185. 1833. *Graphorkis campestris* (Wall. ex Lindl.) Kuntze, Revis. Gen. Pl. 2: 662. 1891 (as "*Graphorchis*").
SYNTYPES: India, Oude, Bhurtapore [Protappur], Mar.
1825, *Wallich 7367* (syntypes, K-LINDL, K-WALL);
Himalayas, *Royle s.n.* (syntype, K-WALL).
Eulophia hemileuca Lindl., J. Proc. Linn. Soc., Bot. 3: 25.
1859 [1858]. TYPE: India, plains of Robilcund, *Thomson 219* (not found).
Eulophia hormusjii Duthie, Ann. Roy. Bot. Gard. (Calcutta)
9(2): 125. 1906. SYNTYPES: Dehra Dun, *Duthie (Mackinnon) 22708a; 22724*; Kheri District of N. Oudh, *Duthie's collector 22797*; Mussoorie Range, 6000 ft, 23 Mar.
1899, *Duthie (Mackinnon) 22708* (syntypes, DD, K);
Bashahr, *Lace* 880; Hazara near Abbottabad, Siran River,
Barrett s.n.; Chital Distr., Gujar valley, *Gatacre s.n.*

Distribution.   Andaman & Nicobar Islands, Aru-
nachal Pradesh, Bihar, Haryana, Himachal Pradesh,
Jammu & Kashmir, Jharkhand, Maharashtra, Manipur,
Odisha, Punjab, Sikkim, Uttar Pradesh, Uttarakhand,
West Bengal [Afghanistan, Bangladesh, Bhutan, China,
Myanmar, Nepal, Pakistan].

Habit.   Terrestrial herb.

Remarks.   Vij et al. (2013) suggest that *Eulophia
hormusjii* is a distinct species, which is distinguished
from *E. dabia* by the bright yellow mid-lobe of the lip
and the compressed, somewhat bilobed spur. The herbarium material we have seen seems quite variable and
almost uniformly lacks color notes; as a result we are at
present unable to form a solid opinion on this matter.
Kumar (2015) considers *E. hormusjii* to fall within the
range of variation of *E. dabia*.

Eulophia densiflora Lindl., J. Proc. Linn. Soc., Bot.
3: 25. 1859 [1858]. *Cyrtopera densiflora* (Lindl.)
Rchb. f., Otia Bot. Hamburg. 68. 1878. *Graphorkis densiflora* (Lindl.) Kuntze, Revis. Gen. Pl.
2: 662. 1891 (as "*Graphorchis*"). TYPE: Bhutan,
Griffith s.n. [6] (holotype, K-LINDL).

Distribution.   ?Sikkim [Bhutan, Nepal].

Habit.   Terrestrial herb.

Remarks.   Recorded from Sikkim Himalaya by
Hooker (1890b), occurring "at the foot of the hills,"
which is rather vague, and the species may not actually
occur in the state of Sikkim.

Eulophia diffusiflora M. W. Chase, Kumar & Schuit.,
Phytotaxa 491(1): 52. 2021. *Geodorum laxiflorum*
Griff., Calcutta J. Nat. Hist. 5: 356. 1845. TYPE:
India, Assam, *Jenkins s.n.* (not found, possibly
same collection as for *G. rariflorum*).

Geodorum rariflorum Lindl., Fol. Orchid. 6/7(*Geodorum*):2.
1855. TYPE: India, Assam, *Jenkins s.n.* (holotype,
K-LINDL).

Distribution.   Assam, Chhattisgarh, Gujarat, Jharkhand, Odisha. Endemic.

Habit.   Terrestrial herb.

Eulophia epidendraea (J. Koenig) C. E. C. Fisch.,
Fl. Madras 1434. 1928. *Serapias epidendraea*
J. Koenig, Observ. Bot. (Retzius) 6: 65. 1791. *Limodorum epidendroides* Willd., Sp. Pl. 4: 124.
1805, nom. superfl. *Eulophia epidendroides* (Willd.)
Schltr., Orchideen Beschreib. Kult. Zücht.: 346.
1914, nom. superfl. TYPE: India, near Madras
and Trangambar, *Koenig s.n.* (syntypes, BM, C).

Limodorum virens Roxb., Pl. Coromandel 1: 32. 1795. *Eulophia virens* (Roxb.) Spreng., Syst. Veg., ed. 16 (Sprengel)
3: 720. 1826. TYPE: India, icon. *Limodorum virens*,
Roxburgh 238 (syntypes, CAL, K).
Limodorum carinatum Willd., Sp. Pl., ed. 4, 4: 124. 1805.
Eulophus carinatus (Willd.) R. Br., Bot. Reg. 7: t. 573.
1821. *Eulophia carinata* (Willd.) Lindl., Gen. Sp. Orchid. Pl.: 183. 1833. TYPE: Icon. Rheede, Hort. Malab.
12: 51, t. 26. 1692.

Distribution. Andaman & Nicobar Islands, Andhra Pradesh, Chhattisgarh, Karnataka, Kerala, Maharashtra, Odisha, Tamil Nadu, Uttarakhand, West Bengal [Bangladesh, Sri Lanka].

Habit. Terrestrial herb.

Eulophia explanata Lindl., Gen. Sp. Orchid. Pl.: 180. 1833. TYPE: "Sri Lanka, *Macrae s.n.*" According to Hooker (1891), the type is actually: Nepal, Morung [Morang], Maghada, *Buchanan-Hamilton s.n.* (holotype, K-LINDL).

Distribution. Andhra Pradesh, Bihar, Chhattisgarh, Jharkhand, Madhya Pradesh, Odisha, Uttar Pradesh, Uttarakhand, West Bengal [Nepal].

Habit. Terrestrial herb.

Eulophia flava (Lindl.) Hook. f., Fl. Brit. India [J. D. Hooker] 6(17): 7. 1890. *Cyrtopera flava* Lindl., Gen. Sp. Orchid. Pl. 189. 1833. *Cyrtopodium flavum* (Lindl.) Benth., J. Linn. Soc., Bot. 18: 320. 1881, nom. illeg., non Link & Otto ex Rchb., 1830. *Graphorkis flava* (Lindl.) Kuntze, Revis. Gen. Pl. 2: 662. 1891 (as "*Graphorchis*"). *Lissochilus flavus* (Lindl.) Schltr., Repert. Spec. Nov. Regni Veg. Beih. 4: 260. 1919. SYNTYPES: India, Morung [Morang] Hills, 6 May 1810, *Wallich 7364 (leg. Buchanan-Hamilton)* (syntypes, K-LINDL, K-WALL); India, Himalayan valleys below Surkunda, Shalma, and Mussoorie, *Royle s.n.* (probable syntype [without locality], K).

Cyrtopera cullenii Wight, Icon. Pl. Ind. Orient. [Wight] 5(1): 21, t. 1754. 1851. *Eulophia cullenii* (Wight) Blume, Coll. Orchid.: 182. 1859. TYPE: India, Travancore, *Wight s.n.* (not found).
Eulophia cullenii (Wight) Blume var. *minor* C. E. C. Fisch., Fl. Madras 3(8): 1435. 1928. TYPE: India, Pulney Hills, alt. 6000 ft, *Van Malderen s.n.* (not found).

Distribution. Andhra Pradesh, Bihar, Chhattisgarh, Jharkhand, Kerala, Madhya Pradesh, Maharashtra, Tamil Nadu, Uttar Pradesh, Uttarakhand [China, Myanmar, Nepal].

Habit. Terrestrial herb.

Eulophia graminea Lindl., Gen. Sp. Orchid. Pl.: 182. 1833. *Graphorkis graminea* (Lindl.) Kuntze, Revis. Gen. Pl. 2: 662. 1891 (as "*Graphorchis*"). TYPE: Singapore, 1822, *Wallich 7372.C (leg. T. Lobb)* (syntypes, BM, CAL, K-LINDL, K-WALL).

Eulophia decipiens Kurz, J. Asiat. Soc. Bengal, Pt. 2, Nat. Hist. 45: 155. 1876. TYPE: India, Nicobar Islands, Feb. 1875, *C. B. Clarke 26009 (leg. V. Kurz)* (syntypes, CAL, K).

Eulophia ucbii Malhotra & Balodi, Bull. Bot. Surv. India 26(1–2): 92, fig. 5. 1984 [1985]. TYPE: India, Pithoragarh, Gori Valley, Garjia, Apr. 1984, *B. Balodi 75561A* (holotype, CAL).

Distribution. Andaman & Nicobar Islands, Andhra Pradesh, Arunachal Pradesh, Assam, Bihar, Chhattisgarh, Jharkhand, Karnataka, Kerala, Madhya Pradesh, Maharashtra, Manipur, Meghalaya, Mizoram, Nagaland, Odisha, Punjab, Sikkim, Tamil Nadu, Tripura, Uttar Pradesh, Uttarakhand, West Bengal [Bangladesh, Bhutan, China, Myanmar, Nepal, Pakistan, Sri Lanka].

Habit. Terrestrial herb.

Eulophia herbacea Lindl., Gen. Sp. Orchid. Pl.: 182. 1833. *Graphorkis herbacea* (Lindl.) Lyons, Pl. Nam. (Ed. 2) 215. 1907. SYNTYPES: Sri Lanka, *Macrae s.n.* (not found); NW India, Mussoorie, *Royle s.n.* (syntype, K).

Limodorum bicolor Roxb., Fl. Ind. ed. 1832, 3: 469. 1832. *Geodorum bicolor* (Roxb.) Voigt, Hort. Suburb. Calcutt.: 628. 1845. *Graphorkis bicolor* (Roxb.) Kuntze, Revis. Gen. Pl. 2: 663. 1891 (as "*Graphorchis*"). *Eulophia vera* Royle, Ill. Bot. Himal. Mts. [Royle] 1: 366. 1839, nom. inval. *Eulophia albiflora* Edgew. ex Lindl., J. Proc. Linn. Soc., Bot. 3: 24. 1859 [1858], nom. inval. TYPE: India, West Bengal, icon. *Limodorum bicolor*, *Roxburgh 2342* (syntypes, CAL, K).
Eulophia brachypetala Lindl., J. Proc. Linn. Soc., Bot. 3: 24. 1859 [1858]. TYPE: India, Gurwhal, Paw, 2000–3000 ft, June 1845, *Thomson 216 [1322]* (holotype, K).

Distribution. Chhattisgarh, Gujarat, Himachal Pradesh, Karnataka, Madhya Pradesh, Maharashtra, Manipur, Mizoram, Odisha, Tripura, Uttarakhand [Bangladesh, China, Myanmar, Nepal].

Habit. Terrestrial herb.

Eulophia kamarupa Sud. Chowdhury, J. Orchid Soc. India 7(1–2): 49. 1993. TYPE: India, Assam, Kamrup District, Jalukbari, Gopinath Bordoloi Nagar, Gauhati University Campus, 26°12′N, 91°45′E, 9 Jan. 1970, *R. Katita 453* (holotype, CAL; isotype, ASSAM).

Distribution. Assam. Endemic.

Habit. Terrestrial herb.

Eulophia mackinnonii Duthie, J. Asiat. Soc. Bengal, Pt. 2, Nat. Hist. 71: 40. 1902 (as "*mackinnoni*"). SYNTYPES: India, Dehra Dun, *Mackinnon s.n.* (syntype, DD); Siwalik range, *Vicary s.n.* (syntypes, CAL, DD); Bahraich District, *Duthie's collector s.n.*; Raipur District; *J. Marten s.n.* (syntypes, DD, K).

Distribution. ?Assam, Chhattisgarh, Jharkhand, Madhya Pradesh, Uttar Pradesh, Uttarakhand [Bangladesh, Nepal].

Habit. Terrestrial herb.

Eulophia macrobulbon (C. S. P. Parish & Rchb. f.) Hook. f., Fl. Brit. India [J. D. Hooker] 6(17): 7. 1890. *Cyrtopera macrobulbon* C. S. P. Parish & Rchb. f., Trans. Linn. Soc. London 30: 144. 1874. *Graphorkis macrobulbon* (C. S. P. Parish & Rchb. f.) Kuntze, Revis. Gen. Pl. 2: 662. 1891 (as "*Graphorchis*"). TYPE: Myanmar, Tenasserim, Moulmein [Mawlamyine], *C. S. P. Parish s.n. [37]* (holotype, K).

Distribution. Manipur, ?Mizoram, ?Nagaland [Myanmar].

Habit. Terrestrial herb.

Eulophia mannii (Rchb. f.) Hook. f., Fl. Brit. India [J. D. Hooker] 6(17): 4. 1890. *Cyrtopera mannii* Rchb. f., Flora 55: 274. 1872. *Graphorkis mannii* (Rchb. f.) Kuntze, Revis. Gen. Pl. 2: 662. 1891 (as "*Graphorchis*"). TYPE: India, Assam, *Mann s.n.* [*12*] (lectotype, W, designated by V. Kumar, 2015).

Distribution. Assam, Manipur, Meghalaya, Sikkim, West Bengal. Endemic.

Habit. Terrestrial herb.

Eulophia nicobarica N. P. Balakr. & N. G. Nair, Bull. Bot. Surv. India 15(3–4): 271. 1973 [1976]. TYPE: India, Nicobar Island, Passa, 15 June 1974, *N. G. Nair 1647A* (holotype, CAL).

Distribution. Andaman & Nicobar Islands. Endemic.

Habit. Terrestrial herb.

Eulophia nuda Lindl., Gen. Sp. Orchid. Pl.: 180. 1833. *Cyrtopera nuda* (Lindl.) Rchb. f., Flora 55: 274. 1872. *Graphorkis nuda* (Lindl.) Kuntze, Revis. Gen. Pl. 2: 662. 1891 (as "*Graphorchis*"). TYPE: Nepal, Morung [Morang] Hills, 27 Apr. 1810, *Wallich 7371 (leg. Buchanan-Hamilton)* (holotype, K-WALL).

Wolfia spectabilis Dennst., Schlüssel Hortus Malab. 38. 1818, nom. inval. (based on Rheede, Hort. Malab. 11: t. 36. 1692). *Eulophia spectabilis* Suresh in D. H. Nicolson, C. R. Suresh & K. S. Manilal, Interpret. Van Rheede's Hort. Malab.: 300. 1988, nom. superfl.
Cyrtopera plicata Lindl., Gen. Sp. Orch. Pl.: 190. 1833. TYPE: Nepal, Morung [Morang] Hills, 19 Apr. 1810, *Wallich 7362 (leg. Buchanan-Hamilton)* (holotype, K-WALL).

Cyrtopera fusca Wight, Icon. Pl. Ind. Orient. [Wight] 5(1): 11, t. 1690. 1851. TYPE: India, Nilgiri Hills, Kartairy Falls near Kaitie, May–June, *Wight s.n.* (holotype, K).
Eulophia bicolor Dalzell, Hooker's J. Bot. Kew Gard. Misc. 3: 343. 1851. TYPE: India, Bombay, Syhadrad, *Dalzell s.n.* (holotype, K).
Cyrtopera mysorensis Lindl., J. Proc. Linn. Soc., Bot. 3: 32. 1859 [1858]. TYPE: India, Mysore, *Stocks 56 (leg. Law)* (holotype, K-LINDL).
Eulophia nuda Lindl. var. *andersonii* Hook. f., Ann. Roy. Bot. Gard. (Calcutta) 5: 33, t. 50. 1895 (as "*andersoni*"). TYPE: India, Darjeeling, May 1862, fid. in H. B. C. May 1867 from Darjeeling, *Anderson s.n.* (not found).

Distribution. Andaman & Nicobar Islands, Andhra Pradesh, Arunachal Pradesh, Assam, Bihar, Chhattisgarh, Goa, Gujarat, Jharkhand, Karnataka, Kerala, Madhya Pradesh, Maharashtra, Manipur, Meghalaya, Mizoram, Nagaland, Odisha, Punjab, Sikkim, Tamil Nadu, Tripura, Uttar Pradesh, Uttarakhand, West Bengal [Bangladesh, Bhutan, China, Myanmar, Nepal, Sri Lanka].

Habit. Terrestrial herb.

Eulophia obtusa (Lindl.) Hook. f., Fl. Brit. India [J. D. Hooker] 6(17): 3. 1890. *Cyrtopera obtusa* Lindl., Gen. Sp. Orchid. Pl.: 190. 1833. *Graphorkis obtusa* (Lindl.) Kuntze, Revis. Gen. Pl. 2: 662. 1891 (as "*Graphorchis*"). *Lissochilus obtusus* (Lindl.) Schltr., Repert. Spec. Nov. Regni Veg. Beih. 4: 260. 1919. TYPE: India, Uttarakhand, Tonse River near Deokhutul, *J. F. Royle s.n.* (holotype, K-LINDL; isotype, K).

Eulophia campanulata Duthie, J. Asiat. Soc. Bengal, Pt. 2, Nat. Hist. 71: 39. 1902. SYNTYPES: India, Dehra Dun, Karwapani, *W. Bell s.n.* (syntype, DD); *P. W. Mackinnon's collector s.n.* (syntypes, DD, ?K [*Mackinnon 22722*]); Gonda District, Chandanpur, *Duthie's collectors s.n.* (syntype, DD).

Distribution. ?Assam, ?Chhattisgarh, Uttar Pradesh, Uttarakhand [Bangladesh, ?Nepal].

Habit. Terrestrial herb.

Remarks. This showy species from swampy lowland grassland has rarely been seen in the last hundred years or so and a conservation assessment is urgently needed; its only known site in Bangladesh has recently been all but destroyed (Sourav et al., 2017). Its occurrence in Assam and Nepal seems doubtful and requires confirmation. A tentative record from Chhattisgarh is based on a painting by Hormusji at Kew, which differs in details of the callus structure from *Eulophia obtusa*. It is not clear how significant these differences are or how accurate the painting is.

In 2020, several populations of *Eulophia obtusa* were found in the Dudhwa Tiger Reserve in Uttar Pradesh (Sourav, pers. comm.).

Eulophia ochreata Lindl., J. Proc. Linn. Soc., Bot. 3: 24. 1859 [1858]. *Graphorkis ochreata* (Lindl.) Kuntze, Revis. Gen. Pl. 2: 662. 1891 (as "*Graphorchis*"). SYNTYPES: India, Concan, *Law s.n.* (syntypes, CAL, K); Canara, Dharwar, and Canara Districts, *Stocks 71* (syntype, K).

Distribution. Andhra Pradesh, Chhattisgarh, Gujarat, Jharkhand, Karnataka, Madhya Pradesh, Maharashtra, Meghalaya, Odisha, Rajasthan, West Bengal. Endemic.

Habit. Terrestrial herb.

Eulophia picta (R. Br.) Ormerod, Checkl. Papuasian Orchids: 293. 2017. *Cymbidium pictum* R. Br., Prodr. Fl. Nov. Holland.: 331. 1810. *Geodorum pictum* (R. Br.) Lindl., Gen. Sp. Orchid. Pl.: 175. 1833. TYPE: Australia, "North Coast of New Holland," Dec. 1802 – Mar. 1803, *R. Brown s.n.* *[5507]* (lectotype, BM, designated by Clements, Austral. Orchid Res. 1: 82, 1989).

Limodorum densiflorum Lam., Encycl. (Lamarck) 3(2): 516. 1792. *Geodorum densiflorum* (Lam.) Schltr., Repert. Spec. Nov. Regni Veg. Beih. 4: 259. 1919. TYPE: India, Malabar, *Rheede*, icon. Hort. Malab. 2: 69, t. 35. 1692.
Limodorum nutans Roxb., Pl. Coromandel 1: 33, t. 40. 1795. *Cymbidium nutans* (Roxb.) Sw., Nova Acta Regiae Soc. Sci. Upsal. 6: 77. 1799. *Malaxis nutans* (Roxb.) Willd., Sp. Pl., ed. 4 [Willdenow] 4(1): 93. 1805. *Geodorum purpureum* R. Br., Hort. Kew., ed. 2 [W. T. Aiton] 5: 207. 1813, nom. superfl. TYPE: India, Coromandel, icon. *Roxburgh, loc. cit.*
Malaxis cernua Willd., Sp. Pl., ed. 4, 4: 93. 1805. *Otandra cernua* (Willd.) Salisb., Trans. Hort. Soc. London 1: 298. 1812. *Cistella cernua* (Willd.) Blume, Bijdr. Fl. Ned. Ind.: 293. 1825. *Ortmannia cernua* (Willd.) Opiz, Flora 17: 592. 1834. *Eulophia cernua* (Willd.) T. C. Hsu, Illustr. Fl. Taiwan 2: 36. 2016. *Eulophia cernua* (Willd.) M. W. Chase, Kumar & Schuit., Phytotaxa 491(1): 51. 2021, isonym. TYPE: India, Malabar, *Rheede*, icon. Hort. Malab. 2: 69, t. 35. 1692.
Geodorum pallidum D. Don, Prodr. Fl. Nepal.: 31. 1825. TYPE: Nepal, 1818, *Wallich s.n.* (syntypes, BM, K).
Limodorum candidum Roxb., Fl. Ind. ed. 1832, 3: 470. 1832. *Geodorum candidum* (Roxb.) Lindl., Fol. Orchid. 6/7 (*Geodorum*): 8. 1855. TYPE: India, Sylhet, icon. *Limodorum candidum, Roxburgh 2343* (syntypes, CAL, K).
Geodorum appendiculatum Griff., Calcutta J. Nat. Hist. 5: 357. 1845. TYPE: India, Assam, *cult. Calcutta Bot. Garden s.n.* (not found).
Geodorum rariflorum Lindl., Fol. Orchid. 6/7(*Geodorum*):2. 1855. TYPE: India, Assam, *Jenkins s.n.* (holotype, K-LINDL).
Geodorum dilatatum sensu Hook. f., Fl. Brit. India [J. D. Hooker] 6(17): 17. 1890, non R. Br., 1813.
Geodorum densiflorum (Lam.) Schltr. var. *kalimpongense* R. Yonzone, Lama & Bhujel, McAllen Int. Orchid Soc. J. 13(6): 7. 2012. TYPE: India, West Bengal, Darjeeling Distr.: Kalimpong Sub-Division, alt. 780–1400 m, *R. Yonzone, D. Lama, R. B. Bhujel & Samuel Rai s.n.* (holotype, CAL; isotype, NBU, C, CWC).

Distribution. Andaman & Nicobar Islands, Andhra Pradesh, Arunachal Pradesh, Assam, Bihar, Chhattisgarh, Goa, Jharkhand, Karnataka, Kerala, Madhya Pradesh, Maharashtra, Manipur, Meghalaya, Mizoram, Nagaland, Odisha, Sikkim, Tamil Nadu, Tripura, Uttarakhand, West Bengal [Bangladesh, Bhutan, China, Myanmar, Nepal, Sri Lanka].

Habit. Terrestrial herb.

Remarks. Paul Ormerod (pers. comm.) has pointed out that *Limodorum densiflorum* Lam. and *Malaxis cernua* Willd. are based on the same type and that therefore the latter is an illegitimate name.

Eulophia pratensis Lindl., J. Proc. Linn. Soc., Bot. 3: 25. 1859 [1858]. *Graphorkis pratensis* (Lindl.) Kuntze, Revis. Gen. Pl. 2: 662. 1891 (as "*Graphorchis*"). TYPE: India, Deccan, *Stocks 22 bis* (holotype, K-LINDL).

Distribution. Gujarat, Karnataka, Kerala, Maharashtra, Tamil Nadu. Endemic.

Habit. Terrestrial herb.

Eulophia promensis Lindl., Gen. Sp. Orchid. Pl.: 181. 1833. *Tainia promensis* (Lindl.) Hook. f., Hooker's Icon. Pl. 19: t. 1883. 1889. *Ascotainia promensis* (Lindl.) Schltr., Repert. Spec. Nov. Regni Veg. Beih. 4: 246. 1919. *Ania promensis* (Lindl.) Senghas, Orchideen (Schlechter) 1(14): 863. 1984. TYPE: Myanmar, Pegu, Prome Hills, Sep.–Oct. 1826, *Wallich 7365* (holotype, K-WALL).

Eulophia geniculata King & Pantl., J. Asiat. Soc. Bengal, Pt. 2, Nat. Hist. 64 (2): 337. 1895 [1896]. TYPE: India, "Sikkim" [West Bengal], Teesta Valley, 1000 ft, Aug., *Pantling s.n.* [?282] (?holotype, K [Rungeet Valley, 1500 ft, Sep. 1894]).

Distribution. Manipur, Nagaland, West Bengal [Myanmar].

Habit. Terrestrial herb.

Eulophia pulchra (Thouars) Lindl., Gen. Sp. Orchid. Pl.: 182. 1833. *Limodorum pulchrum* Thouars, Hist. Orchid. t. 43. 1822. *Graphorkis pulchra* (Thouars) Kuntze, Revis. Gen. Pl. 2: 662. 1891 (as "*Graphorchis*"). *Lissochilus pulcher* (Thouars) H. Perrier, Fl. Madag. 49(2): 41. 1941, nom. illeg., non Schltr., 1915. *Eulophidium pulchrum* (Thouars) Summerh., Bull. Jard. Bot. État Bruxelles 27: 400. 1957. *Oeceoclades pulchra* (Thouars) P. J. Cribb & M. A. Clem., Austral. Orchid Res. 1: 99. 1989. TYPE: Borbonia [Réunion], icon. *Thouars Hist. Orchid. t. 43. 1822.*

Eulophia macrostachya Lindl., Gen. Sp. Orchid. Pl.: 183. 1833. TYPE: Sri Lanka, *Macrae s.n.* (not found).

Distribution.   Karnataka, Kerala, Tamil Nadu [Sri Lanka].

Habit.   Terrestrial herb.

Eulophia recurva (Roxb.) M. W. Chase, Kumar & Schuit., Phytotaxa 491(1): 53. 2021. *Limodorum recurvum* Roxb., Pl. Coromandel 1: 33, t. 39. 1795. *Geodorum recurvum* (Roxb.) Alston in H. Trimen, Handb. Fl. Ceylon 6(Suppl.): 276. 1931. TYPE: India, coast of Coromandel, moist valleys among hills, icon. *Roxburgh, loc. cit.*

Geodorum dilatatum R. Br., Hort. Kew., ed. 2 [W. T. Aiton] 5: 207. 1813. TYPE: *cult. Banks s.n.* (not found).

Distribution.   Andhra Pradesh, Assam, Jharkhand, Madhya Pradesh, Maharashtra, Manipur, Odisha [China, Myanmar].

Habit.   Terrestrial herb.

Eulophia santapaui S. G. Panigrahi & Kataki, Sci. & Cult. 33: 124. 1967. TYPE: India, Meghalaya, Umsaw, 15 Oct. 1954, *Deka s.n.* (holotype, ASSAM, 36002).

Distribution.   Meghalaya. Endemic.

Habit.   Terrestrial herb.

Remarks.   This species is known only from the type locality. A review of its taxonomy and a conservation assessment are needed.

Eulophia zollingeri (Rchb. f.) J. J. Sm., Orch. Java 6: 228. 1905. *Cyrtopera zollingeri* Rchb. f., Bonplandia 5(3): 38. 1857. TYPE: Indonesia, Sumatra, Lampung ("Lampong"), Sep. 1845, *Zollinger 585* (holotype, W).

Cyrtopera sanguinea Lindl., J. Proc. Linn. Soc., Bot. 3: 32. 1859 [1858]. *Eulophia sanguinea* (Lindl.) Hook. f., Fl. Brit. India [J. D. Hooker] 6(17): 8. 1890. *Graphorkis sanguinea* (Lindl.) Kuntze, Revis. Gen. Pl. 2: 662. 1891 (as "*Graphorchis*"). SYNTYPES: India, Sikkim, *Cathcart s.n.* (not found); Sikkim, alt. 4000–5000 ft, *J. D. Hooker 223* (syntypes, K, K-LINDL); *J. D. Hooker 361* (syntype, K). *Eulophia macrorhizon* Hook. f., Fl. Brit. India [J. D. Hooker] 6(17): 7. 1890. TYPE: India, Sikkim Himalaya, Ryang, alt. 2000 ft, *King s.n. [4816]* (holotype, CAL). *Eulophia emilianae* C. J. Saldanha, Indian Forester 100(9): 566. 1974. TYPE: India, Karnataka, Hassan District, Shiradi Ghat, *C. J. Saldanha 16923* (holotype, St. Joseph's College, Bangalore Herbarium).

Distribution.   Andaman & Nicobar Islands, Arunachal Pradesh, Assam, Karnataka, Kerala, Manipur, Meghalaya, Nagaland, Sikkim [China, Sri Lanka].

Habit.   Holomycotrophic terrestrial herb.

———

Galearis Raf., Herb. Raf.: 71. 1833.

TYPE: *Galearis spectabilis* (L.) Raf. (*Orchis spectabilis* L.).

Galearis roborovskii (Maxim.) S. C. Chen, P. J. Cribb & S. W. Gale, Fl. China 25: 92. 2009 (as "*roborowskyi*"). *Orchis roborovskii* Maxim., Bull. Acad. Imp. Sci. Petersb. 31: 104. 1887. *Galeorchis roborovskii* (Maxim.) Nevski, in Kom., Fl. URSS 4: 670. 1935 (as "*roborovskyi*"). *Chusua roborovskii* (Maxim.) P. F. Hunt, Kew Bull. 26: 175. 1975 (as "*roborowskyi*"). *Aorchis roborovskii* (Maxim.) Seidenf., Nordic J. Bot. 2: 9. 1982. TYPE: China, Tangut, *Przewalski s.n.*, 1880.

Orchis stracheyi Hook. f., Fl. Brit. India [J. D. Hooker] 6(17): 128. 1890. *Gymnadenia stracheyi* Hook. f., Fl. Brit. India [J. D. Hooker] 6(17): 128. 1890, in syn. *Galeorchis stracheyi* (Hook. f.) Soó, Acta Bot. Acad. Sci. Hung. 12: 352. 1966. *Galearis stracheyi* (Hook. f.) P. F. Hunt, Kew Bull. 26: 172. 1971. TYPE: India, Garwhal, near Rogile, alt. 11,000 ft, *Strachey & Winterbottom 35*.

Distribution.   Arunachal Pradesh, Uttarakhand [Bhutan, China, Nepal].

Habit.   Terrestrial herb.

Galearis spathulata (Lindl.) P. F. Hunt, Kew Bull. 26(1): 172. 1971. *Gymnadenia spathulata* Lindl., Gen. Sp. Orchid. Pl.: 280. 1835. *Habenaria spathulata* (Lindl.) Benth., J. Linn. Soc., Bot. 18: 355. 1881. *Orchis spathulata* (Lindl.) Rchb. f. ex Hook. f., Fl. Brit. India [J. D. Hooker] 6(17): 127. 1890, nom. illeg. non L. f., 1782. *Galeorchis spathulata* (Lindl.) Soó, Acta Bot. Acad. Sci. Hung. 12: 351. 1966. *Aorchis spathulata* (Lindl.) Verm., Jahresber. Naturwiss. Vereins Wuppertal 25: 33. 1972. TYPE: India, Kedarkanta, *Royle s.n.* (holotype, K-LINDL).

Distribution.   Himachal Pradesh, Sikkim, Uttarakhand [Bhutan, China, Myanmar, Nepal].

Habit.   Terrestrial herb.

Galearis tschiliensis (Schltr.) P. J. Cribb, S. W. Gale & R. M. Bateman, Ann. Bot. (Oxford) 104: 439. 2009. *Aceratorchis tschiliensis* Schltr., Repert. Spec. Nov. Regni Veg. Beih. 12: 329. 1922. *Orchis tschiliensis* (Schltr.) Soó, Ann. Hist.-Nat. Mus. Natl. Hung. 26: 351. 1929. TYPE: China, Hebei [Hubei], Hsiao Wu tai shan, Hsi tai, *Limpricht 3039* (holotype, WRSL, ?lost; isotype, B, lost).

Distribution. Arunachal Pradesh [China].

Habit. Terrestrial herb.

———

Galeola Lour., Fl. Cochinch. 2: 520. 1790.

TYPE: *Galeola nudifolia* Lour.

Galeola cathcartii Hook. f., Fl. Brit. India [J. D. Hooker] 6(17): 89. 1890. TYPE: India, Sikkim Himalaya, *Cathcart s.n.* (holotype, CAL, copy K).

Distribution. Sikkim, West Bengal [also in Thailand].

Habit. Holomycotrophic terrestrial climber.

Galeola nudifolia Lour., Fl. Cochinch. 2: 521. 1790. TYPE: Indochina, *Loureiro s.n.* (holotype, BM).

Galeola hydra Rchb. f., Xenia Orchid. 2: 77. 1865. TYPE: Malaysia: Malacca, *Griffith s.n.* (holotype, K-LINDL).

Distribution. Arunachal Pradesh, Assam, Sikkim [Bhutan, China, Myanmar].

Habit. Holomycotrophic terrestrial climber.

———

Gastrochilus D. Don, Prodr. Fl. Nepal.: 32. 1825.

TYPE: *Gastrochilus calceolaris* (Buch.-Ham. ex Sm.) D. Don.

Gastrochilus acaulis (Hook. f.) Kuntze, Revis. Gen. Pl. 2: 661. 1891. *Cleisostoma acaule* Lindl., Gen. Sp. Orchid. Pl.: 227. 1833 (as "*acaulis*"). *Saccolabium acaule* (Lindl.) Hook. f., Fl. Brit. India [J. D. Hooker] 6(17): 61. 1890. TYPE: Sri Lanka, *Macrae s.n.* (not found).

Vanda pulchella Wight, Icon. Pl. Ind. Orient. [Wight] 5(1): 9, t. 1671. 1851 (as "*pulchilla*"). *Saccolabium nilagiricum* Hook. f., Fl. Brit. India [J. D. Hooker] 6(17): 60. 1890, nom. superfl. *Gastrochilus nilagiricus* Kuntze, Revis. Gen. Pl. 2: 661. 1891, nom. superfl. *Gastrochilus pulchellus* (Wight) Schltr., Repert. Spec. Nov. Regni Veg. 12: 315. 1913, nom. illeg., non Ridley, 1906. *Saccolabium pulchellum* (Wight) C. E. C. Fisch., Fl. Madras 3(8): 1446. 1928. *Gastrochilus indicus* Garay, Bot. Mus. Leafl. 23(4): 180. 1972. TYPE: India, Nilgiri Hills, on the banks of the Kaitairy River, *Wight s.n.* (holotype, K).

Distribution. Karnataka, Kerala, Odisha, Tamil Nadu [Myanmar, Nepal, Sri Lanka].

Habit. Epiphytic herb.

Gastrochilus acutifolius (Lindl.) Kuntze, Revis. Gen. Pl. 2: 661. 1891. *Saccolabium acutifolium* Lindl.,

Gen. Sp. Orchid. Pl.: 223. 1833. TYPE: India Orientali, icon. *Wallich 1137* (holotype, BM).

Distribution. Arunachal Pradesh, Assam, Manipur, Meghalaya, Nagaland, Sikkim, Uttarakhand, West Bengal [Bhutan, ?Myanmar, Nepal].

Habit. Epiphytic herb.

Gastrochilus affinis (King & Pantl.) Schltr., Repert. Spec. Nov. Regni Veg. 12: 314. 1913. *Saccolabium affine* King & Pantl., Ann. Roy. Bot. Gard. (Calcutta) 8: 228, t. 304. 1898. SYNTYPES: India, Sikkim, above Pemiongtsi [Pamayangtse], alt. 8000 ft, June, *Pantling 444* (not found); Sikkim, eastern frontier [Lachong (Lachung) Valley, 8000 ft, July 1897], *Pantling 444* (syntype, BM, CAL, K).

Distribution. Arunachal Pradesh, Sikkim, West Bengal [Bhutan, Nepal].

Habit. Epiphytic herb.

Gastrochilus arunachalensis A. N. Rao, J. Econ. Taxon. Bot. 16(3): 723. 1992. TYPE: India, Arunachal Pradesh, West Kameng District, Dimachang, alt. 500 ft, *A. N. Rao 24220* (holotype, Orchid Herbarium, Tipi).

Distribution. Arunachal Pradesh [Myanmar].

Habit. Epiphytic herb.

Gastrochilus bellinus (Rchb. f.) Kuntze, Rev. Gen. Pl. 2: 661. 1891. *Saccolabium bellinum* Rchb. f., Gard. Chron., n.s., 21: 174. 1884. TYPE: Myanmar, *cult. Low (leg. Boxall) s.n.* (holotype, W).

Distribution. Manipur [China, Myanmar].

Habit. Epiphytic herb.

Gastrochilus calceolaris (Buch.-Ham. ex Sm.) D. Don, Prodr. Fl. Nepal.: 32. 1825. *Aerides calceolaris* Buch.-Ham. ex Sm., Cycl. (Rees) 39: no. 11. 1818 (as "*calceolare*"). *Epidendrum calceolare* (Buch.-Ham. ex Sm.) D. Don, Prodr. Fl. Nepal.: 32. 1825. *Saccolabium calceolare* (Buch.-Ham. ex Sm.) Lindl., Gen. Sp. Orchid. Pl.: 223. 1833. TYPE: Nepal, Bagmati Zone, Kathmandu, Narainhetty, *Buchanan-Hamilton s.n.*

Distribution. Andaman & Nicobar Islands, Andhra Pradesh, Arunachal Pradesh, Assam, Himachal Pradesh, Karnataka, Kerala, Manipur, Meghalaya, Mizoram, Nagaland, Sikkim, Tamil Nadu, Tripura, Uttar Pradesh, Uttarakhand, West Bengal [Bhutan, China, Myanmar, Nepal].

Habit. Epiphytic herb.

Gastrochilus carnosus Z. H. Tsi, Guihaia 16: 137. 1996. TYPE: India, Assam, Jaintia Hills, Sundai, Apr. 1899, *Prain's collector 167* (holotype, AMES).

Distribution. Meghalaya. Endemic.

Habit. Epiphytic herb.

Gastrochilus corymbosus A. P. Das & S. Chanda, J. Econ. Taxon. Bot. 12(2): 401. 1988 [1989]. TYPE: India, West Bengal, Darjeeling, Jalaphar, alt. 7000 ft, 29 Oct. 1982, *A. P. Das 823* (holotype, CAL, not found).

Distribution. West Bengal [Myanmar].

Habit. Epiphytic herb.

Remarks. This was reduced to the synonymy of *Gastrochilus pseudodistichus* (King & Pantl.) Schltr. by Singh et al. (2019). Kumar et al. (2014) pointed out that the type material of *G. corymbosus* could not be found at CAL and based on the protologue considered it distinct.

Gastrochilus dasypogon (Sm.) Kuntze, Revis. Gen. Pl. 2: 661. 1891. *Aerides dasypogon* Sm., Cycl. (Rees) 39: Aerides no. 10. 1818. *Saccolabium dasypogon* (Sm.) Lindl., Gen. Sp. Orchid. Pl.: 222. 1833. TYPE: Nepal, *Buchanan-Hamilton s.n.*

Distribution. Arunachal Pradesh, Assam, Karnataka, Kerala, ?Meghalaya, Mizoram, Sikkim, West Bengal [Nepal].

Habit. Epiphytic herb.

Gastrochilus distichus (Lindl.) Kuntze, Revis. Gen. Pl. 2: 661. 1891. *Saccolabium distichum* Lindl., J. Proc. Linn. Soc., Bot. 3: 36. 1859 [1858]. SYNTYPES: India, Sikkim, alt. 6000–8000 ft, *J. D. Hooker 206* (holotype, K); Khasi Hills, alt. 5000–6000 ft, *J. D. Hooker 83* (syntype, K-LINDL).

Distribution. Himachal Pradesh, Mizoram, Sikkim, Tripura, Uttarakhand, West Bengal [Bhutan, China, Myanmar, Nepal].

Habit. Epiphytic herb.

Gastrochilus flabelliformis (Blatt. & McCann) C. J. Saldanha in C. J. Saldanha & D. H. Nicolson, Fl. Hassan Dist. 830. 1976. *Saccolabium flabelliforme* Blatt. & McCann, J. Bombay Nat. Hist. Soc. 35(4): 722, t. 1. 1932. TYPE: India, North Kanara, Devimane Ghat, *Sedgwick & T. R. Bell 6975* (holotype, BLATT).

Distribution. Goa, Karnataka, Kerala, Maharashtra, Tamil Nadu. Endemic.

Habit. Epiphytic herb.

Gastrochilus garhwalensis Z. H. Tsi, Guihaia 16: 138. 1996. TYPE: India, Uttarakhand, Garhwal, Paravá Kotah Range, 15 June 1902, *Inayat s.n.* (holotype, AMES).

Distribution. Uttarakhand. Endemic.

Habit. Epiphytic herb.

Gastrochilus intermedius (Griff. ex Lindl.) Kuntze, Revis. Gen. Pl. 2: 661. 1891. *Saccolabium intermedium* Griff. ex Lindl., J. Proc. Linn. Soc., Bot. 3: 33. 1859 [1858]. TYPE: India, Khasi Hills, Musmai, Oct. 1835, *Griffith s.n.* (holotype, K-LINDL; isotype, K).

Distribution. Arunachal Pradesh, Meghalaya [also in Vietnam].

Habit. Epiphytic herb.

Remarks. In Swami (2017), a different species is illustrated as *Gastrochilus intermedius*, possibly *G. bellinus* (Rchb. f.) Kuntze.

Gastrochilus linearifolius Z. H. Tsi & Garay, Guihaia 16: 138. 1996. TYPE: India, Sikkim, Nepal Frontier, *Pantling 356* (holotype, AMES; isotype, BM, CAL, K, W).

Distribution. Meghalaya, Sikkim [China].

Habit. Epiphytic herb.

Gastrochilus obliquus (Lindl.) Kuntze, Revis. Gen. Pl. 2: 661. 1891. *Saccolabium obliquum* Lindl., Gen. Sp. Orchid. Pl.: 223. 1833. TYPE: Myanmar, Tung Dong, Nov. 1826, *Wallich 7304* (holotype, K-LINDL; isotype, K-WALL).

Saccolabium bigibbum Rchb. f. ex Hook. f., Bot. Mag. 95: t. 5767. 1869. *Gastrochilus bigibbus* (Rchb. f. ex Hook. f.) Kuntze, Revis. Gen. Pl. 2: 661. 1891. TYPE: Myanmar, Rangoon or Arrakan, *Benson s.n.* (holotype, W).

Distribution. ?Andaman & Nicobar Islands, Assam, Kerala, Manipur, Meghalaya, Nagaland, Sikkim, West Bengal [Bhutan, China, Myanmar].

Habit. Epiphytic herb.

Gastrochilus pseudodistichus (King & Pantl.) Schltr., Repert. Spec. Nov. Regni Veg. 12: 315. 1913. *Saccolabium pseudodistichum* King & Pantl., J. Asiat. Soc. Bengal, Pt. 2, Nat. Hist. 64: 341. 1895

[1896]. *Gastrochilus pseudodistichus* (King & Pantl.) Seidenf., Contr. Rev. Orchid Fl. Cambodia Laos Vietnam 1: 65. 1975 (as "*pseudodisticus*"), isonym. TYPE: India, West Bengal ('Sikkim'), Senchul, alt. 6000–8000 ft, Aug.–Oct. 1892, *Pantling s.n.* [*49*] (lectotype, CAL, designated by P. Kumar et al., 2014; isotype, K).

Distribution. ?Meghalaya, Mizoram, Nagaland, Sikkim, West Bengal [Bhutan, China, ?Myanmar].

Habit. Epiphytic herb.

Gastrochilus rutilans Seidenf., Opera Bot. 95: 293. 1988. TYPE: Thailand, Khao Yai, 2240 ft, *Cumberlege 962* (holotype, C).

Distribution. Arunachal Pradesh [also in Thailand].

Habit. Epiphytic herb.

Gastrochilus sessanicus A. N. Rao, J. Orchid Soc. India 11(1–2): 1. 1997. TYPE: India, Arunachal Pradesh, West Kameng District, Sessa, alt. 3500 ft, *A. N. Rao 29273* (holotype, Orchid Herbarium, Tipi).

Distribution. Arunachal Pradesh. Endemic.

Habit. Epiphytic herb.

Gastrochilus sonamii Lucksom, Orchid Rev. 111 (1253): 278. 2003. TYPE: India, Sikkim, Rachela, 19 Dec. 2002, *Lucksom 455a* (holotype, CAL).

Distribution. Sikkim, West Bengal. Endemic.

Habit. Epiphytic herb.

———

Gastrodia R. Br., Prodr. Fl. Nov. Holland. 330. 1810.

TYPE: *Gastrodia sesamoides* R. Br.

Gastrodia arunachalensis S. N. Hegde & A. N. Rao, Orchid Rev. 93: 171. 1985. TYPE: India, Arunachal Pradesh, West Kameng District, Sessa, alt. 3500 ft, *S. N. Hegde 9250* (holotype, Orchid Herbarium, Tipi).

Distribution. Arunachal Pradesh. Endemic.

Habit. Holomycotrophic terrestrial herb.

Gastrodia dyeriana King & Pantl., J. Asiat. Soc. Bengal, Pt. 2, Nat. Hist. 64: 342. 1895 [1896]. TYPE: India, Sikkim, alt. 7000 ft, Aug. 1893, *Pantling s.n.* [*293*] (holotype, CAL; isotype, BM, K, W).

Distribution. Arunachal Pradesh, Sikkim. Endemic.

Habit. Holomycotrophic terrestrial herb.

Gastrodia elata Blume, Mus. Bot. 2: 174. 1856. TYPE: Japan, Ten-ma, *Bürger s.n.* (holotype, L).

Gastrodia mairei Schltr., Repert. Spec. Nov. Regni Veg. 12: 105. 1913. TYPE: Yunnan: Long-Ky, *Maire 13* (holotype, B).

Distribution. Assam, Sikkim, West Bengal [Bhutan, China, Myanmar, Nepal].

Habit. Holomycotrophic terrestrial herb.

Gastrodia exilis Hook. f., Fl. Brit. India [J. D. Hooker] 6(17): 123. 1890. TYPE: India, Khasi Hills, Amwee, alt. 3000 ft, 24 Sep. 1850, *Hooker & Thomson s.n.* [*2379, orchid 320*] (holotype, K).

Distribution. Arunachal Pradesh, Assam, Kerala, Meghalaya [also in Thailand, etc.].

Habit. Holomycotrophic terrestrial herb.

Remarks. A different species, not listed here, is illustrated as *Gastrodia exilis* in Swami (2016, 2017).

Gastrodia falconeri D. L. Jones & M. A. Clem., Orchadian 12(8): 350. 1998.

Gamoplexis orobanchoides Falc., Trans. Linn. Soc. London 20(2): 293, t. 13. 1847. *Gastrodia orobanchoides* (Falc.) Hook. f., Hooker's Icon. Pl. 19: t. 1852. 1889, nom. illeg., non F. Muell., 1873. TYPE: Himalaya (Emodus Mtns.), *Falconer 1083* (holotype, K).

Distribution. Himachal Pradesh, Jammu & Kashmir, Uttarakhand [Pakistan].

Habit. Holomycotrophic terrestrial herb.

Gastrodia mishmensis A. N. Rao, Harid. & S. N. Hegde, Arunachal Pradesh Forest News 9(1): 10. 1991. TYPE: India, Arunachal Pradesh, Melinja, 2000 m, 30 Aug. 1986, *Haridasan 3529* (holotype, Arunachal Forest Herbarium, Chessa).

Distribution. Arunachal Pradesh. Endemic.

Habit. Holomycotrophic terrestrial herb.

Gastrodia silentvalleyana C. S. Kumar, P. C. S. Kumar, Sibi & S. Anil Kumar, Rheedea 18: 109. 2008. TYPE: India, Kerala, Palakkad District, Silent Valley National Park, ca. 1040 m, *Sibi 43300* (holotype, TBGT).

Distribution. Kerala. Endemic.

Habit. Holomycotrophic terrestrial herb.

Gennaria Parl., Fl. Ital. 3: 404. 1860.

TYPE: *Gennaria diphylla* (Link) Parl. (*Satyrium diphyllum* Link).

Gennaria griffithii (Hook. f.) X. H. Jin & D. Z. Li, Biodivers. Sci. 23: 240. 2015. *Habenaria griffithii* Hook. f., Fl. Brit. India [J. D. Hooker] 6(17): 197. 1890. *Habenaria decipiens* Hook. f., Fl. Brit. India [J. D. Hooker] 6(17): 165. 1890, nom. inval., non Wight, 1851. *Diphylax griffithii* (Hook. f.) Kraenzl., Orchid. Gen. Sp. 1: 599. 1899. *Dithrix decipiens* Soó, Ann. Hist.-Nat. Mus. Natl. Hung. 26: 369. 1929, nom. superfl. *Dithrix griffithii* (Hook. f.) Ormerod & Gandhi, Phytoneuron 2012-61: 3. 2012. *Nujiangia griffithii* (Hook. f.) X. H. Jin & D. Z. Li, J. Syst. Evol. 50: 68. 2012. SYNTYPES: Northwest India, Synj, 3000 ft, 1844, *Edgeworth s.n.* (syntype, K); India, Himachal Pradesh, Lahul, alt. 4000–5000 ft, *Thomson s.n.*; Afghanistan, *Griffith s.n.* [*424*] *(East India Company Herbarium 5326)* (syntypes, CAL, K [2×]); Pakistan, Kurrum Valley, Shalizan, 7000 ft, 18 May 1879, *Aitchison 322* (syntype, K).

Distribution. Himachal Pradesh, Uttarakhand [Afghanistan, Pakistan].

Habit. Terrestrial herb.

Goodyera R. Br., Hort. Kew., ed. 2 [W. T. Aiton] 5: 197. 1813.

TYPE: *Goodyera repens* (L.) R. Br. (*Satyrium repens* L.).

Goodyera biflora (Lindl.) Hook. f., Fl. Brit. India [J. D. Hooker] 6(17): 114. 1890.

Georchis biflora Lindl., Gen. Sp. Orchid. Pl.: 496. 1840. *Epipactis biflora* (Lindl.) A. A. Eaton, Proc. Biol. Soc. Wash. 21: 63. 1908. TYPE: Nepal, *Wallich 7379* (holotype, K-LINDL; isotype, K-WALL).
Goodyera macrantha Maxim., Gartenflora 16: 36. 1867. TYPE: Japan, cultivated at Yedo, *Maximowicz s.n.* (lectotype, LE, designated by Bhattacharjee & Chowdhery, 2018).

Distribution. Arunachal Pradesh, Himachal Pradesh, Uttarakhand, West Bengal [China, Nepal].

Habit. Terrestrial herb.

Goodyera foliosa (Lindl.) Benth. ex C. B. Clarke, J. Linn. Soc., Bot. 25: 73. 1889. *Georchis foliosa* Lindl., Gen. Sp. Orchid. Pl.: 496. 1840. *Orchiodes*

foliosum (Lindl.) Kuntze, Revis. Gen. Pl. 2: 675. 1891. *Epipactis foliosa* (Lindl.) A. A. Eaton, Proc. Biol. Soc. Wash. 21: 64. 1908. TYPE: Myanmar, *Griffith s.n.* (holotype, K-LINDL).

Goodyera secundiflora Griff., Not. Pl. Asiat. 3: 393. 1851. *Orchiodes secundiflorum* (Griff.) Kuntze, Revis. Gen. Pl. 2: 675. 1891. TYPE: India, Khasia Hills, Surureem to Moflong, 1844, *Griffith s.n.* [*35*] *in Assam Herb. 188* (holotype, K-LINDL; isotype, CAL, K).
Goodyera andersonii King & Pantl., J. Asiat. Soc. Bengal, Pt. 2, Nat. Hist. 65: 127. 1896. TYPE: India, Sikkim, Bucheem, 8000 ft, Nov., *Anderson 1228* (holotype, CAL).

Distribution. Arunachal Pradesh, Manipur, Meghalaya, Mizoram, Nagaland, Sikkim, Uttarakhand, West Bengal [Bhutan, China, Myanmar, Nepal].

Habit. Terrestrial herb.

Goodyera fumata Thwaites, Enum. Pl. Zeyl. [Thwaites]: 314. 1861. *Orchiodes fumatum* (Thwaites) Kuntze, Revis. Gen. Pl. 2: 675. 1891. *Salacistis fumata* (Thwaites) T. C. Hsu, Illustr. Fl. Taiwan 2: 184. 2016. *Salacistis fumata* (Thwaites) M. C. Pace, Brittonia 72: 265. 2020, isonym. TYPE: Sri Lanka, Central Province, *Thwaites s.n.* [*Ceylon Plants 3668*] (holotype, K).

Distribution. Arunachal Pradesh, Kerala, Nagaland, Odisha, Sikkim, Tamil Nadu [China, Myanmar, Sri Lanka].

Habit. Terrestrial herb.

Goodyera fusca (Lindl.) Hook. f., Fl. Brit. India [J. D. Hooker] 6(17): 112. 1890. *Hetaeria fusca* Lindl., Gen. Sp. Orchid. Pl.: 491. 1840 (as "*Ætheria*"). *Cystorchis fusca* (Lindl.) Benth. & Hook. f., Gen. Pl. 3: 599. 1883. *Orchiodes fuscum* (Lindl.) Kuntze, Revis. Gen. Pl. 2: 675. 1891. *Epipactis fusca* (Lindl.) A. A. Eaton, Proc. Biol. Soc. Wash. 21: 64. 1908. TYPE: Origin not known, *Wallich 7395* (lectotype, K-LINDL, designated by Bhattacharjee & Bhakat, 2010; isotype, K-WALL).

Distribution. Arunachal Pradesh, Himachal Pradesh, Jammu & Kashmir, Nagaland, Sikkim, Uttarakhand [Bhutan, China, Myanmar, Nepal].

Habit. Terrestrial herb.

Goodyera hemsleyana King & Pantl., J. Asiat. Soc. Bengal, Pt. 2, Nat. Hist. 64: 342. 1895 [1896]. *Epipactis hemsleyana* (King & Pantl.) A. A. Eaton, Proc. Biol. Soc. Wash. 21: 64. 1908. TYPE: India, Sikkim, Senchal, alt. 7000 ft, July, *Pantling 215A* (lectotype, CAL, designated by Bhattacharjee & Bhakat, 2010; isotype, AMES, BM, CAL, K, P, W).

Distribution. Arunachal Pradesh, Sikkim, Uttarakhand, West Bengal [Bhutan, China, Myanmar, Nepal].

Habit. Terrestrial herb.

Goodyera hispida Lindl., J. Proc. Linn. Soc., Bot. 1: 183. 1857. *Orchiodes hispidum* (Lindl.) Kuntze, Revis. Gen. Pl. 2: 675. 1891. *Epipactis hispida* (Lindl.) A. A. Eaton, Proc. Biol. Soc. Wash. 21: 64. 1908. TYPE: India, Khasi Hills, Chuma, alt. 4000 ft, 18 Aug. 1850, *J. D. Hooker 2110* (holotype, K).

Distribution. Arunachal Pradesh, Assam, Meghalaya, Nagaland, Odisha, Sikkim, West Bengal [Nepal].

Habit. Terrestrial herb.

Goodyera marginata Lindl., Gen. Sp. Orchid. Pl.: 493. 1840. *Goodyera repens* (L.) R. Br. var. *marginata* (Lindl.) Tang & F. T. Wang, Acta Phytotax. Sin. 1: 68. 1951. SYNTYPES: India, Kumaon, without date, *Wallich (leg. Royle s.n.) 7394A* (syntypes, K, K-LINDL [2×]); Nepal, Gosainthan [Shishapangma], July 1821, *Wallich (leg. Emodi s.n.) 7394B* (syntypes, K, K-WALL).

Distribution. Himachal Pradesh, Uttarakhand [China, Nepal].

Habit. Terrestrial herb.

Remarks. The distribution in India is unclear, due to confusion with *Goodyera repens*. *Goodyera marginata* is distinguished by the subglabrous (vs. densely glandular hairy) flowers with relatively narrower and longer sepals.

Goodyera procera (Ker Gawl.) Hook., Exot. Fl. 1: t. 39. 1823. *Neottia procera* Ker Gawl., Bot. Reg. 8: t. 639. 1822. *Leucostachys procera* (Ker Gawl.) Hoffmanns., Verz. Orchid.: 26. 1842. *Orchiodes procerum* (Ker Gawl.) Kuntze, Revis. Gen. Pl. 2: 675. 1891. *Epipactis procera* (Ker Gawl.) A. A. Eaton, Proc. Biol. Soc. Wash. 21: 65. 1908. *Peramium procerum* (Ker Gawl.) Makino, J. Jap. Bot. 6: 36. 1929. *Cionisaccus procerus* (Ker Gawl.) M. C. Pace, Brittonia 72: 263. 2020 (as "*procera*"). TYPE: Nepal, *cult. Colville Nursery s.n.* (holotype, BM-LAMBERT).

Cordylestylis foliosa Falc., J. Bot. (Hooker) 4: 75. 1841. *Cordylestylis himalayensis* D. Dietr., Syn. Pl. 5:165. 1852, nom. superfl. TYPE: India, Rajpoor, 3500 ft, *cult. Saharunpoor s.n.*
Goodyera carnea A. Rich., Ann. Sci. Nat., Bot., sér. 2, 15: 80. 1841. TYPE: India, Nilgiri Hills, Konoor, Apr., *Perrottet s.n.* (holotype, P).

Distribution. Andaman & Nicobar Islands, Andhra Pradesh, Arunachal Pradesh, Assam, Bihar, Jharkhand, Karnataka, Kerala, Madhya Pradesh, Manipur, Meghalaya, Mizoram, Nagaland, Odisha, Sikkim, Tamil Nadu, Tripura, Uttar Pradesh, Uttarakhand, West Bengal [Bangladesh, Bhutan, China, Myanmar, Nepal, Sri Lanka].

Habit. Terrestrial herb.

Goodyera recurva Lindl., J. Proc. Linn. Soc., Bot. 1: 183. 1857. *Orchiodes recurvum* (Lindl.) Kuntze, Revis. Gen. Pl. 2: 675. 1891. *Epipactis recurva* (Lindl.) A. A. Eaton, Proc. Biol. Soc. Wash. 21: 65. 1908. TYPE: India, Khasi Hills, Mofling, alt. 5000–6000 ft, 2 Aug. 1850 & 26 Oct. 1850, *Hooker & Thomson [L.]345 [2007] & s.n.* (syntype, K [2×], K-LINDL).

Goodyera prainii Hook. f., Fl. Brit. India [J. D. Hooker] 6(17): 112. 1890. *Epipactis prainii* (Hook. f.) A. A. Eaton, Proc. Biol. Soc. Wash. 21: 65. 1908. *Goodyera recurva* Lindl. var. *prainii* (Hook. f.) Pradhan, Indian Orchids: Guide Identif. & Cult. 1: 106. 1976. TYPE: India, Naga Hills, Pulinabadya, alt. 7200 ft, Dec. 1856, *Prain 59* (holotype, K).

Distribution. Arunachal Pradesh, Assam, Meghalaya, Nagaland, Sikkim, West Bengal [Bhutan, China, Nepal].

Habit. Terrestrial or epiphytic herb.

Goodyera repens (L.) R. Br., Hort. Kew., ed. 2 [W. T. Aiton] 5: 198. 1813. *Satyrium repens* L., Sp. Pl. 2: 945. 1753. *Epipactis repens* (L.) Crantz, Stirp. Austr. Fasc., ed. 2, 2(6): 473. 1769. *Serapias repens* (L.) Vill., Hist. Pl. Dauphiné (Villars) 2: 53. 1787. *Neottia repens* (L.) Sw., Kongl. Vetensk. Acad. Nya Handl. 21: 226. 1800. *Orchis repens* (L.) Eyster ex Poir., Encycl. (Lamarck) 6: 581. 1805. *Peramium repens* (L.) Salisb., Trans. Hort. Soc. London 1: 301. 1812. *Gonogona repens* (L.) Link, Enum. Hort. Berol. Alt. 2: 369. 1822. *Goodyera pubescens* R. Br. var. *repens* (L.) Alph. Wood, Class-book Bot., ed. 2a: 537. 1847. *Elasmatium repens* (L.) Dulac, Fl. Hautes-Pyrénées 121. 1867. *Orchiodes repens* (L.) Kuntze, Revis. Gen. Pl. 2: 675. 1891. TYPE: icon. "*Orchis radice repente*" in Camerarius, Hortus Med. Phil., 111, t. 35. 1588 (lectotype, designated by Baumann et al., 1989).

Distribution. Arunachal Pradesh, Himachal Pradesh, Jammu & Kashmir, Nagaland, Sikkim, Uttarakhand [Bhutan, China, Nepal].

Habit. Terrestrial herb.

Remarks. Indian records, especially those from Himachal Pradesh and Uttarakhand, are in part referable to *Goodyera marginata*.

Goodyera robusta Hook. f., Fl. Brit. India [J. D. Hooker] 6(17): 113. 1890. *Epipactis robusta* (Hook. f.) A. A. Eaton, Proc. Biol. Soc. Wash. 21: 65. 1908. *Goodyera schlechtendaliana* Rchb. f. var. *robusta* (Hook. f.) Av. Bhattacharjee & H. J. Chowdhery, Kew Bull. 67: 506. 2012. TYPE: India, Khasi Hills, alt. 4000 ft, *Hooker & Thomson s.n.* [*orchid 328*] (holotype, K).

Goodyera dongchenii Lucksom, J. Indian Bot. Soc. 72(1–2): 191. 1993. TYPE: India, Sikkim, Fambong Lho Wildlife Sanctuary, *S. Z. Lucksom 210* (holotype, CAL; isotype, Herb. Forest Dept. Gangtok 210).

Distribution. Meghalaya, Sikkim [China].

Habit. Terrestrial herb.

Goodyera rubicunda (Blume) Lindl., Edwards's Bot. Reg. 25(Misc.): 61. 1839. *Neottia rubicunda* Blume, Bijdr. Fl. Ned. Ind. 8: 408. 1825. *Georchis rubicunda* (Blume) Rchb. f., Bonplandia (Hannover) 5: 35. 1857. *Rhamphidia rubicunda* (Blume) F. Muell., Fragm. (Mueller) 7: 30. 1869, nom. illeg., non (Rchb. f.) Rchb. f., 1868. *Orchiodes rubicundum* (Blume) Kuntze, Revis. Gen. Pl. 2: 675. 1891. *Epipactis rubicunda* (Blume) A. A. Eaton, Proc. Biol. Soc. Wash. 21: 65. 1908. *Salacistis rubicunda* (Blume) T. C. Hsu, Illustr. Fl. Taiwan 2: 184. 2016. *Salacistis rubicunda* (Blume) M. C. Pace, Brittonia 72: 266. 2020, isonym. SYNTYPES: Indonesia, Java, Mt. Salak & Mt. Gede, *Blume s.n.* (syntypes, BM, L [2×], P).

Goodyera grandis King & Pantl., Ann. Roy. Bot. Gard. (Calcutta) 8: 284, t. 379. 1898, nom. illeg., non (Blume) Lindl. ex D. Dietr., 1852. *Goodyera clavata* N. Pearce & P. J. Cribb, Edinburgh J. Bot. 58: 116. 2001. *Salacistis clavata* (N. Pearce & P. J. Cribb) M. C. Pace, Brittonia 72: 263. 2020. TYPE: India, "Sikkim" [West Bengal], valley of the Teesta, Rumtek, 4000 ft, 24 Aug. 1896, *Pantling 460* (holotype, CAL; isotype, BM, K).

Distribution. Arunachal Pradesh, Assam, Nagaland, West Bengal [China].

Habit. Terrestrial herb.

Goodyera schlechtendaliana Rchb. f., Linnaea 22: 861. 1849[1850]. *Georchis schlechtendaliana* (Rchb. f.) Rchb. f., Bonplandia (Hannover) 5: 36. 1857. *Orchiodes schlechtendalianum* (Rchb. f.) Kuntze, Revis. Gen. Pl. 2: 675. 1891. *Epipactis schlechtendaliana* (Rchb. f.) A. A. Eaton, Proc. Biol. Soc. Wash. 21: 66. 1908. *Peramium schlechtendalianum* (Rchb. f.) Makino, J. Jap. Bot. 6: 37. 1929. TYPE: Japan, *Göring s.n.* (holotype, W).

Distribution. Arunachal Pradesh, Assam, Manipur, Meghalaya, Mizoram, Nagaland, Sikkim, West Bengal [Bangladesh, Bhutan, China, Myanmar].

Habit. Terrestrial herb.

Goodyera thailandica Seidenf., Bot. Tidsskr. 65: 109. 1969. *Paorchis thailandica* (Seidenf.) M. C. Pace, Brittonia 72(3): 263. 2020. TYPE: Thailand: Khao Yai, Feb. 1965, *Cumberlege 1330* (holotype, K, spirit mat.).

Distribution. Odisha [also in Thailand].

Habit. Terrestrial herb.

Remarks. Pace (2020) doubts that Indian material of this species was correctly identified.

Goodyera viridiflora (Blume) Lindl. ex D. Dietr., Syn. Pl. (D. Dietrich) 5: 165. 1852. *Neottia viridiflora* Blume, Bijdr. Fl. Ned. Ind. 8: 408. 1825. *Physurus viridiflorus* (Blume) Lindl., J. Proc. Linn. Soc., Bot. 1: 180. 1857. *Goodyera viridiflora* (Blume) Blume, Coll. Orchid. 41. 1858, isonym. *Georchis viridiflora* (Blume) F. Muell., Fragm. (Mueller) 8: 29. 1873. *Orchiodes viridiflorum* (Blume) Kuntze, Revis. Gen. Pl. 2: 675. 1891. *Erythrodes viridiflora* (Blume) Schltr., Nachtr. Fl. Schutzgeb. Südsee [Schumann & Lauterbach] 87. 1905. *Epipactis viridiflora* (Blume) Ames, Orchidaceae 2: 61. 1908, nom. illeg., non (Hoffm.) Krock., 1814, non Rupr., 1847. *Eucosia viridiflora* (Blume) M. C. Pace, Brittonia 72: 264. 2020. TYPE: Indonesia, Java, Mt. Salak, *Blume s.n.* (lectotype, P, designated by Bhattacharjee & Bhakat, 2010).

Georchis cordata Lindl., Gen. Sp. Orchid. Pl.: 496. 1840. *Orchiodes cordatum* (Lindl.) Kuntze, Revis. Gen. Pl. 2: 675. 1891. *Epipactis cordata* (Lindl.) A. A. Eaton, Proc. Biol. Soc. Wash. 21: 64. 1908, nom. illeg., non (L.) All., 1785. *Goodyera cordata* (Lindl.) G. Nicholson, Ill. Dict. Gard. 2: 81. 1885. *Eucosia cordata* (Lindl.) T. C. Hsu, Illustr. Fl. Taiwan 2: 14. 2016. TYPE: Sri Lanka, *Macrae s.n.* (holotype, K-LINDL).

Distribution. Arunachal Pradesh, Assam, Meghalaya, Nagaland, Uttarakhand [Bhutan, China, Myanmar, Nepal].

Habit. Terrestrial herb.

Goodyera vittata (Lindl.) Benth. ex Hook. f., Fl. Brit. India [J. D. Hooker] 6(17): 113. 1890. *Georchis vittata* Lindl., J. Proc. Linn. Soc., Bot. 1: 184. 1857. *Orchiodes vittatum* (Lindl.) Kuntze, Revis. Gen. Pl. 2: 675. 1891. TYPE: India, Sikkim, hot valleys, *J. D. Hooker 336* (holotype, K-LINDL).

Distribution. Arunachal Pradesh, Sikkim, Uttara-khand, West Bengal [Bhutan, China, Nepal].

Habit. Terrestrial herb.

———

Grosourdya Rchb. f., Bot. Zeitung (Berlin) 22: 297. 1864.

TYPE: *Grosourdya elegans* Rchb. f.

Grosourdya appendiculata (Blume) Rchb. f., Xenia Orchid. 2: 123. 1868. *Dendrocolla appendiculata* Blume, Bijdr. Fl. Ned. Ind. 7: 289. 1825. *Thrix-spermum appendiculatum* (Blume) Kuntze, Revis. Gen. Pl. 2: 682. 1891. *Sarcochilus appendiculatus* (Blume) J. J. Sm., Orch. Java 564. 1905. *Ptero-ceras appendiculatum* (Blume) Holttum, Kew Bull. 14: 269. 1960. TYPE: Indonesia, Java, Kuripan, *Blume [1818]* (holotype, L).

Distribution. Andaman & Nicobar Islands [Myanmar].

Habit. Epiphytic herb.

Grosourdya muscosa (Rolfe) Garay, Bot. Mus. Leafl. 23(4): 181. 1972. *Sarcochilus muscosus* Rolfe, Bull. Misc. Inform. Kew 1893: 7. 1893. TYPE: India, Andaman Islands, *cult. Kew s.n. (leg. Mann s.n.) Dec. 1892* (holotype, K).

Sarcochilus maculatus Carr, Gard. Bull. Straits Settlem. 59 (1–2): 26. 1929, non (Dalzell) Benth. ex Pfitzer, 1882. *Sarcochilus carrii* L. O. Williams, Bot. Mus. Leafl. 5(4): 57. 1937. *Pteroceras carrii* (L. O. Williams) Holttum, Kew Bull. 14: 269. 1960. TYPE: Malaysia, Tembeling, Mentakab, Krambit, Pahang, described from living plants cultivated at Tembeling, Pahang, *Carr s.n.* (holotype, SING).

Distribution. Andaman & Nicobar Islands [also in Malaysia].

Habit. Epiphytic herb.

———

Gymnadenia R. Br., Hort. Kew., ed. 2 [W. T. Aiton] 5: 191. 1813.

TYPE: *Gymnadenia conopsea* (L.) R. Br. (*Orchis conopsea* L.).

Gymnadenia orchidis Lindl., Gen. Sp. Orchid. Pl.: 278. 1835 var. **orchidis**. *Platanthera orchidis* Lindl., Numer. List [Wallich] n. 7039. 1832, nom. inval. *Habenaria orchidis* (Lindl.) Hook. f., Fl.

Brit. India [J. D. Hooker] 6(17): 142. 1890. *Peristylus orchidis* (Lindl.) Kraenzl., Orchid. Gen. Sp. 1: 515. 1898. *Orchis habenarioides* King & Pantl., Ann. Roy. Bot. Gard. (Calcutta) 8: 302. 1898, nom. superfl. TYPE: Nepal, BagmatiIndia, Kumoon [Kumaon], *Wallich 7039B* (lectotype, K-LINDL, designated by Pearce et al., 2001; isotype, K, K-WALL).

Gymnadenia cylindrostachya Lindl., Gen. Sp. Orchid. Pl.: 278. 1835. *Orchis cylindrostachya* (Lindl.) Kraenzl., Repert. Spec. Nov. Regni Veg. 5: 197. 1908. TYPE: Origin unknown, *Wallich 7056 (leg. Heyne)* (holotype, K-LINDL; isotype, K-WALL).
Gymnadenia violacea Lindl., Gen. Sp. Orchid. Pl.: 278. 1835. TYPE: East Indies, Mussun, *Royle 59* (syntypes, icon K-LINDL, LIV).
Habenaria stoliczkae Kraenzl., Bot. Jahrb. Syst. 16: 215. 1892. TYPE: India, Himachal Pradesh, Narkanda, *Stoliczka 866* (holotype, B).

Distribution. Arunachal Pradesh, Himachal Pradesh, Sikkim, Uttarakhand [Bhutan, China, Myanmar, Nepal, Pakistan].

Habit. Terrestrial herb.

Gymnadenia orchidis Lindl. var. **pantlingii** Renz, Edinburgh J. Bot. 58: 112. 2001. TYPE: India, Sikkim, Tankra-la, 13,000 ft, July 1897, *Pantling 404 p.p.* (holotype, BM; isotype, K K000881637).

Distribution. Sikkim [Bhutan].

Habit. Terrestrial herb.

Remarks. This is considered a synonym of *Gymnadenia orchidis* Lindl. by Singh et al. (2019).

———

Habenaria Willd., Sp. Pl., ed. 4. 4: 5, 44. 1805.

TYPE: *Habernaria macroceratitis* Willd. (*Orchis habenaria* L.).

Habenaria acuifera Wall. ex Lindl., Gen. Sp. Orchid. Pl.: 325. 1835. *Platanthera acuifera* Lindl., Numer. List [Wallich] n. 7045. 1832, nom. inval. *Pecteilis acuifera* (Wall. ex Lindl.) M. A. Clem. & D. L. Jones, Austral. Orchid Rev. 83(6): 51. 2018. TYPE: Myanmar, Tavoy [Dawei], 25 Aug. 1827, *Wallich 7045 (leg. W. Gomez 1195)* (holotype, K-LINDL; isotype, K-WALL).

Distribution. Arunachal Pradesh, Assam, Manipur, Meghalaya, Mizoram, Nagaland [China, Myanmar].

Habit. Terrestrial herb.

Habenaria acuminata (Thwaites) Trimen, Syst. Cat.
Fl. Pl. Ceylon 91. 1885. *Ate acuminata* Thwaites,
Enum. Pl. Zeyl. [Thwaites]: 309. 1861. TYPE: Sri
Lanka, Rambodde and above Galagama, 4000–
5000 ft, Feb. 1861, *Thwaites s.n. [Ceylon Plants
514]* (syntypes, K [2×], G).

Distribution.   Kerala, Tamil Nadu [Sri Lanka].

Habit.   Terrestrial herb.

Habenaria agasthyamalaiana Jalal, Jayanthi & P.
Suresh Kumar, Lankesteriana 19(2): 93. 2019.
TYPE: India, Kerala, Kollam District, Shendur-
ney Wildlife Sanctuary, on the way to Rosemala,
550 m, 12 Oct. 2018, *J. S. Jalal 197753* (holotype, BSI; isotype, BSI).

Distribution.   Kerala. Endemic.

Habit.   Terrestrial herb.

Habenaria aitchisonii Rchb. f., Trans. Linn. Soc.
London, Bot. Ser. 2. 3: 113. 1888. TYPE: Afghanistan, Kuram District, Darban Valley, alt. 7500 ft,
9 Aug. 1880, *J. E. T. Aitchison 413* (holotype, K;
isotype, AMES).

Distribution.   Himachal Pradesh, Sikkim, Uttarakhand [Afghanistan, Bhutan, China, Myanmar, Nepal,
Pakistan].

Habit.   Terrestrial herb.

Habenaria andamanica Hook. f., Fl. Brit. India
[J. D. Hooker] 6(17): 134. 1890. *Medusorchis andamanica* (Hook. f.) Szlach., Orchidee (Hamburg)
55(4): 489. 2004. TYPE: India, Andaman Islands,
C. S. P. Parish s.n. (not found).

Distribution.   Andaman & Nicobar Islands. Endemic.

Habit.   Terrestrial herb.

Habenaria arietina Hook. f., Fl. Brit. India [J. D.
Hooker] 6(17): 138. 1890. *Habenaria pectinata*
D. Don var. *arietina* (Hook. f.) Kraenzl., Orchid.
Gen. Sp. 1: 405. 1898. *Habenaria intermedia*
D. Don var. *arietina* (Hook. f.) Finet, Rev. Gén.
Bot. 13: 530. 1901. *Ochyrorchis arietina* (Hook. f.)
Szlach., Richardiana 4(2): 53. 2004. SYNTYPES:
India, Sikkim, alt. 5000–8000 ft & 10,000 ft,
without collector [*probably Hooker & Thomson s.n.*]
(probable syntype, K); Khasi Hills, alt. 5000–6000
ft, *without collector* [*probably Hooker & Thomson
1251*] (probable syntype, K); India, Sylhet, Pundua Hill, June 1823, *Wallich 7029C (leg. De Silva
1245)* (syntype, K-WALL K001126612); India,

Kumaon, *Wallich 7029B p.p. (leg. R. Blinkworth
s.n.)* (syntype, K-WALL K001126611).

Habenaria pectinata D. Don var. *gigantea* Pradhan, Indian
Orchids: Guide Identif. & Cult. 2: 683. 1979. *Habenaria
ensifolia* Lindl. var. *gigantea* (Pradhan) P. K. Sarkar,
J. Econ. Taxon. Bot. 5: 1008. 1984. TYPE: India, Khasi
Hills, Kala Pani, 24 June 1850, *Hooker & Thompson
[1251] [L.]260* (holotype, K).
Habenaria pectinata D. Don var. *khasiensis* Pradhan, Indian
Orchids: Guide Identif. & Cult. 2: 684. 1979. *Habenaria
ensifolia* Lindl. var. *khasiensis* (Pradhan) P. K. Sarkar,
J. Econ. Taxon. Bot. 5: 1008. 1984. TYPE: India, Khasi
Hills, *Griffith s.n. (East India Company Herbarium
5307)* (holotype, K).

Distribution.   Arunachal Pradesh, Assam, Manipur, Meghalaya, Nagaland, Sikkim, Uttarakhand, West
Bengal [Bhutan, China, Nepal].

Habit.   Terrestrial herb.

Habenaria barbata Wight ex Hook. f., Fl. Brit. India
[J. D. Hooker] 6(17): 133. 1890. TYPE: India,
Travancore, Pulney and Dindygal Mountains, *Wal-
lich 7034 (leg. Wight)* (missing).

Ate virens Lindl., Gen. Sp. Orchid. Pl.: 327. 1835. *Habenaria
virens* (Lindl.) Abeyw., Ceylon J. Sci., Biol. Sci. 2(2): 83.
1959, nom. inval. et illeg., non A. Rich. & Galeotti,
1845. *Habenaria virens* (Lindl.) P. F. Hunt & Summerh.,
Kew Bull. 20(1): 51. 1966, nom. illeg., non A. Rich. &
Galeotti, 1845. TYPE: India, Dindigul, alt. 4000 ft,
Wight s.n. [2084] (holotype, K-LINDL; isotype, K).

Distribution.   Andhra Pradesh, Karnataka, Kerala,
Odisha, Tamil Nadu [Sri Lanka].

Habit.   Terrestrial herb.

Habenaria barnesii Summerh. ex C. E. C. Fisch., Fl.
Madras 3(11): 1887. 1936. SYNTYPES: India,
Nilgiri Hills, Gudalurmalai, 20 Sep. 1934, *Barnes
890* (syntype, K, also spirit mat.); Travancore,
Nemakad Gap, Sep. 1934, *Barnes 814* (syntype,
K, spirit mat.).

Distribution.   Kerala, Tamil Nadu. Endemic.

Habit.   Terrestrial herb.

Habenaria brachyphylla (Lindl.) Aitch., J. Linn.
Soc., Bot. 19: 188. 1882. *Platanthera brachyphylla* Lindl., Gen. Sp. Orchid. Pl.: 293. 1835.
TYPE: India, *Wight s.n.* (holotype, K-LINDL; isotype, P).

Habenaria crassifolia A. Rich., Ann. Sci. Nat., Bot. sér. 2, 15:
72. 1841. SYNTYPES: India, Nilgiri Hills, near Otacamund and Avalanchy, July–Sep., *Perrottet s.n.* (syntype, P).

Distribution. Karnataka, Kerala, Maharashtra, Odisha, Tamil Nadu. Endemic.

Habit. Terrestrial herb.

Habenaria cephalotes Lindl., Gen. Sp. Orchid. Pl.: 322. 1835. *Plantaginorchis cephalotes* (Lindl.) Szlach., Richardiana 4(2): 64. 2004. *Pecteilis cephalotes* (Lindl.) M. A. Clem. & D. L. Jones, Austral. Orchid Rev. 83(6): 51. 2018. TYPE: India, *Wight s.n.* (holotype, K-LINDL; isotype, CAL, G, K).

Distribution. Kerala, Tamil Nadu. Endemic.

Habit. Terrestrial herb.

Habenaria commelinifolia (Roxb.) Wall. ex Lindl., Gen. Sp. Orchid. Pl.: 325. 1835. *Orchis commelinifolia* Roxb., Fl. Ind. ed. 1832, 3: 451. 1832 (as "*commelinaefolia*"). *Platanthera commelinifolia* Lindl., Numer. List [Wallich] n. 7037. 1832, nom. inval. *Pecteilis commelinifolia* (Roxb.) M. A. Clem. & D. L. Jones, Austral. Orchid Rev. 83(6): 51. 2018. TYPE: India, Bengal, cult. Calcutta s.n., icon. *Orchis commelinaefolia* [sic], *Roxburgh 2334* (syntypes, CAL, K).

Distribution. Andhra Pradesh, Assam, Bihar, Chhattisgarh, Himachal Pradesh, Jammu & Kashmir, Jharkhand, Karnataka, Kerala, Madhya Pradesh, Maharashtra, Odisha, Punjab, Uttar Pradesh, Uttarakhand [Myanmar, Nepal].

Habit. Terrestrial herb.

Habenaria crinifera Lindl., Gen. Sp. Orchid. Pl.: 323. 1835. *Synmeria crinifera* (Lindl.) Szlach., Orchidee (Hamburg) 56(3): 331. 2005. *Pecteilis crinifera* (Lindl.) M. A. Clem. & D. L. Jones, Austral. Orchid Rev. 83(6): 51. 2018. TYPE: Sri Lanka, *Macrae s.n.* (holotype, K-LINDL).

Habenaria schizochilus J. Graham, Cal. Pl. Bombay: 252. 1839. *Synmeria schizochilus* (J. Graham) Nimmo in J. Graham, Cat. Pl. Bombay, Add.: s. p. 1839. TYPE: India, Ram Ghaut, *Law s.n.* (not found).

Distribution. Goa, Karnataka, Kerala, Maharashtra, Tamil Nadu [Sri Lanka].

Habit. Terrestrial or more frequently epiphytic herb.

Habenaria dentata (Sw.) Schltr., Repert. Spec. Nov. Regni Veg. Beih. 4: 125. 1919. *Orchis dentata* Sw., Kongl. Vetensk. Acad. Nya Handl. 21: 207. 1800. *Platanthera dentata* (Sw.) Lindl., Gen. Sp. Orchid. Pl.: 296. 1835. *Plantaginorchis dentata* (Sw.) Szlach., Richardiana 4: 64. 2004. *Pecteilis*

dentata (Sw.) M. A. Clem. & D. L. Jones, Austral. Orchid Rev. 83(6): 51. 2018. TYPE: China, without further data.

Habenaria geniculata D. Don, Prodr. Fl. Nepal.: 25. 1825. TYPE: Nepal, Bagmati Zone, Narainhetty, *Buchanan-Hamilton s.n.*

Distribution. Andhra Pradesh, Arunachal Pradesh, Assam, Chhattisgarh, Jharkhand, Manipur, Meghalaya, Mizoram, Nagaland, Odisha, Sikkim, Uttar Pradesh, Uttarakhand, West Bengal [Bhutan, China, Myanmar, Nepal].

Habit. Terrestrial herb.

Habenaria denticulata Rchb. f., Linnaea 19: 376. 1847 [1846]. TYPE: India, Nilgiri Hills, 1838, *Adolphe Delessert s.n.* (holotype, W).

Distribution. Tamil Nadu [Sri Lanka].

Habit. Terrestrial herb.

Habenaria dichopetala Thwaites, Enum. Pl. Zeyl. [Thwaites]: 309. 1861. TYPE: Sri Lanka, Bintenne District, *Thwaites s.n.* [*Ceylon Plants 3564*] (holotype, K).

Distribution. ?Tamil Nadu [Sri Lanka].

Habit. Terrestrial herb.

Remarks. This species is listed as "excluded" in Singh et al. (2019).

Habenaria digitata Lindl., Gen. Sp. Orchid. Pl.: 307. 1835. *Habenaria graveolens* Duthie, Ann. Roy. Bot. Gard. (Calcutta) 9: 177. 1906, sphalm. [*H. digitata* intended]. TYPE: India, Pundua Hills, Apr. 1823, *Wallich 7063 (leg. F. De Silva 1219)* (holotype, K-LINDL; isotype, K-WALL).

Bonatea benghalensis Griff., Calcutta J. Nat. Hist. 4: 382. 1844. TYPE: India, Serampore, *cult. Calcutta Botanical Garden s.n.* (not found).
Habenaria trinervia Wight, Icon. Pl. Ind. Orient. [Wight] 5(1): 12, t. 1701. 1851. TYPE: India, Belgam, *J. S. Law s.n.* (holotype, K-LINDL).
Habenaria lindleyana Wight, Icon. Pl. Ind. Orient. [Wight] 3(2): t. 922. 1844, nom. illeg., non Steud., 1841. *Habenaria travancorica* Hook. f., Fl. Brit. India [J. D. Hooker] 6(17): 135. 1890. *Habenaria digitata* Lindl. var. *travancorica* (Hook. f.) C. E. C. Fisch., Fl. Madras 3(10): 1469. 1928. *Habenaria digitata* Lindl. var. *travancorica* (Hook. f.) Pradhan, Indian Orchids: Guide Identif. & Cult. 2: 683. 1979, isonym. TYPE: India, Travancore, Pulney Hills, alt. 4000–7000 ft, *Wight s.n.* [*3014*] (holotype, K; isotype, CAL).

Distribution. Andhra Pradesh, Arunachal Pradesh, Assam, Chhattisgarh, Gujarat, Himachal Pradesh, Jharkhand, Karnataka, Kerala, Madhya Pradesh, Maharashtra, Meghalaya, Nagaland, Odisha, Rajasthan, Tamil Nadu, Uttar Pradesh, Uttarakhand [Bangladesh, Myanmar, Nepal, Pakistan].

Habit.   Terrestrial herb.

Habenaria diphylla Dalzell, Hooker's J. Bot. Kew Gard. Misc. 2: 262. 1850. TYPE: India, Maharashtra, South Concan, July, *Dalzell s.n.* (syntype, CAL, K, K-LINDL).

Habenaria jerdoniana Wight, Icon. Pl. Ind. Orient. [Wight] 5(1): 14, t. 1715. 1851. TYPE: India, Malabar, *Jerdon s.n.* (holotype, K).
Habenaria sutteri Rchb. f., Linnaea 25: 229. 1852. SYNTYPES: India, Mangalore, 1847, *Sutter s.n.* (syntype, W); Canara, *Metz 142a* (syntype, W).
Planthera canarensis Lindl. ex Rchb. f., Linnaea 41: 100. 1877 [1876], nom. inval.

Distribution.   Andhra Pradesh, Bihar, Chhattisgarh, Goa, Himachal Pradesh, Jharkhand, Karnataka, Kerala, Maharashtra, Odisha, Tamil Nadu, Uttar Pradesh, Uttarakhand [Bangladesh, Myanmar, Nepal].

Habit.   Terrestrial herb.

Remarks.   *Liparis diphyllos* Nimmo is traditionally included in the synonymy of *Habenaria diphylla* Dalzell, often even taken as the basionym for the latter. Not only did Dalzell not refer to *Liparis diphyllos* when he published *H. diphylla*, Nimmo's description, which mentions ovate, acute, plaited leaves and a large round lip, does not fit *H. diphylla* but is indeed applicable to a *Liparis* species. As already suggested by Santapau and Kapadia (1962), *L. diphyllos* could be an earlier name for *L. deflexa* Hook. f., but in the absence of type material, the description is too vague to be certain. *Habenaria diphylla* and *L. deflexa* both occur near Mumbai (formerly Bombay), where the type of *H. diphylla* was collected.

Due to confusion with *Habenaria josephi* the distribution of *H. diphylla* is at present unclear. See remarks under *H. josephi*.

Habenaria elliptica Wight, Icon. Pl. Ind. Orient. [Wight] 5(1): 13, t. 1706. 1851. *Plantaginorchis elliptica* (Wight) Szlach., Richardiana 4: 64. 2004. *Pecteilis elliptica* (Wight) M. A. Clem. & D. L. Jones, Austral. Orchid Rev. 83(6): 51. 2018. TYPE: India, Pulney Hills, Sep., *Wight s.n.* (holotype, K).

Distribution.   Karnataka, Kerala, Tamil Nadu. Endemic.

Habit.   Terrestrial herb.

Habenaria elwesii Hook. f., Bot. Mag. 122: t. 7478. 1896. *Ate elwesii* (Hook. f.) Szlach., Orchidee

(Hamburg) 56(3): 329. 2005. TYPE: India, Nilgiri Mountains, *Proudlock s.n.* (holotype, K).

Distribution.   Goa, Karnataka, Kerala, Maharashtra, Tamil Nadu. Endemic.

Habit.   Terrestrial herb.

Habenaria ensifolia Lindl., Gen. Sp. Orchid. Pl.: 321. 1835. *Habenaria pectinata* D. Don subsp. *ensifolia* (Lindl.) Soó, Ann. Hist.-Nat. Mus. Natl. Hung. 26: 373. 1920. *Ochyrorchis ensifolia* (Lindl.) Szlach., Richardiana 4: 53. 2004. TYPE: Nepal, Bagmati Zone, Rashuwa District, Gosainthan, *Wallich 7030 p.p.* (K-WALL, not found).

Distribution.   Arunachal Pradesh, Himachal Pradesh, Meghalaya, Nagaland, Sikkim, Uttarakhand, West Bengal [Bhutan, Nepal].

Habit.   Terrestrial herb.

Remarks.   We follow Vij et al. (2013) in keeping *Habenaria ensifolia* separate from *H. pectinata* D. Don.

Habenaria flabelliformis Summerh. ex C. E. C. Fisch., in J. S. Gamble, Fl. Madras 3(11): 1887. 1936. TYPE: India, Travancore, Amaimudi slopes, alt. 7500 ft, *Barnes 629* (holotype, K, spirit coll.).

Distribution.   Kerala. Endemic.

Habit.   Terrestrial herb.

Habenaria foliosa A. Rich., Ann. Sci. Nat., Bot. sér. 2, 15: 71. 1841. *Habenaria digitata* Lindl. var. *foliosa* (A. Rich.) Hook. f., Fl. Brit. India [J. D. Hooker] 6(17): 135. 1890. *Habenaria gibsonii* Hook. f. var. *foliosa* (A. Rich.) Santapau & Kapadia, J. Bombay Nat. Hist. Soc. 56: 194. 1959. TYPE: India, Nilgiri Hills near Avalanchy and Otacamund, July–Aug., *Perrottet s.n.* (holotype, P).

Habenaria spencei Blatt. & McCann, J. Bombay Nat. Hist. Soc. 36: 17. 1932. TYPE: India, Western Ghats, Mahableshwar, Fitzgerald Ghat, alt. 4000 ft, *McCann 3026*.

Distribution.   Andhra Pradesh, Goa, Gujarat, Jharkhand, Karnataka, Kerala, Madhya Pradesh, Maharashtra, ?Manipur, Meghalaya, ?Odisha, Tamil Nadu. Endemic.

Habit.   Terrestrial herb.

Remarks.   The distribution of this species is unclear, as *Habenaria gibsonii* and *H. gibsonii* var. *foetida* Blatt. & McCann were often included in it.

Habenaria furcifera Lindl., Gen. Sp. Orchid. Pl.: 319. 1835. *Pecteilis furcifera* (Lindl.) M. A. Clem.

& D. L. Jones, Austral. Orchid Rev. 83(6): 51. 2018. TYPE: India, Mussoorie, *Royle s.n.* (holotype, K-LINDL).

Habenaria hamigera Griff., Calcutta J. Nat. Hist. 4: 380. 1844. TYPE: India, Uttar Pradesh, Goruckpore, *cult. Calcutta (leg. Vicary) s.n.* (not found).

Distribution. Andhra Pradesh, Arunachal Pradesh, Assam, Bihar, Chhattisgarh, Gujarat, Himachal Pradesh, Jammu & Kashmir, Jharkhand, Karnataka, Kerala, Madhya Pradesh, Maharashtra, Manipur, Mizoram, Nagaland, Odisha, Punjab, Tamil Nadu, Tripura, Uttar Pradesh, Uttarakhand, West Bengal [Bangladesh, Bhutan, China, Myanmar, Nepal, Pakistan].

Habit. Terrestrial herb.

Habenaria gibsonii Hook. f., Fl. Brit. India [J. D. Hooker] 6(17): 135. 1890 (as "*gibsoni*") var. **gibsonii**. *Habenaria foliosa* A. Rich. var. *gibsonii* (Hook. f.) Bennett, J. Econ. Taxon. Bot. 5(2): 452. 1984. *Habenaria digitata* Lindl. var. *gibsonii* (Hook. f.) C. E. C. Fisch. in J. S. Gamble, Fl. Madras 3: 1469. 1928. TYPE: India, the Concan, Kyreswur and Kandalla [Khandala], *Gibson s.n.* (holotype (syntype?), K).

Distribution. Karnataka, Maharashtra. Endemic.

Habit. Terrestrial herb.

Habenaria gibsonii Hook. f. var. **foetida** Blatt. & McCann, J. Bombay Nat. Hist. Soc. 36: 16. 1932. *Habenaria foliosa* A. Rich. var. *foetida* (Blatt. & McCann) Bennet, J. Econ. Taxon. Bot. 5: 452. 1984. TYPE: India, Maharastra, Khandala, Monkey Hill, *Hallberg s.n.* (holotype, BLAT).

Distribution. Jharkhand, Karnataka, Kerala, Maharashtra, Rajasthan, Tamil Nadu, West Bengal [Myanmar].

Habit. Terrestrial herb.

Habenaria grandifloriformis Blatt. & McCann, J. Bombay Nat. Hist. Soc. 36: 17. 1932. TYPE: India, Western Ghats, Panchgani, in grassland on Tableland, *Blatter P. 20* (holotype, BLAT).

Habenaria rotundifolia Lindl., Gen. Sp. Orchid. Pl.: 306. 1835, nom. illeg., non (Banks ex Pursh) Richardson, 1823. *Habenaria grandiflora* Lindl., Numer. List [Wallich] n. 7032. 1832, nom. inval. *Habenaria grandiflora* Lindl. ex Dalzell & Gibson, Bombay Fl. 267. 1861, nom. illeg., non (Bigelow) Torr. ex L. C. Beck, 1823. TYPE: India, locality not recorded, 23 Sep. 1816, *Wallich 7032 (leg. Heyne s.n.)* (holotype, K-LINDL; isotype, K-WALL). *Habenaria grandifloriformis* Blatt. & McCann var. *aequiloba* Blatt. & McCann, J. Bombay Nat. Hist. Soc. 36: 18. 1932. TYPE: India, Panchgani, July 1925, *Blatter P. 20a* (holotype, BLAT).

Distribution. Chhattisgarh, Goa, Gujarat, Karnataka, Kerala, Madhya Pradesh, Maharashtra, Odisha, Tamil Nadu. Endemic.

Habit. Terrestrial herb.

Habenaria hallbergii Blatt. & McCann, J. Bombay Nat. Hist. Soc. 36: 24. 1932. TYPE: India, Western ghats, Khandala, ravine, *Hallberg s.n.* (St. Xavier College Herbarium).

Distribution. ?Goa, Karnataka, Kerala, Tamil Nadu. Endemic.

Habit. Terrestrial herb.

Remarks. This is considered a synonym of *Habenaria ovalifolia* Wight by Singh et al. (2019).

Habenaria heyneana Lindl., Gen. Sp. Orchid. Pl.: 320. 1835. *Platanthera heyneana* Lindl., Gen. Sp. Orchid. Pl.: 320. 1835, nom. inval. TYPE: India, locality not recorded, 23 Sep. 1826, *Wallich 7044 (leg. Heyne s.n.)* (holotype, K-LINDL; isotype, K-WALL).

Habenaria glabra A. Rich., Ann. Sci. Nat., Bot., sér. 2, 15: 75. 1841. TYPE: India, Nilgiri Hills, Otacamund, Aug., *Perrottet s.n.* (holotype, P). *Habenaria subpubens* A. Rich., Ann. Sci. Nat., Bot. sér. 2, 15: 75. 1841. *Habenaria heyneana* Lindl. var. *subpubens* (A. Rich.) Pradhan, Indian Orchids: Guide Identif. & Cult. 2: 684. 1979. TYPE: India, Nilgiri Hills, near Otacamund, Aug., *Perrottet s.n.* (holotype, P). *Habenaria candida* Dalzell, Hooker's J. Bot. Kew Gard. Misc. 2: 262. 1850. TYPE: India, Bombay, *Dalzell s.n.* *Habenaria cerea* Blatt. & McCann, J. Bombay Nat. Hist. Soc. 36: 21. 1932. TYPE: India, Panchgani, Tableland, alt. 4400 ft, *Blatter P. 73* (holotype, BLAT). *Habenaria cerea* Blatt. & McCann var. *polyantha* Blatt. & McCann, J. Bombay Nat. Hist. Soc. 36: 22. 1932. TYPE: India, Panchgani, Third Tableland, *Blatter 255* (holotype, BLAT).

Distribution. Andhra Pradesh, Goa, Karnataka, Kerala, Maharashtra, Tamil Nadu. Endemic.

Habit. Terrestrial herb.

Habenaria hollandiana Santapau, Fl. Purandhar: 126. 1958. *Habenaria affinis* Wight, Icon. Pl. Ind. Orient. [Wight] 5(1): t. 1707. 1851, non D. Don, 1825. *Habenaria malleifera* Hook. f. var. *hollandiana* (Santapau) Pradhan, Indian Orchids: Guide Identif. & Cult. 2: 685. 1979. *Habenaria indica* C. S. Kumar & Manilal, Taxon 35(4): 719. 1986, nom. superfl. TYPE: India, ?Belgaum, *?J. S. Law s.n.* (?holotype, K [without collector or locality; Wight was uncertain about these data]).

Distribution. Andhra Pradesh, Jharkhand, Karnataka, Maharashtra, Tamil Nadu. Endemic.

Habit.   Terrestrial herb.

Habenaria intermedia D. Don, Prodr. Fl. Nepal.: 24. 1825. *Kryptostoma intermedium* (D. Don) Olszewski & Szlach., Ann. Bot. Fenn. 37(4): 299. 2000 (as *"intermedia"*). *Ochyrorchis intermedia* (D. Don) Szlach., Richardiana 4(2): 55. 2004. TYPE: Nepal, Bagmati Zone, Rashuwa District, Gosainthan [Shishapangma], *Wallich 7030* (holotype, K-WALL).

Distribution.   Himachal Pradesh, Meghalaya, Nagaland, Uttarakhand [China, Myanmar, Nepal, Pakistan].

Habit.   Terrestrial herb.

Habenaria josephi Rchb. f., Trans. Linn. Soc. London, Bot. 3: 114. 1888. *Habenaria aitchisonii* Rchb. f. var. *josephi* (Rchb. f.) Hook. f., Fl. Brit. India [J. D. Hooker] 6(17): 152. 1890. *Habenaria diphylla* Dalzell var. *josephi* (Rchb. f.) N. Pearce & P. J. Cribb, Edinburgh J. Bot. 58: 114. 2001. TYPE: India, Sikkim, Tungu, 12,000–19,000 ft, 23 July 1849, *Hooker [L.]253 [42]* (holotype, K-LINDL; isotype, AMES, K, ?LE, P).

Habenaria clarkei Kraenzl., Bot. Jahrb. Syst. 16: 148. 1893. TYPE: India, Sikkim, Tungu, 23 July 1849, 12,000– 19,000 ft, *Hooker 42* (lectotype, K-K000247480, designated by Pandey & Jin, 2021; isotype, AMES, K, ?LE, P).

Distribution.   Sikkim [Bhutan, China].

Habit.   Terrestrial herb.

Remarks.   *Habenaria josephi* is considered distinct from *H. diphylla* by Pandey and Jin (2021), differing in the pubescent (vs. glabrous) rachis and ovaries and in the absence of peduncle-scales. The distribution of *H. diphylla* and *H. josephi* in India needs to be reassessed.

Habenaria keralensis K. Prasad, Webbia 74(1): 63. 2019. TYPE: India, Kerala, Thiruvananthapuram district, Agasthyakoodam, 1528 m, 31 Oct. 2014, *K. Prasad 006440* (holotype, CAL; isotype, BSID).

Distribution.   Kerala. Endemic.

Habit.   Terrestrial herb.

Habenaria khasiana Hook. f., Fl. Brit. India [J. D. Hooker] 6(17): 151. 1890. *Habenaria graminea* Lindl., Gen. Sp. Orchid. Pl.: 318. 1835, nom. illeg., non Sprengel, 1826. *Platanthera linifolia* Lindl., Numer. List [Wallich] n. 7041. 1832, nom. inval. SYNTYPES: India, Sylhet, Pundua Hills, Aug. 1820; Cherrapunji, Aug. 1830; Thuar[?], 16 Sep. 1827, *Wallich 7041 (leg. De Silva & Gomez)* (syntypes, K-WALL [3×]).

Distribution.   ?Manipur, Meghalaya, Mizoram, Uttar Pradesh [also in Cambodia, Laos, Thailand].

Habit.   Terrestrial herb.

Habenaria longicorniculata J. Graham, Cat. Pl. Bombay: 202. 1839. *Pecteilis longicorniculata* (J. Graham) M. A. Clem. & D. L. Jones, Austral. Orchid Rev. 83(6): 52. 2018. SYNTYPES: India, South Concan, Kandalla [Khandala], Sir Herbert Compton's Bungalow, *J. S. Law s.n.* (syntypes, K, P); Pulney Hills, *Wight s.n.* (syntypes, K, P).

Habenaria longecalcarata A. Rich., Ann. Sci. Nat., Bot. sér. 2, 15: 71. 1841. TYPE: India, Nilgiri Hills, Konoor, Otacamund, July–Aug., *Perrottet s.n.* (holotype, P).
Habenaria longecalcarata A. Rich. var. *viridis* Blatt. & McCann, J. Bombay Nat. Hist. Soc. 36: 20. 1932 (as *"longicalcarata"*). TYPE: India, Western Ghats, Khandala, *Hallberg s.n.* (holotype, St. Xavier College Herbarium; isotype, K).

Distribution.   Andhra Pradesh, Bihar, Chhattisgarh, Goa, Gujarat, Jharkhand, Karnataka, Kerala, Madhya Pradesh, Maharashtra, Odisha, Rajasthan, Tamil Nadu [Sri Lanka].

Habit.   Terrestrial herb.

Habenaria longicornu Lindl., Gen. Sp. Orchid. Pl.: 322. 1835 (as *"longicornis"*). *Blephariglottis longicornu* (Lindl.) Raf., Fl. Tellur. 2: 38. 1836 [1837] (as *"Blephariglotis longicornis"*). *Plantaginorchis longicornu* (Lindl.) Szlach. & Kras-Lap., Richardiana 6(1): 32. 2006. TYPE: India, locality not recorded, 17 Aug. 1818, *Wallich 7027 (leg. Heyne s.n.)* (holotype, K-LINDL; isotype, K-WALL).

Habenaria montana A. Rich., Ann. Sci. Nat., Bot. sér. 2, 15: 73. 1841. TYPE: India, Nilgiri Hills, *Perrottet s.n.*
Habenaria montana A. Rich. var. *major* A. Rich., Ann. Sci. Nat., Bot. sér. 2, 15: 73.1841. TYPE: India, Nilgiri Hills, Kulhuty, 1840, *Perrottet s.n.* (holotype, P).
Habenaria decipiens Wight, Icon. Pl. Ind. Orient. [Wight] 5(1): 14. 1851. TYPE: India, Pulney Hills, Sep. 1836, *Wight s.n.* (holotype, K)

Distribution.   Andhra Pradesh, Karnataka, Kerala, Tamil Nadu [Bangladesh, Nepal].

Habit.   Terrestrial herb.

Habenaria longifolia Buch.- Ham. ex Lindl., Gen. Sp. Orchid. Pl.: 324. 1835. *Plantaginorchis longifolia* (Buch.-Ham. ex Lindl.) Szlach., Richardiana 4: 64. 2004. *Pecteilis longifolia* (Buch.- Ham. ex Lindl.) M. A. Clem. & D. L. Jones, Austral. Orchid Rev. 83(6): 52. 2018. SYNTYPES: India, Nathpur, *Buchanan-Hamilton s.n.* (syntype, K-LINDL); Mussoree & Portu Kheree, *Royle s.n.*

Distribution. Bihar, Uttarakhand [Bangladesh, ?Nepal].

Habit. Terrestrial herb.

Habenaria macrostachya Lindl., Gen. Sp. Orchid. Pl.: 307. 1835. TYPE: Sri Lanka, *Macrae s.n.* (holotype, K-LINDL; isotype, K).

Distribution. Kerala, Tamil Nadu [Sri Lanka].

Habit. Terrestrial herb.

Habenaria malintana (Blanco) Merr., Sp. Blancoan. 112. 1918. *Thelymitra malintana* Blanco, Fl. Filip. 642. 1837. *Kraenzlinorchis malintana* (Blanco) Szlach., Orchidee (Hamburg) 55: 58. 2004. *Pecteilis malintana* (Blanco) M. A. Clem. & D. L. Jones, Austral. Orchid Rev. 83(6): 52. 2018. TYPE: Philippines, Malinta, *Blanco s.n.*

Habenaria geniculata D. Don var. *ecalcarata* King & Pantl., Ann. Roy. Bot. Gard. (Calcutta) 8: 310. 1898. *Habenaria dentata* (Sw.) Schltr. var. *ecalcarata* (King & Pantl.) Hand.-Mazz., Symb. Sin. 7: 1336. 1936. *Habenaria dentata* (Sw.) Schltr. f. *ecalcarata* (King & Pantl.) Tuyama, Fl. E. Himalaya 1: 438. 1966. *Habenaria dentata* (Sw.) Schltr. subsp. *ecalcarata* (King & Pantl.) Panigrahi & Murti in Murti & Panigrahi, Fl. Bilaspur Distr. 2: 589. 1999. TYPE: India, Sikkim, *Pantling s.n.*
Habenaria pelorioides C. S. P. Parish & Rchb. f., Trans. Linn. Soc. London 30: 139. 1874. TYPE: Myanmar, 1862, *Parish 327 (leg. Amherst)* (holotype, K; isotype, W).
Odisha cleistantha S. Misra, Orchids India: 252. 2007. TYPE: India, Odisha, near Gudugudia, *S. Misra 2434* (holotype, CAL).

Distribution. Assam, Jharkhand, Kerala, Manipur, Nagaland, Odisha, West Bengal [China, Myanmar].

Habit. Terrestrial herb.

Remarks. Habenaria malintana is probably a peloric form of *H. dentata* (Sw.) Schltr., as suggested by Seidenfaden (1977a). The existence of intermediates, which carry a short spur, described as *H. paragenicu-lata* Tang & F. T. Wang and *Odisha cleistantha* S. Misra, seems to support this. Kumar et al. (2018) reduced the latter to *H. malintana*, with which we concur.

Habenaria malleifera Hook. f., Fl. Brit. India [J. D. Hooker] 6(17): 143. 1890. SYNTYPES: India, Sikkim, Senadah, 6000 ft, *King s.n.* (syntype, CAL); Khasi Hills, *T. Lobb s.n.*; Khasi Hills, Myrung, alt. 5000 ft, *Hooker & Thomson 257* (syntypes, K, K-LINDL).

Habenaria furfuracea Hook. f., Fl. Brit. India [J. D. Hooker] 6(17): 144. 1890. TYPE: India, Khasi Hills, Nunklow, alt. 3000–4000 ft, *Hooker & Thomson 1798* (holotype, K).

Distribution. Meghalaya, Mizoram, Nagaland, Sikkim, West Bengal [China, Nepal].

Habit. Terrestrial herb.

Habenaria mandersii Collett & Hemsl., J. Linn. Soc. Bot. 28: 133. 1890. TYPE: Myanmar, Shan Hills, 4400 ft, *Manders s.n.* (holotype, K).

Distribution. Assam, Manipur, Meghalaya, Nagaland [Myanmar].

Habit. Terrestrial herb.

Habenaria marginata Colebr. in W. J. Hooker, Exot. Fl. 2: t. 136. 1824. *Platanthera marginata* (Colebr.) Lindl., Numer. List [Wallich] n. 7038. 1832. *Pecteilis marginata* (Colebr.) M. A. Clem. & D. L. Jones, Austral. Orchid Rev. 83(6): 52. 2018. TYPE: India, July 1814, *cult. Calcutta s.n.* (probably not preserved and the illustration in Exot. Fl. 2: t. 136, 1824, must then be considered as the type).

Habenaria flavescens Hook. f., Fl. Brit. India [J. D. Hooker] 6(17): 150. 1890. *Habenaria marginata* Colebr. var. *flavescens* (Hook. f.) T. Cooke, Fl. Bombay 2(4): 721. 1907. *Habenaria marginata* Colebr. f. *flavescens* (Hook. f.) Blatt. & McCann, J. Bombay Nat. Hist. Soc. 36: 24. 1932. TYPE: India, the Concan, *Law s.n.* (holotype, K-LINDL; isotype, CAL, K).
Habenaria fusifera Hook. f., Fl. Brit. India [J. D. Hooker] 6(17): 147. 1890. *Habenaria marginata* Colebr. var. *fusifera* (Hook. f.) Santapau & Kapadia, J. Bombay Nat. Hist. Soc. 56: 199. 1959. TYPE: India, Travancore, Annamallay Hills, *Beddome s.n.*

Distribution. Andhra Pradesh, Assam, Bihar, Chhattisgarh, Dadra & Nagar Haveli, Daman & Diu, Goa, Gujarat, Himachal Pradesh, Jammu & Kashmir, Jharkhand, Karnataka, Kerala, Madhya Pradesh, Maharashtra, Meghalaya, Odisha, Punjab, Rajasthan, Sikkim, Tamil Nadu, Uttarakhand, West Bengal [Bhutan, Myanmar].

Habit. Terrestrial herb.

Habenaria multicaudata Sedgw., Rec. Bot. Surv. India 6: 352. 1919. TYPE: India, North Kanara, Near Kaswar, Gudihalli, Sep. 1917, *T. R. D. Bell s.n.*

Distribution. Goa, Karnataka, Kerala, Maharashtra, Tamil Nadu. Endemic.

Habit. Terrestrial herb.

Habenaria nicobarica Murugan, Alappatt, S. Prabhu & Arisdason, Bangladesh J. Plant Taxon. 21: 77. 2014. TYPE: India, Andaman & Nicobar Islands: South Nicobar, Little Nicobar Tribal Reserve,

Pulopaha (E), 10 m, 25 Nov. 2008, *C. Murugan 26630* (holotype, CAL; isotype, PBL).

Distribution.   Andaman & Nicobar Islands. Endemic.

Habit.   Terrestrial herb.

Remarks.   This species seems hardly different from *Habenaria stenopetala* Lindl., with which it should be critically compared.

Habenaria osmastonii Karthig., Maina, Sumathi, Jayanthi & Jalal, Phytotaxa 166: 151. 2014. TYPE: India, South Andaman, Rutland Island, Dyer Point, 11 Oct. 1905, *Osmaston 21* (holotype, CAL).

Distribution.   Andaman & Nicobar Islands. Endemic.

Habit.   Terrestrial herb.

Habenaria ovalifolia Wight, Icon. Pl. Ind. Orient. [Wight] 5(1): 13, t. 1708. 1851. TYPE: India, Malabar and Anamally Hills, July–Aug., *Wight 3017* (syntype, K).

Habenaria modesta Dalzell, Hooker's J. Bot. Kew Gard. Misc. 2: 262. 1850. TYPE: India, Bombay Dist., Salsette Island, *Dalzell s.n.* (not found).

Distribution.   Karnataka, Kerala, Maharashtra, Tamil Nadu. Endemic.

Habit.   Terrestrial herb.

Habenaria pallideviridis Seidenf. ex K. M. Matthew, Kew Bull. 48(4): 764. 1993. TYPE: India, Tamil Nadu, Dindigul District, Palni Hills, Kukkal, Bhoothanachiamman temple Hill, alt. 6800 ft, 19 Oct. 1987, *K. M. Matthew RHT 50795* (holotype, C).

Distribution.   Tamil Nadu. Endemic.

Habit.   Terrestrial herb.

Habenaria panigrahiana S. Misra, Blumea 27(1): 213, fig. 1. 1981. TYPE: India, Odisha, Ganjam District, Mohana, 19°26′N, 84°17′E, alt. 1600 ft, on foot hills under light cover in rocky loamy soil, 3 Oct. 1975, *Sarat Misra 122* (holotype, CAL).

Habenaria panigrahiana S. Misra var. *parviloba* S. Misra, Blumea 27: 214, fig. 1. 1981. TYPE: India, Odisha, Ganjam District, Bhanjanagar, 19°57′N, 84°35′E, alt. 650 ft, on reservoir under scrubs, in loamy soil, 25 Jan. 1976, *Sarat Misra 140* (holotype, CAL).
Habenaria ramayyana Ram. Chary & J. J. Wood, Kew Bull. 36(2): 235, fig. 1. 1981. TYPE: India, Andhra Pradesh, Amrabad Forest Reserve, Bahrapur, alt. 2300 ft, 9 Nov. 1979, *Ramachandra Chary 642* (holotype, HY; isotype, CAL, K).

Distribution.   Andhra Pradesh, Gujarat, Kerala, Maharashtra, Odisha, Tamil Nadu. Endemic.

Habit.   Terrestrial herb.

Habenaria pantlingiana Kraenzl., Orchid. Gen. Sp. 1: 892. 1900. TYPE: India, Sikkim, Namgarh, 5000 ft, Aug. 1895, *Pantling 415* (holotype, K; isotype, BM, W).

Habenaria stenopetala Lindl. var. *polytricha* Hook. f., Ann. Roy. Bot. Gard. (Calcutta) 5: 64. 1895. *Habenaria polytricha* (Hook. f.) Pradhan, Indian Orchids: Guide Identif. & Cult. 1: 68. 1976, nom. illeg., non Rolfe, 1896. *Habenaria seshagiriana* A. N. Rao, J. Econ. Taxon. Bot. 6(1): 223. 1985. *Habenaria polytrichoides* Aver., Bot. Zhurn. (Moscow & Leningrad) 73: 432. 1988, nom. superfl. SYNTYPES: India, Sikkim, Chungtan, alt. 6000 ft, *J. D. Hooker s.n.* (syntype, K-LINDL); Naga Hills, *Prain s.n.* (syntype, K).

Distribution.   Arunachal Pradesh, Sikkim, West Bengal [China, Myanmar, Nepal].

Habit.   Terrestrial herb.

Habenaria pectinata D. Don, Prodr. Fl. Nepal.: 24. 1825. *Orchis pectinata* Sm., Exot. Bot. 2: 77. 1806, nom. illeg., non Thunb., 1794. *Habenaria gerardiana* Lindl., Numer. List [Wallich] n. 7031. 1832, nom. inval. *Kryptostoma pectinatum* (D. Don) Olszewski & Szlach., Ann. Bot. Fenn. 37: 299. 2000 (as "*pectinata*"). *Ochyrorchis pectinata* (D. Don) Szlach., Richardiana 4(2): 55. 2004. SYNTYPES: Nepal, Bagmati Zone, Rashuwa District, Gosainthan [Shishapangma], *Wallich s.n.* [*?7029A*] (?syntype, K-WALL [Nepal, Sheopore [Shivapuri], Sep. 1821]); Nepal, *Buchanan-Hamilton s.n.* (syntype, LINN Herb. no. 1381. 22. 1).

Distribution.   Arunachal Pradesh, Himachal Pradesh, Meghalaya, Nagaland, Sikkim, Uttarakhand, West Bengal [China, Myanmar, Nepal, Pakistan].

Habit.   Terrestrial herb.

Habenaria periyarensis Sasidh., K. P. Rajesh & Augustine, Rheedea 8(2): 167. 1998. TYPE: India, Kerala, Idukki District, Periyar Tiger Reserve, alt. 4000 ft, 5 Aug. 1996, *Jomy Augustine 17842* (holotype, KFRI).

Distribution.   Kerala. Endemic.

Habit.   Terrestrial herb.

Habenaria perrottetiana A. Rich., Ann. Sci. Nat., Bot. sér. 2, 15: 74. 1841. *Plectoglossa perrottetiana* (A. Rich.) K. Prasad & Venu, Rheedea 25: 88. 2015. TYPE: India, Nilgiri Hills, Otacamund to Avalanchy, Aug. 1840, *Perrottet s.n.* [*1129 in G*] (holotype, P; isotype, G).

Platanthera lutea Wight, Icon. Pl. Ind. Orient. [Wight] 3(2): t. 919. 1844. *Habenaria lutea* (Wight) Benth., J. Linn. Soc., Bot. 18: 354. 1881. TYPE: India, Pulney Hills, *Wight s.n.* (holotype, K; isotype, P).

Distribution. Karnataka, Kerala, Maharashtra, Tamil Nadu. Endemic.

Habit. Terrestrial herb.

Habenaria plantaginea Lindl., Gen. Sp. Orchid. Pl.: 323. 1835. *Plantaginorchis plantaginea* (Lindl.) Szlach., Richardiana 4(2): 65. 2004. *Pecteilis plantaginea* (Lindl.) M. A. Clem. & D. L. Jones, Austral. Orchid Rev. 83(6): 53. 2018. SYNTYPES: India, betw. Tenevelly & Travancore, Oct. 1814, *Wallich 7053A (leg. Röttler)* (syntypes, K-LINDL, K-WALL [2×]); without locality, Nov. 1816, *Wallich 7053B (leg. Heyne s.n.)*; Mungger [cited as Monghir by Lindley], 10 Sep. 1811, *Wallich 7053C (leg. Buchanan-Hamilton s.n.)* (syntype, K-WALL).

Distribution. Andhra Pradesh, Arunachal Pradesh, Assam, Bihar, Chhattisgarh, Goa, Haryana, Himachal Pradesh, Jammu & Kashmir, Jharkhand, Karnataka, Kerala, Madhya Pradesh, Maharashtra, Meghalaya, Odisha, Punjab, Tamil Nadu, Uttar Pradesh, Uttarakhand, West Bengal [Bangladesh, Myanmar, Nepal, Sri Lanka].

Habit. Terrestrial herb.

Habenaria platantheropsis Kraenzl., Bot. Jahrb. Syst. 28: 172. 1900. SYNTYPES: India, Nilgiri and Kurg [Coorg] Mountains, *Thomson 25.*

Distribution. Tamil Nadu. Endemic.

Habit. Terrestrial herb.

Remarks. This obscure species seems very close to *Habenaria brachyphylla* (Lindl.) Aitch.

Habenaria polyodon Hook. f., Fl. Brit. India [J. D. Hooker] 6(17): 139. 1890. *Habenaria fimbriata* Wight, Icon. Pl. Ind. Orient. [Wight] 5(1): 14, t. 1712. 1851, nom. illeg., non (Aiton) R. Br., 1813. TYPE: India, Nilgiri Hills, autumn, *?Wight s.n.* (holotype, K).

Distribution. Karnataka, Kerala, Tamil Nadu. Endemic.

Habit. Terrestrial herb.

Habenaria pubescens Lindl., Gen. Sp. Orchid. Pl.: 322. 1835. TYPE: India, Sabathoo, *Royle s.n.* (holotype, K).

Distribution. Himachal Pradesh, Uttarakhand [Nepal].

Habit. Terrestrial herb.

Habenaria rangatensis M. C. Naik & K. Prasad, Phytotaxa 442(1): 28. 2020. TYPE: India, Andaman & Nicobar Islands, Middle Andamans, Rangat hills, 51 m, 29 Sep. 2014, *K. Prasad 5201* (holotype, CAL; isotype, CAL).

Distribution. Andaman & Nicobar Islands. Endemic.

Habit. Terrestrial herb.

Habenaria rariflora A. Rich., Ann. Sci. Nat., Bot. sér. 2, 15: 70. 1841. TYPE: India, Nilgiri Hills near Konoor, July, *Perrottet s.n.* (holotype, P).

Habenaria uniflora Dalzell, Hooker's J. Bot. Kew Gard. Misc. 3: 344. 1851, nom. illeg., non Buch.-Ham. ex D. Don, 1825. TYPE: India, South Concan, *Dalzell s.n.* (syntype, K).
Habenaria rariflora A. Rich. var. *latifolia* Blatt. & McCann, J. Bombay Nat. Hist. Soc. 36: 17. 1932. TYPE: India, Panchgani, third milestone towards Wai, July 1925, *Frenchman 213* (holotype, BLAT).

Distribution. Andhra Pradesh, Goa, Karnataka, Kerala, Maharashtra, Tamil Nadu. Endemic.

Habit. Lithophytic herb.

Habenaria reniformis (D. Don) Hook. f., Fl. Brit. India [J. D. Hooker] 6(17): 152. 1890. *Listera reniformis* D. Don, Prodr. Fl. Nepal.: 28. 1825. *Neottia reniformis* (D. Don) Spreng., Syst. Veg. (ed. 16) [Sprengel] 3: 707. 1826. *Herminium reniforme* (D. Don) Lindl., Numer. List [Wallich] n. 7067. 1832, nom. inval. *Aopla reniformis* (D. Don) Lindl., Edwards's Bot. Reg. 20: t. 1701. 1835. TYPE: Nepal, Sankoo, Oct. 1821 *Wallich 7067* (holotype, K-WALL).

Distribution. Assam, Jharkhand, Manipur, Meghalaya, Odisha [China, Nepal].

Habit. Terrestrial herb.

Habenaria rhodocheila Hance, Ann. Sci. Nat., Bot. sér. 5, 5: 243. 1866. *Smithanthe rhodocheila* (Hance) Szlach. & Marg., Orchidee (Hamburg) 55(2): 174. 2004. *Pecteilis rhodocheila* (Hance) M. A. Clem. & D. L. Jones, Austral. Orchid Rev. 83(6): 53. 2018. TYPE: China, Canton, along North River, 15 July 1864, *Sampson 11332* (holotype, BM).

Distribution. Mizoram [China, Myanmar].

Habit. Lithophytic herb.

Habenaria rhynchocarpa (Thwaites) Trimen, Syst. Cat. Fl. Pl. Ceylon 91. 1885. *Platanthera rhynchocarpa* Thwaites, Enum. Pl. Zeyl. [Thwaites]: 310. 1861. *Plantaginorchis rhynchocarpa* (Thwaites) Szlach., Richardiana 4(2): 65. 2004. TYPE: Sri Lanka, Galagama, 4000 ft, *Ceylon Plants 3058* (not found).

Habenaria stenopetala Lindl., Gen. Sp. Orchid. Pl.: 324. 1835, nom. illeg., non Lindl.: 319. 1835. TYPE: India, Sri Lanka, *Macrae s.n.* (holotype, K-LINDL).

Distribution.   Karnataka [Sri Lanka].

Habit.   Terrestrial herb.

Habenaria richardiana Wight, Icon. Pl. Ind. Orient. [Wight] 5(1): 14, t. 1713. 1851. TYPE: India, Nilgiri Hills and Anamallies, *Wight s.n.* (lectotype, K K000247430, left hand specimen, designated by Ravichandran et al., 2019).

Distribution.   Kerala, Tamil Nadu. Endemic.

Habit.   Terrestrial herb.

Habenaria roxburghii Nicolson in C. J. Saldanha & D. H. Nicolson, Fl. Hassan Distr.: 834. 1976. *Orchis plantaginea* Roxb., Pl. Coromandel 1: 32, t. 37. 1795. *Orchis platyphyllos* Sw. ex Willd., Sp. Pl., ed. 4. 4: 11. 1805, nom. superfl. *Orchis roxburghii* Pers., Syn. Pl. (Persoon) 2: 503. 1807, nom. superfl. *Habenaria platyphylla* Spreng., Syst. Veg. (ed. 16) [Sprengel] 3: 690. 1826, nom. superfl. *Gymnadenia plantaginea* (Roxb.) Lindl., Numer. List [Wallich] n. 7053. 1832. *Habenaria platyphylloides* M. R. Almeida, Fl. Maharashtra 5A: 61. 2009, nom. superfl. *Pecteilis roxburghii* (Nicolson) M. A. Clem. & D. L. Jones, Austral. Orchid Rev. 83(6): 53. 2018. TYPE: India, Coromandel coast, icon. *Roxburgh s.n., loc. cit.*

Distribution.   Andhra Pradesh, Chhattisgarh, Jharkhand, Karnataka, Kerala, Madhya Pradesh, Maharashtra, Odisha, Tamil Nadu [Sri Lanka].

Habit.   Terrestrial herb.

Habenaria sahyadrica K. M. P. Kumar, Nirmesh, V. B. Sreek. & Kumar, Phytotaxa 244(2): 196. 2016. TYPE: India, Kerala: Palakkad district, Muthikulam, way to Elival hills, 1700 m, 22 Nov. 2013, *Nirmesh & Prabhukumar 28501* (holotype, KFRI; isotype, CMPR, CALI).

Distribution.   Kerala. Endemic.

Habit.   Terrestrial herb.

Habenaria stenopetala Lindl., Gen. Sp. Orchid. Pl.: 319. 1835. TYPE: India, towards Cashmir, *Royle s.n.* (holotype, K-LINDL).

Distribution.   Andaman & Nicobar Islands, Arunachal Pradesh, Assam, Himachal Pradesh, Maharashtra, Meghalaya, Mizoram, Nagaland, Odisha, Sikkim, Uttar Pradesh, Uttarakhand, West Bengal [Bhutan, China, Myanmar, Nepal].

Habit.   Terrestrial herb.

Habenaria suaveolens Dalzell, Hooker's J. Bot. Kew Gard. Misc. 2: 263. 1850. *Plantaginorchis suaveolens* (Dalzell) Szlach. & Kras-Lap., Richardiana 6(1): 32. 2006. *Pecteilis suaveolens* (Dalzell) M. A. Clem. & D. L. Jones, Austral. Orchid Rev. 83(6): 53. 2018. TYPE: India, betw. Vingula & Malwan, *Dalzell s.n.* (holotype, K).

Habenaria variabilis Blatt. & McCann, J. Bombay Nat. Hist. Soc. 36: 19. 1932, nom. illeg., non Ridl., 1886. *Habenaria panchganiensis* Santapau & Kapadia, J. Bombay Nat. Hist. Soc. 54: 478. 1957. TYPE: India, Western Ghats, third milestone from Panchgani to Wai, *Frenchman P. 21.*

Distribution.   Goa,   Karnataka,   Maharashtra. Endemic.

Habit.   Terrestrial herb.

Habenaria trichosantha Wall. ex Lindl., Gen. Sp. Orchid. Pl.: 324. 1835. *Fimbrorchis trichosantha* (Wall. ex Lindl.) Szlach., Orchidee (Hamburg) 55: 491. 2004. *Pecteilis trichosantha* (Wall. ex Lindl.) M. A. Clem. & D. L. Jones, Austral. Orchid Rev. 83(6): 53. 2018. TYPE: Myanmar, Mt. Tong Dong, Nov. 1826, *Wallich 7028* (holotype, K-LINDL; isotype, K-WALL).

Distribution.   Manipur [Bhutan, Myanmar].

Habit.   Terrestrial herb.

Habenaria trifurcata Hook. f., Fl. Brit. India [J. D. Hooker] 6(17): 148. 1890. TYPE: India, Khasi Hills, Nowgong, collector not cited (holotype, CAL).

Distribution.   Assam, Meghalaya, ?Tamil Nadu. Endemic.

Habit.   Terrestrial herb.

Habenaria viridiflora (Sw.) R. Br. ex Spreng., Syst. Veg., ed. 16, 3: 691. 1826. *Orchis viridiflora* Sw., Kongl. Vetensk. Acad. Nya Handl. 21: 206. 1800. TYPE: India Orientali, *Rottler s.n.*

Habenaria graminea A. Rich., Ann. Sci. Nat., Bot., sér. 2, 15: 73. 1841, nom. illeg., non Lindl., 1835, non (Thouars) Spreng., 1826. TYPE: India, Nilgiri Hills, Kaity, *Perrottet s.n.* (holotype, P).
Habenaria tenuis Griff., Calcutta J. Nat. Hist. 4: 379. 1844. TYPE: India, Serampore, *cult. Calcutta s.n.* (not found).
Coeloglossum luteum Dalzell, Hooker's J. Bot. Kew Gard. Misc. 2: 263. 1850. *Habenaria viridiflora* (Sw.) R. Br. ex Spreng. var. *dalzellii* Hook. f., Fl. Brit. India [J. D. Hooker] 6(17): 150. 1890. TYPE: India, Bombay Res., Malwan, Aug., *Dalzell s.n.*

Distribution. Assam, Karnataka, Kerala, Maharashtra, Tamil Nadu, West Bengal [Bangladesh, Myanmar, Sri Lanka].

Habit. Terrestrial herb.

———

Hemipilia Lindl., Gen. Sp. Orchid. Pl.: 296. 1835.

TYPE: *Hemipilia cordifolia* Lindl.

Hemipilia cordifolia Lindl., Gen. Sp. Orchid. Pl.: 296. 1835. *Platanthera cordifolia* Lindl., Numer. List [Wallich] n. 7049. 1832, nom. inval. TYPE: Nepal, Bagmati Zone, Rashuwa District, Gosainthan [Shishapangma], Aug. 1821, *Wallich 7049* (holotype, K-LINDL; isotype, K, K-WALL).

Distribution. Himachal Pradesh, Uttarakhand [Bhutan, Myanmar, Nepal].

Habit. Terrestrial herb.

Hemipilia purpureopunctata (K. Y. Lang) X. H. Jin, Schuit. & W. T. Jin, Molec. Phylogen. Evol. 77: 50. 2014. *Habenaria purpureopunctata* K. Y. Lang, Acta Phytotax. Sin. 16(4): 127. 1978. *Hemipiliopsis purpureopunctata* (K. Y. Lang) Y. B. Luo & S. C. Chen, Novon 13: 450. 2003. TYPE: China, Tibet, Bomi, 2500 m, 19 July 1965, *Y. T. Chang & K. Y. Lang 384* (holotype, PE).

Distribution. Arunachal Pradesh [China].

Habit. Terrestrial herb.

———

Herminium L., Opera Var. 251. 1758.

TYPE: *Herminium monorchis* (L.) R. Br. (*Ophrys monorchis* L. [as "*monochris*"]).

Herminium albomarginatum (King & Pantl.) X. H. Jin, Schuit., Raskoti & L. Q. Huang, Cladistics 32(2): 210. 2015. *Habenaria albomarginata* King

& Pantl., Ann. Roy. Bot. Gard. (Calcutta) 8: 322. 1898. *Platanthera albomarginata* (King & Pantl.) Kraenzl., Orchid. Gen. Sp. 1: 939. 1901. *Peristylus albomarginatus* (King & Pantl.) K. Y. Lang, Acta Phytotax. Sin. 34: 640. 1996. *Bhutanthera albomarginata* (King & Pantl.) Renz, Edinburgh J. Bot. 58: 101. 2001. TYPE: India, Sikkim, Singalelah, 13,000 ft, July 1896, *Pantling 450* (holotype, K; isotype, BM, E, L, PE, W).

Distribution. Sikkim [Bhutan, Nepal].

Habit. Terrestrial herb.

Herminium albovirens (Renz) X. H. Jin, Schuit., Raskoti & L. Q. Huang, Cladistics 32(2): 210. 2015. *Bhutanthera albovirens* Renz, Edinburgh J. Bot. 58(1): 102. 2001. TYPE: Bhutan, Tongsa District, Thita Tso, Rinchen Chu, 13 July 1937, *Ludlow & Sherriff 3441* (holotype, BM).

Distribution. Sikkim [Bhutan].

Habit. Terrestrial herb.

Remarks. This species was recently photographed in Sikkim (Swami, 2016, as *Bhutanthera albovirens*).

Herminium clavigerum (Lindl.) X. H. Jin, Schuit., Raskoti & L. Q. Huang, Cladistics 32(2): 210. 2015. *Platanthera clavigera* Lindl., Gen. Sp. Orchid. Pl.: 289. 1835. *Habenaria clavigera* (Lindl.) Dandy, J. Bot. 68: 246. 1930. *Habenella clavigera* (Lindl.) Szlach. & Kras-Lap., Richardiana 6: 34. 2006. TYPE: India, Simla, Aug. 1831, *Dalhousie s.n.* (holotype, K-LINDL; isotype, K, LIV).

Habenaria densa Wall. ex Lindl., Gen. Sp. Orchid. Pl.: 326. 1835. *Platanthera densa* Lindl., Numer. List [Wallich] n. 7046. 1832, nom. inval. *Platantheroides densa* (Wall. ex Lindl.) Szlach., Richardiana 4: 106. 2004. TYPE: Nepal, Sheopore [Shivapuri], 1821, *Wallich 7046* (holotype, K-WALL; isotype, K-LINDL).

Distribution. Arunachal Pradesh, Himachal Pradesh, Sikkim, Uttarakhand, West Bengal [Bhutan, China, Nepal].

Habit. Terrestrial herb.

Herminium edgeworthii (Hook. f. ex Collett) X. H. Jin, Schuit., Raskoti & L. Q. Huang, Cladistics 32(2): 210. 2015. *Habenaria edgeworthii* Hook. f. ex Collett, Fl. Siml. 504. 1902. *Platanthera edgeworthii* (Hook. f. ex Collett) R. K. Gupta, Fl. Nainital. 349. 1968. *Platantheroides edgeworthii* (Hook. f. ex Collett) Szlach., Richardiana 4(3): 106. 2004. *Habenella edgeworthii* (Hook. f. ex Collett) Szlach. & Kras-Lap., Richardiana 6(1):

35. 2006. TYPE: India, Simla, Banatar, Aug. 1834, *Edgeworth s.n.* (holotype, K).

Distribution. Arunachal Pradesh, Himachal Pradesh, Jammu & Kashmir, Uttar Pradesh, Uttarakhand [Bhutan, Myanmar, Nepal, Pakistan].

Habit. Terrestrial herb.

Herminium elisabethae (Duthie) Tang & F. T. Wang, Bull. Fan Mem. Inst. Biol. Bot. 7: 129. 1936. *Habenaria elisabethae* Duthie, J. Asiat. Soc. Bengal, Pt. 2, Nat. Hist. 71: 44. 1902. *Peristylus elisabethae* (Duthie) R. K. Gupta, Fl. Nainital. 351. 1968. TYPE: India, Uttarakhand, Mussoorie, 6000–7000 ft, Aug. 1899, *P. W. Mackinnon 22990* (lectotype, K, designated by Raskoti et al., 2017; isotype, K, P).

Peristylus kumaonensis Renz, J. Orchid Soc. India 1(1–2): 23. 1987. TYPE: India, Uttar Pradesh, Kumaon, from Nainital to Garmpani, old path, 2 Aug. 1983, *J. Renz, P. S. Pangtey & B. S. Kalokoti 13587* (holotype, RENZ).

Distribution. Himachal Pradesh, Meghalaya, Sikkim, Uttarakhand, West Bengal [Bhutan, China, Myanmar, Nepal].

Habit. Terrestrial herb.

Remarks. This species is illustrated as *Peristylus duthiei* (Hook. f.) Deva & H. B. Naithani in Swami (2016).

Herminium fallax (Lindl.) Hook. f., Fl. Brit. India [J. D. Hooker] 6(17): 129. 1890. *Peristylus fallax* Lindl., Gen. Sp. Orchid. Pl.: 298. 1835. *Habenaria fallax* (Lindl.) King & Pantl., Ann. Roy. Bot. Gard. (Calcutta) 8: 325. 1898. *Platanthera fallax* (Lindl.) Schltr., Repert. Spec. Nov. Regni Veg. Beih. 4: 111. 1919. *Monorchis fallax* (Lindl.) O. Schwarz, Mitt. Thüring. Bot. Ges. 1: 95. 1949. TYPE: Nepal, 1821, *Wallich 7412* (holotype, K-LINDL; isotype, K-WALL).

Peristylus fallax Lindl. var. *dwarikae* Deva & H. B. Naithani, Orch. Fl. N. W. Himal. 187. 1986 (as "*dwarikii*"). TYPE: India, Uttarakhand, Dodital, Uttarkashi, alt. 9200 ft, Aug. 1974, *Dwarika Prasad s.n.* (holotype, DD).

Distribution. Arunachal Pradesh, Meghalaya, Nagaland, Sikkim, Uttarakhand, West Bengal [Bhutan, China, Myanmar, Nepal].

Habit. Terrestrial herb.

Herminium gracile King & Pantl., J. Asiat. Soc. Bengal, Pt. 2, Nat. Hist. 65: 131. 1896. *Androcorys gracilis* (King & Pantl.) Schltr., Notizbl. Bot. Gart. Berlin-Dahlem 7(68): 397. 1920. *Monorchis grac-*

ilis (King & Pantl.) O. Schwarz, Mitt. Thüring. Bot. Ges. 1: 95. 1949. TYPE: India, Sikkim, Lachen Valley, July 1895, *Pantling 397* (lectotype, CAL CAL0000000673, designated by Lawkush & Chowdhery, 2014; isotype, CAL, K, L, P).

Distribution. Sikkim [Bhutan, China].

Habit. Terrestrial herb.

Herminium handelii X. H. Jin, Schuit., Raskoti & L. Q. Huang, Cladistics 32(2): 210. 2015. *Habenaria alpina* Hand.-Mazz., Symb. Sin. 7: 1336. 1936. *Bhutanthera alpina* (Hand.-Mazz.) Renz, Edinburgh J. Bot. 58: 102. 2001. TYPE: China, Yunnan, betw. Mekong & Salwin, alt. 4200–4300 m, 6 Aug. 1916, *Handel-Mazzetti 9716* (holotype, WU; isotype, AMES, E).

Distribution. Sikkim [Bhutan, China, Nepal].

Habit. Terrestrial herb.

Remarks. This is not *Herminium alpinum* (Jacq.) Sweet, or *Herminium alpinum* Lindl.

Herminium jaffreyanum King & Pantl., J. Asiat. Soc. Bengal, Pt. 2, Nat. Hist. 65: 130. 1896. *Monorchis jaffreyana* (King & Pantl.) O. Schwarz, Mitt. Thüring. Bot. Ges. 1: 95. 1949. *Androcorys jaffreyana* (King & Pantl.) Lawkush & Vik. Kumar, Indian J. Forest. 37: 483. 2014. TYPE: India, Sikkim, near the top of Sinchal, alt. 8600 ft, Aug. 1892, *Pantling 237* (holotype, CAL; isotype, AMES ["8000 ft"], K ["8000 ft"]).

Distribution. Arunachal Pradesh, Sikkim, ?Uttarakhand, West Bengal [Nepal].

Habit. Terrestrial herb.

Herminium josephi Rchb. f., Flora 55: 276. 1872. *Herminium grandiflorum* Lindl. ex Hook. f., Fl. Brit. India [J. D. Hooker] 6(17): 129. 1890, nom. inval. *Androcorys josephi* (Rchb. f.) Agrawala & H. J. Chowdhery, Kew Bull. 65(1): 106. 2010. TYPE: India, Sikkim, *J. D. Hooker s.n.* [*264*] (holotype, K-LINDL; isotype, AMES, K, P, W).

Herminium duthiei Hook. f., Fl. Brit. India [J. D. Hooker] 6(17): 130. 1890. *Monorchis duthiei* (Hook. f.) O. Schwarz, Mitt. Thüring. Bot. Ges. 1: 95. 1949. *Peristylus duthiei* (Hook. f.) Deva & H. B. Naithani, Orchid Fl. N. W. Himalaya 181. 1986. TYPE: India, Uttarakhand, Garwhal, Khasi pass, alt. 11,000–12,000 ft, 10 Sep. 1885, *Duthie s.n.* [*4424*] (holotype, K; isotype, BM, CAL).

Distribution. Arunachal Pradesh, Sikkim, Uttarakhand [Bhutan, China, Nepal].

Habit. Terrestrial herb.

Herminium kalimpongensis Pradhan, Indian Orchids: Guide Identif. & Cult. 2: 680. 1979. *Androcorys kalimpongensis* (Pradhan) Agrawala & H. J. Chowdhery, Kew Bull. 65(1): 106. 2010. TYPE: India, West Bengal, Kalimpong, 2000 m, 16 Aug. 1974, *U. C. Pradhan's collector* (holotype, U. C. Pradhan Herbarium).

Distribution. West Bengal. Endemic.

Habit. Terrestrial herb.

Remarks. The generic position of this species is uncertain; the cordate leaf is unlike any other species of *Herminium*.

Herminium kamengense A. N. Rao, J. Econ. Taxon. Bot. 25(2): 287. 2001. TYPE: India, Arunachal Pradesh, West Kameng District, Bomdila, alt. 8000 ft, *A. N. Rao 30600A* (holotype, Orchid Herbarium, Tipi).

Distribution. Arunachal Pradesh [China, Myanmar, Nepal].

Habit. Terrestrial herb.

Herminium kumaunense Deva & H. B. Naithani, Orchid Fl. N. W. Himalaya: 159, fig. 81. 1986 (as "*kumaunensis*"). TYPE: India, Uttarakhand, Pithoragrah District, Chalik Byans, alt. 10,500–11,500 ft, 23 July 1886, *Duthie 6003* (holotype, DD).

Distribution. Uttar Pradesh. Endemic.

Habit. Terrestrial herb.

Remarks. In Raskoti et al. (2017) a different collection is wrongly cited as the holotype.

Herminium lanceum (Thunb. ex Sw.) Vuijk, Blumea 11(1): 228. 1961. *Ophrys lancea* Thunb. ex Sw., Kongl. Vetensk. Acad. Nya Handl. 21: 223. 1800. *Satyrium lanceum* (Thunb. ex Sw.) Pers., Syn. Pl. (Persoon) 2(2): 507. 1807. *Aceras lanceum* (Thunb. ex Sw.) Steud., Nomencl. Bot., ed. 2 (Steudel) 1: 12. 1840. *Spiranthes lancea* (Thunb. ex Sw.) Backer, Bakh. f. & Steenis, Blumea 6(2): 361. 1950. TYPE: Indonesia, Java, *Thunberg s.n.* [*21289*] (holotype, UPS).

Aceras angustifolium Lindl., Edwards's Bot. Reg. 18: t. 1525. 1832 (as "*angustifolia*"). *Herminium angustifolium* (Lindl.) Benth. ex Hook. f., Fl. Brit. India [J. D. Hooker] 6(17): 129. 1890, non (Lindl.) Benth. ex Clarke, 1889. TYPE: Nepal, Bagmati Zone, Rasuwa District, Gosainthan [Shishapangma], Aug. 1821, *Wallich 7061* (holotype, K-LINDL; isotype, K, K-WALL [2×], L).

Distribution. Arunachal Pradesh, Assam, Himachal Pradesh, Karnataka, Manipur, Meghalaya, Mizoram, Nagaland, Sikkim, Tamil Nadu, Tripura, Uttarakhand, West Bengal [Bhutan, China, Myanmar, Nepal, Pakistan].

Habit. Terrestrial herb.

Herminium latilabre (Lindl.) X. H. Jin, Schuit., Raskoti & L. Q. Huang, Cladistics 32(2): 210. 2015. *Platanthera latilabris* Lindl., Gen. Sp. Orchid. Pl.: 289. 1835. *Habenaria latilabris* (Lindl.) Hook. f., Fl. Brit. India [J. D. Hooker] 6 (17): 153. 1890. *Platantheroides latilabris* (Lindl.) Szlach., Richardiana 4(3): 107. 2004. *Habenella latilabris* (Lindl.) Szlach. & Kras-Lap., Richardiana 6(1): 36. 2006. TYPE: India, Kumaon, *Wallich 7040B p.p. (leg. Blinkworth)* (holotype, K-LINDL; isotype, K-WALL).

Platanthera acuminata Lindl., Gen. Sp. Orchid. Pl.: 289. 1835. SYNTYPES: Nepal, Sep. 1821, *Wallich 7040A* (syntypes, K-LINDL, K-WALL); India, Kumaon, *Wallich 7040B p.p. (leg. Blinkworth)* (syntype K-WALL); India, *icon Wallich 167* (syntype, K).
Habenaria cumminsiana King & Pantl., J. Asiat. Soc. Bengal, Pt. 2, Nat. Hist. 64: 343. 1895 [1896]. *Platanthera cumminsiana* (King & Pantl.) Renz, Edinburgh J. Bot. 58: 117. 2001. *Platantheroides cumminsiana* (King & Pantl.) Szlach., Richardiana 4(3): 106. 2004. *Habenella cumminsiana* (King & Pantl.) Szlach. & Kras-Lap., Richardiana 6(1): 35. 2006. TYPE: India, Sikkim, Gnatong, 11,000 ft, Aug. 1894, *Pantling 329* (holotype, CAL; isotype, K).

Distribution. Arunachal Pradesh, Himachal Pradesh, Sikkim, Uttarakhand, West Bengal [Bhutan, China, Nepal, Pakistan].

Habit. Terrestrial herb.

Herminium longilobatum S. N. Hegde & A. N. Rao, Himalayan Pl. J. 1(2): 47. 1982. TYPE: India, Arunachal Pradesh, West Kameng District, Towang, 9000 ft, 6 Sep. 1980, *S. N. Hegde 3208A* (holotype, Orchid Herbarium, Tipi).

Distribution. Arunachal Pradesh, Nagaland [China, Nepal].

Habit. Epiphytic herb.

Herminium mackinnonii Duthie, J. Asiat. Soc. Bengal, Pt. 2, Nat. Hist. 71: 44. 1902 (as "*mackinnoni*"). *Monorchis mackinonii* (Duthie) O. Schwarz, Mitt. Thüring. Bot. Ges. 1: 95. 1949. TYPE: India, Uttarakhand, Lehari-Garhwal, near Mussoorie, 6000 ft, 30 Aug. 1901, *P. W. Mackinnon 25421* (holotype, K; isotype, AMES).

Distribution.   Uttarakhand [Nepal].

Habit.   Terrestrial herb.

Herminium macrophyllum (D. Don) Dandy, J. Bot. 70: 328. 1932. *Neottia macrophylla* D. Don, Prodr. Fl. Nepal.: 27. 1825. *Spiranthes macrophylla* (D. Don) Spreng., Syst. Veg. (ed. 16) [Sprengel] 3: 708. 1826. *Peristylus macrophyllus* (D. Don) Lawkush, Vik. Kumar & Bankoti, Indian J. Forest. 36: 388. 2013. TYPE: Nepal, 1819, *Wallich s.n.* (holotype, BM).

Herminium congestum Lindl., Edwards's Bot. Reg. 18: sub t. 1499. 1832. TYPE: Nepal, Gosainthan [Shishapangma], Aug. 1821, *Wallich 7068* (holotype, K-LINDL; isotype, BM, E, K, K-WALL).
Peristylus duthiei (Hook. f.) Deva & H. B. Naithani var. *inayatii* S. Deva & H. B. Naithani, Orch. Fl. N. W. Himal. 185. 1986. TYPE: India, Uttarakhand, Pithoragarh District, Kumaon, Ralam Valley, 19 Aug. 1900, *Inayat 24103* (holotype, DD).

Distribution.   Arunachal Pradesh, Nagaland, Sikkim, Uttarakhand, West Bengal [Bhutan, China, Nepal, Pakistan].

Habit.   Terrestrial herb.

Herminium mannii (Rchb. f.) Tang & F. T. Wang, Bull. Fan Mem. Inst. Biol. Bot. 7: 128. 1936. *Coeloglossum mannii* Rchb. f., Linnaea 41: 54. 1877 [1876]. *Peristylus mannii* (Rchb. f.) Mukerjee, Notes Roy. Bot. Gard. Edinburgh 21(3): 153. 1953. *Platanthera mannii* (Rchb. f.) Schltr., Repert. Spec. Nov. Regni Veg. Beih. 4: 114. 1919. TYPE: India, Meghalaya, Khasi Hills, alt. 5000 ft, *Mann s.n.* (holotype, W).

Habenaria gracillima Hook. f., Fl. Brit. India [J. D. Hooker] 6 (17): 163. 1890. *Peristylus gracillimus* (Hook. f.) Kraenzl., Orchid. Gen. Sp. 1: 513. 1898. TYPE: India, Khasi Hills, Munnipore, alt. 4000–5000 ft, *C. B. Clarke s.n.* (not found).

Distribution.   ?Andaman & Nicobar Islands, Assam, Manipur, Meghalaya, Nagaland [China, Myanmar, Nepal].

Habit.   Terrestrial herb.

Herminium monophyllum (D. Don) P. F. Hunt & Summerh., Kew Bull. 20: 51. 1966. *Neottia monophylla* D. Don, Prodr. Fl. Nepal.: 27. 1825. *Spiranthes monophylla* (D. Don) Spreng., Syst. Veg. (ed. 16) [Sprengel] 3: 709. 1826. *Monorchis monophylla* (D. Don) O. Schwarz, Mitt. Thüring. Bot. Ges. 1: 95. 1949. *Androcorys monophylla* (D. Don) Agrawala & H. J. Chowdhery, Kew Bull. 65(1): 106. 2010. *Herminium gramineum* Lindl., Gen.

Sp. Orchid. Pl.: 305. 1835, nom. superfl. TYPE: Nepal, 1819, *Wallich s.n.* (holotype, BMK-WALL).

Distribution.   Arunachal Pradesh, Himachal Pradesh, Uttarakhand [Bhutan, Nepal].

Habit.   Terrestrial herb.

Herminium monorchis (L.) R. Br., Hort. Kew., ed. 2 [W. T. Aiton] 5: 191. 1813. *Ophrys monorchis* L., Sp. Pl. 2: 947. 1753. *Orchis monorchis* (L.) Crantz, Stirp. Austr. Fasc., ed. 2. 6: 478. 1769. *Epipactis monorchis* (L.) F. W. Schmidt in Mayer, Phys. Aufs. 1: 246. 1795. *Arachnites monorchis* (L.) Hoffm., Deutschl. Fl., ed. 2, 2: 179. 1804. *Satyrium monorchis* (L.) Pers., Syn. Pl. (Persoon) 2: 507. 1807. *Monorchis herminium* O. Schwarz, Mitt. Thüring. Bot. Ges. 1: 95. 1949. TYPE: Europe, *Herb. Linn. No. 1056. 22, middle specimen* (lectotype, LINN, designated by Baumann et al., 1989).

Herminium haridasanii A. N. Rao, J. Econ. Taxon. Bot. 16(3): 725. 1992. TYPE: India, Arunachal Pradesh, Lohit District, Hot Spring, alt. 9600 ft, *Haridasan 3674A* (holotype, Orchid Herbarium, Tipi).

Distribution.   Arunachal Pradesh, Himachal Pradesh, Nagaland, Sikkim, Uttarakhand [China, Nepal, Pakistan].

Habit.   Terrestrial herb.

Remarks.   This species is illustrated as *Peristylus elisabethae* (Duthie) R. K. Gupta in Swami (2016).

Herminium pugioniforme Lindl. ex Hook. f., Fl. Brit. India [J. D. Hooker] 6(17): 130. 1890. *Monorchis pugioniformis* (Lindl. ex Hook. f.) O. Schwarz, Mitt. Thüring. Bot. Ges. 1: 96. 1949. *Androcorys pugioniformis* (Lindl. ex Hook. f.) K. Y. Lang, Guihaia 16(2): 105. 1996. SYNTYPES: India, Kashmir, alt. 12,000 ft, Aug. 1877, *J. E. T. Aitchison s.n.* [106] (syntype, K); Garwhal, Dudu Glacier under Srikanta, alt. 14,000–15,000 ft, 10 Aug. 1883, *Duthie s.n.* (syntypes, CAL, K); Sikkim, Samdong, 15,000–16,000 ft, 11 Sep. 1849, *J. D. Hooker s.n.* (syntype, K).

Distribution.   Himachal Pradesh, Sikkim, Uttarakhand [Bhutan, China, Nepal].

Habit.   Terrestrial herb.

Remarks.   This species probably holds the altitudinal record in Orchidaceae, occurring up to at least 5200 m.

Herminium quinquelobum King & Pantl., J. Asiat. Soc. Bengal, Pt. 2, Nat. Hist. 65: 130. 1896. *Monor-*

chis quinqueloba (King & Pantl.) O. Schwarz, Mitt. Thüring. Bot. Ges. 1: 96. 1949. TYPE: India, Sikkim, Tendong, alt. 7000 ft, Aug. 1894, *Pantling 339* (holotype, CAL; isotype, K).

Distribution. Sikkim, West Bengal [China, Nepal].

Habit. Terrestrial or sometimes epiphytic herb.

———

Herpysma Lindl., Edwards's Bot. Reg. 19: t. 1618. 1833.

TYPE: *Herpysma longicaulis* Lindl.

Herpysma longicaulis Lindl., Gen. Sp. Orchid. Pl.: 506. 1840. TYPE: Nepal, Mar. 1821, *Wallich 7389* (holotype, K-LINDL; isotype, AMES, C, CAL, K, K-WALL).

Distribution. Arunachal Pradesh, Assam, Meghalaya, Mizoram, Sikkim, West Bengal [Bhutan, China, Myanmar, Nepal].

Habit. Terrestrial herb.

———

Hetaeria Blume, Bijdr. Fl. Ned. Ind. 8: 409. 1825 ("*Etaeria*"), nom. et orth. cons.

TYPE: *Hetaeria oblongifolia* Blume (typ. cons.).

Hetaeria affinis (Griff.) Seidenf. & Ormerod, Oasis, Suppl. 2: 9. 2001. *Goodyera affinis* Griff., Not. Pl. Asiat. 3: 391. 1851. TYPE: India, Assam, near Negrogam, 18 Jan. 1836, *Griffith in Assam Herb. 315* (holotype, CAL [s.n.]); Khasi Hills, Mamlu, 610 m, Jan. 1897, *Pantling s.n.* (epitype, CAL, designated by Bhattacharjee & Chowdhery, 2013).

Cerochilus rubens Lindl., Gard. Chron. 1854: 87. 1854. *Rhamphidia rubens* (Lindl.) Lindl., J. Proc. Linn. Soc., Bot. 1: 182. 1857. *Hetaeria rubens* (Lindl.) Benth. ex Hook. f., Fl. Brit. India [J. D. Hooker] 6(17): 115. 1890. TYPE: Unknown locality, *R. Hanbury s.n.* (holotype, K-LINDL).

Distribution. Arunachal Pradesh, Assam, Meghalaya, Nagaland, Tripura [Bhutan, Myanmar].

Habit. Terrestrial herb.

Hetaeria anomala Lindl., J. Proc. Linn. Soc., Bot. 1: 185. 1857. TYPE: India, Assam, forest at Tingree, among tea trees, *Griffith s.n.* (holotype, K-LINDL).

Distribution. Assam, Sikkim [Myanmar].

Habit. Terrestrial herb.

Hetaeria obliqua Blume, Coll. Orchid. 104. 1859. TYPE: Indonesia, South Borneo, near Lake Babay, *Korthals s.n.* (holotype, L).

Distribution. Andaman & Nicobar Islands [also in Indonesia].

Habit. Terrestrial herb.

Hetaeria oblongifolia Blume, Bijdr. Fl. Ned. Ind. 8: 410. 1825 (as "*etaeria*"). TYPE: Indonesia, Java, Tjanjor Prov., Solassie Valley, *Blume s.n.* (holotype, L; isotype, K).

Goodyera ovalifolia Wight, Icon. Pl. Ind. Orient. [Wight] 5(1): t. 1730. 1851. *Rhamphidia ovalifolia* (Wight) Lindl., J. Proc. Linn. Soc., Bot. 1: 181. 1857. *Hetaeria ovalifolia* (Wight) Benth. ex Hook. f., Fl. Brit. India [J. D. Hooker] 6(17): 115. 1890. TYPE: India, Tamil Nadu, Courtallum, Aug.–Sep., *Wight s.n.* (syntypes, K [2×]).
Rhamphidia gardneri Thwaites, Enum. Pl. Zeyl. [Thwaites]: 313. 1861. *Hetaeria gardneri* (Thwaites) Benth. ex Hook. f., Fl. Brit. India [J. D. Hooker] 6(17): 115. 1890. TYPE: Sri Lanka, Central Province, *Gardner s.n.* [*Ceylon Plants 3425*] (holotype, K).
Hetaeria helferi Hook. f., Fl. Brit. India [J. D. Hooker] 6(17): 115. 1890. TYPE: Myanmar, Tenasserim, Misses Eider Islands, *Helfer s.n.* (holotype, K).

Distribution. Andaman & Nicobar Islands, Karnataka, Kerala, Tamil Nadu [Myanmar, Sri Lanka].

Habit. Terrestrial herb.

———

Holcoglossum Schltr., Repert. Spec. Nov. Regni Veg. Beih. 4: 285. 1919.

TYPE: *Holcoglossum quasipinifolium* (Hayata) Schltr. (*Saccolabium quasipinifolium* Hayata).

Holcoglossum amesianum (Rchb. f.) Christenson, Notes Roy. Bot. Gard. Edinburgh 44: 255. 1987. *Vanda amesiana* Rchb. f., Gard. Chron., ser. 3, 1: 764. 1887. TYPE: East Indies, *S. Low s.n.* (cult. Messrs. Hugh Low & Co.) (holotype, W).

Distribution. Assam, Manipur, Meghalaya [China, Myanmar].

Habit. Epiphytic herb.

Holcoglossum himalaicum (Deb, Sengupta & Malick) Aver., Bot. Zhurn. (Moscow & Leningrad) 73: 432. 1988. *Saccolabium himalaicum* Deb, Sengupta & Malick, Bull. Bot. Soc. Bengal 22(2): 213. 1968. *Ascocentrum himalaicum* (Deb, Sengupta & Malick) Christenson, Notes Roy. Bot. Gard. Edinburgh 44: 256. 1987. *Vanda himalaica* (Deb, Sengupta & Malick) L. M. Gardiner, Phytotaxa 61:

50. 2012. *Pendulorchis himalaica* (Deb, Sengupta & Malick) Z. J. Liu, K. Wei Liu & X. J. Xiao, PLoS ONE 8(4): art. e60097. 2013. TYPE: Myanmar, Sima, alt. 6500 ft, Jan. 1900, *Shaik Mokim 13* (holotype, CAL).

Distribution. Arunachal Pradesh, Assam, Manipur, Sikkim, West Bengal [Bhutan, Myanmar].

Habit. Epiphytic herb.

Holcoglossum nagalandense (Phukan & Odyuo) X. H. Jin, PLoS ONE 7(12): art. e52050. 2012 (as *"nagalandensis"*). *Penkimia nagalandensis* Phukan & Odyuo, Orchid Rev. 114(1272): 331. 2006. TYPE: India, Nagaland, Kiphirie District, Penkim village, 5100 ft, 3 Feb. 2004, *N. Odyuo 102808* (holotype, CAL).

Distribution. Nagaland [China].

Habit. Epiphytic herb.

———

Hsenhsua X. H. Jin, Schuit. & W. T. Jin, Molec. Phylogen. Evol. 77: 48. 2014.

TYPE: *Hsenhsua chrysea* (W. W. Sm.) X. H. Jin, Schuit., W. T. Jin & L. Q. Huang (*Habenaria chrysea* W. W. Sm.).

Hsenhsua chrysea (W. W. Sm.) X. H. Jin, Schuit., W. T. Jin & L. Q. Huang, Molec. Phylogen. Evol. 77: 48. 2014. *Habenaria chrysea* W. W. Sm., Notes Roy. Bot. Gard. Edinburgh 13: 204. 1921. *Orchis chrysea* (W. W. Sm.) Schltr., Repert. Spec. Nov. Regni Veg. 19: 372. 1924. *Ponerorchis chrysea* (W. W. Sm.) Soó, Acta Bot. Acad. Sci. Hung. 12: 353. 1966. *Chusua chrysea* (W. W. Sm.) P. F. Hunt, Kew Bull. 26: 174. 1971. TYPE: China: Tibet, Tsarong, on Doker-la, Mekong-Salween divide, 28°20′ N, 12,000–13,000 ft, Aug. 1917, *Forrest 14738* (holotype, E; isotype, K).

Distribution. Arunachal Pradesh [Bhutan, China].

Habit. Terrestrial herb.

———

Ipsea Lindl., Gen. Sp. Orchid. Pl.: 124. 1831.

TYPE: *Ipsea speciosa* Lindl.

Ipsea malabarica (Rchb. f.) Hook. f., Fl. Brit. India [J. D. Hooker] 5(16): 812. 1890. *Pachystoma malabaricum* Rchb. f., Ann. Bot. Syst. (Walpers) 6: 462. 1862. *Spathoglottis malabarica* (Rchb. f.) Pradhan, Indian Orchids: Guide Identif. & Cult. 2: 702. 1979 (as *"malbarica"*). TYPE: India, Malabar, *Jerdon s.n.* (holotype, W).

Distribution. Kerala. Endemic.

Habit. Terrestrial herb.

———

Lecanorchis Blume, Mus. Bot. 2: 188. 1856.

TYPE: *Lecanorchis javanica* Blume.

Lecanorchis sikkimensis N. Pearce & P. J. Cribb, Edinburgh J. Bot. 56(2): 280. 1999. TYPE: India, Sikkim, above Rungbee (Rimbi Chhu), alt. 7000 ft, 13 June 1902, *Prain's collector s.n.* (holotype, K).

Distribution. Sikkim [Bhutan, Myanmar].

Habit. Holomycotrophic terrestrial herb.

Lecanorchis taiwaniana S. S. Ying, Quart. J. Chin. Forest. 20(4): 133. 1987. TYPE: Taiwan, Wulai District, Chiaku-lin to Chia-ku-liao, 780 m, 2 Aug. 1987, *S. S. Ying s.n.* (holotype, NTUF F00000200; isotype, NTUF F00008292); Taiwan, Wulai District, Pataoerhshan, 1000 m, 15 Aug. 2009, *T. C. Hsu 2259* (epitype, TAIF, designated by Suetsugu et al., 2016).

Distribution. Assam [also in Japan, Laos, Taiwan].

Habit. Holomycotrophic terrestrial herb.

———

Liparis Rich., De Orchid. Eur. 21, 30, 38. 1817, nom. cons.

TYPE: *Liparis loeselii* (L.) Rich. (*Ophrys loeselii* L.).

Liparis acuminata Hook. f., Fl. Brit. India [J. D. Hooker] 5(16): 696. 1890. *Leptorkis acuminata* (Hook. f.) Kuntze, Revis. Gen. Pl. 2: 671. 1891 (as *"Leptorchis"*). TYPE: India, Khasi Hills, *Griffith s.n.* (holotype, K-LINDL).

Distribution. Meghalaya [also in Vietnam].

Habit. Terrestrial herb.

Liparis alata A. Rich., Ann. Sci. Nat., Bot. sér. 2, 15: 17. 1841. *Leptorkis alata* (A. Rich.) Kuntze, Revis. Gen. Pl. 2: 671. 1891 (as *"Leptorchis"*). TYPE: India, Nilgiri Hills, Waterfall ("Water-Fat") not far from Kaiti, *Perrottet s.n.*

Distribution. Tamil Nadu. Endemic.

Habit. Terrestrial herb.

Liparis assamica King & Pantl., Ann. Roy. Bot. Gard. (Calcutta) 8: 36. 1898. *Platystyliparis assamica* (King & Pantl.) Marg., Richardiana 7: 38. 2007. SYNTYPES: India, Khasi Hills and Assam, *Griffith 5082* (syntypes, CAL, K); *Mann s.n.* (syntypes, CAL [*21*], K); *Wallich s.n.*

Distribution. Arunachal Pradesh, Assam, Meghalaya, Nagaland, West Bengal [China].

Habit. Epiphytic herb.

Liparis atropurpurea Lindl., Gen. Sp. Orchid. Pl.: 28. 1830. *Platystylis atropurpurea* (Lindl.) Lindl., Gen. Sp. Orchid. Pl.: 18. 1830. TYPE: Sri Lanka, *Macrae s.n.* (holotype, K-LINDL).

Malaxis nilgiriensis T. Muthuk., A. Rajendran, Priyadh. & Sarval., Webbia 70: 65. 2015. *Crepidium nilgiriensis* (T. Muthuk., A. Rajendran, Priyadh. & Sarval.) Sushil K. Singh, Agrawala & Jalal, Orchids India: 172. 2019. TYPE: India, Tamil Nadu, Nilgiri District, Mukurthi National Park, *Muthukumar 230* (holotype, MH; isotype, BUH).

Distribution. Karnataka, Kerala, Madhya Pradesh, Tamil Nadu [Sri Lanka].

Habit. Terrestrial herb.

Liparis atrosanguinea Ridl., J. Straits Branch Roy. Asiat. Soc. 39: 71. 1903. TYPE: Malaysia, Perak, Taiping Hills, The Gap, *Curtis & Derry s.n.* (holotype, SING).

Distribution. Andaman & Nicobar Islands [Sri Lanka].

Habit. Terrestrial herb.

Liparis barbata Lindl., Gen. Sp. Orchid. Pl.: 27. 1830. *Empusa barbata* (Lindl.) T. C. Hsu, Illustr. Fl. Taiwan 2: 14. 2016. TYPE: Sri Lanka, *Macrae s.n.* (holotype, K-LINDL).

Liparis wrayi Hook. f., Fl. Brit. India [J. D. Hooker] 6(17): 181. 1890 (as "*wrayii*"). *Leptorkis wrayi* (Hook. f.) Kuntze, Revis. Gen. Pl. 2: 671. 1891 (as "*Leptorchis*"). *Diteilis wrayi* (Hook. f.) M. A. Clem. & D. L. Jones, Orchadian 15(1): 41. 2005. TYPE: Malaysia: Upper Perak, alt. 300 ft, *Wray s.n.* [*3713*] (holotype, CAL; isotype, K).
Liparis indirae Manilal & C. S. Kumar, Pl. Syst. Evol. 145 (1–2): 155. 1984 (as "*indiraii*"). TYPE: India, Kerala, Silent Valley, Valiayaparathode, *C. Sathish Kumar SV 10585* (holotype, CAL).

Distribution. Arunachal Pradesh, Kerala [Myanmar, Sri Lanka].

Habit. Terrestrial herb.

Liparis beddomei Ridl., J. Linn. Soc., Bot. 22: 268. 1886. *Leptorkis beddomei* (Ridl.) Kuntze, Revis. Gen. Pl. 2: 671. 1891 (as "*Leptorchis*"). TYPE: India, Shembanganoor, Pulneys Mountains, alt. 5000 ft, *Beddome s.n.* (holotype, BM).

Distribution. Tamil Nadu. Endemic.

Habit. Terrestrial herb.

Liparis biloba Wight, Icon. Pl. Ind. Orient. [Wight] 5(1): 4, t. 1633. 1851. *Leptorkis biloba* (Wight) Kuntze, Revis. Gen. Pl. 2: 671. 1891 (as "*Leptorchis*"). TYPE: India, Nilgiri Hills, July–Aug., *Wight s.n.* (holotype, K).

Distribution. Karnataka, Tamil Nadu. Endemic.

Habit. Epiphytic herb.

Liparis bistriata C. S. P. Parish & Rchb. f., Trans. Linn. Soc. London 30: 155. 1874. *Leptorkis bistriata* (C. S. P. Parish & Rchb. f.) Kuntze, Revis. Gen. Pl. 2: 671. 1891 (as "*Leptorchis*"). *Stichorkis bistriata* (C. S. P. Parish & Rchb. f.) Marg., Szlach. & Kulak, Acta Soc. Bot. Poloniae 77: 37. 2008. TYPE: Myanmar, Moulmein [Mawlamyine], *C. S. P. Parish s.n.* [*80*] (holotype, K).

Liparis bistriata C. S. P. Parish & Rchb. f. var. *robusta* Hook. f., Fl. Brit. India [J. D. Hooker] 5(16): 702. 1890. TYPE: India, Naga Hills, Kohima, *Prain s.n.* [*39*] (holotype, CAL).
Liparis hookeri Ridl., J. Linn. Soc., Bot. 22: 288. 1886. TYPE: India, Khasi Hills, *Hooker & Thomson 17 p.p.* (holotype, K).

Distribution. Arunachal Pradesh, Manipur, ?Meghalaya, Nagaland, Sikkim, Tripura, West Bengal [China, Myanmar].

Habit. Epiphytic herb.

Liparis bootanensis Griff., Not. Pl. Asiat. 3: 278. 1851. *Leptorkis bootanensis* (Griff.) Kuntze, Revis. Gen. Pl. 2: 671. 1891 (as "*Leptorchis*"). *Stichorkis bootanensis* (Griff.) Marg., Szlach. & Kulak, Acta Soc. Bot. Poloniae 77: 37. 2008. *Cestichis bootanensis* (Griff.) T. C. Hsu, Illustr. Fl. Taiwan 1: 334. 2016. TYPE: India, Khasi Hills, Duranga, alt. 500 ft, *Griffith in Bootan Herb. 1460* (holotype, K-LINDL; isotype, K).

Liparis lancifolia Hook. f., Fl. Brit. India [J. D. Hooker] 5(16): 700. 1890. SYNTYPES: India, Khasi Hills, Chuurra and Pomrang, 4000–6000 ft, *Hooker & Thomson Liparis no. 18*; Shillong, alt. 6000 ft, *C. B. Clarke s.n.* (syntype, K); Khasi Hills, 5000 ft, *Mann s.n.* (syntypes, CAL, K).
Liparis pachypus C. S. P. Parish & Rchb. f., Trans. Linn. Soc. London 30: 155. 1874. TYPE: Myanmar, Moulmein [Mawlamyine], *C. S. P. Parish s.n.* (holotype, W).

Distribution. Arunachal Pradesh, Assam, Manipur, Meghalaya, Mizoram, Nagaland, Sikkim, Tripura, West Bengal [Bhutan, China, Myanmar].

Habit. Epiphytic herb.

Liparis cathcartii Hook. f., Hooker's Icon. Pl. 19: t. 1808. 1889. *Leptorkis cathcartii* (Hook. f.) Kuntze, Revis. Gen. Pl. 2: 671. 1891 (as "*Leptorchis*"). SYNTYPES: India, Sikkim Himalaya, alt. 8000–10,000 ft, *J. D. Hooker* 103. (syntypes, K, K-LINDL); Darjeeling, Sandakphu, *C. B. Clarke 35033* (syntypes, BM, K, W); Darjeeling, Tonglu, *C. B. Clarke 35037* (syntype, K).

Distribution. Arunachal Pradesh, West Bengal [Bhutan, China, Nepal].

Habit. Terrestrial herb.

Liparis cespitosa (Lam.) Lindl., Bot. Reg. 11: sub t. 882. 1825 (as "*caespitosa*"). *Epidendrum cespitosum* Lam., Encycl. (Lamarck) 1: 187. 1783. *Cestichis cespitosa* (Lam.) Ames, Orchidaceae 2: 132. 1908 (as "*caespitosa*"). *Malaxis cespitosa* (Lam.) Thouars, Hist. Orchid. t. 90. 1822. *Stichorkis cespitosa* (Lam.) Thouars ex Marg. in H. B. Margońska & D. L. Szlachetko, Orchid. Tahiti: 61. 2010. TYPE: Bourbon [Réunion], *Jussieu s.n.* (holotype, P).

Liparis auriculata Rchb. f., Flora 55: 277. 1872, nom. illeg., non Blume ex Miq., 1866. *Liparis pusilla* Ridl., J. Linn. Soc., Bot. 22: 294. 1886. *Stichorkis pusilla* (Ridl.) Marg., Szlach. & Kulak, Acta Soc. Bot. Poloniae 77: 39. 2008. TYPE: India, Khasi Hills, *Hooker & Thomson s.n.* (not found).
Liparis duthiei Hook. f., Hooker's Icon. Pl. 19: t. 1857b. 1889. *Leptorkis duthiei* (Hook. f.) Kuntze, Revis. Gen. Pl. 2: 671. 1891 (as "*Leptorchis*"). TYPE: Western Himalaya: Kumaon, Gori Valley, alt. 2000–3000 ft, 18 Aug. 1886, *Duthie s.n. [5991]* (holotype, K; isotype, CAL).
Liparis obscura Hook. f., Fl. Brit. India [J. D. Hooker] 5(16): 701. 1890. TYPE: Sri Lanka, Central Province, alt. 2000–5000 ft, *Thwaites s.n.* (holotype, K).
Liparis prainii Hook. f., Fl. Brit. India [J. D. Hooker] 5(16): 700. 1890. TYPE: India, Upper Assam, Naga Hills, *Prain 78* (holotype, K).

Distribution. Arunachal Pradesh, Assam, Kerala, Manipur, Meghalaya, Mizoram, Nagaland, Sikkim, Tamil Nadu, Uttarakhand, West Bengal [Bhutan, China, Myanmar, Nepal, Sri Lanka].

Habit. Epiphytic herb.

Remarks. *Liparis duthiei* is still occasionally referred to as a distinct species, but we failed to find consistent differences that would justify this.

Liparis chungthangensis Lucksom, Orchid Rev. 112(1255): 14. 2004. TYPE: India, Sikkim,

Chungthang, 706–914 m, 12 June 2003, *S. Z. Lucksom 314* (holotype, CAL).

Distribution. Sikkim, West Bengal. Endemic.

Habit. Terrestrial herb.

Liparis cordifolia Hook. f., Hooker's Icon. Pl. 19: t. 1811. 1889. *Leptorkis cordifolia* (Hook. f.) Kuntze, Revis. Gen. Pl. 2: 671. 1891 (as "*Leptorchis*"). SYNTYPES: India, Sikkim, Yoksun, alt. 6000 ft, *C. B. Clarke s.n.*; Khasi Hills, alt. 3600 ft, *Griffith 35* (syntype, K-LINDL); *Hooker & Thomson, Liparis No. 8* (syntypes, K, K-LINDL, LE).

Distribution. Arunachal Pradesh, Assam, Meghalaya, Mizoram, Nagaland, Sikkim, Uttarakhand, West Bengal [Bhutan, China, Nepal].

Habit. Terrestrial herb.

Liparis deflexa Hook. f., Fl. Brit. India [J. D. Hooker] 5(16): 697. 1890. *Leptorkis deflexa* (Hook. f.) Kuntze, Revis. Gen. Pl. 2: 671. 1891 (as "*Leptorchis*"). TYPE: India, Sikkim Himalaya, Darjeeling, 1844, *Griffith's collector 5367* (holotype, K; isotype, K-LINDL).

?*Liparis diphyllos* Nimmo in J. Graham, Cat. Pl. Bombay: 252. 1839. TYPE: India, South Concan, collector not named.
Liparis prazeri King & Pantl., J. Asiat. Soc. Bengal, Pt. 2, Nat. Hist. 66: 582. 1898 [1897]. SYNTYPES: Myanmar, Kendat, Aug., *Calcutta Botanic Garden collector s.n.* (not found [Two specimens at Kew, collected in June 1892 by Prazer and Abdul Huk respectively, are probably types, as they were received in March 1897 from G. King labeled "*Liparis prazeri n. spec.*" However, the type locality Kendat is not mentioned on the labels and the specimens were collected in June, not August, which is in conflict with the protologue.]); *Kurz 345* (sterile, not found).
Liparis flavoviridis Blatt. & McCann, J. Bombay Nat. Hist. Soc. 35(2): 260. 1931. TYPE: India, North Kanara, right hand side of the road from Yellapur to Karwar, Aug. 1912, *T. R. Bell 4217*.

Distribution. Andhra Pradesh, Assam, Chhattisgarh, Goa, Karnataka, Kerala, Odisha, Tamil Nadu, Uttarakhand, West Bengal [Myanmar, Nepal].

Habit. Terrestrial herb.

Remarks. As suggested by Santapau and Kapadia (1962), *Liparis diphyllos* is possibly an earlier name for this species, but in the absence of type material the description is too vague to be certain.

Liparis delicatula Hook. f., Hooker's Icon. Pl. 19: t. 1889. 1889. *Leptorkis delicatula* (Hook. f.) Kuntze, Revis. Gen. Pl. 2: 671. 1891 (as "*Leptorchis*"). *Platystyliparis delicatula* (Hook. f.) Marg., Richardiana 7: 39. 2007. SYNTYPES: India, Khasi Hills, alt. 4000–5000 ft, *T. Lobb s.n.* (syntype, K); *Hooker & Thomson s.n.* [*1109*] (syntypes, CAL, K); India, Assam, Mishmi Hills, *Griffith s.n.* (syntypes, K, K-LINDL).

Distribution. Arunachal Pradesh, Assam, Meghalaya, Mizoram, Nagaland, Sikkim, Uttarakhand, West Bengal [China].

Habit. Epiphytic herb.

Liparis distans C. B. Clarke, J. Linn. Soc., Bot. 25: 71. 1889. *Leptorkis distans* (C. B. Clarke) Kuntze, Revis. Gen. Pl. 2: 671. 1891 (as "*Leptorchis*"). *Stichorkis distans* (C. B. Clarke) Marg., Szlach. & Kulak, Acta Soc. Bot. Poloniae 77: 38. 2008. SYNTYPES: India, Assam, Kohima, alt. 6000 ft, *C. B. Clarke 41105* (syntype, K); *41071* (syntype, K); *41574* (syntype, K); *41099* (syntypes, CAL, K).

Liparis macrantha Hook. f., Hooker's Icon. Pl. 19: t. 1854. 1889. TYPE: India, Assam, Naga Hills, *Prain 44.*

Distribution. Arunachal Pradesh, Assam, Manipur, ?Meghalaya, Nagaland [China].

Habit. Epiphytic herb.

Liparis dongchenii Lucksom, Indian J. Forest. 23(1): 113. 2000. TYPE: India, Sikkim, Lingchey, 706–1500 m, June, *Lucksom 316* (holotype, CAL).

Distribution. Sikkim, West Bengal. Endemic.

Habit. Terrestrial herb.

Liparis downii Ridl., J. Straits Branch Roy. Asiat. Soc. 49: 27. 1907 (as "*Siparis*"). TYPE: Thailand, *cult. Singapore Botanic Garden (leg. Down s.n.)*, June 1905.

Distribution. Odisha [Myanmar].

Habit. Terrestrial herb.

Liparis elegans Lindl., Gen. Sp. Orchid. Pl.: 30. 1830. *Leptorkis elegans* (Lindl.) Kuntze, Revis. Gen. Pl. 2: 671. 1891. *Cestichis elegans* (Lindl.) M. A. Clem. & D. L. Jones, Orchadian 15: 39. 2005. *Stichorkis elegans* (Lindl.) Marg., Szlach. & Kulak, Acta Soc. Bot. Poloniae 77: 38. 2008. TYPE: Malaysia, Penang, Oct. 1823, *Wallich 1943 (leg. G. Porter)* (lectotype, K-LINDL, designated by Tetsana et al., 2013).

Distribution. Andaman & Nicobar Islands [also in Thailand, etc.].

Habit. Epiphytic herb.

Liparis elliptica Wight, Icon. Pl. Ind. Orient. [Wight] 5(1): 17, t. 1735. 1851. *Liparis wightii* Rchb. f., Ann. Bot. Syst. (Walpers) 6: 218. 1861, nom. superfl. *Leptorkis elliptica* (Wight) Kuntze, Revis. Gen. Pl. 2: 671. 1891 (as "*Leptorchis*"). *Cestichis elliptica* (Wight) M. A. Clem. & D. L. Jones, Orchadian 15(1): 39. 2005. *Stichorkis elliptica* (Wight) Marg., Szlach. & Kulak, Acta Soc. Bot. Poloniae 77: 38. 2008. TYPE: India, Nilgiri Hills, Coonoor, Aug.–Oct., *Wight s.n.* (holotype, K).

Distribution. Andhra Pradesh, Arunachal Pradesh, Assam, Karnataka, Kerala, Manipur, Meghalaya, Odisha, Tamil Nadu, West Bengal [China, Nepal, Sri Lanka].

Habit. Epiphytic herb.

Liparis gamblei Hook. f., Hooker's Icon. Pl. 19: t. 1812. 1889. *Leptorkis gamblei* (Hook. f.) Kuntze, Revis. Gen. Pl. 2: 671. 1891 (as "*Leptorchis*"). TYPE: India, Sikkim Himalaya, Rungbee, *C. B. Clarke 12450* (syntype, K), Sinchul, alt. 6500 ft, *Gamble 3988A* (syntype, K).

Distribution. Sikkim, West Bengal [Bangladesh].

Habit. Terrestrial herb.

Liparis gigantea C. L. Tso, Sunyatsenia 1: 136. 1933. *Empusa gigantea* (C. L. Tso) T. C. Hsu, Illustr. Fl. Taiwan 2: 16. 2016. TYPE: China, Guangdong, Xinyi, 31 Mar. 1931, *Ko 51255* (holotype, SYS; isotype, PE).

Liparis nigra Seidenf., Bot. Tidsskr. 65: 129. 1969. *Diteilis nigra* (Seidenf.) M. A. Clem. & D. L. Jones, Orchadian 15: 41. 2005. TYPE: Thailand, Ban Langka, north-west of Lomsak, 1200–1300 m, *Seidenfaden & Smitinand GT 5401* (holotype, C).

Distribution. Arunachal Pradesh [China].

Habit. Terrestrial herb.

Liparis glossula Rchb. f., Linnaea 41: 44. 1877 [1876]. *Leptorkis glossula* (Rchb. f.) Kuntze, Revis. Gen. Pl. 2: 671. 1891 (as "*Leptorchis*"). TYPE: Not designated ("India orientalis").

Distribution. Himachal Pradesh, Sikkim, Uttarakhand, West Bengal [China, Nepal].

Habit. Terrestrial herb.

Liparis luteola Lindl., Gen. Sp. Orchid. Pl.: 32. 1830. *Leptorkis luteola* (Lindl.) Kuntze, Revis. Gen. Pl. 2: 671. 1891 (as "*Leptorchis*"). *Stichorkis luteola* (Lindl.) Marg., Szlach. & Kułak, Acta Soc. Bot. Poloniae 77(1): 38. 2008. TYPE: India, Pundua

Hills, Nov. 1824, *Wallich 1944 (leg. De Silva 1462)* (holotype, K-LINDL; isotype, BM, G, K, K-WALL, P).

Distribution.   Arunachal Pradesh, Assam, Meghalaya, Nagaland [China, Myanmar].

Habit.   Epiphytic herb.

Liparis lydiae Lucksom, J. Bombay Nat. Hist. Soc. 89: 105. 1992 (as "*lydiaii*"). TYPE: India, Sikkim, Bhusuk Valley, *Lucksom 198a* (holotype, CAL; isotype, Gangtok, Forest Department Herbarium).

Distribution.   Sikkim. Endemic.

Habit.   Epiphytic herb.

Liparis mannii Rchb. f., Flora 55: 275. 1872. *Leptorkis mannii* (Rchb. f.) Kuntze, Revis. Gen. Pl. 2: 671. 1891 (as "*Leptorchis*"). *Stichorkis mannii* (Rchb. f.) Marg., Szlach. & Kulak, Acta Soc. Bot. Poloniae 77: 38. 2008. *Cestichis mannii* (Rchb. f.) T. C. Hsu, Illustr. Fl. Taiwan 1: 339. 2016. TYPE: India, Assam, *Mann 27* (holotype, W, Herb. No. 46235; isotype, G).

Liparis tenuifolia Hook. f., Hooker's Icon. Pl. 21: t. 2013. 1890. *Stichorkis tenuifolia* (Hook. f.) Marg., Szlach. & Kulak, Acta Soc. Bot. Poloniae 77: 39. 2008. TYPE: India, Upper Assam, Mikir Hills, alt. 1000 ft, *Mann 30* (holotype, K).

Distribution.   Arunachal Pradesh, Assam, Sikkim, West Bengal [China, Myanmar].

Habit.   Epiphytic herb.

Liparis nervosa (Thunb.) Lindl., Gen. Sp. Orchid. Pl.: 26. 1830 var. **nervosa**. *Ophrys nervosa* Thunb., Fl. Jap. (Thunberg): 27. 1784. *Epidendrum nervosum* (Thunb.) Thunb., Trans. Linn. Soc. London 2: 327. 1794. *Cymbidium nervosum* (Thunb.) Sw., Nova Acta Regiae Soc. Sci. Upsal. 6: 76. 1799. *Malaxis nervosa* (Thunb.) Sw., Kongl. Vetensk. Acad. Nya Handl. 21: 235. 1800. *Iebine nervosa* (Thunb.) Raf., Fl. Tellur. 4: 39. 1836 [1838]. *Sturmia nervosa* (Thunb.) Rchb. f., Bonplandia (Hannover) 3: 250. 1855. *Leptorkis nervosa* (Thunb.) Kuntze, Revis. Gen. Pl. 2: 671. 1891 (as "*Leptorchis*"). *Diteilis nervosa* (Thunb.) M. A. Clem. & D. L. Jones, Orchadian 15(1): 40. 2005. *Empusa nervosa* (Thunb.) T. C. Hsu, Illustr. Fl. Taiwan 2: 18. 2016. TYPE: Japan, Osacca [Osaka] & Iedo, *Thunberg s.n.* (holotype, UPS).

Cymbidium bituberculatum Hook., Exot. Fl. 2: t. 116. 1824. *Liparis bituberculata* (Hook.) Lindl., Bot. Reg. 11: t. 882. 1825. *Sturmia bituberculata* (Hook.) Rchb. f., Bonplandia (Hannover) 2: 22. 1854. *Leptorkis bituberculata*

(Hook.) Kuntze, Revis. Gen. Pl. 2: 671. 1891. TYPE: Nepal, *Carey s.n.* (holotype, K).
Liparis macrocarpa Hook. f., Fl. Brit. India [J. D. Hooker] 5(16): 696. 1890. *Leptorkis macrocarpa* (Hook. f.) Kuntze, Revis. Gen. Pl. 2: 671. 1891. SYNTYPES: India, Sikkim Himalaya and Khasi Hills, *J. D. Hooker s.n.* (syntypes, K).

Distribution.   Andhra Pradesh, Arunachal Pradesh, Assam, Bihar, ?Goa, Karnataka, Madhya Pradesh, Meghalaya, Nagaland, Odisha, Sikkim, Tamil Nadu, Uttarakhand, West Bengal [China, Myanmar, Nepal].

Habit.   Terrestrial herb.

Remarks.   This species was recently recorded from Nagaland as *Liparis formosana* Rchb. f. by Odyuo et al. (2017), but the stated differences with *L. nervosa* all fall within the range of variation of the latter as circumscribed by Tetsana et al. (2019).

Liparis nervosa (Thunb.) Lindl. var. **khasiana** (Hook. f.) P. K. Sarkar, J. Econ. Taxon. Bot. 5(5): 1008. 1984. *Liparis bituberculata* (Hook.) Lindl. var. *khasiana* Hook. f., Fl. Brit. India [J. D. Hooker] 5(16): 696. 1890. *Liparis khasiana* (Hook. f.) Tang & F. T. Wang, Acta Phytotax. Sin. 1(1): 76. 1951. SYNTYPES: India, Khasi Hills, Myrung, *Griffith s.n.* [*5068*] (syntype, K), "*etc.*"

Liparis breviscapa A. P. Das & Lama, J. Econ. Taxon. Bot. 16(1): 226. 1992. TYPE: : West Bengal, Darjeeling, Birch Hill, alt. 6400 ft, 4 Aug. 1981, *A. P. Das 603* (holotype, CAL).

Distribution.   Arunachal Pradesh, Assam, Meghalaya [Bhutan].

Habit.   Terrestrial herb.

Liparis odorata (Willd.) Lindl., Gen. Sp. Orchid. Pl.: 26. 1830. *Malaxis odorata* Willd., Sp. Pl., ed. 4 [Willdenow] 4(1): 91. 1805. *Leptorkis odorata* (Willd.) Kuntze, Revis. Gen. Pl. 2: 671. 1891 (as "*Leptorchis*"). *Empusa odorata* (Willd.) T. C. Hsu, Illustr. Fl. Taiwan 2: 19. 2016. TYPE: India, Malabar, icon. of "la-Poulou-Mararava" in Rheede, Hort. Malabar. 12: 53, t. 27.

Empusa paradoxa Lindl., Bot. Reg. 10: sub t. 825. 1824. *Liparis paradoxa* (Lindl.) Rchb. f., Ann. Bot. Syst. (Walpers) 6: 218. 1861. TYPE: Nepal, *Lambert s.n.* (not found).
Liparis dalzellii Hook. f., Fl. Brit. India [J. D. Hooker] 5(16): 698. 1890. *Leptorkis dalzellii* (Hook. f.) Kuntze, Revis. Gen. Pl. 2: 671. 1891 (as "*Leptorchis*"). TYPE: India, South Concan, *Dalzell s.n.* (holotype, K).
Liparis vestita Rchb. f. subsp. *seidenfadenii* S. Misra, Nordic J. Bot. 6(1): 26, fig. 2. 1986. *Liparis espeevijii* S. Misra, Orchids Orissa 370. 2004. TYPE: India, Odisha State, Keonjhar District, Rebana reserve forests, Kendughata, alt. 2000 ft, 13 July 1982, *Misra 753* (holotype, CAL).

Liparis udaii S. Misra, J. Orchid Soc. India 23: 87. 2009. TYPE: India, Odisha, Sundargada district, Khajuridihi forest block, ca. 800 m, *Misra 2494* (holotype, CAL).

Liparis tortilis P. M. Salim & J. Mathew, Species (India) 20: 36. 2019. TYPE: India, Kerala, Wayanad district, Aranamala, altitude 895 m, 10 July 2018, *P. M. Salim & J. Mathew 4464* (holotype, KUBH; isotype, KUBH).

Distribution. Andhra Pradesh, Arunachal Pradesh, Assam, Bihar, Goa, Himachal Pradesh, Jharkhand, Karnataka, Kerala, Maharashtra, Manipur, Meghalaya, Mizoram, Nagaland, Odisha, Sikkim, Tamil Nadu, Uttarakhand, West Bengal [Bangladesh, Bhutan, China, Myanmar, Nepal, Sri Lanka].

Habit. Terrestrial herb.

Remarks. The sharply defined notch in the lip apex of the recently described *Liparis tortilis* appears to be an artifact and its eponymous twisted inflorescence may well be a developmental abnormality. In other respects this species does not differ significantly from the widespread *L. odorata*.

Liparis perpusilla Hook. f., Hooker's Icon. Pl. 19: t. 1856 B. 1889. *Leptorkis perpusilla* (Hook. f.) Kuntze, Revis. Gen. Pl. 2: 671. 1891 (as "*Leptorchis*"). *Platystyliparis perpusilla* (Hook. f.) Marg., Richardiana 7(1): 39. 2007. SYNTYPES: India, Sikkim Himalaya, on trees, *J. D. Hooker 110* (syntypes, K, K-LINDL); Darjeeling, *C. B. Clarke 26872* (syntypes, K, W); Darjeeling, Senchal, *Gamble 9421* (syntypes, K, K-LINDL).

Liparis togashii Tuyama in H. Hara, Fl. E. Himalaya 1: 441. 1966. TYPE: Nepal: Hati Sar, Minchin Dhap, alt. 8500–9500 ft, 28 Oct. 1963, *H. Hara et al. s.n.* (holotype, TI).

Distribution. Arunachal Pradesh, Sikkim, West Bengal [Bhutan, China, Nepal].

Habit. Epiphytic herb.

Liparis petiolata (D. Don) P. F. Hunt & Summerh., Kew Bull. 20(1): 52. 1966. *Acianthus petiolatus* D. Don, Prodr. Fl. Nepal.: 29. 1825. TYPE: Nepal, June 1821, *Wallich s.n.* [*1945*] (holotype, BM; isotype, C, E, G, K-LINDL, K-WALL, P, PR).

Malaxis cordifolia Sm. in A. Rees, Cycl. 22: art. Malaxis no. 12. 1812. TYPE: Nepal, Narainhetty, 2 Oct. 1802, *Buchanan s.n.* (holotype, LINN).

Liparis nepalensis Lindl., Bot. Reg. 11: t. 882. 1825. TYPE: Nepal, June 1821, *Wallich s.n.* [*1945*] (holotype, K-LINDL; isotype, BM, C, E, G, K-WALL, P, PR).

Liparis rupestris Griff. var. *purpurascens* Ridl., J. Linn. Soc., Bot. 22: 268. 1886. TYPE: India, Sikkim, Lachen, *J. D. Hooker s.n.* (not found).

Liparis pulchella Hook. f., Hooker's Icon. Pl. 19: t. 1810. 1889. *Leptorkis pulchella* (Hook. f.) Kuntze, Revis. Gen. Pl. 2: 671. 1891. SYNTYPES: India, Khasi Hills,

Myrung, Moflong, and Surureem, alt. 5000–6000 ft, *J. D. Hooker & T. Thomson s.n.* [& *1272*, 24 June 1850] (syntypes, CAL, K, K-LINDL, P); Assam, Naga Hills, *Prain 58* (syntype, K).

Distribution. Assam, Meghalaya, Nagaland, Sikkim, West Bengal [Bhutan, China, Myanmar, Nepal].

Habit. Terrestrial herb.

Remarks. More than one species may be included here.

Liparis plantaginea Lindl., Gen. Sp. Orchid. Pl.: 29. 1830. *Leptorkis plantaginea* (Lindl.) Kuntze, Revis. Gen. Pl. 2: 671. 1891 (as "*Leptorchis*"). *Stichorkis plantaginea* (Lindl.) Marg., Szlach. & Kulak, Acta Soc. Bot. Poloniae 77: 39. 2008. TYPE: India, icon. *Wallich 633* (holotype, BM).

Liparis selligera Rchb. f., Linnaea 41: 42. 1877 [1876]. TYPE: India, without locality, *Saunders s.n.* (holotype, W).

Distribution. Arunachal Pradesh, Assam, Manipur, Meghalaya, Nagaland, Sikkim, West Bengal [Bhutan, Myanmar, Nepal].

Habit. Epiphytic herb.

Liparis platyphylla Ridl., J. Linn. Soc., Bot. 22: 264. 1886. *Leptorkis platyphylla* (Ridl.) Kuntze, Revis. Gen. Pl. 2: 671. 1891 (as "*Leptorchis*"). TYPE: India, Anamallays, alt. 3000 ft, *Beddome s.n.* (holotype, BM).

Distribution. Karnataka, Kerala, Tamil Nadu. Endemic.

Habit. Terrestrial herb.

Liparis platyrachis Hook. f., Hooker's Icon. Pl. 19: t. 1890. 1889. *Leptorkis platyrachis* (Hook. f.) Kuntze, Revis. Gen. Pl. 2: 671. 1891 (as "*Leptorchis*"). *Platystyliparis platyrachis* (Hook. f.) Marg., Richardiana 7(1): 39. 2007. SYNTYPES: India, Sikkim Himalaya, alt. 4000–5000 ft, *Treutler 967* (syntype, K); *C. B. Clarke s.n.* (syntype, K).

Distribution. Nagaland, Sikkim, Uttarakhand, West Bengal [China, Nepal].

Habit. Epiphytic herb.

Liparis pygmaea King & Pantl., Ann. Roy. Bot. Gard. (Calcutta) 8: 34, t. 44. 1898. TYPE: India, Sikkim, below Jongri, alt. 13,000 ft, *Pantling 449* (holotype, CAL; isotype, E, K, LE, W).

Liparis nana Rolfe, Bull. Misc. Inform. Kew 1913: 28. 1913. TYPE: Vietnam, Annam, Sep. 1912, cult. *G. Wilson s.n* (holotype, K).

Liparis meniscophora Gagnep., Bull. Soc. Bot. France 79: 166. 1932. TYPE: Vietnam, Annam, Dalat, waterfalls of Camly, *Evrard 1029* (holotype, P).

Distribution. Sikkim, Uttarakhand [China, Nepal].

Habit. Terrestrial herb.

Liparis resupinata Ridl., J. Linn. Soc., Bot. 22: 290. 1886. *Leptorkis resupinata* (Ridl.) Kuntze, Revis. Gen. Pl. 2: 671. 1891 (as "*Leptorchis*"). *Platystyliparis resupinata* (Ridl.) Marg., Richardiana 7(1): 39. 2007. TYPE: India, Khasi Hills, *T. Lobb 122* (lectotype, K-LINDL, designated by Tetsana et al., 2019).

Liparis ridleyi Hook. f., Hooker's Icon. Pl. 19: t. 1887. 1889. *Leptorkis ridleyi* (Hook. f.) Kuntze, Revis. Gen. Pl. 2: 671. 1891 (as "*Leptorchis*"). *Liparis resupinata* Ridl. var. *ridleyi* (Hook. f.) King & Pantl., Ann. Roy. Bot. Gard. (Calcutta) 8: 37. 1898. *Platystyliparis resupinata* (Ridl.) Marg. var. *ridleyi* (Hook. f.) Marg., Richardiana 7: 39. 2007. SYNTYPES: India, Sikkim Himalaya, alt. 4000–6000 ft, *Griffith's collectors* (syntype, K, Distribution 5081); *Hooker 95* (syntype, K-LINDL).

Distribution. Arunachal Pradesh, Assam, Manipur, Meghalaya, Mizoram, Nagaland, ?Odisha, Sikkim, Tamil Nadu, Uttarakhand, West Bengal [Bhutan, China, Nepal].

Habit. Epiphytic herb.

Liparis rostrata Rchb. f., Linnaea 41: 44. 1877 [1876]. *Leptorkis rostrata* (Rchb. f.) Kuntze, Revis. Gen. Pl. 2: 671. 1891 (as "*Leptorchis*"). TYPE: Without origin, 1872, *cult. Balfour s.n.* (holotype, W).

Liparis diodon Rchb. f., Linnaea 41: 43. 1877 [1876]. *Leptorkis diodon* (Rchb. f.) Kuntze, Revis. Gen. Pl. 2: 671. 1891 (as "*Leptorchis*"). TYPE: India, Dehra Dun ("Derha Doon"), *Wilson Saunders s.n.*

Distribution. Himachal Pradesh, Uttarakhand [China, Nepal, Pakistan].

Habit. Terrestrial herb.

Liparis rupestris Griff., Not. Pl. Asiat. 3: 276. 1851. *Leptorkis rupestris* (Griff.) Kuntze, Revis. Gen. Pl. 2: 671. 1891 (as "*Leptorchis*"). TYPE: India, Assam, Nunklow, 15 Nov. 1835, *Griffith in Assam Herb. 257* (not found).

Distribution. Assam, Meghalaya, Sikkim, West Bengal. Endemic.

Habit. Terrestrial herb.

Liparis sanamalabarica P. M. Salim, Taiwania 62(4): 345. 2017. TYPE: India, Kerala, Wayanad Dis-trict, Kattimattom Hills, 1392 m, Sep. 2011, *P. M. Salim 0416* (holotype, MSSRF; isotype, MSSRF).

Distribution. Kerala. Endemic.

Habit. Terrestrial herb.

Liparis stricklandiana Rchb. f., Gard. Chron., n.s., 13: 232. 1880. *Leptorkis stricklandiana* (Rchb. f.) Kuntze, Revis. Gen. Pl. 2: 671. 1891 (as "*Leptorchis*"). *Stichorkis stricklandiana* (Rchb. f.) Marg., Szlach. & Kułak, Acta Soc. Bot. Poloniae 77(1): 39. 2008. TYPE: India, Assam, *Strickland (leg. Bull) s.n.* (holotype, K).

Liparis griffithii Ridl., J. Linn. Soc., Bot. 22: 285. 1886. TYPE: Bhutan, *Griffith 5069* (holotype, K).

Liparis dolabella Hook. f., Fl. Brit. India [J. D. Hooker] 6(17): 183. 1890. TYPE: India, Khasi Hills, alt. 4000 ft, *Mann 25* (holotype, K).

Distribution. Arunachal Pradesh, ?Meghalaya, Nagaland, Sikkim, West Bengal [Bhutan, China, Nepal].

Habit. Epiphytic herb.

Liparis tigerhillensis A. P. Das & S. Chanda, J. Econ. Taxon. Bot. 12(2): 403. 1988 [1989]. TYPE: India, West Bengal, Darjeeling, Tiger Hill, alt. 8000 ft, 16 July 1982, *A. P. Das 1051* (holotype, CAL; isotype, CAL, Herb. Presidency College, Calcutta, Herb A. P. Das).

Distribution. West Bengal. Endemic.

Habit. Terrestrial herb.

Liparis torta Hook. f., Hooker's Icon. Pl. 21: t. 2014. 1890. *Leptorkis torta* (Hook. f.) Kuntze, Revis. Gen. Pl. 2: 671. 1891 (as "*Leptorchis*"). *Stichorkis torta* (Hook. f.) Marg., Szlach. & Kulak, Acta Soc. Bot. Poloniae 77: 39. 2008. TYPE: India, Khasi Hills, alt. 3000 ft, *Mann s.n.* (holotype, K).

Distribution. Meghalaya. Endemic.

Habit. Epiphytic herb.

Liparis vestita Rchb. f., Flora 55: 274. 1872. *Leptorkis vestita* (Rchb. f.) Kuntze, Revis. Gen. Pl. 2: 671. 1891 (as "*Leptorchis*"). *Stichorkis vestita* (Rchb. f.) Marg., Szlach. & Kulak, Acta Soc. Bot. Poloniae 77: 39. 2008. TYPE: India, Assam, *Mann s.n.* (holotype, W; isotype, C).

Distribution. Assam, Meghalaya. Endemic.

Habit. Epiphytic herb.

Liparis viridiflora (Blume) Lindl., Gen. Sp. Orchid. Pl.: 31. 1830. *Malaxis viridiflora* Blume, Bijdr. Fl.

Ned. Ind. 8: 392, t. 54. 1825. *Leptorkis viridiflora* (Blume) Kuntze, Revis. Gen. Pl. 2: 671. 1891 (as "*Leptorchis*"). *Stichorkis viridiflora* (Blume) Marg., Szlach. & Kulak, Acta Soc. Bot. Poloniae 77: 39. 2008. *Cestichis viridiflora* (Blume) T. C. Hsu, Illustr. Fl. Taiwan 1: 339. 2016. TYPE: Indonesia, Java, Tjanjor Prov, Solassie Valley, *Blume s.n.* [*1297*] (holotype, L).

Liparis longipes Lindl., Pl. Asiat. Rar. (Wallich) 1: 31, t. 35. 1829. *Sturmia longipes* (Lindl.) Rchb. f., Bonplandia (Hannover) 3: 250. 1855. *Leptorkis longipes* (Lindl.) Kuntze, Revis. Gen. Pl. 2: 670. 1891 (as "*Leptorchis*"). *Cestichis longipes* (Lindl.) Ames, Orchidaceae 1: 75. 1905. SYNTYPES: India, Sylhet, *cult. Calcutta s.n. (leg. M. R. Smith)* (syntype, K-WALL); Sri Lanka, in the mountains, *Macrae s.n.*
Liparis pendula Lindl., Edwards's Bot. Reg. 24(Misc.): 94. 1838. TYPE: India, *cult. Loddiges s.n.* (holotype, K-LINDL).
Liparis spathulata Lindl., Edwards's Bot. Reg. 26(Misc.): 81. 1840. *Liparis longipes* Lindl. var. *spathulata* (Lindl.) Ridl., J. Linn. Soc., Bot. 22: 294. 1886. *Liparis viridiflora* (Blume) Lindl. var. *spathulata* (Lindl.) A. N. Rao, Bull. Arunachal Forest Res. 26: 103. 2010, nom. inval. SYNTYPES: India, *cult. Loddiges s.n.*; Burma, *Griffith 772.*
Liparis stachyurus Rchb. f., Flora 55: 274. 1872. *Leptorkis stachyurus* (Rchb. f.) Kuntze, Revis. Gen. Pl. 2: 671. 1891 (as "*Leptorchis*"). TYPE: India, Assam, without further details.
Liparis sikkimensis Lucksom & S. Kumar, J. Indian Bot. Soc. 73: 159. 1994. TYPE: India, Sikkim, Dikchu, 10 Nov. 1992, *Lucksom & S. Kumar 197* (holotype, CAL; isotype, Gangtok Forest Dept. Herbarium).

Distribution.   Andhra Pradesh, Arunachal Pradesh, Assam, Karnataka, Kerala, Manipur, Meghalaya, Mizoram, Nagaland, Odisha, Sikkim, Tamil Nadu, Tripura, Uttarakhand, West Bengal [Bangladesh, Bhutan, China, Myanmar, Nepal, Sri Lanka].

Habit.   Epiphytic herb.

Liparis walakkadensis M. Kumar & Sequiera, J. Orchid Soc. India 13(1–2): 29, fig. 1. 1999. TYPE: India, Kerala, Palghat, Silent Valley National Park, Walakkad, 3500 ft, 26 Aug. 1995, *Stephen 007877* (holotype, KFRI).

Distribution.   Kerala. Endemic.

Habit.   Terrestrial herb.

Liparis walkerae Graham, Edinburgh New Philos. J. 20: 194. 1836 (as "*walkeriae*"). *Leptorkis walkerae* (Graham) Kuntze, Revis. Gen. Pl. 2: 671. 1891 (as "*Leptorchis walkeriae*"). *Diteilis walkerae* (Graham) M. A. Clem. & D. L. Jones, Orchadian 15: 41. 2005 (as "*walkeriae*"). TYPE: Sri Lanka, 1834, *Mrs. Colonel Walker s.n.* (holotype, K).

Distribution.   Kerala, Tamil Nadu [Sri Lanka].

Habit.   Terrestrial herb.

Liparis wightiana Thwaites, Enum. Pl. Zeyl. [Thwaites]: 295. 1861. *Leptorkis wightiana* (Thwaites) Kuntze, Revis. Gen. Pl. 2: 671. 1891 (as "*Leptorchis*"). *Diteilis wightiana* (Thwaites) M. A. Clem. & D. L. Jones, Orchadian 15(1): 41. 2005. *Liparis atropurpurea* Wight, Icon. Pl. Ind. Orient. [Wight] 3(2): 9, t. 904. 1845, nom. illeg., non Lindl., 1830. TYPE: Sri Lanka, Central Province, alt. 3000–5000 ft, *Thwaites s.n.* [*Ceylon Plants 3179*] (holotype, BM; isotype, K, P).

Distribution.   Assam, Karnataka, Kerala, Tamil Nadu [Sri Lanka].

Habit.   Terrestrial herb.

——

Luisia Gaudich., Voy. Uranie, Bot. 426. 1826 [1829].

TYPE: *Luisia teretifolia* Gaudich.

Luisia abrahamii Vatsala in Abraham & Vatsala, Intr. Orchids 489. 1981 (as "*abrahami*"). TYPE: India, Thenmalai-Aryankavu Range, alt. 1000–2000 ft, 11 Apr. 1973, *P. Vatsala 207* (holotype, University of Kerala, Botany Department Herbarium).

Distribution.   Kerala. Endemic.

Habit.   Epiphytic herb.

Luisia antennifera Blume, Rumphia 4: 50. 1849. TYPE: Indonesia, Borneo, *Korthals s.n.* (holotype, L).

Distribution.   Manipur [also in Indonesia, Malaysia, Thailand, Vietnam].

Habit.   Epiphytic herb.

Luisia balakrishnanii S. Misra, Nelumbo 52: 152. 2010 (as "*balakrishnani*"). TYPE: India, South Andaman, near Jarwa on Andaman Trunk Road, *S. Misra 2498* (holotype, CAL).

Distribution.   Andaman & Nicobar Islands. Endemic.

Habit.   Epiphytic herb.

Luisia brachystachys (Lindl.) Blume, Rumphia 4: 50. 1849. *Mesoclastes brachystachys* Lindl., Gen. Sp. Orchid. Pl.: 44. 1830. TYPE: Bangladesh or India, Sylhet, *Wallich 1994 (leg. De Silva)* (holotype, K-LINDL; isotype, K-WALL).

Luisia indivisa King & Pantl., Ann. Roy. Bot. Gard. (Calcutta) 8: 201, t. 269. 1898. TYPE: India, Sikkim, at the base of the range, *Anderson s.n.* (Pantling's drawing no. 470) (?holotype, K).

Distribution. Andaman & Nicobar Islands, Arunachal Pradesh, Assam, Jharkhand, Meghalaya, Nagaland, Odisha, Sikkim, Tripura, Uttar Pradesh, Uttarakhand, West Bengal [Bangladesh, Bhutan, China, Myanmar, Nepal].

Habit. Epiphytic herb.

Luisia diglipurensis Sanjay Mishra & Jalal, Phytotaxa 453(3): 255. 2020. TYPE: India, Andaman & Nicobar Islands, North Andaman, Diglipur, Shyam Nagar, 7 m, 15 Mar. 2016, *Mishra 32425* (holotype, PBL; isotype, PBL).

Distribution. Andaman & Nicobar Islands. Endemic.

Habit. Epiphytic herb.

Luisia filiformis Hook. f., Fl. Brit. India [J. D. Hooker] 6(17): 23. 1890. TYPE: Bangladesh, Sylhet, Terrya Ghat, *Mann s.n.* (holotype, K).

Luisia grovesii Hook. f., Fl. Brit. India [J. D. Hooker] 6(17): 25. 1890. TYPE: India, eastern Bengal, Lushai Hills, *G. B. Groves s.n.* (holotype, CAL).

Distribution. Arunachal Pradesh, Bihar, Manipur, Meghalaya, Mizoram, Sikkim, West Bengal [Bangladesh, China].

Habit. Epiphytic herb.

Luisia inconspicua (Hook. f.) Hook. f. ex King & Pantl., Ann. Roy. Bot. Gard. (Calcutta) 8: 203, t. 272. 1898. *Saccolabium inconspicuum* Hook. f., Fl. Brit. India [J. D. Hooker] 6(17): 56. 1890. *Gastrochilus inconspicuus* (Hook. f.) Kuntze, Revis. Gen. Pl. 2: 661. 1891. *Luisiopsis inconspicua* (Hook. f.) C. S. Kumar & P. C. S. Kumar, Rheedea 15(1): 48. 2005. TYPE: India, Lower Assam, *Jenkins s.n.*, icon. in *Herb. Calcutta* (holotype, CAL; isotype, K).

Luisia micrantha Hook. f., Fl. Brit. India [J. D. Hooker] 6(17): 23. 1890. SYNTYPES: India, Assam, *Griffith 5186* (syntype, K); Khasi Hills, alt. 3000–4000 ft, *Mann 12* (syntype, K).

Distribution. Andaman & Nicobar Islands, Arunachal Pradesh, Assam, Jharkhand, Manipur, Meghalaya, Nagaland, Odisha, Sikkim, Tripura, Uttarakhand, West Bengal [Bangladesh, Bhutan, Nepal].

Habit. Epiphytic herb.

Luisia jarawana Sanjay Mishra & Jalal, Phytotaxa 453(3): 260. 2020. TYPE: India, Andaman &

Nicobar Islands, Middle Andaman, on the way to Sagwan nallah, 16 m, 16 Mar. 2018, *Mishra 33199* (holotype, PBL; isotype, PBL).

Distribution. Andaman & Nicobar Islands. Endemic.

Habit. Epiphytic herb.

Luisia jonesii J. J. Sm., Blumea 5: 311. 1943. TYPE: Peninsular Malaysia, *cult. Singapore Botanic Garden s.n. (leg. Jones)* (holotype, L).

Distribution. Manipur [also in Malaysia].

Habit. Epiphytic herb.

Luisia macrantha Blatt. & McCann, J. Bombay Nat. Hist. Soc. 35(3): 492, t. 10. 1932. SYNTYPES: India, North Kanara, Yellapur, in forest, *T. R. Bell 5397 & 5400* (syntypes, BLAT).

Distribution. Karnataka, Kerala. Endemic.

Habit. Epiphytic herb.

Luisia macrotis Rchb. f., Gard. Chron. 1869: 1110. 1869. TYPE: India, Assam, *J. Day s.n.* (holotype, W).

Distribution. Assam [China, Myanmar].

Habit. Epiphytic herb.

Luisia megamalaiana Karupp. & V. Ravich., Phytotaxa 387(4): 295. 2019. TYPE: India, Tamil Nadu, Theni District, Megamalai Wildlife Sanctuary, near Kardana Estate, 1400–1600 m, 10 Nov. 2016, *Karuppusamy & Ravichandran 864* (holotype, MH; isotype, The Madura College, Madurai).

Distribution. Tamil Nadu. Endemic.

Habit. Epiphytic herb.

Luisia microptera Rchb. f., Gard. Chron. 1870: 1503. 1870. TYPE: India, Assam, *Benson s.n.* (holotype, W).

Distribution. Assam. Endemic.

Habit. Epiphytic herb.

Luisia primulina C. S. P. Parish & Rchb. f., Trans. Linn. Soc. London 30: 144, t. 10. 1874. TYPE: Myanmar, Moulmein [Mawlamyine], *C. S. P. Parish s.n.* (holotype, W).

Distribution. ?Odisha [Myanmar].

Habit. Epiphytic herb.

Remarks. This species is listed as "excluded" in Singh et al. (2019).

Luisia psyche Rchb. f., Bot. Zeitung (Berlin) 21(12): 98. 1863. TYPE: not designated.

Distribution. Arunachal Pradesh, Assam, Meghalaya, Nagaland [Myanmar].

Habit. Epiphytic herb.

Luisia recurva Seidenf., Dansk Bot. Ark. 27(4): 31. 1971. TYPE: Thailand, Phu Krading, Loei, alt. 5000 ft, *Kerr 0140* (holotype, K).

Distribution. Andaman & Nicobar Islands [also in Thailand].

Habit. Epiphytic herb.

Luisia secunda Seidenf., Dansk Bot. Ark. 27(4): 32. 1971. TYPE: Thailand, Muang Lan, Ranong, *Seidenfaden & Smitinand GT 6210* (holotype, C).

Distribution. West Bengal [also in Thailand].

Habit. Epiphytic herb.

Remarks. This is considered a synonym of *Luisia filiformis* Hook. f. by Singh et al. (2019).

Luisia tenuifolia Blume, Rumphia 4: 50. 1849. *Cymbidium tenuifolium* Lindl., Gen. Sp. Orchid. Pl.: 167. 1833, nom. illeg., non (L.) Willd., 1805. *Luisia laurifolia* M. R. Almeida, Fl. Maharashtra 5A: 66 (2009), nom. superfl. TYPE: Sri Lanka, Peradenia, *Macrae s.n.* [67] (holotype, K-LINDL).

Birchea teretifolia A. Rich., Ann. Sci. Nat., Bot. sér. 2. 15: 67 1841. *Luisia birchea* Blume, Mus. Bot. 1: 64. 1849. *Birchea nilgherrensis* D. Dietr., Syn. Pl. 5:118. 1852, nom. superfl. TYPE: India, Nilgiri Hills, near Avalanchy, *Perrottet s.n.* (holotype, P).
Luisia evangelinae Blatt. & McCann, J. Bombay Nat. Hist. Soc. 35(3): 493. 1932. *Luisia tenuifolia* Blume var. *evangelinae* (Blatt. & McCann) Santapau & Kapadia, J. Bombay Nat. Hist. Soc. 59: 829. 1962. *Luisia birchea* Blume var. *evangelinae* (Blatt. & McCann) P. K. Sarkar, J. Econ. Taxon. Bot. 5(5): 1008. 1984. *Luisia laurifolia* M. R. Almeida var. *evangelinae* (Blatt. & McCann) M. R. Almeida, Fl. Maharashtra 5A: 66. 2009. TYPE: India, North Kanara, Astoli & Chandwadi, Apr. 1911, *T. R. Bell s.n.*
Luisia pseudotenuifolia Blatt. & McCann, J. Bombay Nat. Hist. Soc. 35(3): 492. 1932. TYPE: India, North Kanara, in forest, *T. R. Bell 5401.*
Luisia pulniana Vatsala in Abraham & Vatsala, Intr. Orchids: 486. 1981. TYPE: India, Kodaikanal Hill near Shembaganur, alt. 6200–8000 ft, May 1973, *V. S. Manickam 205* (holotype, University of Kerala, Botany Department Herbarium).

Distribution. Goa, Karnataka, Kerala, Maharashtra, Tamil Nadu [Sri Lanka].

Habit. Epiphytic herb.

Remarks. Abraham and Vatsala (1981) may be correct in recognizing more than one species in this complex. A revision of Indian *Luisia* is much needed.

Luisia trichorrhiza (Hook.) Blume, Mus. Bot. 1: 63. 1849 (as "*trichorhiza*"). *Vanda trichorrhiza* Hook., Exot. Fl. 1: t. 72. 1823 (as "*trichorhiza*"). TYPE: Nepal, cult. Liverpool *(leg. Wallich s.n.)* (not found).

Luisia trichorrhiza (Hook.) Blume var. *flava* Gogoi, Richardiana 16: 324. 2016. TYPE: India, Assam, Tinsukia Dist., Daisajan, 118 m, 16 Apr. 2016, *K. Gogoi 0764* (holotype, CAL; isotype, DU, TOSEHIM [The Orchid Society of Eastern Himalaya]).

Distribution. Andhra Pradesh, Arunachal Pradesh, Assam, Jharkhand, Maharashtra, Meghalaya, Nagaland, Odisha, Sikkim, Tamil Nadu, Uttarakhand, West Bengal [Bangladesh, Myanmar, Nepal].

Habit. Epiphytic herb.

Remarks. The photographs of *Luisia trichorrhiza* var. *flava* in the protologue show that the inflorescence is few-flowered, not 9- to 10-flowered as stated in the description. The only other difference with the typical variety is in the color, and we have no doubt that the variety is merely a color form.

Luisia volucris Lindl., Fol. Orchid. 4(*Luisia*): 1. 1853. TYPE: India, Khasi Hills, *T. Lobb s.n.* (holotype, K-LINDL; isotype, K).

Distribution. Assam, Meghalaya, Mizoram, Sikkim, West Bengal [Bangladesh].

Habit. Epiphytic herb.

Luisia zeylanica Lindl., Fol. Orchid. 4(*Luisia*): 3. 1853. TYPE: Sri Lanka, *Macrae 50* (holotype, K-LINDL).

Luisia truncata Blatt. & McCann, J. Bombay Nat. Hist. Soc. 35(3): 491. 1932. SYNTYPES: India, North Kanara, Yellapur, May–June 1911, *T. R. Bell s.n.*; Western Ghats, Castle Rock, *T. R. Bell s.n.*
Luisia indica Khuraijam & R. K. Roy, Biodivers. J. 6(3): 699. 2015. TYPE: India, Bihar, West Champaran, Valmiki Tiger Reserve, 13 Mar. 2015, *J. S. Khuraijam 101206* (holotype, LWG).

Distribution. Andaman & Nicobar Islands, Andhra Pradesh, Arunachal Pradesh, Assam, Bihar, Goa, Himachal Pradesh, Jharkhand, Karnataka, Kerala, Madhya Pradesh, Maharashtra, Meghalaya, Nagaland, Odisha, Sikkim, Tamil Nadu, Uttar Pradesh, Uttarakhand, West Bengal [Bhutan, China, Myanmar, Nepal, Sri Lanka].

Habit. Epiphytic herb.

Remarks. This species has been reported from India erroneously as *Luisia teretifolia* Gaudich. and *L. tristis* (G. Forst.) Hook. f. Kumar et al. (2018) synonymized *L. indica.*

Luisia zollingeri Rchb. f., Ann. Bot. Syst. (Walpers) 6: 622. 1863. SYNTYPES: India, Ganges Delta, *Roxburgh s.n.*; Sylhet, *Wallich s.n.*; Khasi Hills, *T. Lobb s.n.*; Indonesia, Borneo and Java, Mt. Salak, Mar. 1848, *Zollinger 1265* (syntypes, W).

Distribution. Andaman & Nicobar Islands [also in Indochina, Indonesia, Thailand].

Habit. Epiphytic herb.

———

Macropodanthus L. O. Williams, Bot. Mus. Leafl. 6(4): 103. 1938.

TYPE: *Macropodanthus philippinensis* L. O. Williams.

Macropodanthus alatus (Holttum) Seidenf. & Garay, Opera Bot. 95: 261, fig. 166, pl. 29-C. 1988. *Sarcochilus alatus* Holttum, Gard. Bull. Singapore 14: 5. 1953. *Pteroceras alatum* (Holttum) Holttum, Kew Bull. 14: 269. 1960. TYPE: Malaya: Fraser's Hill, *Holttum S. F. N. 39467* (holotype, SING).

Distribution. Andaman & Nicobar Islands [also in Malaysia].

Habit. Epiphytic herb.

Macropodanthus berkeleyi (Rchb. f.) Seidenf. & Garay, Opera Bot. 95: 261, fig. 167. 1988. *Thrixspermum berkeleyi* Rchb. f., Gard. Chron., n.s., 17: 557. 1882. *Sarcochilus berkeleyi* (Rchb. f.) Hook. f., Fl. Brit. India [J. D. Hooker] 6(17): 37. 1890. *Pteroceras berkeleyi* (Rchb. f.) Holttum, Kew Bull. 14: 269. 1960. TYPE: India, Nicobar Islands, *E. G. Berkeley 8* (holotype, W).

Distribution. Andaman & Nicobar Islands. Endemic.

Habit. Epiphytic herb.

———

Malaxis Sol. ex Sw., Prodr. [O. P. Swartz] 8, 119. 1788.

TYPE: *Malaxis spicata* Sw.

Malaxis cylindrostachya (Lindl.) Kuntze, Revis. Gen. Pl. 2: 673. 1891. *Dienia cylindrostachya* Lindl.,

Gen. Sp. Orchid. Pl.: 22. 1830. *Microstylis cylindrostachya* (Lindl.) Rchb. f., Ann. Bot. Syst. (Walpers) 6: 207. 1861. TYPE: Nepal, Sheopore [Shivapuri], July 1821, *Wallich 1934* (lectotype, K-LINDL, designated by Margońska, 2012; isotype, G, K-WALL).

Distribution. Himachal Pradesh, Sikkim, Uttarakhand, West Bengal [Bhutan, China, Myanmar, Nepal, Pakistan].

Habit. Terrestrial herb.

Malaxis muscifera (Lindl.) Kuntze, Revis. Gen. Pl. 2: 673. 1891. *Dienia muscifera* Lindl., Gen. Sp. Orchid. Pl.: 23. 1830. *Microstylis muscifera* (Lindl.) Ridl., J. Linn. Soc., Bot. 24: 333. 1888. TYPE: Nepal, Bagmati Zone, Kathmandu District, Gosainthan [Shishapangma], 27 Oct. 1821, *Wallich 1935* (lectotype, K-LINDL, designated by Margońska, 2012; isotype, BM, BR, E, G, K-WALL, L).

Distribution. Arunachal Pradesh, Himachal Pradesh, Sikkim, Uttar Pradesh, Uttarakhand, West Bengal [Bhutan, China, Myanmar, Nepal, Pakistan].

Habit. Terrestrial herb.

———

Micropera Lindl., Edwards's Bot. Reg. 18: t. 1522. 1832.

TYPE: *Micropera pallida* (Roxb.) Lindl. (*Aerides pallida* Roxb.).

Micropera mannii (Hook. f.) Tang & F. T. Wang, Acta Phytotax. Sin. 1(1): 94. 1951. *Sarcochilus mannii* Hook. f., Fl. Brit. India [J. D. Hooker] 6(17): 36. 1890. *Thrixspermum mannii* (Hook. f.) Kuntze, Revis. Gen. Pl. 2: 682. 1891. *Camarotis mannii* (Hook. f.) King & Pantl., Ann. Roy. Bot. Gard. (Calcutta) 8: 239. 1898. TYPE: India, Khasi Hills, alt. 2000–3000 ft, June 1878, *Mann 11/80* (holotype, K).

Distribution. Arunachal Pradesh, Assam, Meghalaya, Nagaland, Sikkim, West Bengal [Bhutan].

Habit. Epiphytic herb.

Micropera obtusa (Lindl.) Tang & F. T. Wang, Acta Phytotax. Sin. 1(1): 94. 1951. *Camarotis obtusa* Lindl., Edwards's Bot. Reg. 30(Misc.): 73. 1844. *Sarcochilus obtusus* (Lindl.) Benth. ex Hook. f., Fl. Brit. India [J. D. Hooker] 6(17): 36. 1890, non (Lindl.) Rchb. f., 1863. TYPE: India, icon. *cult. Loddiges (leg. Bateman) s.n.* (holotype, K-LINDL).

Distribution. Arunachal Pradesh, Assam, ?Meghalaya, Nagaland, Sikkim, Tripura, West Bengal [Bangladesh, Bhutan, Myanmar, Nepal].

Habit. Epiphytic herb.

Micropera pallida (Roxb.) Lindl., Edwards's Bot. Reg. 18: t. 1522. 1832. *Aerides pallida* Roxb., Fl. Ind. ed. 1832, 3: 475. 1832 (as *"pallidum"*). *Camarotis pallida* (Roxb.) Lindl., J. Proc. Linn. Soc., Bot. 3: 37. 1859 [1858]. *Sarcochilus roxburghii* Hook. f., Fl. Brit. India [J. D. Hooker] 6(17): 36. 1890. SYNTYPES: India, West Bengal, eastern parts, and Bangladesh, Chittagong, collector not named, icon. *Aerides pallidum* [sic], *Roxburgh 2349* [?origin] (syntypes, CAL, K).

Distribution. Assam, Meghalaya, Nagaland, Odisha, Tripura, West Bengal [Bangladesh, Myanmar].

Habit. Epiphytic herb.

Micropera rostrata (Roxb.) N. P. Balakr., J. Bombay Nat. Hist. Soc. 67: 66. 1970 (as *"rostratum"*). *Aerides rostrata* Roxb., Fl. Ind. ed. 1832, 3: 474. 1832 (as *"rostratum"*). *Camarotis rostrata* (Roxb.) Rchb. f., Ann. Bot. Syst. (Walpers) 6: 881. 1864. TYPE: Bangladesh or India, Sylhet, icon. *Aerides rostratum* [sic], *Roxburgh 2348* (syntypes, CAL, K).

Camarotis purpurea Lindl., Gen. Sp. Orchid. Pl.: 219. 1833. *Sarcochilus purpureus* (Lindl.) Benth. ex Hook. f., Fl. Brit. India [J. D. Hooker] 6(17): 36. 1890. *Micropera purpurea* (Lindl.) Pradhan, Indian Orchids: Guide Identif. & Cult. 2: 619. 1979. TYPE: Bangladesh or India, Sylhet, Purarooak, Mar. 1828, *Wallich 7329 (leg. W. Gomez 185)* (holotype, K-LINDL; isotype, K-WALL).

Distribution. Arunachal Pradesh, Assam, Manipur, Meghalaya, Mizoram, Nagaland [Bangladesh, ?Myanmar].

Habit. Epiphytic herb.

———

Mycaranthes Blume, Bijdr. Fl. Ned. Ind.: 352. 1825.

TYPE: *Mycaranthes lobata* Blume.

Mycaranthes floribunda (D. Don) S. C. Chen & J. J. Wood, Fl. China 25: 348. 2009. *Dendrobium floribundum* D. Don, Prodr. Fl. Nepal.: 34. 1825. *Callista floribunda* (D. Don) Kuntze, Revis. Gen. Pl. 2: 654. 1891. TYPE: Nepal, *Wallich s.n.* (holotype, BM).

Eria paniculata Lindl., Pl. Asiat. Rar. (Wallich) 1: 32, t. 36. 1830. *Pinalia paniculata* (Lindl.) Kuntze, Revis. Gen. Pl. 2: 679. 1891. *Mycaranthes paniculata* (Lindl.)

Schuit., Y. P. Ng & H. A. Pedersen, Bot. J. Linn. Soc. 186: 196. 2018. SYNTYPES: Nepal, Noakote, *Wallich 1971* (not found); India, Pundua, Mar. 1824, *Wallich 1971 (leg. De Silva 1348)* (syntypes, BM, E, K, K-LINDL, K-WALL, W).

Distribution. Arunachal Pradesh, Assam, Manipur, Meghalaya, Mizoram, Nagaland, Sikkim, Tripura, West Bengal [Bangladesh, Bhutan, China, Myanmar, Nepal].

Habit. Epiphytic herb.

———

Myrmechis (Lindl.) Blume, Coll. Orchid. 76. 1858 [1859]. *Anoectochilus* sect. *Myrmechis* Lindl., Gen. Sp. Orchid. Pl.: 500. 1840.

TYPE: *Myrmechis gracilis* (Blume) Blume (*Anoectochilus gracilis* Blume).

Myrmechis bakhimensis D. Maity, N. Pradhan & Maiti, Acta Phytotax. Sin. 45(3): 321. 2007. TYPE: India, Sikkim, Bakhim to Dzongri, 10,800 ft, 25 July 1999, *D. Maity 21921* (holotype, CAL; isotype, BSHC, PE).

Distribution. Sikkim. Endemic.

Habit. Terrestrial herb.

Myrmechis pumila (Hook. f.) Tang & F. T. Wang, Acta Phytotax. Sin. 1(1): 69. 1951. *Odontochilus pumilus* Hook. f., Fl. Brit. India [J. D. Hooker] 6(17): 99. 1890. *Cystopus pumilus* (Hook. f.) Kuntze, Revis. Gen. Pl. 2: 658. 1891. *Zeuxine pumila* (Hook. f.) King & Pantl., Ann. Roy. Bot. Gard. (Calcutta) 8: 291, t. 389. 1898. *Anoectochilus pumilus* (Hook. f.) Seidenf. & Smitinand, Orchids Thailand (Prelim. List): 89. 1959. TYPE: India, Sikkim Himalaya, alt. 8000–10,000 ft, *J. D. Hooker 325A* (lectotype, K-LINDL, designated by Seidenfaden, 1978).

Cheirostylis franchetiana King & Pantl., J. Asiat. Soc. Bengal, Pt. 2, Nat. Hist. 64: 341. 1895 [1896]. *Zeuxine franchetiana* (King & Pantl.) King & Pantl., Ann. Roy. Bot. Gard. (Calcutta) 8: 292, t. 398. 1898. *Myrmechis franchetiana* (King & Pantl.) Schltr., Repert. Spec. Nov. Regni Veg. Beih. 4: 174. 1919. TYPE: India, Sikkim, above Sureil, alt. 6500 ft, Aug. [1894], *Pantling 338B* (holotype, CAL; isotype, BM, K, W).

Distribution. Arunachal Pradesh, Assam, Manipur, Nagaland, Sikkim, West Bengal [Bhutan, China, Myanmar, Nepal].

Habit. Terrestrial herb.

———

Neottia Guett., Hist. Acad. Roy. Sci. Mem. Math. Phys. (Paris, 4) 1750: 374. 1754, nom. cons.

TYPE: *Neottia nidus-avis* (L.) Rich. (*Ophrys nidus-avis* L.).

Neottia acuminata Schltr., Acta Horti Gothob. 1: 141. 1924. SYNTYPES: China, Sichuan, Dongrergo, below Huang-lung-ssü, ca. 12,300 ft, 23 July 1922, *Harry Smith 3859* (syntypes, GB, UPS); *3940* (syntype, UPS); Karlong, 10,500 ft, *Harry Smith 2997* (syntype, UPS).

Aphyllorchis parviflora King & Pantl., J. Asiat. Soc. Bengal, Pt. 2, Nat. Hist. 65: 128. 1896. *Neottia parviflora* (King & Pantl.) Schltr., Acta Horti Gothob. 1: 142. 1924, nom. illeg., non Sm., 1813. TYPE: India, Sikkim, Lachung Valley, alt. 10,000 ft, July 1895, *Pantling 383* (holotype, CAL; isotype, AMES, BM, E, G, K, W [most presumed isotypes do not agree in all collecting details with the protologue, differing in locality, elevation, and/or collecting month]).

Distribution. Arunachal Pradesh, Sikkim, Uttarakhand, West Bengal [Bhutan, China, Nepal].

Habit. Holomycotrophic terrestrial herb.

Neottia alternifolia (King & Pantl.) Szlach., Fragm. Florist. Geobot. Suppl. 3: 117. 1995. *Listera alternifolia* King & Pantl., J. Asiat. Soc. Bengal, Pt. 2, Nat. Hist. 65: 126. 1896. TYPE: India, Sikkim, Lachen Valley, alt. 10,000 ft, July 1895, *Pantling 390* (holotype, CAL; isotype, AMES, E, K, W).

Distribution. Arunachal Pradesh, Sikkim [China].

Habit. Terrestrial herb.

Neottia brevicaulis (King & Pantl.) Szlach., Fragm. Florist. Geobot. Suppl. 3: 117. 1995. *Listera brevicaulis* King & Pantl., J. Asiat. Soc. Bengal, Pt. 2, Nat. Hist. 65: 126. 1896. TYPE: India, Sikkim, Lachen Valley, alt. 9000 ft, July 1895, *Pantling 392* (holotype, CAL; isotype, CAL, BM, K, L, W).

Distribution. Sikkim [Bhutan, China].

Habit. Terrestrial herb.

Neottia chandrae Raskoti, J. J. Wood & Ale, Nordic J. Bot. 30: 187. 2012. TYPE: Nepal, Rasuwa District, Lauribinayak, Lantang National Park, 3800 m, 20 July 2008, *Raskoti 280* (holotype, KATH; isotype, TUCH).

Distribution. Uttarakhand [Nepal].

Habit. Terrestrial herb.

Neottia confusa Bhaumik, Edinburgh J. Bot. 69: 381. 2012. TYPE: India, Arunachal Pradesh, Upper

Siang District, Sitoma camp-Ruitala camp, 3200–3800 m, 22 July 2010, *Bhaumik 36891* (holotype, CAL; isotype, ARUN).

Distribution. Arunachal Pradesh. Endemic.

Habit. Terrestrial herb.

Neottia dentata (King & Pantl.) Szlach., Fragm. Florist. Geobot. Suppl. 3: 117. 1995. *Listera dentata* King & Pantl., Ann. Roy. Bot. Gard. (Calcutta) 8: 257, t. 342. 1898. TYPE: India, Sikkim, Jongri, alt. 13,000 ft, 8 July 1896, *Pantling 452* (holotype, CAL; isotype, BM, K, W).

Distribution. Sikkim [Myanmar].

Habit. Terrestrial herb.

Neottia dihangensis Bhaumik, Edinburgh J. Bot. 69: 379. 2012. TYPE: India, Arunachal Pradesh, Upper Siang District, Sitoma camp-Ruitala camp, 3600 m, 22 July 2010, *Bhaumik 36874* (holotype, CAL; isotype, ARUN).

Distribution. Arunachal Pradesh. Endemic.

Habit. Terrestrial herb.

Neottia divaricata (Panigrahi & P. Taylor) Szlach., Fragm. Florist. Geobot. Suppl. 3: 117. 1995. *Listera divaricata* Panigrahi & P. Taylor, Kew Bull. 30(3): 559. 1975. TYPE: India, Arunachal Pradesh, Lohit Frontier District, Delei Valley, 28°15′N, 96°35′E, 9300–10,500 ft, 23 Aug. 1928, *F. Kingdon Ward 8572* (holotype, K).

Distribution. Arunachal Pradesh [China].

Habit. Terrestrial herb.

Neottia inayatii (Duthie) Schltr., Bot. Jahrb. Syst. 45: 387. 1911 (as "*inayati*"). *Listera inayatii* Duthie, J. Asiat. Soc. Bengal, Pt. 2, Nat. Hist. 71: 41. 1902 (as "*inayati*"). TYPE: Pakistan, Bhurju, Kangan Valley, Hazara, July 1897, *Inayat Khan 22596* (syntypes, CAL, K).

Listera kashmiriana Duthie, Ann. Roy. Bot. Gard. (Calcutta) 9: 153, t. 118. 1906. *Neottia kashmiriana* (Duthie) Schltr., Bot. Jahrb. Syst. 45: 387. 1911. TYPE: India, Kashmir, Liddar Valley, betw. 8000–9000 ft, *Inayat Khan 25372* (?holotype, K).

Distribution. Himachal Pradesh, Jammu & Kashmir [Pakistan].

Habit. Holomycotrophic terrestrial herb.

Neottia karoana Szlach., Fragm. Florist. Geobot. Suppl. 3: 117. 1995. *Listera micrantha* Lindl., J.

Proc. Linn. Soc., Bot. 1: 176. 1857. *Diphryllum micranthum* (Lindl.) Kuntze, Revis. Gen. Pl. 2: 659. 1891. TYPE: India, Sikkim, alt. 10,000 ft, *J. D. Hooker 353* (holotype, K-LINDL).

Distribution. Sikkim [Bhutan, China, Myanmar].

Habit. Terrestrial herb.

Neottia listeroides Lindl., Ill. Bot. Himal. Mts. [Royle] 1: 368. 1839. *Nidus listeroides* (Lindl.) Kuntze, Revis. Gen. Pl. 2: 674. 1891 (as "*listerodes*"). TYPE: India, Mussoorie, *Royle s.n.* (holotype, K-LINDL).

Neottia lindleyana Decne. in Jacquem., Voy. Inde 4: 163, t. 163. 1844. *Listera lindleyana* (Decne.) King & Pantl., Ann. Roy. Bot. Gard. (Calcutta) 8: 258, t. 343. 1898. TYPE: India, near Semlah, alt. 7000 ft, *Jacquemont s.n.*

Distribution. Arunachal Pradesh, Himachal Pradesh, Jammu & Kashmir, Nagaland, Sikkim, Uttarakhand [Bhutan, China, Nepal, Pakistan].

Habit. Holomycotrophic terrestrial herb.

Neottia longicaulis (King & Pantl.) Szlach., Fragm. Florist. Geobot. Suppl. 3: 117. 1995. *Listera longicaulis* King & Pantl., J. Asiat. Soc. Bengal, Pt. 2, Nat. Hist. 65: 126. 1896. TYPE: India, Sikkim, Lachen Valley, alt. 7000 ft, July, *Pantling 391* (holotype, CAL; isotype, BM, ?K [July 1897], L, W).

Distribution. Sikkim, Uttarakhand [Bhutan, China, Myanmar, Nepal].

Habit. Terrestrial herb.

Neottia mackinnonii Deva & H. B. Naithani, Orch. Fl. N. W. Himal. 75. 1986. TYPE: India, Uttarakhand, East of Tehri, Tehri Garhwal, Sep. 1901, *Mackinnon 2542a* (holotype, DD).

Distribution. Uttarakhand. Endemic.

Habit. Holomycotrophic terrestrial herb.

Neottia microglottis (Duthie) Schltr., Bot. Jahrb. Syst. 45: 387. 1911. *Listera microglottis* Duthie, J. Asiat. Soc. Bengal, Pt. 2, Nat. Hist. 71: 42. 1902. *Archineottia microglottis* (Duthie) S. C. Chen, Acta Phytotax. Sin. 17(2): 14. 1979. *Holopogon microglottis* (Duthie) S. C. Chen, Acta Phytotax. Sin. 35(2): 179. 1997. TYPE: India, Western Himalaya, Tehri-Garhwal, alt. 5000–6000 ft, Sep. 1901, *P. W. Mackinnon s.n.* (holotype, CAL).

Distribution. ?Himachal Pradesh, Uttarakhand. Endemic.

Habit. Holomycotrophic terrestrial herb.

Neottia mucronata (Panigrahi & J. J. Wood) Szlach., Fragm. Florist. Geobot. Suppl. 3: 118. 1995. *Listera mucronata* Panigrahi & J. J. Wood, Kew Bull. 29(4): 731. 1974 [1975]. TYPE: Nepal: Gandaki Zone, on the ridge north of Gandrung, 7000 ft, 30 May 1971, *Barclay & Synge 2410* (holotype, K).

Distribution. Meghalaya, Uttarakhand [China, Nepal].

Habit. Terrestrial herb.

Remarks. This is considered a synonym of *Neottia pinetorum* (Lindl.) Szlach. by Singh et al. (2019).

Neottia nandadeviensis (Hajra) Szlach., Fragm. Florist. Geobot. Supp. 3: 118. 1995. *Listera nandadeviensis* Hajra, Bull. Bot. Surv. India 25(1–4): 181, fig. 1. 1983 [1985]. TYPE: India, Uttar Pradesh, Chamoli District, Nandadevi National Park, Himtoli, *P. K. Hajra 73202A* (holotype, CAL).

Distribution. Uttarakhand. Endemic.

Habit. Terrestrial herb.

Neottia ovata (L.) Bluff & Fingerh., Comp. Fl. German., ed. 2, 2: 435. 1838. *Ophrys ovata* L., Sp. Pl.: 946. 1753. *Epipactis ovata* (L.) Crantz, Stirp. Austr. Fasc., ed. 2, 2: 473. 1769. *Listera ovata* (L.) R. Br., Hort. Kew., ed. 2 [W. T. Aiton] 5: 201. 1813. TYPE: Germany, near Tübingen, icon. *L. Fuchs, Hist. Stirp.: 566, 1542* (lectotype, designated by Baumann et al., 1989).

Distribution. Jammu & Kashmir [Pakistan].

Habit. Terrestrial herb.

Neottia pantlingii (W. W. Sm.) Tang & F. T. Wang, Acta Phytotax. Sin. 1: 66. 1951. *Aphyllorchis pantlingii* W. W. Sm., Rec. Bot. Surv. India 4: 243. 1911. *Archineottia pantlingii* (W. W. Sm.) S. C. Chen, Acta Phytotax. Sin. 17(2): 14. 1979. *Holopogon pantlingii* (W. W. Sm.) S. C. Chen, Acta Phytotax. Sin. 35(2): 179. 1997. SYNTYPES: India, Sikkim, Zemu Valley, *Smith & Cave 1020* (syntype, K); Lachen, 8500 ft, 27 July 1909, *Smith & Cave 2657* (syntype, CAL, drawing K).

Distribution. Sikkim [Bhutan].

Habit. Holomycotrophic terrestrial herb.

Neottia pinetorum (Lindl.) Szlach., Fragm. Florist. Geobot. Suppl. 3: 118. 1995. *Listera pinetorum* Lindl., J. Proc. Linn. Soc., Bot. 1: 175. 1857. *Diphryllum pinetorum* (Lindl.) Kuntze, Revis. Gen. Pl.

2: 659. 1891. TYPE: India, Sikkim, near Lachen, alt. 10,000–11,000 ft, *J. D. Hooker 355* (holotype, K-LINDL; isotype, CAL).

Distribution.   Sikkim, Uttarakhand [Bhutan, China, Myanmar, Nepal].

Habit.   Terrestrial herb.

Neottia tenuis (Lindl.) Szlach., Fragm. Florist. Geobot. Suppl. 3: 119. 1995. *Listera tenuis* Lindl., J. Proc. Linn. Soc., Bot. 1: 176. 1857. TYPE: India, Sikkim, alt. 11,000–12,000 ft, *J. D. Hooker 354* (holotype, K-LINDL).

Distribution.   Himachal Pradesh, Sikkim, Uttarakhand [China, Nepal].

Habit.   Terrestrial herb.

———

Nephelaphyllum Blume, Bijdr. Fl. Ned. Ind. 8: 372. 1825.

TYPE: *Nephelaphyllum pulchrum* Blume.

Nephelaphyllum cordifolium (Lindl.) Blume, Fl. Javae, n.s., 1: 145. 1858. *Cytheris cordifolia* Lindl., Gen. Sp. Orchid. Pl.: 129. 1831. *Tainia cordifolia* (Lindl.) Gagnep., Bull. Mus. Natl. Hist. Nat., Ser. 2. 4: 706. 1932, nom. illeg., non Hook. f., 1889. TYPE: India, Sylhet Mountains, Aug. 1822, *Wallich 3750 (leg. De Silva 1135)* (holotype, K-LINDL; isotype, K-WALL).

Distribution.   Assam, Meghalaya, Nagaland [Bangladesh, Myanmar].

Habit.   Terrestrial herb.

Nephelaphyllum nudum Hook. f., Fl. Brit. India [J. D. Hooker] 6(17): 192. 1890. TYPE: India, Sikkim Himalaya, *King s.n.* (holotype, CAL).

Distribution.   Sikkim. Endemic.

Habit.   Terrestrial herb.

Remarks.   This species is apparently only known from the type.

Nephelaphyllum pulchrum Blume, Bijdr. Fl. Ned. Ind.: 373. 1825 var. **pulchrum.** TYPE: Indonesia, Java, Mt. Salak, *?Blume s.n.* (not found).

Distribution.   Andaman & Nicobar Islands [Myanmar].

Habit.   Terrestrial herb.

Nephelaphyllum pulchrum Blume var. **sikkimensis** Hook. f., Fl. Brit. India [J. D. Hooker] 5(16): 819. 1890. *Nephelaphyllum sikkimensis* (Hook. f.) Karthik. in S. Karthikeyan et al., Fl. Ind. ser. 4, 1 (Monocotyledon): 154. 1989. TYPE: ?Bhutan, *Griffith s.n. (Kew Distr. 5370)* (holotype, K).

Distribution.   Arunachal Pradesh, Assam, Manipur, Sikkim, West Bengal [Bhutan].

Habit.   Terrestrial herb.

———

Nervilia Comm. ex Gaudich., Voy. Uranie, Bot. 421, t. 35. 1826 [1829], nom. cons.

TYPE: *Nervilia aragoana* Gaudich. (= *Nervilia concolor* (Blume) Schltr.) (typ. cons.).

Nervilia concolor (Blume) Schltr., Bot. Jahrb. Syst. 45: 404. 1911. *Cordyla concolor* Blume, Bijdr. Fl. Ned. Ind.: 416. 1825. *Roptrostemon concolor* (Blume) Lindl., Gen. Sp. Orchid. Pl.: 453. 1840. *Pogonia concolor* (Blume) Blume, Mus. Bot. 1: 32. 1849. TYPE: Indonesia, Java, Mt. Salak, *Blume s.n.* (?holotype, L).

Nervilia aragoana Gaudich., Voy. Uranie, Bot. 422, t. 35. 1826 [1829]. *Pogonia flabelliformis* Lindl., Gen. Sp. Orchid. Pl.: 415. 1840, nom. superfl. TYPE: Mariana Islands (Gaum), *Gaudichaud s.n.* (holotype, P).
Epipactis carinata Roxb., Fl. Ind. ed. 1832, 3: 454. 1832. *Nervilia carinata* (Roxb.) Schltr., Bot. Jahrb. Syst. 45: 404. 1911. *Pogonia carinata* (Roxb.) Lindl., Gen. Sp. Orchid. Pl.: 414. 1840. TYPE: India, Bengal, cult. Calcutta s.n., icon. *Epipactis carinata*, *Roxburgh 2092* (syntypes, CAL, K).
Pogonia scottii Rchb. f., Flora 55: 276. 1872. *Nervilia scottii* (Rchb. f.) Schltr., Bot. Jahrb. Syst. 45: 404. 1911. TYPE: India, West Bengal, Darjeeling Himalaya, *Scott s.n.* (holotype, W).

Distribution.   Andaman & Nicobar Islands, Andhra Pradesh, Arunachal Pradesh, Assam, Bihar, Gujarat, Goa, Himachal Pradesh, Jharkhand, Karnataka, Kerala, Madhya Pradesh, Maharashtra, Manipur, Meghalaya, Nagaland, Odisha, Rajasthan, Sikkim, Tamil Nadu, Uttar Pradesh, Uttarakhand, West Bengal [Bangladesh, Bhutan, China, Myanmar, Nepal, Pakistan].

Habit.   Terrestrial herb.

Nervilia falcata (King & Pantl.) Schltr., Bot. Jahrb. Syst. 45: 402. 1911. *Pogonia falcata* King & Pantl., J. Asiat. Soc. Bengal, Pt. 2, Nat. Hist. 65: 129. 1896. TYPE: India, West Bengal, Duar, E of Jaldacca River, Apr., *Pantling 439* (holotype, CAL).

Distribution. Assam, Jharkhand, Manipur, Odisha, Uttar Pradesh, Uttarakhand, West Bengal [Bangladesh, Bhutan].

Habit. Terrestrial herb.

Nervilia gammieana (Hook. f.) Pfitzer, Nat. Pflanzenfam. [Engler & Prantl] 2(6): 56. 1888. *Pogonia gammieana* Hook. f., Bot. Mag. 109: t. 6671. 1883. SYNTYPES: India, May & July 1881, *cult. Kew (leg. Gammie) s.n.* (syntype, K); India, hot valleys below Darjeeling, *J. D. Hooker s.n.* (syntype, K); India, Kumaon, *R. Strachey & J. E. Winterbottom 19* (syntype, K).

Distribution. Arunachal Pradesh, Himachal Pradesh, Jammu & Kashmir, Odisha, Sikkim, Uttarakhand, West Bengal [Nepal, Pakistan].

Habit. Terrestrial herb.

Nervilia gleadowii A. N. Rao, Indian Forester 118 (11): 846. 1992. TYPE: India, Uttarnchal, Tehri Garhwal, Bamsu Valley below Lusitach, alt. 7000 ft, 9 May 1900, *F. Gleadow 23940* (holotype, DD).

Distribution. Uttarakhand. Endemic.

Habit. Terrestrial herb.

Nervilia hispida Blatt. & McCann, J. Bombay Nat. Hist. Soc. 35(4): 728. 1932. TYPE: India, North Kanara, Yellapur, *T. R. Bell s.n.*

Distribution. Karnataka. Endemic.

Habit. Terrestrial herb.

Nervilia hookeriana (King & Pantl.) Schltr., Bot. Jahrb. Syst. 45: 405. 1911. *Pogonia hookeriana* King & Pantl., J. Asiat. Soc. Bengal, Pt. 2, Nat. Hist. 65: 129. 1896. TYPE: India, Sikkim, alt. 3000 ft, Aug., *G. King 2153* (holotype, CAL).

Distribution. Arunachal Pradesh, Meghalaya, Nagaland, Sikkim, West Bengal. Endemic.

Habit. Terrestrial herb.

Nervilia infundibulifolia Blatt. & McCann, J. Bombay Nat. Hist. Soc. 35(4): 725. 1932. TYPE: India, North Kanara, Yellapur, June 1911, icon. *T. R. Bell s.n.* (holotype, BLAT).

Nervilia hallbergii Blatt. & McCann, J. Bombay Nat. Hist. Soc. 35(4): 726. 1932. TYPE: India, Western Ghat, Kuna near Khandala, June 1917, *Hallberg s.n.* (holotype, BLAT).

Distribution. Andhra Pradesh, Arunachal Pradesh, Jharkhand, Karnataka, Kerala, Maharashtra, Odisha,

Sikkim, Tamil Nadu, Uttar Pradesh, Uttarakhand [Bhutan, China].

Habit. Terrestrial herb.

Nervilia juliana (Roxb.) Schltr., Bot. Jahrb. Syst. 45: 402. 1911. *Epipactis juliana* Roxb., Fl. Ind. ed. 1832, 3: 453. 1832. *Pogonia juliana* (Roxb.) Wall ex Lindl., Gen. Sp. Orchid. Pl.: 414. 1840. TYPE: India, West Bengal, vicinity of Calcutta, icon. *Epipactis juliana, Roxburgh 2091* (syntypes, CAL, K).

Distribution. Arunachal Pradesh, Assam, Karnataka, Mizoram, West Bengal [Bangladesh, ?Myanmar, ?Sri Lanka].

Habit. Terrestrial herb.

Nervilia khasiana (King & Pantl.) Schltr., Bot. Jahrb. Syst. 45: 402. 1911. *Pogonia khasiana* King & Pantl., J. Asiat. Soc. Bengal, Pt. 2, Nat. Hist. 66: 597. 1898 [1897]. TYPE: India, Jaintia Hills, Jharain, South Jawai, alt. 3000 ft, June, *Pantling 626* (not found).

Distribution. Assam, Meghalaya. Endemic.

Habit. Terrestrial herb.

Nervilia mackinnonii (Duthie) Schltr., Bot. Jahrb. Syst. 45: 402. 1911. *Pogonia mackinnonii* Duthie, J. Asiat. Soc. Bengal, Pt. 2, Nat. Hist. 71: 43. 1902. *Nervilia macroglossa* (Hook. f.) Schltr. var. *mackinnonii* (Duthie) Pradhan, Indian Orchids: Guide Identif. & Cult. 1: 148. 1976. SYNTYPES: Western Himalaya: Musssoorie, alt. 4500–6000 ft, June (flowers), July & Sep. (leaves) 1899, *P. W. Mackinnon s.n.* [*22705*] (syntypes, CAL, K).

Distribution. Kerala, Tamil Nadu, Uttar Pradesh, Uttarakhand [China, Myanmar, Nepal].

Habit. Terrestrial herb.

Nervilia macroglossa (Hook. f.) Schltr., Bot. Jahrb. Syst. 45: 402. 1911. *Pogonia macroglossa* Hook. f., Fl. Brit. India [J. D. Hooker] 6(17): 120. 1890. SYNTYPES: India, Sikkim Himalaya, hot valleys, *J. D. Hooker s.n.* (syntype, K); Tomlong, alt. 6500 ft, *C. B. Clarke 27725* (syntype, K).

Distribution. Arunachal Pradesh, Assam, Jharkhand, Meghalaya, Sikkim, West Bengal [Bhutan, Myanmar, Nepal].

Habit. Terrestrial herb.

Nervilia pangteyana Jalal, Kumar & G. S. Rawat, Nordic J. Bot. 30: 407. 2012. TYPE: India, Uttarakhand, Nainital, Sitabani, 900 m.a.s.l., 16 Aug. 2009, *Jalal 15051* (holotype, WII).

Distribution. Uttarakhand. Endemic.

Habit. Terrestrial herb.

Nervilia plicata (Andrews) Schltr., Bot. Jahrb. Syst. 45: 403. 1911. *Arethusa plicata* Andrews, Bot. Repos. 5: t. 321. 1803. *Pogonia plicata* (Andrews) Lindl., Gen. Sp. Orchid. Pl.: 415. 1840. TYPE: East Indies, 1803, *Aylmer Bourke Lambert s.n.* ("figure is made from the flower in the hot-house of J. Vere, Esquire, Kensington Gore").

Epipactis plicata Roxb., Fl. Ind. ed. 1832, 3: 454. 1832. TYPE: India, West Bengal, in the vicinity of Calcutta, icon. *Epipactis plicata*, *Roxburgh 1647* (syntypes, CAL, K).
Pogonia biflora Wight, Icon. Pl. Ind. Orient. [Wight] 5(1): 22, t. 1758-2. 1851. *Nervilia biflora* (Wight) Schltr., Bot. Jahrb. Syst. 45: 403. 1911. TYPE: India, Wynad, *Jerdon s.n.* (not found).
Cordyla discolor Blume, Bijdr. Fl. Ned. Ind. 8: 416. 1825. *Roptrostemon discolor* (Blume) Lindl., Gen. Sp. Orchid. Pl.: 453. 1840. *Nervilia discolor* (Blume) Schltr., Bot. Jahrb. Syst. 45: 403. 1911. TYPE: Indonesia, Java, Mt. Salak, *Blume s.n.* (holotype, ?L [*Blume 652*]).

Distribution. Andaman & Nicobar Islands, Andhra Pradesh, Arunachal Pradesh, Assam, Bihar, Gujarat, Himachal Pradesh, Jammu & Kashmir, Jharkhand, Karnataka, Kerala, Maharashtra, Manipur, Meghalaya, Mizoram, Odisha, Sikkim, Tamil Nadu, Uttarakhand, West Bengal [Bangladesh, Bhutan, Myanmar, Nepal, Pakistan].

Habit. Terrestrial herb.

Nervilia punctata (Blume) Makino, Bot. Mag. (Tokyo) 16: 199. 1902. *Pogonia punctata* Blume, Mus. Bot. 1: 32. 1849. *Aplostellis punctata* (Blume) Ridl., Fl. Malay Penins. 4: 204. 1924. SYNTYPES: Indonesia, Sumatra, *Korthals 653* (syntype, L); West Java, *Kuhl & van Hasselt s.n.* (syntype, L).

Distribution. Andaman & Nicobar Islands, Mizoram, Sikkim, West Bengal [also in Thailand].

Habit. Terrestrial herb.

Remarks. It remains to be confirmed that the true *Nervilia punctata* occurs in continental Asia, where several closely related new species have been identified recently, with perhaps more to follow.

Nervilia simplex (Thouars) Schltr., Bot. Jahrb. Syst. 45: 401. 1911. *Arethusa simplex* Thouars, Hist. Orchid.: t. 24. 1822. *Pogonia simplex* (Thouars) Rchb. f., Xenia Orchid. 2: 92. 1865. TYPE: icon.

Thouars, Hist. Orchid.: t. 24. 1822 (lectotype, designated by Pettersson, 1990).

Bolborchis crociformis Zoll. & Moritzi, Syst. Verz. Java [Moritzi] 89. 1846. *Nervilia crociformis* (Zoll. & Moritzi) Seidenf., Dansk Bot. Ark. 32(2): 151, fig. 92. 1978. TYPE: Indonesia, Java, Tjikoya, 11 Oct. 1812, *Zollinger 762*.
Pogonia crispata Blume, Mus. Bot. 1: 32. 1849. *Nervilia crispata* (Blume) Schltr. ex K. Schum. & Lauterb., Fl. Schutzgeb. Südsee: 240. 1900. TYPE: Indonesia, Java, Bantam Prov., Mt. Batu-auwel, *van Hasselt s.n.* (holotype, ?L, not found, painting W).
Pogonia prainiana King & Pantl., J. Asiat. Soc. Bengal, Pt. 2, Nat. Hist. 65: 129. 1896. *Nervilia prainiana* (King & Pantl.) Seidenf. & Smitinand, Orchids Thailand 4, 2: 730. 1965. TYPE: India, Sikkim, Lachung Valley, alt. 6500 ft, June 1895, *Pantling 372* (holotype, CAL; isotype, BM, G, K).
Nervilia monantha Blatt. & McCann, J. Bombay Nat. Hist. Soc. 35(4): 724. 1932. TYPE: India, North Kanara, Yellapur, June 1911, *T. R. Bell 5428a* (holotype, BLAT).

Distribution. Andhra Pradesh, Arunachal Pradesh, Goa, Jharkhand, Karnataka, Kerala, Madhya Pradesh, Maharashtra, Manipur, Nagaland, Odisha, Sikkim, Tamil Nadu, Uttarakhand, West Bengal [Nepal].

Habit. Terrestrial herb.

Oberonia Lindl., Gen. Sp. Orchid. Pl.: 15. 1830, nom. cons.

TYPE: *Oberonia iridifolia* Lindl., nom. superfl. (*Malaxis ensiformis* Sm. = *Oberonia ensiformis* (Sm.) Lindl.) (typ. cons.).

Oberonia acaulis Griff., Not. Pl. Asiat. 3: 275. 1851. SYNTYPES: Bhutan, *Griffith in Bootan Herb. 1130* (not found), India, Khasi Hills, Churra, *Griffith s.n.* (syntype, K-LINDL).

Oberonia myriantha Lindl., Fol. Orchid. 8(*Oberonia*): 4. 1859. *Malaxis myriantha* (Lindl.) Rchb. f., Ann. Bot. Syst. (Walpers) 6: 213. 1861. *Iridorkis myriantha* (Lindl.) Kuntze, Revis. Gen. Pl. 2: 669. 1891 (as "*Iridorchis*"). TYPE: India, Khasi Hills, *Hooker & Thomson 113* (lectotype, K-LINDL; designated by Ansari & Balakrishnan, 1990; isotype, L, MH, ?P [s.n.]).
Oberonia sikkimensis Lindl., Fol. Orchid. 8(*Oberonia*): 4. 1859. *Malaxis sikkimensis* (Lindl.) Rchb. f., Ann. Bot. Syst. (Walpers) 6: 212. 1861. TYPE: India, Sikkim, low Valleys, *J. D. Hooker 114* (syntypes, AMES, K, K-LINDL, P).
Oberonia ritae King & Pantl., J. Asiat. Soc. Bengal, Pt. 2, Nat. Hist. 66: 579. 1898 [1897] (as "*ritaii*"). TYPE: India, Khasi Hills, Jowai, alt. 2000–3000 ft, Aug., *S. E. Rita s.n.* [*599*] (lectotype, CAL, designated by Ansari & Balakrishnan, 1990).
Oberonia acaulis Griff. var. *latipetala* Chowlu, Nanda & A. N. Rao, Bangladesh J. Pl. Taxon. 21(1): 93. 2014. TYPE: India, Manipur, Senapati District, Hengbung, 1298 m,

Chowlu 00368 (holotype, Herb. Centre Orchid Gene Conservation E. Himalayan Region).

Distribution. Arunachal Pradesh, Assam, Manipur, Meghalaya, Mizoram, Nagaland, Sikkim, Uttarakhand, West Bengal [Bhutan, China, Myanmar, Nepal].

Habit. Epiphytic herb.

Remarks. *Oberonia ritae* is considered a synonym by Geiger (pers. comm.).

Oberonia agastyamalayana C. S. Kumar in C. Sathish Kumar & K. S. Manilal, Cat. Indian Orchids 57. 1994. TYPE: India, Kerala, Trivandrum District, Agastyamala, alt. 3200 ft, *C. S. Kumar 1398* (TBGT).

Oberonia longifolia M. Kumar & Sequiera, J. Orchid Soc. India 12(1–2): 29, fig. 1. 1998, nom. illeg., non Ridl., 1908. TYPE: India, Kerala, Palghat District, Silent Valley National Park, Sispara, alt. 5700 ft, *Stephen 007856* (holotype, KFRI).

Distribution. Kerala. Endemic.

Habit. Epiphytic herb.

Oberonia anamalayana J. Joseph, J. Indian Bot. Soc. 42: 222. 1964. TYPE: India, Tamil Nadu, Coimbatore District, Anamalais, Waverly Estate Reserve Forests, alt. 1500 ft, 16 Jan. 1961, *J. Joseph 13537A* (holotype, BSI; isotype, CAL).

Distribution. Kerala, Tamil Nadu. Endemic.

Habit. Epiphytic herb.

Oberonia angustifolia Lindl., Fol. Orchid. 8(*Oberonia*): 5. 1859. *Malaxis angustifolia* (Lindl.) Rchb. f., Ann. Bot. Syst. (Walpers) 6: 213. 1861. *Oberonia iridifolia* Lindl. var. *angustifolia* (Lindl.) Hook. f., Fl. Brit. India [J. D. Hooker] 5(15): 676. 1888. *Iridorkis angustifolia* (Lindl.) Kuntze, Revis. Gen. Pl. 2: 669. 1891 (as "*Iridorchis*"). *Oberonia denticulata* Wight var. *angustifolia* (Lindl.) S. Misra, J. Orchid Soc. India 3: 70. 1989. *Oberonia smisrae* Panigrahi var. *angustifolia* (Lindl.) Panigrahi in S. K. Murti & G. Panigrahi, Fl. Bilaspur Distr. 2: 593. 1999. SYNTYPES: India, Sikkim, Darjeeling, *Griffith s.n.* (syntype, K-LINDL); Khasya, *Griffith 1223* (syntype, K-LINDL), alt. 4000–5000 ft, *Hooker & Thomson 111* (syntype, K-LINDL).

Distribution. Assam, Meghalaya, Sikkim, West Bengal. Endemic.

Habit. Epiphytic herb.

Oberonia anthropophora Lindl., Gen. Sp. Orchid. Pl.: 16. 1830. *Malaxis anthropophora* (Lindl.)

Rchb. f., Ann. Bot. Syst. (Walpers) 6: 215. 1861. *Iridorkis anthropophora* (Lindl.) Kuntze, Revis. Gen. Pl. 2: 669. 1891 (as "*Iridorchis*"). TYPE: Myanmar, Tavoy [Dawei], 25 Aug. 1827, *Wallich 1951* (lectotype, K-LINDL, designated by Seidenfaden, 1968; isotype, G, K-WALL).

Oberonia mannii Hook. f., Fl. Brit. India [J. D. Hooker] 6(17): 180. 1890. TYPE: India, Jyntea Hills, north of Sylhet, alt. 3000 ft, *Mann s.n.* (lectotype, K, designated by Seidenfaden, 1968; isotype, CAL).
Oberonia falcata King & Pantl., J. Asiat. Soc. Bengal, Pt. 2, Nat. Hist. 64: 329. 1895 [1896]. TYPE: India, Sikkim, ?Labha [= Chungthang], alt. ?6000 ft, *King & Pantling s.n.* [presumably one of *Pantling 218*, collected in 1892, 1893, and 1895] (lectotype, BM, designated by Seidenfaden, 1968; isotype, BM, CAL, G, K, L, P, W).

Distribution. Arunachal Pradesh, Assam, Meghalaya, Nagaland, Sikkim, West Bengal [Bangladesh, Bhutan, China, Myanmar, Nepal].

Habit. Epiphytic herb.

Remarks. *Oberonia falcata* and *O. mannii* are considered synonyms of *O. anthropophora* by Geiger (pers. comm.).

Oberonia balakrishnanii R. Ansari, Orchid Monogr. 4: 16. 1990. TYPE: India, Tamil Nadu, Church Cliff, 27 Oct. 1897, *Bourne 1837* (holotype, MH).

Distribution. Tamil Nadu. Endemic.

Habit. Epiphytic herb.

Remarks. Geiger (2019) has suggested that *Oberonia balakrishnanii* could be a synonym of *O. nayarii* R. Ansari & N. P. Balakr.

Oberonia bicornis Lindl., Gen. Sp. Orchid. Pl.: 16. 1830. *Malaxis bicornis* (Lindl.) Rchb. f., Ann. Bot. Syst. (Walpers) 6: 211. 1861. *Iridorkis bicornis* (Lindl.) Kuntze, Revis. Gen. Pl. 2: 669. 1891 (as "*Iridorchis*"). TYPE: Bangladesh or India, Sylhet, Nov. 1823, *Wallich 1949 (leg. De Silva 1293)* (holotype, K-WALL; isotype, E).

Oberonia tenuis Lindl., Fol. Orchid. 8(*Oberonia*): 3. 1859. TYPE: Sri Lanka, Hittawaka, *Thwaites [2654]* (holotype, K).
Oberonia umbonata Blatt. & McCann, J. Bombay Nat. Hist. Soc. 35(2): 259. 1931. TYPE: India, Maharashtra, North Kanara, Siddhapur, alt. 1400 ft, *Bell & Sedgwick 7270* (holotype, BLAT).

Distribution. Assam, Karnataka, Kerala, Maharashtra, Manipur, Meghalaya, Mizoram, Tamil Nadu [Bangladesh, Sri Lanka].

Habit. Epiphytic herb.

Oberonia bopannae Chowlu & Kumar, Phytotaxa
316(3): 287. 2017. TYPE: India, Arunachal Pradesh, Namsai District, Tengapani, 2 Oct. 2014,
K. Chowlu 40001 (holotype, CAL).

Distribution.   Arunachal Pradesh. Endemic.

Habit.   Epiphytic herb.

Oberonia brachyphylla Blatt. & McCann, J. Bombay Nat. Hist. Soc. 35(2): 257. 1931. TYPE:
India, North Kanara, *T. R. Bell s.n.* (not found).
NEOTYPE: *Kapadia 2855* (neotype, BLAT, designated by Santapau & Kapadia, 1960).

Oberonia arunachalensis A. N. Rao, Rheedea 7(1): 130. 1997.
TYPE: India, Arunachal Pradesh, Lohit District, Kamlang Reserve Forest on way to Chamba-glowlake area,
alt. 1600 ft, *A. N. Rao 28967* (holotype, Orchid Herbarium, Tipi).
Oberonia kamlangensis A. N. Rao, J. Econ. Taxon. Bot. 24(2):
267, t. 1. 2000. TYPE: India, Arunachal Pradesh, Lohit
District, Kamlang sanctuary, alt. 480 ft, *A. N. Rao
28283-A* (holotype, Orchid Herbarium, Tipi).

Distribution.   Arunachal Pradesh, Goa, Karnataka,
Kerala. Endemic.

Habit.   Epiphytic herb.

Remarks.   *Oberonia arunachalensis* and *O. kamlangensis* are considered synonyms by Geiger (pers.
comm.).

Oberonia brachystachys Lindl., Sert. Orchid. 2: t.
8B. 1838. *Iridorkis brachystachys* (Lindl.) Kuntze,
Revis. Gen. Pl. 2: 669. 1891 (as "*Iridorchis*").
SYNTYPES: "Burmese Empire," *Griffith 697* (not
found); *Griffith 778* (syntypes, K-LINDL, P); *Griffith s.n.* (possible syntype, K).

Oberonia demissa Lindl., Fol. Orchid. 8(*Oberonia*): 4. 1859.
TYPE: India, Sikkim, above Terai, *J. D. Hooker 121* (holotype, K).

Distribution.   Meghalaya, Sikkim, West Bengal
[Myanmar, Nepal].

Habit.   Epiphytic herb.

Oberonia brunoniana Wight, Icon. Pl. Ind. Orient.
[Wight] 5(1): 3, t. 1622. 1851. *Malaxis brunoni-
ana* (Wight) Rchb. f., Ann. Bot. Syst. (Walpers) 6:
209. 1861. *Iridorkis brunoniana* (Wight) Kuntze,
Revis. Gen. Pl. 2: 669. 1891 (as "*Iridorchis*").
TYPE: India, Tamil Nadu, Iyamally Hills near
Coimbatore, Mt. Agamullu, June–July, *Wight s.n.*
(lectotype, K, designated by Geiger, 2019).

Oberonia lindleyana Wight, Icon. Pl. Ind. Orient. [Wight]
5(1): 3, t. 1624. 1851, nom. illeg., non Brogn., 1834.

Malaxis lindleyana Rchb. f., Ann. Bot. Syst. (Walpers)
6: 210. 1861. *Iridorkis lindleyana* (Rchb. f.) Kuntze,
Revis. Gen. Pl. 2: 669. 1891 (as "*Iridorchis*"). *Oberonia
santapaui* Kapadia, J. Bombay Nat. Hist. Soc. 57: 265.
1960. TYPE: India, Tamil Nadu, Iyamally Hills, near
Coimbatore, Aug.–Sep., *Wight s.n.* (holotype, K).
Oberonia wallichii Hook. f., Fl. Brit. India [J. D. Hooker]
5(15): 681. 1888. *Iridorkis wallichii* (Hook. f.) Kuntze,
Revis. Gen. Pl. 2: 669. 1891 (as "*Iridorchis*"). TYPE:
Bangladesh or India, Sylhet, Dec. 1820, *Wallich 1948.2
(leg. De Silva s.n.)* (syntypes, K, K-WALL).
Oberonia swaminathanii Ratheesh, Manudev & Sujanapal,
Nordic J. Bot. 28: 713. 2010. TYPE: India, Kerala,
Wayanad District, Kurichiarmala, ca 1500 m, *M. K.
Ratheesh Narayanan MSSH 1693* (holotype, MH; iso-
type, CALI, KFRI).
Oberonia saintberchmansii Kad. V. George & J. Mathew, Spe-
cies (India) 20: 112. 2019. TYPE: India, Kerala, Idukki
District, way to Nedukandam, Cardomom Hills, 3rd
mile, 1310 m, Jan. 2014, *K. V. George & S. Antony 0126*
(holotype, RHK [SB College Herbarium, Changanassery,
Kerala]; isotype, MSSRF).

Distribution.   Andhra Pradesh, Assam, Dadra &
Nagar Haveli, Goa, Karnataka, Kerala, Maharashtra,
Meghalaya, Tamil Nadu. Endemic.

Habit.   Epiphytic herb.

Remarks.   Geiger (2019) has demonstrated that
Oberonia santapaui Kapadia is not significantly differ-
ent from *O. brunoniana*; he is also of the opinion that
O. wallichii, *O. saintberchmansii*, and probably *O. swa-
minathanii* are additional synonyms of *O. brunoniana*
(Geiger, 2020 and pers. comm.).

Oberonia caulescens Lindl., Gen. Sp. Orchid. Pl.:
15. 1830. *Malaxis caulescens* (Lindl.) Rchb. f.,
Ann. Bot. Syst. (Walpers) 6: 215. 1861. *Iridorkis
caulescens* (Lindl.) Kuntze, Revis. Gen. Pl. 2: 669.
1891 (as "*Iridorchis*"). TYPE: Nepal, July, *Wallich
1950* (lectotype, K-LINDL, designated by Ansari
& Balakrishnan, 1990; isotype, K-WALL).

Oberonia longilabris King & Pantl., J. Asiat. Soc. Bengal,
Pt. 2, Nat. Hist. 64: 330. 1895 [1896]. TYPE: India,
Sikkim, Songchongloo, alt. 6000 ft, July [?1896], *King
& Pantling s.n. [?227]* (lectotype, CAL, designated by
Ansari & Balakrishnan, 1990; isotype, K, P, W).
Oberonia auriculata King & Pantl., Ann. Roy. Bot. Gard. (Cal-
cutta) 8: 13, t. 16A. 1898. TYPE: India, Sikkim, Sureil
& Runglee, alt. 6000 ft, July 1897, *Pantling 166* (lecto-
type, CAL, designated by Ansari & Balakrishnan, 1990).
Oberonia katakiana A. N. Rao, J. Econ. Taxon. Bot. 20: 711.
1996. TYPE: India, Arunachal Pradesh, West Kameng
District, Tipi, alt. 640 ft, *A. N. Rao 26077-A* (holotype,
Orchid Herbarium, Tipi).

Distribution.   Arunachal Pradesh, Assam, Meghalaya, Nagaland, Sikkim, Uttar Pradesh, ?Uttarakhand,
West Bengal [Bhutan, China, Nepal].

Habit.   Epiphytic herb.

Remarks. Oberona katakiana was recently synonymized by Geiger (2019).

Oberonia cavaleriei Finet, Bull. Soc. Bot. France 55: 334. 1908. SYNTYPES: China, Guizhou and Kou-Tchéou, South Tin-fan, *Cavalerie 1904* (syntypes, E, P).

Distribution. Manipur, Meghalaya, Uttarakhand [China, Myanmar, Nepal].

Habit. Epiphytic herb.

Remarks. Indian records of *Oberonia myosurus* (G. Forst.) Lindl. (= *Phreatia matthewsii* Rchb. f.) belong here.

Oberonia chandrasekharanii V. J. Nair, V. S. Ramach. & R. Ansari, Blumea 28(2): 361, fig. 1. 1983. TYPE: India, Kerala State, Cannanore District, Chandanathode, 2500 ft, 15 Aug. 1980, *Ramachandran 66948* (holotype, CAL; isotype, K, MH).

Oberonia rangannaiana Kesh. Murthy, Yogan. & K. V. Nair, Curr. Sci. 56(12): 621. 1987. TYPE: India, Karnataka, Bhagamandala to Mercara, alt. 3900 ft, 8 Aug. 1983, *K. R. Keshava Murthy et al. 4233A* (holotype, RRCBI).

Distribution. Arunachal Pradesh, Karnataka, Kerala, Tamil Nadu. Endemic.

Habit. Epiphytic herb.

Oberonia clarkei Hook. f., Hooker's Icon. Pl. 18: t. 1779A. 1888. *Iridorkis clarkei* (Hook. f.) Kuntze, Revis. Gen. Pl. 2: 669. 1891 (as "*Iridorchis*"). TYPE: India, Khasi Hills, Shillong, alt. 5000 ft, *C. B. Clarke 5818* (holotype, K).

Distribution. Assam, Manipur, Meghalaya, Mizoram, Nagaland, Tamil Nadu. Endemic.

Habit. Epiphytic herb.

Remarks. This is considered a synonym of *Oberonia jenkinsiana* Griff. ex Lindl. by Singh et al. (2019).

Oberonia emarginata King & Pantl., Ann. Roy. Bot. Gard. (Calcutta) 8: 6, t. 2A. 1898. TYPE: India, Sikkim, Namgah, alt. 6000 ft, Sep. 1895, *Pantling 423* (lectotype, K, designated by Seidenfaden, 1968; isotype, AMES, BM, CAL, L, W).

Oberonia micrantha King & Pantl., Ann. Roy. Bot. Gard. (Calcutta) 8: 6, t. 5. 1898, nom. illeg., non A. Rich., 1833. TYPE: India, Sikkim, Tendong, 6000 ft, July, *Pantling 324* (lectotype, K, designated by Seidenfaden, 1968; isotype, CAL, P).

Distribution. Arunachal Pradesh, Assam, Meghalaya, Nagaland, Sikkim, West Bengal [Bhutan, Nepal].

Habit. Epiphytic herb.

Oberonia ensiformis (Sm.) Lindl., Fol. Orchid. 8(*Oberonia*): 4. 1859. *Malaxis ensiformis* Sm., Cycl. (Rees) 22: Malaxis no. 14. 1812. *Oberonia iridifolia* Lindl., Gen. Sp. Orchid. Pl.: 15. 1830, nom. superfl. *Iridorkis ensiformis* (Sm.) Kuntze, Revis. Gen. Pl. 2: 669. 1891 (as "*Iridorchis*"). *Iridorkis iridifolia* Kuntze, Revis. Gen. Pl. 2: 669. 1891 (as "*Iridorchis*"), nom. superfl. *Malaxis iridifolia* Rchb. f., Ann. Bot. Syst. (Walpers) 6: 208. 1861, nom. superfl. *Oberonia smisrae* Panigrahi in S. K. Murti & G. Panigrahi, Fl. Bilaspur Distr. 2: 593. 1999, nom. superfl. TYPE: Nepal, Bagmati Zone, Kathmandu, Narayanhetty, on trees, 12 Nov. 1802, *Buchanan-Hamilton s.n.* (lectotype, LINN sheet LINN-HS 1396.11.1–2, designated by Ansari & Balakrishnan, 1990 and Geiger, 2020; isolectotype, K).

Oberonia trilobata Griff., Not. Pl. Asiat. 3: 273. 1851. TYPE: India, Khasi Hills, Nera Nowgong, 18 Nov. 1835, *Assam Herb. 269* (not found).

Distribution. Andaman & Nicobar Islands, Andhra Pradesh, Arunachal Pradesh, Assam, Karnataka, Kerala, Maharashtra, Manipur, Meghalaya, Mizoram, Nagaland, Odisha, Sikkim, Tamil Nadu, Tripura, Uttarakhand, West Bengal [China, Myanmar, Nepal].

Habit. Epiphytic herb.

Oberonia falconeri Hook. f., Hooker's Icon. Pl. 18: t. 1780. 1888. SYNTYPES: Kumaon, Dehradun, Apr. 1825, *Wallich s.n.* [*1948.3*] (syntypes, K, K-WALL); *Falconer 1008* (syntypes, AMES, CAL, K, W); Bihar, *J. D. Hooker 120* (syntype, K-LINDL); Chota Nagpur, *C. B. Clarke 21457* (syntype, K); Concan, *J. S. Law s.n.* (syntype, K).

Distribution. Andhra Pradesh, Arunachal Pradesh, Assam, Bihar, Chhattisgarh, Jharkhand, Karnataka, Kerala, Madhya Pradesh, Maharashtra, Meghalaya, Odisha, Sikkim, Uttarakhand, West Bengal [Bangladesh, China, Nepal].

Habit. Epiphytic herb.

Oberonia forcipata Lindl., Fol. Orchid. 8(*Oberonia*): 2. 1859. *Malaxis forcipata* (Lindl.) Rchb. f., Ann. Bot. Syst. (Walpers) 6: 208. 1861. *Iridorkis forcipata* (Lindl.) Kuntze, Revis. Gen. Pl. 2: 669. 1891 (as "*Iridorchis*"). SYNTYPES: Sri Lanka, Hewahette District, on trees (syntype, K-LINDL); *Thwaites 2511* (syntypes, AMES, BM, K-LINDL, P, W).

Distribution.   Kerala [Sri Lanka].

Habit.   Epiphytic herb.

Oberonia griffithiana Lindl., Sert. Orchid. 2: t. 8B. 1838. *Malaxis griffithiana* (Lindl.) Rchb. f., Ann. Bot. Syst. (Walpers) 6: 208. 1861. *Iridorkis griffithiana* (Lindl.) Kuntze, Revis. Gen. Pl. 2: 669. 1891 (as "*Iridorchis*"). TYPE: Myanmar, Moulmein [Mawlamyine], *Griffith s.n.* [*355*] (lectotype, K-LINDL K000974222, designated by Ansari & Balakrishnan, 1990).

Distribution.   Andaman & Nicobar Islands, Nagaland, Uttarakhand [Myanmar].

Habit.   Epiphytic herb.

Oberonia jenkinsiana Griff. ex Lindl., Fol. Orchid. 8(*Oberonia*): 4. 1859. *Malaxis jenkinsiana* (Griff. ex Lindl.) Rchb. f., Ann. Bot. Syst. (Walpers) 6: 211. 1861. *Iridorkis jenkinsiana* (Griff. ex Lindl.) Kuntze, Revis. Gen. Pl. 2: 669. 1891 (as "*Iridorchis*"). TYPE: India, Assam, Debooro Mook, *Jenkins s.n.* (syntypes, CAL, K-LINDL, K).

Distribution.   Arunachal Pradesh, Assam, Manipur, Meghalaya, Sikkim, West Bengal [China, Myanmar, Nepal].

Habit.   Epiphytic herb.

Oberonia josephi C. J. Saldanha, Indian Forester 100(9): 569, t. 2. 1974. TYPE: India, Karnataka, Hassan District, near Genkalbetta, alt. 3000 ft, 9 Oct. 1969, *C. J. Saldanha 15247-A* (holotype, JCB).

Distribution.   Arunachal Pradesh, Karnataka, Kerala, Tamil Nadu. Endemic.

Habit.   Epiphytic herb.

Oberonia kingii Lucksom, Orchid Rev. 110: 346. 2002. TYPE: India, Sikkim, Bhusuk Valley, 3400–4500 ft, Nov., *Luckson 411a* (holotype, CAL).

Distribution.   Sikkim. Endemic.

Habit.   Epiphytic herb.

Oberonia langbianensis Gagnep., Bull. Soc. Bot. France 79: 168. 1932. TYPE: Vietnam, Annam, Lang-bian, *Evrard 1402* (holotype, P).

Oberonia sulcata J. Joseph & Sud. Chowdhury, J. Bombay Nat. Hist. Soc. 63: 54. 1966. TYPE: India, Arunachal Pradesh, Selari forest, *Joseph 40358* (holotype, CAL; isotype, ASSAM).

Distribution.   Arunachal Pradesh [also in Thailand, Vietnam].

Habit.   Epiphytic herb.

Oberonia longibracteata Lindl., Gen. Sp. Orchid. Pl.: 15. 1830. *Malaxis longibracteata* (Lindl.) Rchb. f., Ann. Bot. Syst. (Walpers) 6: 209. 1861. *Iridorkis longibracteata* (Lindl.) Kuntze, Revis. Gen. Pl. 2: 669. 1891 (as "*Iridorchis*"). TYPE: Sri Lanka, *Macrae s.n.* (holotype, K-LINDL; isotype, ?BM [collector not named]).

Distribution.   Kerala, Manipur, Tamil Nadu [China, Sri Lanka].

Habit.   Epiphytic herb.

Oberonia maxima C. S. P. Parish ex Hook. f., Fl. Brit. India [J. D. Hooker] 5(15): 677. 1888. *Iridorkis maxima* (C. S. P. Parish ex Hook. f.) Kuntze, Revis. Gen. Pl. 2: 669. 1891 (as "*Iridorchis*"). TYPE: India, Tenasserim, Moulmein [Mawlamyine], *C. S. P. Parish 287* (holotype, K).

Oberonia orbicularis Hook. f., Fl. Brit. India [J. D. Hooker] 5(15): 677. 1888. *Iridorkis orbicularis* (Hook. f.) Kuntze, Revis. Gen. Pl. 2: 669. 1891 (as "*Iridorchis*"). TYPE: Sikkim Himalaya, Dikeeling, alt. 3000 ft, *C. B. Clarke s.n.* [*9610*] (holotype, K).

Distribution.   Arunachal Pradesh, Manipur, ?Meghalaya, Nagaland, Sikkim [Myanmar].

Habit.   Epiphytic herb.

Oberonia mucronata (D. Don) Ormerod & Seidenf. in G. Seidenfaden, Contr. Orchid. Fl. Thailand 13: 20. 1997. *Stelis mucronata* D. Don, Prodr. Fl. Nepal.: 32. 1825. TYPE: Nepal, *Buchanan-Hamilton s.n.* (holotype, BM).

Oberonia denticulata Wight, Icon. Pl. Ind. Orient. [Wight] 5(1): 3, t. 1625. 1851. *Oberonia iridifolia* Lindl. var. *denticulata* (Wight) Hook. f., Fl. Brit. India [J. D. Hooker] 5(15): 676. 1888. TYPE: India, Tamil Nadu, Coimbatore, Iyamally Hills, July–Aug. [December 1847], *Wight s.n.* [*2939*] (syntypes, CAL, K [2×]).

Cymbidium iridifolium Roxb., Fl. Ind. ed. 1832, 3: 458. 1832. TYPE: India, Sylhet, collector not named (not found).

Oberonia iridifolia Lindl. var. *brevifolia* Hook. f., Fl. Brit. India [J. D. Hooker] 5(15): 676. 1888. SYNTYPES: Moulmein [Mawlamyine], *Griffith s.n.* (syntype, K); *C. S. P. Parish s.n.* (syntype, K).

Oberonia lobulata King & Pantl., J. Asiat. Soc. Bengal, Pt. 2, Nat. Hist. 64(2): 331. 1895 [1896]. TYPE: India, Sikkim, valley of the Teesta, alt. 1000, Oct., *Pantling s.n.* [*199*] (probable holotype, CAL).

Oberonia gammiei King & Pantl., J. Asiat. Soc. Bengal, Pt. 2, Nat. Hist. 66: 578. 1898 [1897]. TYPE: India, Lower Bengal: Sunderbans, *G. A. Gammie & R. L. Heinig 92* (lectotype, CAL, designated by Seidenfaden, 1968; probable isotype, K [*Gammie s.n.*] K000387706).

Oberonia manipurensis Chowlu, Y. N. Devi, A. N. Rao, N. Angela, H. B. Sharma & Akimpou, Nordic J. Bot. 34: 384. 2016. TYPE: India, Manipur, Tamenglong District, Tamenglong, 24°48.78′N, 93°32.77′E, 403 m, *Chowlu 00362* (holotype, CAL).

Distribution. Andaman & Nicobar Islands, Andhra Pradesh, Arunachal Pradesh, Assam, Bihar, Chhattisgarh, Goa, Gujarat, Karnataka, Kerala, Maharashtra, Manipur, Meghalaya, Mizoram, Nagaland, Odisha, Sikkim, Tamil Nadu, Tripura, Uttarakhand, West Bengal [Bangladesh, Bhutan, Myanmar, Nepal].

Habit. Epiphytic herb.

Remarks. *Oberona manipurensis* was recently synonymized by Geiger (2019). He also considers (pers. comm.) *O. gammiei* and *O. lobulata* to be additional synonyms of *O. mucronata*.

Oberonia muthikulamensis K. Prasad, K. M. P. Kumar & P. Sudheshna, Nordic J. Bot. 36(5): art. e01797: 2. 2018. TYPE: India, Kerala, Palakkad district, Muthikulam, way to Elival hills, 10°56′12.6″N, 76°38′26.7″E, 1846 m, Nov., *K. Prasad 008471* (holotype, CAL; isotype, CAL).

Distribution. Kerala. Endemic.

Habit. Epiphytic herb.

Oberonia nayarii R. Ansari & N. P. Balakr., Orchid Monogr. 4: 17. 1990. TYPE: India, Tamil Nadu, Nilgiri Hills, Pykara, 28 Dec. 1900, *C. A. Barber 2687* (holotype, MH).

Distribution. Karnataka, Kerala, Tamil Nadu. Endemic.

Habit. Epiphytic herb.

Oberonia obcordata Lindl., Fol. Orchid. 8(*Oberonia*): 7. 1859. *Malaxis obcordata* (Lindl.) Rchb. f., Ann. Bot. Syst. (Walpers) 6: 216. 1861. *Iridorkis obcordata* (Lindl.) Kuntze, Revis. Gen. Pl. 2: 669. 1891 (as "*Iridorchis*"). SYNTYPES: India, Sikkim, alt. 2000 ft, *Hooker & Thomson 112* (lectotype, K-LINDL, designated by Ansari & Balakrishnan, 1990).

Oberonia obcordata Lindl. var. *bracteata* Hook. f., Fl. Brit. India [J. D. Hooker] 5(15): 684. 1888. TYPE: India, Sikkim, *Thomson s.n.*

Oberonia treutleri Hook. f., Fl. Brit. India [J. D. Hooker] 5(15): 683. 1888. *Iridorkis treutleri* (Hook. f.) Kuntze, Revis. Gen. Pl. 2: 669. 1891 (as "*Iridorchis*"). TYPE: India, Sikkim Himalaya, alt. 6000 ft, *Treutler s.n.* [*1151*] (holotype, K).

Distribution. Arunachal Pradesh, Meghalaya, Nagaland, Sikkim, West Bengal [Bhutan, China, Myanmar, Nepal].

Habit. Epiphytic herb.

Oberonia pachyphylla King & Pantl., Ann. Roy. Bot. Gard. (Calcutta) 8: 5, t. 3. 1898 (as "*pachyrachis*" in the plate). TYPE: India, Sikkim, Salgurra, near Siliguri, alt. 900 ft, Jan. 1896, *Pantling 429* (lectotype, CAL, designated by Seidenfaden, 1968).

Distribution. West Bengal [Bhutan, Nepal].

Habit. Epiphytic herb.

Oberonia pachyrachis Rchb. f. ex Hook. f., Fl. Brit. India [J. D. Hooker] 5(15): 681. 1888. *Iridorkis pachyrachis* (Rchb. f. ex Hook. f.) Kuntze, Revis. Gen. Pl. 2: 669. 1891 (as "*Iridorchis*"). SYNTYPE: Tropical Himalaya, Garwhal, *Falconer s.n.* [*1009*] (lectotype, K, designated by Ansari & Balakrishnan, 1990; isotype, W).

Distribution. Arunachal Pradesh, Assam, Manipur, Meghalaya, Mizoram, Nagaland, Sikkim, Tripura, Uttarakhand, West Bengal [Bhutan, China, Myanmar, Nepal].

Habit. Epiphytic herb.

Oberonia platycaulon Wight, Icon. Pl. Ind. Orient. [Wight] 5(1): 3, t. 1623. 1851. *Malaxis platycaulon* (Wight) Rchb. f., Ann. Bot. Syst. (Walpers) 6: 209. 1861. *Iridorkis platycaulon* (Wight) Kuntze, Revis. Gen. Pl. 2: 669. 1891 (as "*Iridorchis*"). TYPE: India, Pulney Hills, Sep., *Wight s.n.* (syntypes, K, K-LINDL, possibly E [*Wight 921*]).

Oberonia bisaccata Manilal & C. S. Kumar, Kew Bull. 39(1): 121, fig. 1. 1984. TYPE: India, Kerala, Silent Valley, near dam site, 1982, *C. Sathish Kumar SV 10738* (holotype, CALI; isotype, K).

Distribution. Karnataka, Kerala, Tamil Nadu. Endemic.

Habit. Epiphytic herb.

Oberonia prainiana King & Pantl., J. Asiat. Soc. Bengal, Pt. 2, Nat. Hist. 64: 331. 1895 [1896]. TYPE: India, Sikkim, Teesta Valley, alt. 1000 ft, July, *Pantling s.n.* [*probably 225*] (probable holotype, CAL).

Distribution. Arunachal Pradesh, Assam, Sikkim, Uttarakhand, West Bengal [Nepal].

Habit. Epiphytic herb.

Oberonia proudlockii King & Pantl., J. Asiat. Soc.
Bengal, Pt. 2, Nat. Hist. 66: 580. 1898 [1897].
TYPE: India, Nilgiri Hills, Gudalur, Sep. 1896,
Proudlock s.n. (holotype, CAL; isotype, K).

Oberonia sedgwickii Blatt. & McCann, J. Bombay Nat. Hist.
Soc. 35(2): 257. 1931. TYPE: India, Western Ghats,
Castle Rock, alt. 1600 ft, *Sedgwick's collector 5615* (ho-
lotype, BLAT).

Distribution. Karnataka, Kerala, Odisha, Tamil
Nadu. Endemic.

Habit. Epiphytic herb.

Oberonia pumilio Rchb. f., Bonplandia (Hannover)
5: 58. 1857. *Malaxis pumilio* (Rchb. f.) Rchb.
f., Ann. Bot. Syst. (Walpers) 6: 216. 1861. TYPE:
Indonesia, Java, *Zollinger s.n.* (holotype, W).

Oberonia treubii Ridl., J. Linn. Soc., Bot. 32: 219. 1896.
TYPE: Indonesia, Java, Buitenzorg Botanic Gardens,
Treub s.n. (holotype, SING).

Distribution. Arunachal Pradesh [also in Thailand].

Habit. Epiphytic herb.

Oberonia pyrulifera Lindl., Fol. Orchid. 8(*Oberonia*): 3. 1859. *Malaxis pyrulifera* (Lindl.) Rchb. f.,
Ann. Bot. Syst. (Walpers) 6: 211. 1861. *Iridorkis
pyrulifera* (Lindl.) Kuntze, Revis. Gen. Pl. 2: 669.
1891 (as "*Iridorchis*"). TYPE: India, Khasi Hills,
Myrung, *Griffith s.n.* (holotype, K-LINDL).

Oberonia verticillata Wight var. *khasiana* Lindl., Fol. Orchid.
8(*Oberonia*): 3. 1859. *Malaxis verticillata* (Wight) Rchb.
f. var. *khasiana* (Lindl.) Rchb. f., Ann. Bot. Syst. (Walpers) 6: 210. 1861. TYPE: India, Khasi Hills, Bor Pance,
Hooker & Thomson 118 (syntypes, K, K-LINDL); *T. Lobb
s.n.* (syntype, K-LINDL).

Distribution. Arunachal Pradesh, Assam, Manipur,
Meghalaya, Mizoram, Nagaland, Odisha, Sikkim, Uttarakhand, West Bengal [Bhutan, China, Myanmar].

Habit. Epiphytic herb.

Oberonia raoi L. R. Shakya & R. P. Chaudhary,
Rheedea 10(1): 57. 2000 (as "*raoii*"). TYPE: India,
Meghalaya, Khasi Hills, Nongpoh, 4000 ft, Aug.
1897, *Pantling s.n.* (holotype, CAL).

Distribution. Meghalaya. Endemic.

Habit. Epiphytic herb.

Remarks. Geiger (pers. comm.) suspects that
Oberonia raoi is a synonym of *O. brachyphylla* Blatt. &
McCann.

Oberonia recurva Lindl., Edwards's Bot. Reg. 25
(Misc.): 14. 1839. *Malaxis recurva* (Lindl.) Rchb.
f., Ann. Bot. Syst. (Walpers) 6: 212. 1861. *Iridorkis recurva* (Lindl.) Kuntze, Revis. Gen. Pl. 2:
669. 1891 (as "*Iridorchis*"). TYPE: India, Bombay, *Messrs. Loddiges s.n.* (holotype, K-LINDL).

Oberonia setifera Lindl., Fol. Orchid. 8(*Oberonia*): 3. 1859.
TYPE: India, South Conkan, *Dalzell 38* (holotype,
K-LINDL).
Oberonia parvula King & Pantl., J. Asiat. Soc. Bengal, Pt. 2,
Nat. Hist. 64: 330. 1895 [1896]. TYPE: India ("Bho-
tan"), Guru-bathan, alt. 1500 ft, Feb. [1892], *Pantling
s.n.* [presumably *Pantling 203*, Valley of the Teesta, alt.
1500 ft, Feb. 1892] (syntypes, AMES, CAL, K, P).
Oberonia croftiana King & Pantl., Ann. Roy. Bot. Gard. (Cal-
cutta) 8: 7, t. 6A. 1898. TYPE: Sikkim-Bhutan Frontier,
Jaldakha River Banks, alt. 900 ft, Oct. 1893, *Pantling
254* (holotype, CAL; isotype, AMES, K, P, W).
Oberonia lingmalensis Blatt. & McCann, J. Bombay Nat. Hist.
Soc. 35(2): 255. 1931. *Oberonia recurva* Lindl. var. *lingmalensis* (Blatt. & McCann) Santapau & Kapadia, J. Bombay Nat. Hist. Soc. 57: 259. 1960. TYPE: India, Western
Ghats, near Mahableshwar, Lingmala, *Blatter & Hallberg 1681* (holotype, BLAT).

Distribution. Arunachal Pradesh, Goa, Karnataka,
Kerala, Maharashtra, Meghalaya, Nagaland, Sikkim,
West Bengal [Bhutan, China, Nepal, Sri Lanka].

Habit. Epiphytic herb.

Oberonia rufilabris Lindl., Sert. Orchid. 2: t. 8A.
1838. *Malaxis rufilabris* (Lindl.) Rchb. f., Ann.
Bot. Syst. (Walpers) 6: 213. 1861. *Iridorkis rufilabris* (Lindl.) Kuntze, Revis. Gen. Pl. 2: 669.
1891 (as "*Iridorchis*"). TYPE: Myanmar, Mergui
[Myeik], *Griffith 1834* (holotype, K-LINDL; isotype, ?L, P).

Oberonia pantlingiana L. R. Shakya & R. P. Chaudhary, Harvard Pap. Bot. 4: 360. 1999. TYPE: India, Dooars, 175 m,
Pantling 430 (holotype, CAL; isotype, BM, K, W).

Distribution. Arunachal Pradesh, Assam, Meghalaya, Sikkim, West Bengal [Bangladesh, Bhutan, China,
Myanmar, Nepal].

Habit. Epiphytic herb.

Oberonia sebastiana B. V. Shetty & Vivek., Bull.
Bot. Surv. India 17(1–4): 157. 1975 [1978]. TYPE:
India, Kerala, Umaiyamalai, Animudi Slope,
6600 ft, 17 Nov. 1965, *B. V. Shetty & K. Vivekanan-
than 26480 A* (holotype, CAL; isotype, MH).

Distribution. Kerala, Tamil Nadu. Endemic.

Habit. Epiphytic herb.

Oberonia seidenfadeniana J. Joseph & Vajr., Bull.
Bot. Surv. India 13(3–4): 344. 1971 [1974]. TYPE:

India, Tamil Nadu, Eastern side of Anamalai Hills, Andiparaishola, alt. 4200 ft, 22 Sep. 1962, *Joseph 17476-A* (holotype, CAL; isotype, MH).

Distribution. Kerala, Tamil Nadu. Endemic.

Habit. Epiphytic herb.

Oberonia teres Kerr, Bull. Misc. Inform. Kew 1927: 214. 1927. TYPE: Thailand: Mae Tun, Chieng Mai, alt. 2560 ft, *Kerr 484* (holotype, K).

Distribution. Assam, Manipur, Meghalaya, Mizoram [China].

Habit. Epiphytic herb.

Oberonia thwaitesii Hook. f., Fl. Brit. India [J. D. Hooker] 5(15): 678. 1888. TYPE: Sri Lanka, 1853, *Thwaites s.n.* [*Ceylon Plants 2516*] (holotype, K; isotype, MH).

Oberonia verticillata Wight var. *pubescens* Lindl., Fol. Orchid. 8(*Oberonia*): 3. 1859. *Malaxis verticillata* (Wight) Rchb. f. var. *pubescens* (Lindl.) Rchb. f., Ann. Bot. Syst. (Walpers) 6: 210. 1861. TYPE: Sri Lanka, 1853, *Thwaites s.n.* [*Ceylon Plants 2516*] (holotype, K; isotype, MH).

Distribution. Kerala [Sri Lanka].

Habit. Epiphytic herb.

Remarks. Ansari and Balakrishnan (1990) designated *Ceylon Plants 2572* (PDA) as "lectotype." This is not admissible, since Hooker only cited *Ceylon Plants 2516*; this must be considered the type.

Oberonia verticillata Wight, Icon. Pl. Ind. Orient. [Wight] 5(1): 3, t. 1626. 1851. *Malaxis verticillata* (Wight) Rchb. f., Ann. Bot. Syst. (Walpers) 6: 210. 1861. *Iridorkis verticillata* (Wight) Kuntze, Revis. Gen. Pl. 2: 669. 1891 (as "*Iridorchis*"). TYPE: India, Nilgiri Hills, July–Oct., *Wight s.n.* (lectotype, BM, designated by Ansari & Balakrishnan, 1990 [typification incomplete, as there is more than one collection at BM, fide Geiger (pers. comm.)]; isotype, AMES, BM, CAL, K, K-LINDL, SING).

Oberonia spiralis Blatt. & McCann, J. Bombay Nat. Hist. Soc. 35(2): 256. 1931, non Griff., 1851, nec Lindl., 1852. TYPE: India, Yellapur, N Kanara, *Bell 219* (holotype, BLAT).
Oberonia bellii Blatt. & McCann, J. Bombay Nat. Hist. Soc. 35(2): 256. 1931. TYPE: Not designated.

Distribution. Goa, Karnataka, Kerala, Maharashtra, Tamil Nadu. Endemic.

Habit. Epiphytic herb.

Oberonia wightiana Lindl., Edwards's Bot. Reg. 25(Misc.): 14. 1839. *Malaxis wightiana* (Lindl.) Rchb. f., Ann. Bot. Syst. (Walpers) 6: 212. 1861. *Iridorkis wightiana* (Lindl.) Kuntze, Revis. Gen. Pl. 2: 669. 1891 (as "*Iridorchis*"). TYPE: India, Nilgiri & Pulney Hills, Aug.–Sep., *Wight 181* (lectotype, K-LINDL, designated by Ansari & Balakrishnan, 1990; isotype, MH).

Oberonia stachyoides A. Rich., Ann. Sci. Nat., Bot. sér. 2, 15: 15. 1841. SYNTYPES: India, Nilgiri Hills, valleys near Avalanchy and Dodabetta, Apr. and May, *Perrottet s.n.* (syntypes, W).
Oberonia arnottiana Wight, Icon. Pl. Ind. Orient. [Wight] 5(1): 3, t. 1628. 1851. *Oberonia wightiana* Lindl. var. *arnottiana* (Wight) R. Ansari, N. C. Nair & V. J. Nair, J. Econ. Taxon. Bot. 3(1): 118. 1982. TYPE: India, Nilgiri & Pulney Hills, Sep., *Wight s.n.* (lectotype, BM, designated by Ansari & Balakrishnan, 1990; isotype, K, K-LINDL).
Oberonia wightiana Lindl. var. *nilgirensis* R. Ansari, N. C. Nair & V. J. Nair, J. Econ. Taxon. Bot. 3(1): 118. 1982. TYPE: India, Tamil Nadu, Nilgiri Hills, 1 Sep. 1970, *Sharma 35978* (holotype, CAL; isotype, MH).

Distribution. Karnataka, Kerala, Tamil Nadu [Sri Lanka].

Habit. Epiphytic herb.

Oberonia wynadensis Sivad. & R. T. Balakr., Nordic J. Bot. 9(4): 395. 1989 [1990]. TYPE: India, Kerala State, Wynad District, Chembra Peak near Meppadi, alt. ca. 4000 ft, 23 Sep. 1984, *R. T. Balakrishnan CU 40643* (holotype, C; isotype, CAL).

Oberonia pakshipadalensis M. Kumar & Sequiera, J. Orchid Soc. India 12(1–2): 31, fig. 2. 1998. TYPE: India, Kerala, Wayanad District, Pakshipadalam, alt. 4800 ft, *Stephen 008138* (holotype, KFRI).

Distribution. Kerala. Endemic.

Habit. Epiphytic herb.

Oberonia zeylanica Hook. f., Hooker's Icon. Pl. 18: t. 1782A. 1888. *Iridorkis zeylanica* (Hook. f.) Kuntze, Revis. Gen. Pl. 2: 669. 1891 (as "*Iridorchis*"). TYPE: Sri Lanka, Hantani, *Thwaites s.n.* [*Ceylon Plants 543*] (lectotype, designated by Ansari & Balakrishnan, 1990, K).

Distribution. Karnataka, ?Sikkim, Tamil Nadu [Sri Lanka].

Habit. Epiphytic herb.

Remarks. It is not clear why Ansari and Balakrishnan (1990) designated "*C.P. 543*," under which number Thwaites's collection was distributed, as the type. This is a fruiting specimen. Perhaps they were unaware of the existence of the syntype *Beckett s.n.*, which has flowers and was used by Hooker to describe and illustrate the fertile parts.

———

Odontochilus Blume, Coll. Orchid. 79. 1858 [1859].

TYPE: *Odontochilus flavescens* (Blume) Blume (*Anoectochilus flavescens* Blume).

Odontochilus clarkei Hook. f., Fl. Brit. India [J. D. Hooker] 6(17): 100. 1890. *Cystopus clarkei* (Hook. f.) Kuntze, Revis. Gen. Pl. 2: 658. 1891. *Anoectochilus clarkei* (Hook. f.) Seidenf. & Smitinand, Orchids Thailand 1: 88. 1959. TYPE: India, Sikkim, Mongpo, alt. 3000 ft, *C. B. Clarke 26741* (holotype, K).

Distribution. Arunachal Pradesh, Sikkim, West Bengal [Myanmar].

Habit. Terrestrial herb.

Odontochilus crispus (Lindl.) Hook. f., Fl. Brit. India [J. D. Hooker] 6(17): 99. 1890. *Anoectochilus crispus* Lindl., J. Proc. Linn. Soc., Bot. 1: 180. 1857. *Cystopus crispus* (Lindl.) Kuntze, Revis. Gen. Pl. 2: 658. 1891. TYPE: India, Sikkim, icon. *Cathcart s.n.* (holotype, K).

Distribution. Arunachal Pradesh, Assam, Meghalaya, Nagaland, Sikkim, West Bengal [Bhutan, China, Myanmar, Nepal].

Habit. Terrestrial herb.

Odontochilus elwesii C. B. Clarke ex Hook. f., Fl. Brit. India [J. D. Hooker] 6(17): 100. 1890. *Cystopus elwesii* (C. B. Clarke ex Hook. f.) Kuntze, Revis. Gen. Pl. 2: 658. 1891. *Anoectochilus elwesii* (C. B. Clarke ex Hook. f.) King & Pantl., Ann. Roy. Bot. Gard. (Calcutta) 8: 296, t. 394. 1898. TYPE: India, Khasi Hills, Shillong, alt. 6100 ft, *C. B. Clarke* [44609] (lectotype, K, designated by Seidenfaden, 1978).

Distribution. Arunachal Pradesh, Assam, ?Meghalaya, Nagaland, Sikkim, West Bengal [Bhutan, China, Myanmar, Nepal].

Habit. Terrestrial herb.

Odontochilus grandiflorus (Lindl.) Benth. ex Hook. f., Fl. Brit. India [J. D. Hooker] 6(17): 100. 1890. *Anoectochilus grandiflorus* Lindl., J. Proc. Linn. Soc., Bot. 1: 179. 1857. *Cystopus grandiflorus* (Lindl.) Kuntze, Revis. Gen. Pl. 2: 658. 1891. TYPE: India, Sikkim, alt. 4000–6000 ft, *J. D.*

Hooker 329 (lectotype, K-LINDL, designated by Bhattacharjee & Chowdhery, 2018; isotype, K [3×]).

Distribution. Arunachal Pradesh, Assam, Manipur, Meghalaya, Mizoram, Nagaland, Sikkim, West Bengal [Bhutan].

Habit. Terrestrial herb.

Odontochilus lanceolatus (Lindl.) Blume, Coll. Orchid. 81, t. 29, fig. 2. 1858 (as "*lanceolatum*"). *Anoectochilus lanceolatus* Lindl., Gen. Sp. Orchid. Pl.: 499. 1840. *Cystopus lanceolatus* (Lindl.) Kuntze, Revis. Gen. Pl. 2: 658. 1891. SYNTYPES: India, Assam, *Mack s.n.* (syntype, K); Khasi Hills, *Griffith s.n.* (syntype, K).

Anoectochilus luteus Lindl., J. Proc. Linn. Soc., Bot. 1: 179. 1857. *Anoectochilus flavus* Benth. & Hook. f., Gen. Pl. 3: 598. 1883, nom. superfl. (in error, should have been "*luteus*"). *Cystopus flavus* (Benth. & Hook. f.) Kuntze, Revis. Gen. Pl. 2: 658, nom. superfl. 1891. SYNTYPES: India, Sikkim, alt. 5000 ft, *J. D. Hooker 341* [341 p.p.] (syntype, K-LINDL); icon. *Cathcart* (syntype, K).

Distribution. Arunachal Pradesh, Assam, Meghalaya, Mizoram, Nagaland, Sikkim, West Bengal [Bangladesh, Bhutan, China, Myanmar, Nepal].

Habit. Terrestrial herb.

Odontochilus tortus King & Pantl., J. Asiat. Soc. Bengal, Pt. 2, Nat. Hist. 65: 125. 1896. *Anoectochilus tortus* (King & Pantl.) King & Pantl., Ann. Roy. Bot. Gard. (Calcutta) 8: 298, t. 396. 1898. *Pristiglottis torta* (King & Pantl.) Aver., Bot. Zhurn. (Moscow & Leningrad) 81(10): 78. 1996. TYPE: Bhutan, Kumai near Jaldaca river, alt. 4000 ft, [Nov.–]Dec. 1894, *Pantling 354* (lectotype, CAL, designated by Bhattacharjee & Chowdhery, 2018; isotype, BM, CAL, K).

Distribution. Arunachal Pradesh, Assam, Sikkim, West Bengal [Bhutan, China, Myanmar].

Habit. Terrestrial herb.

———

Oreorchis Lindl., J. Proc. Linn. Soc., Bot. 3: 26. 1859 [1858].

TYPE: *Oreorchis patens* (Lindl.) Lindl.

Oreorchis foliosa (Lindl.) Lindl., J. Proc. Linn. Soc., Bot. 3: 27. 1859 [1858]. *Corallorhiza foliosa* Lindl., Gen. Sp. Orchid. Pl.: 535. 1840. *Kitigorchis foliosa* (Lindl.) Maek., Wild Orchids Japan Colour 469. 1971. TYPE: India septentrionali: Mussoorie, *Royle s.n.* (holotype, K-LINDL).

Distribution. Himachal Pradesh, Nagaland, Sikkim, Uttarakhand [Myanmar, Nepal].

Habit. Terrestrial herb.

Oreorchis indica (Lindl.) Hook. f., Fl. Brit. India [J. D. Hooker] 5(16): 709. 1890. *Corallorhiza indica* Lindl., J. Proc. Linn. Soc., Bot. 3: 26. 1859 [1858]. *Oreorchis foliosa* (Lindl.) Lindl. var. *indica* (Lindl.) N. Pearce & P. J. Cribb, J. Orchid Soc. India 10(1–2): 5. 1996. SYNTYPES: India, Simla, Hattu, *Thomson 1724* (syntype, K); Garwhal, *Edgeworth 1844* (syntype, K).

Distribution. Assam, Himachal Pradesh, Sikkim, Uttarakhand, West Bengal [Bhutan, China, Nepal].

Habit. Terrestrial herb.

Oreorchis micrantha Lindl., J. Proc. Linn. Soc., Bot. 3: 27. 1859 [1858]. TYPE: India, Himalayas, 8000–10,000 ft, Yaklul Mountains, Kumaon, *T. Thomson 214* (syntypes, K, K-LINDL).

Oreorchis rolfei Duthie, J. Asiat. Soc. Bengal, Pt. 2, Nat. Hist. 71: 38. 1902. TYPE: Western Himalaya: Nag Tiba, Tehri-Garwal, alt. 9000 ft, June 1900, *Mackinnon's collector s.n.* (holotype, CAL).

Distribution. Arunachal Pradesh, Himachal Pradesh, Jammu & Kashmir, Sikkim, Uttarakhand [Bhutan, China, Myanmar, Nepal].

Habit. Terrestrial herb.

Remarks. This species is illustrated as *Oreorchis indica* (Lindl.) Hook. f. in Swami (2016).

Oreorchis patens (Lindl.) Lindl., J. Proc. Linn. Soc., Bot. 3: 27. 1859 [1858]. *Corallorhiza patens* Lindl., Gen. Sp. Orchid. Pl.: 535. 1840. TYPE: Russia, Siberia, *Prescott s.n.* (holotype, K-LINDL).

Distribution. Uttarakhand, West Bengal [China].

Habit. Terrestrial herb.

Remarks. This species is illustrated as *Oreorchis micrantha* Lindl. in Swami (2016).

———

Pachystoma Blume, Bijdr. Fl. Ned. Ind. 8: 376. 1825.

TYPE: *Pachystoma pubescens* Blume.

Pachystoma hirsutum (J. Joseph & Vajr.) C. S. Kumar & Manilal, Kew Bull. 42(4): 942. 1987 (as "*hirsuta*"). *Eulophia hirsuta* J. Joseph & Vajr., Bull. Bot. Surv. India 17(1–4): 192. 1975 [1978]. TYPE:

India, Kerala, Palghat District, Karasuryamalai-Anamada area, Nemmara Division, 3 Mar. 1976, *Vajravelu 46201A* (holotype, CAL).

Distribution. Kerala, Tamil Nadu. Endemic.

Habit. Terrestrial herb.

Pachystoma pubescens Blume, Bijdr. Fl. Ned. Ind. 8: 376, t. 3, fig. 29. 1825. *Pachychilus pubescens* (Blume) Blume, Mus. Bot. 2: 178. 1856. TYPE: Indonesia, Java, Krawang Prov., Tjiradjas, *Blume s.n.* (holotype, ?L [*Blume 1529*]).

Apaturia senilis Lindl., Gen. Sp. Orchid. Pl.: 130. 1831. *Pachystoma senile* (Lindl.) Rchb. f., Bonplandia 3(18): 251. 1855. *Pachychilus senilis* (Lindl.) Blume, Mus. Bot. 2: 178. 1856. SYNTYPES: India, Sylhet Mountains, Mar. 1829, *Wallich 3739 (leg. H. Bruce 126)* (syntypes, E, K, K-LINDL); Uligapur & Kalegung, 21 & 26 Feb. 1809, *Wallich 3739 (leg. Buchanan-Hamilton s.n.)* (syntypes, E, K-LINDL, K-WALL).
Apaturia lindleyana Wight, Icon. Pl. Ind. Orient. [Wight] 5(1): 8, t. 1662. 1851. TYPE: India, Coorg, Dec.–Jan., *Jerdon s.n.* (holotype, K).

Distribution. Andaman & Nicobar Islands, Arunachal Pradesh, Assam, Bihar, Himachal Pradesh, Karnataka, Kerala, Madhya Pradesh, Maharashtra, Manipur, Meghalaya, Mizoram, Nagaland, Odisha, Punjab, Sikkim, Tamil Nadu, Uttar Pradesh, Uttarakhand, West Bengal [Bangladesh, Bhutan, China, Myanmar, Nepal].

Habit. Terrestrial herb.

———

Paphiopedilum Pfitzer, Morph. Stud. Orchideenbl. 11. 1886, nom. cons.

TYPE: *Paphiopedilum insigne* (Wall. ex Lindl.) Pfitzer (*Cypripedium insigne* Wall. ex Lindl.) (typ. cons.).

Remarks. All species of *Paphiopedilum* (slipper orchids) are currently listed on CITES Appendix I. In India, all are extremely rare and endangered.

Paphiopedilum charlesworthii (Rolfe) Pfitzer, Bot. Jahrb. Syst. 19: 40. 1894. *Cypripedium charlesworthii* Rolfe, Orchid Rev. 1: 303. 1893. *Cordula charlesworthii* (Rolfe) Rolfe, Orchid Rev. 20: 2. 1912. TYPE: East Indies, *Messrs. Charlesworth s.n.* (holotype, K).

Distribution. ?Assam, Mizoram, ?Sikkim [China, Myanmar].

Habit. Lithophytic herb.

Paphiopedilum × crossianum (Rchb.f.) Stein, Orchid.-Buch: 465. 1892. *Cypripedium × crossia-*

num Rchb.f., Gard. Chron. 1873: 877. 1873. TYPE: Artificial hybrid, *cult. Veitch s.n.*

Paphiopedilum × *venustoinsigne* Pradhan, Indian Orchids: Guide Identif. & Cult. 2: 675. 1979. TYPE: India, Meghalaya, *U. C. Pradhan's collector 7* (holotype, Herbarium U. C. Pradhan).

Distribution.   Meghalaya. Endemic.

Habit.   Terrestrial herb.

Remarks.   Although first described and named as an artificial hybrid, this taxon is also a natural hybrid: *Paphiopedilum insigne* × *P. venustum.*

Paphiopedilum druryi (Bedd.) Stein, Orchideenbuch 466. 1892. *Cypripedium druryi* Bedd., Icon. Pl. Ind. Or. (Beddome) 1: 23, t. 112. 1874 (as "*drurii*"). TYPE: India, Calcad Hills, *Beddome s.n.* (lectotype, BM, designated by Cribb, 1998).

Distribution.   Kerala, Tamil Nadu. Endemic.

Habit.   Terrestrial herb.

Paphiopedilum fairrieanum (Lindl.) Stein, Orchideenbuch 467. 1892 (as "*fairieanum*"). *Cypripedium fairrieanum* Lindl., Gard. Chron. 1857: 740. 1857 (as "*fairieanum*"). *Cordula fairrieana* (Lindl.) Rolfe, Orchid Rev. 20: 2. 1912. TYPE: Unknown locality, *cult. Fairrie ("Fairie") s.n.* (holotype, K-LINDL).

Distribution.   Arunachal Pradesh, ?Assam, Sikkim [Bhutan].

Habit.   Terrestrial herb.

Paphiopedilum hirsutissimum (Lindl. ex Hook.) Stein, Orchideenbuch 470. 1892. *Cypripedium hirsutissimum* Lindl. ex Hook., Bot. Mag. 83: t. 4990. 1857. *Cordula hirsutissima* (Lindl. ex Hook.) Rolfe, Orchid Rev. 20: 2. 1912. TYPE: "Indonesia, Java," [certainly wrong] *cult. Parker s.n.* (holotype, K).

Distribution.   ?Assam, ?Manipur, Meghalaya, Mizoram, Nagaland, ?Sikkim, ?West Bengal [China, Myanmar].

Habit.   Lithophytic herb.

Paphiopedilum insigne (Wall. ex Lindl.) Pfitzer, Jahrb. Wiss. Bot. 19: 159. 1888. *Cypripedium insigne* Wall. ex Lindl., Coll. Bot. (Lindley) t. 32. 1821. *Cordula insignis* (Wall. ex Lindl.) Raf., Fl. Tellur. 4: 46. 1836 [1838]. TYPE: ?India, *cult. Cattley (leg. Wallich)* (holotype, K-LINDL).

Distribution.   ?Assam, ?Manipur, ?Nagaland, Meghalaya, ?West Bengal [Bangladesh].

Habit.   Terrestrial herb.

Remarks.   In Wallich's herbarium at Kew, under No. 7022, are two different collections by De Silva, from Cherrapunji (Feb. 1820) and Pundua (Sep. 1820), of *Cypripedium* (=*Paphiopedilum) insigne.* It is probable that the living plant that Wallich sent to England, where it was first grown in the Liverpool Botanic Garden and then passed on to Mr. Cattley, was collected by De Silva in one of these places.

Paphiopedilum × **polystigmaticum** (Rchb. f.) Stein, Orchid.-Buch: 481. 1892. *Cypripedium* × *polystigmaticum* Rchb.f., Gard. Chron., ser. 3, 4: 407. 1888. TYPE: Artificial hybrid, *cult. R. Measures s.n.*

Paphiopedilum × *spicerovenustum* Pradhan, Indian Orchids: Guide Identif. & Cult. 2: 676. 1979. TYPE: Sonai, E Assam, *U. C. Pradhan's collector 8* (holotype, Herb. U. C. Pradhan).

Distribution.   Assam. Endemic.

Habit.   Terrestrial herb.

Remarks.   Although first described and named as an artificial hybrid, this taxon is also a natural hybrid: *Paphiopedilum spicerianum* × *P. venustum.*

Paphiopedilum × **pradhanii** Pradhan, Indian Orchids: Guide Identif. & Cult. 2: 675. 1979. TYPE: *Icones t. 5, 11–14 October 1974* (holotype, Herb. U. C. Pradhan).

Distribution.   Sikkim. Endemic.

Habit.   Terrestrial herb.

Remarks.   This taxon is a natural hybrid: *Paphiopedilum fairrieanum* × *P. venustum.*

Paphiopedilum spicerianum (Rchb. f.) Pfitzer, Jahrb. Wiss. Bot. 19: 144. 1888. *Cypripedium spicerianum* Rchb. f., Gard. Chron., n.s., 13: 40 (also 74 & 363). 1880. *Cordula spiceriana* (Rchb. f.) Rolfe, Orchid Rev. 20: 2. 1912. TYPE: India, Assam, *cult. Veitch s.n. (leg. Spicer)* (holotype, W).

Distribution.   Assam, Manipur, Mizoram, ?Sikkim [Bhutan, China, Myanmar].

Habit.   Terrestrial herb.

Paphiopedilum venustum (Wall. ex Sims) Pfitzer, Jahrb. Wiss. Bot. 19: 165. 1888. *Cypripedium venustum* Wall. ex Sims, Bot. Mag. 47: t. 2129. 1820.

Stimegas venustum (Wall. ex Sims) Raf., Fl. Tellur. 4: 46. 1836 [1838]. *Cordula venusta* (Wall. ex Sims) Rolfe, Orchid Rev. 20: 2. 1912. TYPE: India, Sylhet, Pundua, June 1820, *Wallich 7023 (leg. De Silva s.n.)* (holotype, K-LINDL; isotype, K-WALL).

Paphiopedilum venustum (Wall. ex Sims) Pfitzer var. *rubrum* Pradhan, Indian Orchids: Guide Identif. & Cult. 2: 675. 1979. TYPE: India, Meghalaya, *U. C. Pradhan's collector 6* (holotype, Herb. U. C. Pradhan).

Paphiopedilum venustum (Wall. ex Sims) Pfitzer var. *teestaensis* Pradhan, Indian Orchids: Guide Identif. & Cult. 2: 675. 1979. TYPE: India, West Bengal, Teesta Valley, *K. B. Basnet 3* (holotype, Herb. U. C. Pradhan).

Distribution. ?Arunachal Pradesh, Assam, ?Manipur, Meghalaya, Sikkim, West Bengal [Bangladesh, China, Nepal].

Habit. Terrestrial herb.

Paphiopedilum villosum (Lindl.) Stein, Orchideenbuch 490. 1892. *Cypripedium villosum* Lindl., Gard. Chron. 1854: 135. 1854. *Cordula villosa* (Lindl.) Rolfe, Orchid Rev. 20: 2. 1912. TYPE: Moulmein [Mawlamyine], alt. 5000 ft, *T. Lobb s.n.* (holotype, K-LINDL).

Distribution. Assam, ?Manipur, Mizoram, ?Nagaland [China, Myanmar].

Habit. Epiphytic or lithophytic herb.

Paphiopedilum wardii Summerh., Gard. Chron., ser. 3, 92: 446. 1932. TYPE: Myanmar, in the North, *F. Kingdon Ward s.n.* (holotype, K, spirit mat.).

Distribution. Arunachal Pradesh [China, Myanmar].

Habit. Terrestrial herb.

————

Papilionanthe Schltr., Orchis 9: 78. 1915.

TYPE: *Papilionanthe teres* (Lindl.) Schltr. (*Vanda teres* Lindl.).

Papilionanthe biswasiana (Ghose & Mukerjee) Garay, Bot. Mus. Leafl. 23(10): 371. 1974. *Aerides biswasiana* Ghose & Mukerjee, Orchid Rev. 53: 124. 1945 (as "*biswasianum*"). TYPE: Myanmar, Shan States, Apr. 1942, *cult. Ghose s.n.* (holotype, CAL).

Distribution. ?Nagaland, ?Sikkim, ?West Bengal [Myanmar].

Habit. Epiphytic herb.

Remarks. This species is listed as "excluded" in Singh et al. (2019).

Papilionanthe cylindrica (Lindl.) Seidenf., Descr. Epidendrorum J. G. König: 33. 1995. *Aerides cylindrica* Lindl., Gen. Sp. Orchid. Pl.: 240. 1833 (as "*cylindricum*"). TYPE: India, without locality, *Wallich 7317B (leg. Wight s.n.)* (syntype, K-WALL), *Wallich 7317A (leg. Heyne s.n., 21 Mar. 1817)* (syntypes, K-LINDL, K-WALL).

Distribution. Arunachal Pradesh, Assam, ?Himachal Pradesh, Karnataka, Kerala, ?Manipur, ?Meghalaya, Mizoram, Nagaland, Sikkim, Tamil Nadu, Tripura [Sri Lanka].

Habit. Epiphytic herb.

Remarks. *Papilionanthe cylindrica* is often thought to be a synonym of *P. subulata* (Willd.) Garay. However, that name is likely to be a synonym of *Thrixspermum filiforme* (Hook. f.) Kuntze (Ormerod et al., 2019).

Papilionanthe teres (Roxb.) Schltr., Orchis 9: 78. 1915. *Dendrobium teres* Roxb., Fl. Ind. ed. 1832, 3: 485. 1832. *Vanda teres* (Roxb.) Lindl., Gen. Sp. Orchid. Pl.: 217. 1833. TYPE: Bangladesh, Chittagong and Sylhet, icon. *Dendrobium teres, Roxburgh 2355* (syntypes, CAL, K).

Distribution. Andaman & Nicobar Islands, Andhra Pradesh, Arunachal Pradesh, Assam, Bihar, Manipur, Meghalaya, Mizoram, Nagaland, Odisha, Sikkim, Tamil Nadu, Tripura, West Bengal [Bangladesh, Bhutan, China, Myanmar, Nepal].

Habit. Epiphytic herb.

Papilionanthe uniflora (Lindl.) Garay, Bot. Mus. Leafl. 23(10): 372. 1974. *Mesoclastes uniflora* Lindl., Gen. Sp. Orchid. Pl.: 45. 1830. *Luisia uniflora* (Lindl.) Blume, Rumphia 4: 50. 1849. *Aerides longicornu* Hook. f., Fl. Brit. India [J. D. Hooker] 6(17): 44. 1890, nom. superfl. *Aerides uniflora* (Lindl.) Summerh., Kew Bull. 10(4): 588. 1955 [1956]. TYPE: Nepal, Bagmati Zone, Rashuwa District, Gosainthan [Shishapangma], Oct. 1821, *Wallich 1993* (holotype, K-LINDL; isotype, K-WALL).

Distribution. Arunachal Pradesh, Assam, Meghalaya, Nagaland, Sikkim, West Bengal [Bhutan, Nepal].

Habit. Epiphytic herb.

Papilionanthe vandarum (Rchb. f.) Garay, Bot. Mus. Leafl. 23(10): 372. 1974. *Aerides vandarum* Rchb. f., Gard. Chron. 1867: 997. 1867. *Vanda vandarum* (Rchb. f.) K. Karas., Orchid Atlas 8: 199. 1992. SYNTYPES: Without provenance, *cult. Jenisch s.n.* (syntype, W); *cult. Schiller s.n.* (syntype, W).

Distribution. Arunachal Pradesh, Assam, Manipur, Meghalaya, Mizoram, Sikkim [Bhutan, China, ?Myanmar, Nepal].

Habit. Epiphytic herb.

Pecteilis Raf., Fl. Tellur. 2: 37. 1836 [1837].

TYPE: *Pecteilis susannae* (L.) Raf. (as "*susanna*") (*Orchis susannae* L.).

Pecteilis gigantea (Sm.) Raf., Fl. Tellur. 2: 38. 1836 [1837]. *Orchis gigantea* Sm., Exot. Bot. 2: 79, t. 100. 1806. *Habenaria gigantea* (Sm.) D. Don, Prodr. Fl. Nepal.: 24. 1825. *Platanthera gigantea* (Sm.) Lindl., Numer. List [Wallich] n. 7052. 1832. TYPE: Nepal, *Buchanan-Hamilton s.n.*

Distribution. Andhra Pradesh, Assam, Bihar, Chhattisgarh, Dadra & Nagar Haveli, Goa, Gujarat, Himachal Pradesh, Jammu & Kashmir, Jharkhand, Karnataka, Kerala, Madhya Pradesh, Maharashtra, Nagaland, Odisha, Sikkim, Tamil Nadu, Uttarakhand [Myanmar, Nepal, Pakistan].

Habit. Terrestrial herb.

Pecteilis henryi Schltr., Repert. Spec. Nov. Regni Veg. Beih. 4: 45. 1919. *Pecteilis susannae* (L.) Raf. subsp. *henryi* (Schltr.) So6, Ann. Hist.-Nat. Mus. Natl. Hung. 26: 368. 1929. TYPE: China: Puteng, South of Szemao (Yunnan), alt. 3000 ft, *A. Henry 12534.*

Distribution. Manipur, Odisha [China, Myanmar].

Habit. Terrestrial herb.

Pecteilis korigadensis Jalal & Jayanthi, Phytotaxa 388(2): 169. 2019. TYPE: India, Maharashtra, Pune District, Korigad-Aamby Valley, 640 m elev., 29 Sep. 2018, *Jalal & Jayanthi197752* (holotype, BSI).

Distribution. Maharashtra. Endemic.

Habit. Terrestrial herb.

Pecteilis rawatii Kumar & Veldkamp, Gard. Bull. Singapore 61: 336. 2010, nom. inval. TYPE: India (origin unknown), icon. Ann. Roy. Bot. Gard. (Calcutta) 5: t. 99. 1895 (not an acceptable type, Art. 37.4).

Platanthera triflora (D. Don) Pradhan var. *multiflora* Pradhan, Indian Orchids: Guide Identif. & Cult. 1: 57. 1976, nom. inval. TYPE: Not designated.

Habenaria triflora auct. non D. Don: Hook. f., Ann. Roy. Bot. Gard. Calcutta 5: 66, t. 99. 1895.

Distribution. India (origin unknown). Endemic.

Habit. Terrestrial herb.

Remarks. To validate the name of this species, an actual specimen must be designated as the type. It appears that no specimen is available at present; the species may even be extinct.

Pecteilis susannae (L.) Raf., Fl. Tellur. 2: 38. 1836 [1837] (as "*susanna*"). *Orchis susannae* L., Sp. Pl. 2: 939. 1753. *Habenaria susannae* (L.) R. Br. ex Spreng., Syst. Veg. (ed. 16) [Sprengel] 3: 692. 1826. *Platanthera susannae* (L.) Lindl., Gen. Sp. Orchid. Pl.: 295. 1835. *Hemihabenaria susannae* (L.) Finet, Rev. Gén. Bot. 13: 532. 1902. TYPE: Indonesia, Maluku, Ambon, icon. "*Orchis Amboinensis*" in Hermann, *Parad. Bat.: 209. 1698* (lectotype, designated by Cribb in Cafferty & Jarvis (eds.), 1999).

Distribution. Andhra Pradesh, Arunachal Pradesh, Assam, Chhattisgarh, ?Goa, Karnataka, Kerala, Madhya Pradesh, Manipur, Meghalaya, Mizoram, Nagaland, Sikkim, Tamil Nadu, Uttarakhand, West Bengal [China, Myanmar, Nepal].

Habit. Terrestrial herb.

Pecteilis triflora (D. Don) Tang & F. T. Wang, Acta Phytotax. Sin. 1(1): 62. 1951. *Habenaria triflora* D. Don, Prodr. Fl. Nepal.: 25. 1825. *Platanthera candida* Lindl., Gen. Sp. Orchid. Pl.: 295. 1835, nom. superfl. *Pecteilis candida* Schltr., Repert. Spec. Nov. Regni Veg. Beih. 4: 120. 1919. *Platanthera triflora* (D. Don) Pradhan, Indian Orchids: Guide Identif. & Cult. 2: 680. 1979. TYPE: Nepal, Aug. 1821, *Wallich s.n.* [7035] (syntypes, K-LINDL, K-WALL).

Distribution. Jharkhand, Uttarakhand, West Bengal [Bangladesh, Nepal].

Habit. Terrestrial herb.

Pelatantheria Ridl., J. Linn. Soc., Bot. 32: 371. 1896.

TYPE: *Pelatantheria ctenoglossum* Ridl.

Pelatantheria insectifera (Rchb. f.) Ridl., J. Linn. Soc., Bot. 32: 373. 1896. *Sarcanthus insectifer* Rchb. f., Bot. Zeitung (Berlin) 15(10): 159. 1857. TYPE: India, Calcutta, *cult. Schiller s.n.*

Distribution.   Andaman & Nicobar Islands, Assam, Bihar, Chhattisgarh, Jharkhand, Manipur, Meghalaya, Nagaland, Odisha, Sikkim, Tripura, Uttarakhand, West Bengal [Bangladesh, Myanmar, Nepal].

Habit.   Epiphytic herb.

———

Pennilabium J. J. Sm., Bull. Jard. Bot. Buitenzorg, sér. 2, 13: 47. 1914.

TYPE: *Pennilabium angraecum* (Ridl.) J. J. Sm. (*Saccolabium angraecum* Ridl.).

Pennilabium labanyaeanum C. Deori, Odyuo & A. A. Mao, Gard. Bull. Singapore 67: 144. 2015. TYPE: India, Meghalaya, Laitkyrhong, 5 km from Smith, East Khasi Hills, 1753 m, 23 July 2014, *Deori & Odyuo 131601* (holotype, CAL; isotype, ASSAM).

Distribution.   Meghalaya. Endemic.

Habit.   Epiphytic herb.

Pennilabium proboscideum A. S. Rao & J. Joseph, Bull. Bot. Surv. India 10(2): 232. 1968 [1969]. TYPE: India, Arunachal Pradesh, Gauhati-Shillong Road, 23 July 1966, *A. S. Rao 45622* (holotype, CAL).

Distribution.   Arunachal Pradesh, Assam, Meghalaya, Nagaland, West Bengal. Endemic.

Habit.   Epiphytic herb.

Pennilabium struthio Carr, Gard. Bull. Straits Settlem. 5: 151. 1930. TYPE: Malaysia, Teku River, alt. 500 ft, 1928, *C. E. Carr s.n.* (holotype, SING; ?isotype, [298] K).

Distribution.   Arunachal Pradesh [also in Malaysia].

Habit.   Epiphytic herb.

———

Peristylus Blume, Bijdr. Fl. Ned. Ind. 8: 404. 1825, nom. cons.

TYPE: *Peristylus grandis* Blume.

Peristylus affinis (D. Don) Seidenf., Dansk Bot. Ark. 31(3): 48. 1977. *Habenaria affinis* D. Don, Prodr. Fl. Nepal.: 25. 1825. *Gymnadenia affinis* (D. Don) Rchb. f., Otia Bot. Hamburg. 33. 1878. *Peristylus goodyeroides* Lindl. var. *affinis* (D. Don) T. Cooke, Fl. Bombay 2: 712. 1908. *Phyllomphax affinis*

(D. Don) Schltr., Repert. Spec. Nov. Regni Veg. 16: 286. 1919. TYPE: Nepal, *Wallich s.n.* (not found).

Distribution.   Andhra Pradesh, Arunachal Pradesh, Assam, Himachal Pradesh, Jharkhand, Karnataka, Kerala, Manipur, ?Meghalaya, Nagaland, Sikkim, Uttar Pradesh, Uttarakhand, West Bengal [China, Myanmar, Nepal].

Habit.   Terrestrial herb.

Peristylus aristatus Lindl., Gen. Sp. Orchid. Pl.: 300. 1835. *Habenaria aristata* (Lindl.) Hook. f., Syst. Cat. Fl. Pl. Ceylon 91. 1885. TYPE: Sri Lanka, *Macrae s.n.* (holotype, K-LINDL).

Distribution.   Arunachal Pradesh, Assam, Goa, Karnataka, Kerala, Maharashtra, Meghalaya, Sikkim, Tamil Nadu [Myanmar, Nepal, Sri Lanka].

Habit.   Terrestrial herb.

Peristylus balakrishnanii Karthig., Sumathi & Jayanthi, Kew Bull. 65(3): 491. 2010 [2011]. TYPE: India, South Andaman, Rutland Island, Mt. Ford, 15 July 2003, *Karthigeyan, Sumathi & Jayanthi 6074* (holotype, CAL; isotype, MCCH, PBL).

Distribution.   Andaman & Nicobar Islands. Endemic.

Habit.   Terrestrial herb.

Peristylus biermannianus (King & Pantl.) X. H. Jin, Schuit. & W. T. Jin, Molec. Phylogen. Evol. 77: 51. 2014. *Habenaria biermanniana* King & Pantl., J. Asiat. Soc. Bengal, Pt. 2, Nat. Hist. 64: 343. 1895 [1896]. *Platanthera biermanniana* (King & Pantl.) Kraenzl., Orchid. Gen. Sp. 1: 636. 1899. TYPE: India, Darjeeling, Senchal, 8000 ft, July 1896, *Pantling s.n.* [*332*] (holotype, CAL; isotype, K).

Distribution.   Sikkim, West Bengal [Nepal].

Habit.   Terrestrial herb.

Peristylus brachyphyllus A. Rich., Ann. Sci. Nat., Bot. sér. 2, 15: 70, t. 2 A. 1841. TYPE: India, Nilgiri Hills, near Kulhuty, not far from Otacamund, July, *Perrottet s.n.* (holotype, P; isotype, K-LINDL).

Habenaria malabarica Hook. f., Fl. Brit. India [J. D. Hooker] 6(17): 159. 1890. SYNTYPES: India, Canara, Nilgiri and Bababudun Hills, *Heyne s.n.*; *Perrottet s.n.*; *J. E. Stocks s.n.* (syntypes, K-LINDL, P)

Distribution.   Karnataka, Tamil Nadu. Endemic.

Habit.   Terrestrial herb.

Peristylus caranjensis (Dalzell) Ormerod & C. S. Kumar, Harvard Pap. Bot. 23 (2): 283. 2018. *Habenaria caranjensis* Dalzell, Hooker's J. Bot. Kew Gard. Misc. 2: 262. 1850 (as "*caraujensis*"). TYPE: India, Bombay, Caranja ("Carauja") Island, *Dalzell s.n.* NEOTYPE: India, Dronagheree, July 1848, *J. E. Stocks(?) s.n.* (holotype, K, designated by Ormerod & Kumar, 2018).

Habenaria stocksii Hook. f., Fl. Brit. India [J. D. Hooker] 6(17): 158. 1890. *Peristylus stocksii* (Hook. f.) Kraenzl., Orchid. Gen. Sp. 1: 513. 1898. TYPE: India, Mysore, *J. E. Stocks 173* (lectotype, K, designated by Ormerod & Kumar, 2018).

Distribution. Bihar, Goa, Gujarat, Karnataka, Madhya Pradesh, Maharashtra, Tamil Nadu. Endemic.

Habit. Terrestrial herb.

Peristylus constrictus (Lindl.) Lindl., Gen. Sp. Orchid. Pl.: 300. 1835. *Herminium constrictum* Lindl., Edwards's Bot. Reg. 18: t. 1499. 1832. *Platanthera constricta* Lindl., Numer. List [Wallich] n. 7043. 1832, nom. inval. *Habenaria constricta* (Lindl.) Hook. f., Fl. Brit. India [J. D. Hooker] 6(17): 161. 1890. TYPE: Nepal, *Wallich s.n.* [*7043*] or *icon. Wallich* (not found).

Distribution. Arunachal Pradesh, Assam, Bihar, Chhattisgarh, Himachal Pradesh, Jammu & Kashmir, Jharkhand, Madhya Pradesh, Maharashtra, Manipur, Meghalaya, Nagaland, Odisha, Sikkim, Uttar Pradesh, Uttarakhand [Bangladesh, China, Myanmar, Nepal].

Habit. Terrestrial herb.

Remarks. When he transferred *Herminium constrictum* to *Peristylus*, Lindley cited *Wallich 7043* as the type collection. Under this number there are two sheets in the Wallich Herbarium at Kew, both labeled *Platanthera constricta*, one (7043B) collected on 30 May 1802 by Buchanan-Hamilton in Goyalpara (Goalpara, Assam) and one (7043A) collected in 1827 in Moulmein, with a question mark added. Neither of these are from Nepal. Lindley also mentions a drawing by Wallich, which we have not seen. Perhaps this was based on material from Nepal and it should then be considered the holotype.

Peristylus densus (Lindl.) Santapau & Kapadia, J. Bombay Nat. Hist. Soc. 57: 128. 1960. *Coeloglossum densum* Lindl., Gen. Sp. Orchid. Pl.: 302. 1835. TYPE: India, Sylhet, Pundua, July 1820, *Wallich 7057 (leg. De Silva 115)* (holotype, K-LINDL; isotype, K-WALL).

Habenaria gracilis Colebr. in W. J. Hooker, Exot. Fl. 2: t. 135. 1824. TYPE: India or Bangladesh, Sylhet, *Colebrooke s.n.* (probably not preserved and the illustration in Exot. Fl. 2: t. 135, 1824, must then be considered as the type).

Habenaria peristyloides Wight, Icon. Pl. Ind. Orient. [Wight] 5(1): 13, t. 1702. 1851, nom. illeg., non A. Rich., 1840. *Coeloglossum peristyloides* Rchb. f., Bonplandia (Hannover) 4: 321. 1856. *Peristylus peristyloides* M. R. Almeida, Fl. Maharashtra 5A: 79. 2009. TYPE: India, Pulney Hills, Sep., *Wight s.n.* (holotype, K).

Platanthera stenostachya Lindl., Hooker's J. Bot. Kew Gard. Misc. 7: 37. 1855. *Habenaria stenostachya* (Lindl.) Benth., Fl. Hongk. 362. 1861. *Peristylus stenostachyus* (Lindl.) Kraenzl., Orchid. Gen. Sp. 1: 502. 1898. TYPE: Hong Kong, *Champion s.n.* (holotype, K-LINDL).

Habenaria neglecta King & Pantl., J. Asiat. Soc. Bengal, Pt. 2, Nat. Hist. 66: 603. 1898 [1897]. *Peristylus neglectus* (King & Pantl.) Kraenzl., Orchid. Gen. Sp. 1: 924. 1901. TYPE: Bababudun, *Stocks s.n. (leg. Law)* (holotype, K; isotype, CAL, K, W).

Peristylus xanthochlorus Blatt. & McCann, J. Bombay Nat. Hist. Soc. 35(4): 733. 1932. TYPE: India, Western Ghats, Panchgani, first Tableland, *Blatter & Halberg B1686*.

Distribution. Arunachal Pradesh, Assam, Goa, Karnataka, Kerala, Maharashtra, Manipur, Meghalaya, Nagaland, Sikkim, Tamil Nadu [Bangladesh, China, Myanmar, Nepal, Sri Lanka].

Habit. Terrestrial herb.

Remarks. *Habenaria gracilis* Colebr. probably belongs here, as already suggested by Lindley. It would be the oldest name for this species, but the epithet cannot be used in *Peristylus* because of the existence of *P. gracilis* Blume.

Peristylus exilis Wight, Icon. Pl. Ind. Orient. [Wight] 5(1): 12, t. 1698. 1851. TYPE: India, Pulney Hills, Sep., *Wight s.n.* (holotype, K).

Peristylus aristatus sensu C. E. C. Fisch., Fl. Madras 2(8): 1474. 1928, non Lindl., 1835.

Distribution. Tamil Nadu. Endemic.

Habit. Terrestrial herb.

Remarks. This is considered a synonym of *Peristylus aristatus* Lindl. by Singh et al. (2019).

Peristylus goodyeroides (D. Don) Lindl., Gen. Sp. Orchid. Pl.: 299. 1835. *Habenaria goodyeroides* D. Don, Prodr. Fl. Nepal.: 25. 1825. *Herminium goodyeroides* (D. Don) Lindl., Numer. List [Wallich] n. 7066. 1832. TYPE: Nepal, Toka, July 1821, *Wallich s.n.* [*7066A*] (lectotype, K-LINDL; isotype, K-WALL).

Distribution. ?Andaman & Nicobar Islands, Andhra Pradesh, Arunachal Pradesh, Assam, Bihar, Chhattisgarh, Himachal Pradesh, Jharkhand, Karnataka, Kerala, Madhya Pradesh, Manipur, Meghalaya, Mizoram, Nagaland, Odisha, Sikkim, Tamil Nadu, Tripura, Uttar

Pradesh, Uttarakhand, West Bengal [Bangladesh, Bhutan, China, Myanmar, Nepal, Sri Lanka].

Habit. Terrestrial herb.

Peristylus gracilis Blume, Bijdr. Fl. Ned. Ind. 8: 404. 1825. TYPE: Indonesia, Java, Mt. Seribu, *Blume s.n.* (holotype, L [*Blume 955*]).

Distribution. ?Andaman & Nicobar Islands, Assam, Meghalaya [Myanmar].

Habit. Terrestrial herb.

Peristylus hamiltonianus (Lindl.) Lindl., Gen. Sp. Orchid. Pl.: 299. 1835. *Herminium hamiltonianum* Lindl. Edwards's Bot. Reg. 18: t. 1499. 1832. *Habenaria hamiltoniana* (Lindl.) Hook. f., Fl. Brit. India [J. D. Hooker] 6(17): 160. 1890. TYPE: Nepal, Morung [Morang] Hills, 28 July 1810, *Wallich 7069 (leg. Buchanan-Hamilton s.n.)* (holotype, K-LINDL; isotype, K-WALL).

Distribution. Assam, Meghalaya, Nagaland [Bhutan, Nepal].

Habit. Terrestrial herb.

Peristylus intrudens (Ames) Ormerod, Taiwania 56: 46. 2011. *Habenaria intrudens* Ames, Schedul. Orchid. 6: 1. 1923. TYPE: Philippines, Luzon, Antipoto, 100 m, Oct. 1913, *Bur. Sci. (Ramos) 21999* (holotype, AMES).

Distribution. Andaman & Nicobar Islands [China, Nepal].

Habit. Terrestrial herb.

Peristylus lacertifer (Lindl.) J. J. Sm., Bull. Jard. Bot. Buitenzorg ser. 3. 9: 23. 1927. *Gymnadenia tenuiflora* Lindl., Numer. List [Wallich] n. 7055. 1832, nom. inval. *Coeloglossum lacertiferum* Lindl., Gen. Sp. Orchid. Pl.: 302. 1835. *Habenaria lacertifera* (Lindl.) Benth., Fl. Hongk. 362. 1861. TYPE: Myanmar, Tavoy [Dawei], 19 Aug. 1827, *Wallich 7055 (leg. W. Gomez)* (holotype, K-LINDL; isotype, K-WALL).

Distribution. Assam, Manipur, Meghalaya, Nagaland, Sikkim, West Bengal [China, Myanmar].

Habit. Terrestrial herb.

Peristylus lawii Wight, Icon. Pl. Ind. Orient. [Wight] 5(1): 12, t. 1695. 1851. *Habenaria lawii* (Wight) Hook. f., Fl. Brit. India [J. D. Hooker] 6(17): 162. 1890. TYPE: India, Belgaum, *J. S. Law s.n.[68]* (holotype, K; isotype, CAL).

Distribution. Bihar, Chhattisgarh, Gujarat, Jharkhand, Karnataka, Kerala, Madhya Pradesh, Maharashtra, Odisha, Tamil Nadu, Uttarakhand, West Bengal [Myanmar, Nepal].

Habit. Terrestrial herb.

Peristylus monticola (Ridl.) Seidenf., Dansk Bot. Ark. 31: 35. 1977. *Habenaria monticola* Ridl., J. Linn. Soc., Bot. 32: 413. 1896. SYNTYPES: Malaysia, Mt. Ophir, 3000 ft; Kedak Peak, 3000 ft, *Ridley s.n.* (syntypes, SING).

Distribution. Andaman & Nicobar Islands [also in Indonesia, Malaysia].

Habit. Terrestrial herb.

Peristylus parishii Rchb. f., Trans. Linn. Soc. London 30: 139. 1874. *Habenaria parishii* (Rchb. f.) Hook. f., Fl. Brit. India [J. D. Hooker] 6(17): 161. 1890. TYPE: India, Tenasserim, Moulmein [Mawlamyine], *C. S. P. Parish s.n.* (holotype, K).

Distribution. Andaman & Nicobar Islands, Assam, Madhya Pradesh, ?Meghalaya, Nagaland, Odisha, Sikkim, West Bengal [China, Myanmar, Nepal].

Habit. Terrestrial herb.

Peristylus plantagineus (Lindl.) Lindl., Gen. Sp. Orchid. Pl.: 300. 1835. *Herminium plantagineum* Lindl., Edwards's Bot. Reg. 18: t. 1499. 1832. *Habenaria wightii* Trimen, Syst. Cat. Fl. Pl. Ceylon: 91. 1885. TYPE: Sri Lanka, *Macrae s.n.* (holotype, K-LINDL).

Habenaria goodyeroides sensu Hook., Bot. Mag. 62: t. 3397. 1835.
Peristylus elatus Dalzell, Hooker's J. Bot. Kew Gard. Misc. 3: 344. 1851. *Habenaria elata* (Dalzell) Alston in H. Trimen, Handb. Fl. Ceylon 6(Suppl.): 280. 1931. TYPE: India, Bombay, Malwan prov., *Dalzell s.n.* (holotype, K).

Distribution. Andhra Pradesh, Chhattisgarh, Goa, Gujarat, Jharkhand, Karnataka, Kerala, Madhya Pradesh, Maharashtra, Odisha, Tamil Nadu [Nepal, Sri Lanka].

Habit. Terrestrial herb.

Peristylus prainii (Hook. f.) Kraenzl., Orchid. Gen. Sp. 1: 514. 1898. *Habenaria prainii* Hook. f., Fl. Brit. India [J. D. Hooker] 6(17): 159. 1890. TYPE: India, Assam, Naga Hills, Kohima, Aug. 1886, *Prain 42* (holotype, K).

Distribution. Arunachal Pradesh, Assam, Nagaland, Sikkim, West Bengal [Bangladesh, Bhutan, Myanmar, Nepal].

Habit. Terrestrial herb.

Peristylus pseudophrys (King & Pantl.) Kraenzl., Orchid. Gen. Sp. 1: 925. 1901. *Habenaria pseudophrys* King & Pantl., J. Asiat. Soc. Bengal, Pt. 2, Nat. Hist. 65: 133. 1896. *Peristylus pseudophrys* (King & Pantl.) Pradhan, Indian Orchids: Guide Identif. & Cult. 2: 687. 1979, isonym. TYPE: India, Sikkim, Chungthang, 5500 ft, July, *Pantling 424* (holotype, CAL).

Distribution. Assam, Sikkim, West Bengal. Endemic.

Habit. Terrestrial herb.

Peristylus richardianus Wight, Icon. Pl. Ind. Orient. [Wight] 5(1): 12, t. 1697. 1851. TYPE: India, Nilgiri Hills, *Wight s.n.* (syntypes, K).

Habenaria bicornuta Hook. f., Fl. Brit. India [J. D. Hooker] 6(17): 156. 1890. SYNTYPES: India, Western Ghats, Nilgiri and Pulney Hills, up to 6500 ft, *Wight s.n.* (syntype, K).

Distribution. Arunachal Pradesh, Kerala, Meghalaya, Tamil Nadu [Nepal].

Habit. Terrestrial herb.

Peristylus sahanii Kumar, G. S. Rawat & Jalal, Kew Bull. 65(1): 101. 2010. TYPE: Jharkhand: Latehar, Netarhat, 8 Aug. 2005, *Pankaj Kumar 051039* (holotype, WII).

Distribution. Jharkhand. Endemic.

Habit. Terrestrial herb.

Peristylus secundus (Lindl.) Rathakr., Indian Forester 98(1): 31. 1972. *Gymnadenia secunda* Lindl., Numer. List [Wallich] n. 7054. 1832, nom. inval. *Coeloglossum secundum* Lindl., Gen. Sp. Orchid. Pl.: 303 1835. TYPE: South India or Sri Lanka, without locality, 23 Sep. 1816, *Wallich 7054 (leg. Heyne s.n.)* (holotype, K-LINDL; isotype, K-WALL).

Peristylus lancifolius A. Rich., Ann. Sci. Nat., Bot. sér. 2, 15: 69, t. 2C. 1841. TYPE: India, Nilgiri Hills near Neddoubetta, Sep., *Perrottet s.n.* (holotype, P; isotype, K-LINDL).
Peristylus robustior Wight, Icon. Pl. Ind. Orient. [Wight] 5(1): 12, t. 1699. 1851. *Habenaria robustior* (Wight) Hook. f., Fl. Brit. India [J. D. Hooker] 6(17): 160. 1890. TYPE: India, *Wight s.n.* (holotype, K).

Distribution. Karnataka, Kerala, Tamil Nadu [Sri Lanka].

Habit. Terrestrial herb.

Peristylus spiralis A. Rich., Ann. Sci. Nat., Bot. sér. 2, 15: 69, t. 2B. 1841. *Habenaria torta* Hook. f., Fl. Brit. India [J. D. Hooker] 6(17): 159. 1890. TYPE: India, Nilgiri Hills, Avalanchy, Aug.–Sep., *Perrottet s.n.* (holotype, P).

Distribution. Karnataka, Kerala, ?Maharashtra, Tamil Nadu [Sri Lanka].

Habit. Terrestrial herb.

Peristylus tentaculatus (Lindl.) J. J. Sm., Orch. Java 6: 35. 1905. *Glossula tentaculata* Lindl., Bot. Reg. 10: t. 862. 1824. *Glossaspis tentaculata* (Lindl.) Spreng., Syst. Veg. (ed. 16) [Sprengel] 3: 694. 1826. *Habenaria tentaculata* (Lindl.) Rchb. f., Otia Bot. Hamburg. 34. 1878. TYPE: China, 1824, *John Damper Parks s.n.* (holotype, K-LINDL).

Distribution. Assam, Kerala, ?Meghalaya [China].

Habit. Terrestrial herb.

Peristylus tipulifer (C. S. P. Parish & Rchb. f.) Mukerjee, Notes Roy. Bot. Gard. Edinburgh 21(3): 153. 1953 (as *"tipuliferus"*). *Habenaria tipulifera* C. S. P. Parish & Rchb. f., Trans. Linn. Soc. London 30: 139. 1874. TYPE: Myanmar, *Parish 292* (holotype, K).

Distribution. Assam, Manipur, ?Meghalaya [Myanmar, Nepal].

Habit. Terrestrial herb.

———

Phalaenopsis Blume, Bijdr. Fl. Ned. Ind. 7: 294. 1825.

TYPE: *Phalaenopsis amabilis* (L.) Blume (*Epidendrum amabile* L.).

Phalaenopsis arunachalensis K. Gogoi & Rinya, Lankesteriana 20(3): 275. 2020. TYPE: India, Arunachal Pradesh, Lower Subansiri, Ziro, 1400 m, 16 Oct. 2019, fl. 15 Apr. 2020, *K. Gogoi & K. Rinya 00809* (holotype, Orchid Herbarium, Tipi; isotype, ASSAM, TOSEHIM).

Distribution. Arunachal Pradesh. Endemic.

Habit. Epiphytic herb.

Phalaenopsis cacharensis (Barbhuiya, B. K. Dutta & Schuit.) Kocyan & Schuit., Phytotaxa 161: 67. 2014. *Ornithochilus cacharensis* Barbhuiya, B. K. Dutta & Schuit., Kew Bull. 67: 511. 2012. TYPE:

India, Assam, Borail Wildlife Sanctuary, 130 m, 16 July 2011, *Barbhuiya 665* (holotype, ASSAM).

Distribution. Assam. Endemic.

Habit. Epiphytic herb.

Phalaenopsis cornu-cervi (Breda) Blume & Rchb. f., Hamburger Garten-Blumenzeitung 16: 116. 1860. *Polychilos cornu-cervi* Breda, Gen. Sp. Orchid. Asclep. 1: t. [1.] 1827 [1828]. *Polystylus cornu-cervi* (Breda) Hasselt ex Hassk., Retzia 1: 3. 1855. TYPE: Indonesia, Java, Bantam Prov, *Kuhl & van Hasselt s.n.* (holotype, L, lost; isotype, BO, W).

Distribution. Andaman & Nicobar Islands, Assam, Manipur, Mizoram, West Bengal [Myanmar].

Habit. Epiphytic herb.

Phalaenopsis deliciosa Rchb. f., Bonplandia 2(7): 93. 1854. *Kingidium deliciosum* (Rchb. f.) H. R. Sweet, Amer. Orchid. Soc. Bull. 39(12): 1095. 1970. *Doritis deliciosa* (Rchb. f.) T. Yukawa & K. Kita, Acta Phytotax. Geobot. 56: 156. 2005. TYPE: Indonesia, Java, *Zollinger 1429* (holotype, W).

Phalaenopsis wightii Rchb. f., Bot. Zeitung (Berlin) 20(27): 214. 1862. *Doritis wightii* (Rchb. f.) Benth. & Hook. f., Gen. Pl. 3: 574. 1883. *Kingidium wightii* (Rchb. f.) O. Gruss & Roellke, Orchidee (Hamburg) 46: 23. 1995. TYPE: India, Malabar, *Wight s.n.* (holotype, K). *Kingidium hookerianum* O. Gruss & Roellke, Orchidee (Hamburg) 45(6): 230. 1994. *Phalaenopsis deliciosa* Rchb. f. subsp. *hookeriana* (O. Gruss & Roellke) Christenson, *Phalaenopsis*: 223. 2001, nom. inval. *Doritis deliciosa* (Rchb. f.) T. Yukawa & K. Kita subsp. hookeriana (O. Gruss & Roellke) T. Yukawa & K. Kita, Acta Phytotax. Geobot. 56(2): 156. 2005, nom. inval. *Kingidium deliciosum* (Rchb. f.) H. R. Sweet subsp. *hookerianum* (O. Gruss & Roellke) S. Misra, Orchids India: 259. 2007, nom. inval. TYPE: Myanmar, Tenasserim, Moulmein [Mawlamyine], *Parish 175* (holotype, K).

Distribution. Andaman & Nicobar Islands, Arunachal Pradesh, Assam, Goa, Karnataka, Kerala, Meghalaya, Nagaland, Odisha, Sikkim, Tamil Nadu, Uttarakhand, West Bengal [Bangladesh, Bhutan, China, Myanmar, Nepal, Sri Lanka].

Habit. Epiphytic herb.

Phalaenopsis difformis (Wall. ex Lindl.) Kocyan & Schuit., Phytotaxa 161: 67. 2014. *Aerides difformis* Wall. ex Lindl., Gen. Sp. Orchid. Pl.: 242. 1833 (as "*difforme*"). *Ornithochilus fuscus* Wall. ex Lindl., Gen. Sp. Orchid. Pl.: 242. 1833, nom. inval. *Ornithochilus difformis* (Wall. ex Lindl.) Schltr., Repert. Spec. Nov. Regni Veg. Beih. 4:

277. 1919. *Sarcochilus difformis* (Wall. ex Lindl.) Tang & F. T. Wang, Acta Phytotax. Sin. 1: 92. 1951. *Trichoglottis difformis* (Wall. ex Lindl.) T. B. Nguyen & D. H. Duong in T. B. Nguyen (ed.), Fl. Taynguyen. Enum. 206. 1984 (as "*diformis*"), nom. inval. TYPE: Nepal, icon. *Wallich s.n.* (not found).

Distribution. Arunachal Pradesh, Assam, Manipur, Meghalaya, Mizoram, Nagaland, Sikkim, Uttarakhand, West Bengal [Bhutan, China, Myanmar, Nepal].

Habit. Epiphytic herb.

Phalaenopsis lobbii (Rchb. f.) H. R. Sweet, Gen. Phalaenopsis 53. 1980. *Phalaenopsis parishii* Rchb. f. var. *lobbii* Rchb. f., Refug. Bot. (Saunders) 2: sub t. 85. 1870 [1869]. *Phalaenopsis decumbens* (Griff.) Holttum var. *lobbii* (Rchb. f.) P. F. Hunt, Amer. Orchid Soc. Bull. 40: 1094. 1971. *Polychilos lobbii* (Rchb. f.) Shim, Malayan Nat. J. 36: 24. 1982. *Doritis lobbii* (Rchb. f.) T. Yukawa & K. Kita, Acta Phytotax. Geobot. 56(2): 156. 2005. TYPE: Eastern Himalaya, 1845, *T. Lobb s.n.* (holotype, W).

Distribution. Assam, Sikkim, West Bengal [Bhutan, Myanmar].

Habit. Epiphytic herb.

Phalaenopsis malipoensis Z. J. Liu & S. C. Chen, Acta Bot. Yunnan. 27: 37. 2005. TYPE: China, Yunnan, Malipo County, Xia Jin Chang Xiang, 1200 m, *Z. J. Liu 2890* (holotype, Herb. Shenzhen City Wutongshan Nurseries).

Distribution. Assam [China, Myanmar].

Habit. Epiphytic herb.

Phalaenopsis mannii Rchb. f., Gard. Chron. 1871: 902. 1871. *Polychilos mannii* (Rchb. f.) Shim, Malayan Nat. J. 36: 24. 1982. TYPE: Unknown locality, May 1868, *Gustav Mann s.n.* (holotype, W).

Distribution. Arunachal Pradesh, Assam, Meghalaya, Mizoram, Sikkim, West Bengal [Bhutan, China, Myanmar, Nepal].

Habit. Epiphytic herb.

Phalaenopsis marriottiana (Rchb. f.) Kocyan & Schuit. var. **parishii** (Rchb. f.) Kocyan & Schuit., comb. nov. Basionym: *Vanda parishii* Rchb. f., Xenia Orchid. 2: 138. 1868. *Hygrochilus parishii* (Rchb. f.) Pfitzer, Nat. Pflanzenfam. Nachtr. [Engler & Prantl] 1: 112. 1897. *Vandopsis parishii*

(Rchb. f.) Schltr., Repert. Spec. Nov. Regni Veg. 11: 47. 1912. *Stauropsis parishii* (Rchb. f.) Rolfe, Orchid Rev. 27: 97. 1919. *Phalaenopsis hygrochila* J. M. H. Shaw, Orchid Rev. Suppl., 123(1309): 23. 2015. *Phalaenopsis tigrina* Ming H. Li, O. Gruss & Z. J. Liu, Phytotaxa 275(1): 59. 2016, nom. superfl. TYPE: Myanmar, *Parish s.n.* (holotype, K).

Distribution. Arunachal Pradesh, Assam, Manipur, Mizoram, Nagaland, Sikkim [China, Myanmar].

Habit. Epiphytic herb.

Remarks. Phalaenopsis marriottiana var. *marriottiana*, apparently not known from India, differs in having unspotted, light brown to purplish brown sepals and petals, in contrast to variety *parishii*, which has yellow sepals and petals with red-brown spots. However, intermediate forms are known and in the absence of evident morphological differences there is, in our opinion, no reason to consider these color morphs as different species. The appropriate infraspecific rank is still an open question.

Due to an oversight, the variety *parishii* was not validly published in an earlier paper by Kocyan and Schuiteman (2014); this is here remedied.

Ormerod et al. (2021) mistakenly adopted the later name *Phalaenopsis hygrochila* J. M. H. Shaw for the species.

Phalaenopsis mysorensis C. J. Saldanha, Indian Forester 100(9): 571, t. 3. 1974. *Polychilos mysorensis* (C. J. Saldanha) Shim, Malayan Nat. J. 36: 25. 1982. *Kingidium mysorense* (C. J. Saldanha) C. S. Kumar in C. Sathish Kumar & K. S. Manilal, Cat. Indian Orchids 95. 1994 (as "*mysorensis*"). *Doritis mysorensis* (C. J. Saldanha) T. Yukawa & K. Kita, Acta Phytotax. Geobot. 56: 157. 2005. TYPE: India, Karnataka, Hassan District, Vanagur, alt. 3000 ft, 18 Dec. 1969, *C. J. Saldanha 15915* (holotype, St. Joseph's College, Bangalore Herbarium).

Kingidium niveum C. S. Kumar in C. Sathish Kumar & K. S. Manilal, Cat. Indian Orchids 53, fig. 16. 1994. TYPE: India, Kerala, Palghat District, Walghat, *C. S. Kumar SV 11264* (holotype, CAL).

Distribution. Karnataka, Kerala [Sri Lanka].

Habit. Epiphytic herb.

Phalaenopsis parishii Rchb. f., Bot. Zeitung (Berlin) 23: 146. 1865. *Grafia parishii* (Rchb. f.) A. D. Hawkes, Phytologia 13: 306. 1966. *Polychilos parishii* (Rchb. f.) Shim, Malayan Nat. J. 36: 25. 1982. *Doritis parishii* (Rchb. f.) T. Yukawa & K. Kita, Acta Phytotax. Geobot. 56: 157. 2005. TYPE: Myanmar, *C. S. P. Parish s.n.* (holotype, W).

Aerides decumbens Griff., Not. Pl. Asiat. 3: 365. 1851. *Kingiella decumbens* (Griff.) Rolfe, Orchid Rev. 25: 197. 1917. *Biermannia decumbens* (Griff.) Tang & F. T. Wang ex Merr. & Metcalf, Lingnan Sci. J. 21: 7. 1945. *Phalaenopsis decumbens* (Griff.) Holttum, Gard. Bull. Singapore 11: 286. 1947. *Kingidium decumbens* (Griff.) P. F. Hunt, Kew Bull. 24(1): 97. 1970. TYPE: Myanmar, towards Mogoung, Apr. 1837, *Griffith s.n.* (holotype, K).

Distribution. Arunachal Pradesh, Assam, Karnataka, Kerala, Sikkim, Tamil Nadu, West Bengal [Myanmar, Nepal].

Habit. Epiphytic herb.

Phalaenopsis pulcherrima (Lindl.) J. J. Sm., Repert. Spec. Nov. Regni Veg. 32: 366. 1933. *Doritis pulcherrima* Lindl., Gen. Sp. Orchid. Pl.: 178. 1833. TYPE: Vietnam, Turon Gulf, *Finlayson s.n.* [*521*] (holotype, K).

Phalaenopsis mastersii King & Pantl., J. Asiat. Soc. Bengal, Pt. 2, Nat. Hist. 66: 591. 1898 [1897]. TYPE: India, Assam, Nambur falls, Feb. 1845, *Masters s.n.* (not found).

Distribution. Assam, Manipur, Meghalaya [China, Myanmar].

Habit. Lithophytic herb.

Phalaenopsis taenialis (Lindl.) Christenson & Pradhan, Indian Orchid J. 1(4): 154. 1985. *Aerides taenialis* Lindl., Gen. Sp. Orchid. Pl.: 239. 1833 (as "*taeniale*"). *Doritis taenialis* (Lindl.) Benth. ex Hook. f., Fl. Brit. India [J. D. Hooker] 6(17): 31. 1890. *Kingiella taenialis* (Lindl.) Rolfe, Orchid Rev. 25: 197. 1917. *Biermannia taenialis* (Lindl.) Tang & F. T. Wang, Acta Phytotax. Sin. 1: 96. 1951. *Kingidium taeniale* (Lindl.) P. F. Hunt, Kew Bull. 24(1): 98. 1970 (as "*taenialis*"). *Polychilos taenialis* (Lindl.) Shim, Malayan Nat. J. 36: 28. 1982. TYPE: Nepal, *Wallich s.n.* (holotype, BM; isotype, K).

Doritis braceana Hook. f., Fl. Brit. India [J. D. Hooker] 6(17): 196. 1890. *Phalaenopsis braceana* (Hook. f.) Christenson, Selbyana 9(1): 169. 1986. *Kingidium braceanum* (Hook. f.) Seidenf., Opera Bot. 95: 187. 1988. TYPE: India, Sikkim Himalaya, *Gamble s.n.*, drawing by Brace (holotype, CAL).

Distribution. Arunachal Pradesh, Assam, Manipur, Meghalaya, Mizoram, Nagaland, Sikkim, Uttarakhand, West Bengal [Bhutan, China, ?Myanmar, Nepal].

Habit. Epiphytic herb.

Phalaenopsis tetraspis Rchb. f., Xenia Orchid. 2: 146. 1870. *Phalaenopsis speciosa* Rchb. f. var. *tetraspis* (Rchb. f.) H. R. Sweet, Amer. Orchid. Soc.

Bull. 37(12): 1092, fig. 16, 17a. 1968. TYPE: Himalaya, *T. Lobb s.n.* (holotype, W).

Phalaenopsis speciosa Rchb. f., Gard. Chron., n.s., 15: 562. 1881. *Polychilos speciosus* (Rchb. f.) Shim, Malayan Nat. J. 36: 26. 1982 (as "*speciosa*"). TYPE: Tropical Asia, *Bull s.n.* (holotype, W).

Distribution. Andaman & Nicobar Islands [also in Indonesia].

Habit. Epiphytic herb.

Phalaenopsis yingjiangensis (Z. H. Tsi) Kocyan & Schuit., Phytotaxa 161: 67. 2014. *Ornithochilus yingjiangensis* Z. H. Tsi, Acta Phytotax. Sin. 22: 479. 1984. TYPE: China, Yunnan, Yingjiang, *Tsi 168* (holotype, PE).

Distribution. Mizoram, Nagaland [China].

Habit. Epiphytic herb.

————

Phreatia Lindl., Gen. Sp. Orchid. Pl.: 63. 1830.

TYPE: *Phreatia elegans* Lindl.

Phreatia albofarinosa Ormerod, Taiwania 50(3): 183. 2005. TYPE: Thailand Doi Sutep, 3200 ft, 12 Aug. 1910, *Kerr 259* (holotype, K).

Distribution. Meghalaya [also in Thailand].

Habit. Epiphytic herb.

Phreatia elegans Lindl., Gen. Sp. Orchid. Pl.: 63. 1830. *Thelasis elegans* (Lindl.) Blume, Mus. Bot. 2: 187. 1856. *Eria elegans* (Lindl.) Rchb. f. in B. Seemann, Fl. Vit. 301. 1868. TYPE: Sri Lanka, *Macrae 38* (holotype, K-LINDL; isotype, BM).

Distribution. Arunachal Pradesh, Kerala, Manipur, Meghalaya, Nagaland, Sikkim, Tamil Nadu, West Bengal [Sri Lanka].

Habit. Epiphytic herb.

Phreatia plantaginifolia (J. Koenig) Ormerod, Opera Bot. 124: 22. 1995. *Epidendrum plantaginifolium* J. Koenig, Observ. Bot. (Retzius) 6: 60. 1791. *Cymbidium plantaginifolium* (J. Koenig) Willd., Sp. Pl., ed. 4 [Willdenow] 4: 101. 1805. TYPE: Peninsular Malaysia, Tsing, near Malacca, *J. Koenig s.n.*

Dendrolirium secundum Blume, Bijdr. Fl. Ned. Ind. 7: 350. 1825. *Phreatia secunda* (Blume) Lindl., Gen. Sp. Orchid. Pl.: 64. 1830. TYPE: Indonesia, Java, Pantjar, *Blume s.n.* (holotype, L).

Distribution. Andaman & Nicobar Islands [also in Indochina, Indonesia, Thailand].

Habit. Epiphytic herb.

————

Pinalia Buch.-Ham. ex Lindl., Orchid. Scelet.: 14, 21, 23. 1826.

TYPE: *Pinalia alba* Buch.-Ham. ex Lindl. (= *Pinalia spicata* (D. Don) S. C. Chen & J. J. Wood).

Pinalia acervata (Lindl.) Kuntze, Revis. Gen. Pl. 2: 679. 1891. *Eria acervata* Lindl., Paxton's Fl. Gard. 1: 170. 1850. TYPE: Origin not known, *East India Company s.n.* (not found).

Distribution. Arunachal Pradesh, Assam, Manipur, Meghalaya, Mizoram, Nagaland, Sikkim, Tripura, West Bengal [Bhutan, China, Myanmar].

Habit. Epiphytic herb.

Pinalia acutifolia (Lindl.) Kuntze, Revis. Gen. Pl. 2: 679. 1891. *Eria acutifolia* Lindl., Edwards's Bot. Reg. 28(Misc.): 38. 1842. TYPE: India, *Messrs. Loddiges 209* (not found).

Distribution. India (no precise locality known). Endemic.

Habit. Epiphytic herb.

Remarks. This is an obscure species of uncertain provenance of which no material could be found. Lindley wrote that it should be "classed near" *Eria clavicaulis* (≡ *Cylindrolobus clavicaulis*) but it is said to have racemes as long as the leaves with distant flowers, which is quite unlike *E. clavicaulis*.

Pinalia apertiflora (Summerh.) A. N. Rao, Bull. Arunachal Forest Res. 26: 103. 2010. *Eria apertiflora* Summerh., Bull. Misc. Inform. Kew 1929: 9. 1929. TYPE: India, Assam, Sep. 1928, *cult. Royal Botanical Garden, Kew (leg. Hinde) s.n.* (holotype, K).

Distribution. Arunachal Pradesh, Assam, Meghalaya [Bhutan, Myanmar, Nepal].

Habit. Epiphytic herb.

Remarks. This is listed as a synonym of *Eria bipunctata* Lindl. by Singh et al. (2019).

Pinalia bicolor (Lindl.) Kuntze, Revis. Gen. Pl. 2: 679. 1891. *Eria bicolor* Lindl., Gen. Sp. Orchid. Pl.: 65. 1830. TYPE: Sri Lanka, Maturatam, *Macrae s.n.* (holotype, K-LINDL).

Distribution.   Tamil Nadu [Sri Lanka].

Habit.   Epiphytic herb.

Pinalia bilobulata (Seidenf.) Schuit., Y. P. Ng & H. A. Pedersen, Bot. J. Linn. Soc. 186: 196. 2018. *Eria bilobulata* Seidenf., Opera Bot. 62: 123. 1982. TYPE: Thailand: Mae Sarieng road at Km. 80, 3200 ft, *GT 7728* (holotype, C).

Distribution.   Odisha [also in Thailand].

Habit.   Epiphytic herb.

Pinalia bipunctata (Lindl.) Kuntze, Revis. Gen. Pl. 2: 679. 1891. *Eria bipunctata* Lindl., Edwards's Bot. Reg. 27(Misc.): 83. 1841. TYPE: India, Chiree District, Khasi Hills, *J. Gibson s.n.* (not found).

Distribution.   Arunachal Pradesh, Assam, Meghalaya, Nagaland, Sikkim, Uttarakhand, West Bengal [China, Myanmar, Nepal].

Habit.   Epiphytic herb.

Pinalia bractescens (Lindl.) Kuntze, Revis. Gen. Pl. 2: 679. 1891. *Eria bractescens* Lindl., Edwards's Bot. Reg. 27(Misc.): 18. 1841. SYNTYPES: Singapore, *Messrs. Loddiges 214* (not found); Myanmar, *Griffith 1055* (syntype, K-LINDL).

Eria bractescens Lindl. var. *kurzii* Hook. f., Fl. Brit. India [J. D. Hooker] 5(16): 797. 1890. TYPE: India, Andaman Islands, icon. *T. Anderson s.n.* (holotype, CAL).

Distribution.   Andaman & Nicobar Islands, Arunachal Pradesh, Assam, Meghalaya, Nagaland, Sikkim, West Bengal [Myanmar, Nepal].

Habit.   Epiphytic herb.

Pinalia connata (J. Joseph, S. N. Hegde & Abbar.) Ormerod & E. W. Wood, Harvard Pap. Bot. 15: 351. 2010. *Eria connata* J. Joseph, S. N. Hegde & Abbar., Bull. Bot. Surv. India 24(1–4): 114. 1982 [1983]. TYPE: India, Arunachal Pradesh, Khellong, 1100 m, 7 Aug. 1980, *N. R. Abbareddy 62391A* (holotype, CAL; isotype, ASSAM, MH).

Distribution.   Arunachal Pradesh, Assam [Bhutan].

Habit.   Epiphytic herb.

Pinalia excavata (Lindl.) Kuntze, Revis. Gen. Pl. 2: 679. 1891. *Eria excavata* Lindl., Gen. Sp. Orchid. Pl.: 67. 1830. TYPE: Nepal, Bagmati Zone, Kathmandu District, Sheopore [Shivapuri], May 1821, *Wallich 1974* (holotype, K-LINDL; isotype, K, K-WALL).

Eria flava Lindl. var. *rubida* Lindl., J. Proc. Linn. Soc., Bot. 3: 49. 1859 [1858]. TYPE: India, Sikkim, *Cathcart s.n.* (holotype, K-LINDL).
Eria sphaerochila Lindl., J. Proc. Linn. Soc., Bot. 3: 54. 1859 [1858]. TYPE: India, Khasi Hills, 4000–6000 ft, *Hooker 69* (holotype, K-LINDL).

Distribution.   Arunachal Pradesh, Assam, Manipur, Meghalaya, Mizoram, Nagaland, Sikkim, West Bengal [Bhutan, China, Nepal].

Habit.   Epiphytic herb.

Pinalia globulifera (Seidenf.) A. N. Rao, Bull. Arunachal Forest Res. 26: 103. 2010. *Eria globulifera* Seidenf., Opera Bot. 62: 125. 1982. TYPE: Thailand, Doi Pae Poe, 3300–3800 ft, *Seidenfaden & Smitinand GT 7317* (holotype, C).

Distribution.   Arunachal Pradesh, Manipur, Nagaland, Tripura, Uttarakhand, West Bengal [Myanmar].

Habit.   Epiphytic herb.

Pinalia graminifolia (Lindl.) Kuntze, Revis. Gen. Pl. 2: 679. 1891. *Eria graminifolia* Lindl., J. Proc. Linn. Soc., Bot. 3: 54. 1859 [1858]. TYPE: India, Darjeeling, *Griffith s.n.* (holotype, K-LINDL; isotype, CAL).

Distribution.   Arunachal Pradesh, Nagaland, Sikkim, West Bengal [Bhutan, China, Myanmar, Nepal].

Habit.   Epiphytic herb.

Pinalia leucantha Kuntze, Revis. Gen. Pl. 2: 679. 1891. *Eria alba* Lindl., Gen. Sp. Orchid. Pl.: 67. 1830. *Octomeria alba* Wall. ex Hook. f., Fl. Brit. India [J. D. Hooker] 5(16): 795. 1890, nom. inval. TYPE: Nepal, Bagmati Zone, Kathmandu District, Sheopore [Shivapuri], *Wallich s.n.* (holotype, K-LINDL).

Distribution.   Arunachal Pradesh, Himachal Pradesh, Nagaland, Sikkim, Uttarakhand, West Bengal [Bhutan, Nepal].

Habit.   Epiphytic herb.

Remarks.   This is not *Pinalia alba* Buch.-Ham. ex Lindl.

Pinalia lineata (Lindl.) Kuntze, Revis. Gen. Pl. 2: 678. 1891. *Eria lineata* Lindl., J. Proc. Linn. Soc., Bot. 3: 53. 1858. SYNTYPES: ?Java, icon. *Veitch cult. s.n.* (syntype, K-LINDL), ?Continent of India, *Anonymous cult. s.n.* (not found).

Eria amica Rchb. f., Xenia Orchid. 2: 162. 1870. *Pinalia amica* (Rchb. f.) Kuntze, Revis. Gen. Pl. 2: 679. 1891. TYPE: India, Assam, *cult. J. Day s.n.* (holotype, W).

Eria andersonii Hook. f., Fl. Brit. India [J. D. Hooker] 5(16): 795. 1890 (as *"andersoni"*). *Pinalia andersonii* (Hook. f.) Kuntze, Rev. Gen. Pl. 2: 679. 1891. TYPE: India, Sikkim, Darjeeling, icon. *T. Anderson s.n.* (holotype, CAL).

Eria confusa Hook. f., Hooker's Icon. Pl. 19: t. 1850. 1889. *Pinalia confusa* (Hook. f.) Kuntze, Rev. Gen. Pl. 2: 679. 1891. SYNTYPES: Nepal, *Wallich s.n.* (syntype, K); India, Sikkim, alt. 4000–6000 ft, *J. D. Hooker s.n.* (syntype, K).

Distribution. Arunachal Pradesh, Assam, Manipur, Meghalaya, Mizoram, Nagaland, Sikkim, Uttarakhand, West Bengal [Bhutan, China, Myanmar, Nepal].

Habit. Epiphytic herb.

Pinalia meghasaniensis (S. Misra) Schuit., Y. P. Ng & H. A. Pedersen, Bot. J. Linn. Soc. 186: 197. 2018. *Eria bilobulata* Seidenf. subsp. *meghasaniensis* S. Misra, J. Orchid Soc. India 2(1–2): 49. 1988. *Eria meghasaniensis* (S. Misra) S. Misra, J. Orchid Soc. India 3(1–2): 69. 1989. TYPE: India, Odisha, Similipal forests, Meghasani Parbat, alt. 3500 ft, 19 Aug. 1987, *Sarat Misra SM 1348A* (holotype, CAL).

Distribution. Odisha. Endemic.

Habit. Epiphytic herb.

Pinalia merguensis (Lindl.) Kuntze, Rev. Gen. Pl. 2: 679. 1891. *Eria merguensis* Lindl., J. Proc. Linn. Soc., Bot. 3: 52. 1859 [1858]. *Mycaranthes merguensis* (Lindl.) Rauschert, Feddes Repert. 94: 456. 1983. TYPE: Myanmar, Mergui [Myeik], *Griffith 1034* (holotype, K).

Distribution. Mizoram [Bhutan, Myanmar].

Habit. Epiphytic herb.

Pinalia mysorensis (Lindl.) Kuntze, Revis. Gen. Pl. 2: 679. 1891. *Eria mysorensis* Lindl., J. Proc. Linn. Soc., Bot. 3: 54. 1859 [1858]. TYPE: India, Mysore, *Law s.n.* (holotype, K-LINDL; isotype, K).

Eria pubescens Wight, Icon. Pl. Ind. Orient. [Wight] 5(1): 4, t. 1634. 1851, nom. illeg., non (Hook.) Lindl. ex Loudon, 1830. TYPE: India, western slopes of the Nilgiri Hills, Aug.–Sep., *Wight s.n.* (holotype, K).

Distribution. Karnataka, Kerala, Maharashtra, Tamil Nadu. Endemic.

Habit. Epiphytic herb.

Pinalia obesa (Lindl.) Kuntze, Revis. Gen. Pl. 2: 679. 1891. *Eria obesa* Lindl., Gen. Sp. Orchid. Pl.: 68. 1830. *Trias obesa* (Lindl.) Mason, Burmah, ed. 3: 809. 1860. *Hymeneria obesa* (Lindl.) M. A. Clem.

& D. L. Jones, Orchadian 13(11): 501. 2002. TYPE: Myanmar, Ataran River, Martaban, 29 Jan. 1827, *Wallich 1976* (holotype, K-WALL).

Eria prainii Briq., Annuaire Conserv. Jard. Bot. Genève 4: 210. 1900. TYPE: India, "Sikkim," cult. *Geneva Botanic Garden s.n. (leg. Calcutta Botanic Garden)* (holotype, G).

Distribution. Assam, Mizoram [Myanmar].

Habit. Epiphytic herb.

Pinalia occidentalis (Seidenf.) Schuit., Y. P. Ng & H. A. Pedersen, Bot. J. Linn. Soc. 186: 197. 2018. *Eria occidentalis* Seidenf., Nordic J. Bot. 2(1): 15, fig. 1. 1982. TYPE: India, Kumaon, Pithoragarh, betw. Chowpta & Maitli, alt. 4500 ft, 31 July 1980, *Arora 70806* (holotype, BSD).

Distribution. Uttarakhand. Endemic.

Habit. Epiphytic herb.

Pinalia polystachya (A. Rich.) Kuntze, Revis. Gen. Pl. 2: 679. 1891. *Eria polystachya* A. Rich., Ann. Sci. Nat., Bot. sér. 2, 15: 20, t. 9. 1841. TYPE: India, Nilgiri Hills near Neddoubetta, Sep. 1840, *Perrottet s.n.* (holotype, P).

Distribution. Karnataka, Kerala, Maharashtra, Tamil Nadu. Endemic.

Habit. Epiphytic herb.

Pinalia pumila (Lindl.) Kuntze, Revis. Gen. Pl. 2: 679. 1891. *Eria pumila* Lindl., Gen. Sp. Orchid. Pl.: 68. 1830. TYPE: India, Pundua, *Wallich 1972 (leg. De Silva s.n.)* (holotype, K-LINDL; isotype, K, K-WALL).

Distribution. Arunachal Pradesh, Assam, Meghalaya, Sikkim, West Bengal [Bhutan, Myanmar].

Habit. Epiphytic herb.

Pinalia sharmae (H. J. Chowdhery, G. S. Giri & G. D. Pal) A. N. Rao, Bull. Arunachal Forest Res. 26: 104. 2010. *Eria sharmae* H. J. Chowdhery, G. S. Giri & G. D. Pal, Indian J. Forest. 16(1): 91. 1993. TYPE: India, Arunachal Pradesh, Lower Subansiri District, Itanagar, alt. 1600 ft, 20 Feb. 1992, *H. J. Chowdhery 1726* (holotype, CAL).

Distribution. Arunachal Pradesh. Endemic.

Habit. Epiphytic herb.

Pinalia spicata (D. Don) S. C. Chen & J. J. Wood, Fl. China 25: 354. 2009. *Octomeria spicata* D. Don,

Prodr. Fl. Nepal.: 31. 1825. *Pinalia alba* Buch.-Ham. ex Lindl., Coll. Bot. (Lindley) 8: t. 41B. 1826. *Eria convallarioides* Lindl., Gen. Sp. Orchid. Pl.: 70. 1830, nom. superfl. *Eria spicata* (D. Don) Hand.-Mazz., Symb. Sin. 7(5): 1353. 1936 (as "*spicatae*"). SYNTYPES: Nepal, Bagmati Zone, Narainhetty, *Buchanan-Hamilton s.n.* (not found); Nepal, Toka, Aug. 1821, *Wallich s.n.* [*1975*] (syntypes, K-LINDL, K-WALL).

Distribution.   Arunachal Pradesh, Assam, Himachal Pradesh, Manipur, Meghalaya, Mizoram, Nagaland, Sikkim, Uttar Pradesh, Uttarakhand, West Bengal [Bhutan, China, Myanmar, Nepal].

Habit.   Epiphytic herb.

———

Platanthera Rich., De Orchid. Eur. 20, 26, 35. 1817, nom. cons.

TYPE: *Platanthera bifolia* (L.) Rich. (*Orchis bifolia* L.).

Platanthera bakeriana (King & Pantl.) Kraenzl., Orchid. Gen. Sp. 1: 632. 1899. *Habenaria bakeriana* King & Pantl., J. Asiat. Soc. Bengal, Pt. 2, Nat. Hist. 65: 132. 1896. TYPE: India, Sikkim, Lachen Valley, alt. 9000 ft, July, *Pantling 401* (lectotype, CAL, designated by Efimov, 2016; isotype, CAL, K).

Distribution.   Arunachal Pradesh, Meghalaya, Mizoram, Sikkim, West Bengal [Bhutan, China, Myanmar, Nepal].

Habit.   Terrestrial herb.

Platanthera bhutanica K. Inoue, J. Jap. Bot. 61: 193. 1986. TYPE: Bhutan, Parshary Jiurpe, alt. 13,000 ft, 13 July 1914, *R. E. Cooper & A. K Bulley 2596* (holotype, F; isotype, BM).

Distribution.   Meghalaya [Bhutan, China].

Habit.   Terrestrial herb.

Platanthera calceoliformis (W. W. Sm.) X. H. Jin, Schuit. & W. T. Jin, Molec. Phylogen. Evol. 77: 51. 2014. *Herminium calceoliforme* W. W. Sm., Notes Roy. Bot. Gard. Edinburgh 12: 211. 1921. *Smithorchis calceoliformis* (W. W. Sm.) Tang & F. T. Wang., Bull. Fan Mem. Inst. Biol. Bot. 7: 140. 1936. TYPE: China, Yunnan, Mekong-Yangtze divide, Kari Pass, *Forrest 13110* (lectotype, E 381991, designated by Efimov, 2016; isotype, E 381992).

Distribution.   Sikkim [China].

Habit.   Terrestrial herb.

Platanthera concinna (Hook. f.) Kraenzl., Orchid. Gen. Sp. 1: 621. 1899. *Habenaria concinna* Hook. f., Fl. Brit. India [J. D. Hooker] 6(17): 155. 1890. SYNTYPES: India, Khasi Hills, Kala-pane, alt. 5000 ft, *Hooker & Thomson 283* (lectotype, K-LINDL, designated by Efimov, 2016).

Habenaria dyeriana King & Pantl., J. Asiat. Soc. Bengal, Pt. 2, Nat. Hist. 65: 133. 1896. *Platanthera dyeriana* (King & Pantl.) Kraenzl., Orchid. Gen. Sp. 1: 636. 1899. *Platanthera dyeriana* (King & Pantl.) Pradhan, Indian Orchids: Guide Identif. & Cult. 2: 682. 1979 (as "*dyerana*"), isonym. TYPE: India, Sikkim, Lachen valley, 12,000 ft, Aug. 1895, *Pantling 407* (lectotype, CAL, designated by Singh et al., 2014; isotype, AMES, CAL, G, K, L, LE, W).

Distribution.   Arunachal Pradesh, Meghalaya, Sikkim [China].

Habit.   Terrestrial herb.

Platanthera dulongensis X. H. Jin & Efimov, Nordic J. Bot. 30: 294. 2012. TYPE: China, Yunnan, Gongshan County, *Jin 8386* (holotype, PE).

Platanthera fugongensis Ormerod, Taiwania 58: 29. 2013. TYPE: China, Yunnan, Fugong Xian, Lishadi Xian, Yaduo Cun, above Shidali along the N side of S fork of Yam He, E side of Gaoligong Shan, *Gaoligong Shan Biodiversity Survey Team 28318* (holotype, CAS).

Distribution.   Arunachal Pradesh, West Bengal [China].

Habit.   Terrestrial herb.

Remarks.   This species was recently recorded from West Bengal (Swami, 2016, as *Platanthera fugongensis*).

Platanthera japonica (Thunb.) Lindl., Gen. Sp. Orchid. Pl.: 290. 1835. *Orchis japonica* Thunb., Syst. Veg., ed. 14: 811. 1784. TYPE: Japan, *Thunberg s.n.* (holotype, UPS: Herb. Thunberg 21205).

Platanthera arcuata Lindl., Gen. Sp. Orchid. Pl.: 289. 1835. *Habenaria arcuata* (Lindl.) Hook. f., Fl. Brit. India [J. D. Hooker] 6(17): 155. 1890. TYPE: India, Mussoorie, *Royle s.n.* (holotype, LIV).

Distribution.   Himachal Pradesh, Nagaland, Uttarakhand [Bhutan, China, Nepal].

Habit.   Terrestrial herb.

Platanthera leptocaulon (Hook. f.) Soó, Ann. Hist.-Nat. Mus. Natl. Hung. 26: 360. 1929. *Habenaria leptocaulon* Hook. f., Fl. Brit. India [J. D. Hooker] 6(17): 154. 1890. TYPE: India, Sikkim Himalaya, Lachen Valley, alt. 10,000–11,000 ft, *J. D. Hooker* *s.n.* [*315*] (lectotype, K, barcode 247392, designated by Efimov, 2016; ?isotype, K).

Distribution.   Arunachal Pradesh, Sikkim, Uttarakhand, West Bengal [Bhutan, China, Myanmar, Nepal].

Habit.   Terrestrial herb.

Platanthera nematocaulon (Hook. f.) Kraenzl., Orchid. Gen. Sp. 1: 942. 1901. *Habenaria nematocaulon* Hook. f., Fl. Brit. India [J. D. Hooker] 6(17): 154. 1890. *Peristylus nematocaulon* (Hook. f.) Banerji & Prabha Pradhan, Orchids Nepal Himalaya 106. 1984. TYPE: India, Sikkim Himalaya, Rechila, alt. 10,000–12,000 ft, *J. D. Hooker s.n.* [*14*] (lectotype, K 387513, designated by Efimov & Jin, 2014; isotype, K 387512).

Habenaria juncea King & Pantl., J. Asiat. Soc. Bengal, Pt. 2, Nat. Hist. 65: 132. 1896. *Platanthera juncea* (King & Pantl.) Kraenzl., Orchid. Gen. Sp. 1: 942. 1901. TYPE: Sikkim, Lachen Valley, alt. 11,000 ft, Aug. 1894, *Pantling 406* (syntypes, AMES, CAL, E, L, LE, W).

Distribution.   Meghalaya, Nagaland, Sikkim, West Bengal [Bhutan, China, Myanmar, Nepal].

Habit.   Terrestrial herb.

Platanthera orbicularis (Hook. f.) X. H. Jin, Schuit. & Raskoti, PhytoKeys 79: 72. 2017. *Herminium orbiculare* Hook. f., Fl. Brit. India [J. D. Hooker] 6(17): 130. 1890. *Monorchis orbicularis* (Hook. f.) O. Schwarz, Mitt. Thüring. Bot. Ges. 1: 95. 1949. *Peristylus orbicularis* (Hook. f.) Agrawala, H. J. Chowdhery & S. Choudhury, Kew Bull. 65(1): 106. 2010. TYPE: Sikkim Himalaya, Chumbi Valley, Rungboo, Aug. 1884, *King's collector 164* (holotype, K; isotype, CAL).

Distribution.   Sikkim [Bhutan, China, Myanmar, Nepal].

Habit.   Terrestrial herb.

Platanthera pachycaulon (Hook. f.) Soó, Ann. Hist.-Nat. Mus. Natl. Hung. 26: 364. 1929. *Habenaria pachycaulon* Hook. f., Fl. Brit. India [J. D. Hooker] 6(17): 154. 1890. TYPE: India, Sikkim Himalaya, Nattong, 12 July 1877, *King's collectors 4345* (lectotype, K 247394, designated by Efimov & Jin, 2014; isotype, CAL).

Habenaria oligantha Hook. f., Fl. Brit. India [J. D. Hooker] 6(17): 154. 1890, nom. illeg., non Hochst. ex Engl., 1892. TYPE: India, Sikkim Himalaya, interior valleys, alt. 10,000–12,000 ft, *J. D. Hooker s.n.* [*311*] (lectotype, K 247391, designated by Efimov, 2016; isotype, K).
Platanthera exelliana Soó, Ann. Hist.-Nat. Mus. Natl. Hung. 26: 359. 1929. *Platanthera elachyantha* Tang & F. T.

Wang, Acta Phytotax. Sin. 1: 58. 1951, nom. superfl. TYPE: India, Sikkim, *Hooker 311* (holotype, K 247391, designated by Efimov, 2016; isotype, K-LINDL).

Distribution.   Sikkim, Uttarakhand, West Bengal [Bhutan, China, Myanmar, Nepal].

Habit.   Terrestrial herb.

Platanthera sikkimensis (Hook. f.) Kraenzl., Orchid. Gen. Sp. 1: 621. 1899. *Habenaria sikkimensis* Hook. f., Fl. Brit. India [J. D. Hooker] 6(17): 155. 1890. *Platanthera sikkimensis* (Hook. f.), Pradhan, Indian Orchids: Guide Identif. & Cult. 2: 675. 1979, isonym. TYPE: India, Sikkim Himalaya, Sinchal, alt. 8000–9000 ft, *T. Thomson s.n.* (not found).

Distribution.   Sikkim [China, Nepal].

Habit.   Terrestrial or epiphytic herb.

Platanthera stenantha (Hook. f.) Soó, Ann. Hist.-Nat. Mus. Natl. Hung. 26: 363. 1929. *Habenaria stenantha* Hook. f., Fl. Brit. India [J. D. Hooker] 6(17): 153. 1890. *Hemihabenaria stenantha* (Hook. f.) Finet, Rev. Gén. Bot. 13: 532. 1902. TYPE: India, Sikkim, Jongri, 15 Oct. 1875, *Clarke 25954* (lectotype, K, designated by Efimov & Jin, 2014).

Distribution.   Arunachal Pradesh, Nagaland, Sikkim, Uttarakhand, West Bengal [Bhutan, China, Myanmar, Nepal].

Habit.   Terrestrial herb.

Platanthera stenochila X. H. Jin, Schuit., Raskoti & Lu Q. Huang, Cladistics 32(2): 210. 2015. *Herminium angustilabre* King & Pantl., J. Asiat. Soc. Bengal, Pt. 2, Nat. Hist. 65: 131. 1896. *Monorchis angustilabris* (King & Pantl.) O. Schwarz, Mitt. Thüring. Bot. Ges. 1: 95. 1949. *Androcorys angustilabris* (King & Pantl.) Agrawala & H. J. Chowdhery, Kew Bull. 65(1): 105. 2010. *Platanthera angustilabris* (King & Pantl.) X. H. Jin, Schuit. & W. T. Jin, Molec. Phylogen. Evol. 77: 51. 2014, nom. illeg., non Seidenf., 1995. TYPE: India, Sikkim, Lingtu, 11,000 ft, June, *Pantling 375* (holotype, CAL; isotype, CAL, ?K, W).

Distribution.   Sikkim, West Bengal [China].

Habit.   Terrestrial herb.

Platanthera superantha (J. J. Wood) X. H. Jin, Schuit., Raskoti & Lu Q. Huang, Cladistics 32(2): 210. 2015. *Peristylus superanthus* J. J. Wood, Kew Bull. 41(4): 811, fig. 1. 1986. TYPE: Nepal, *Grey-Wilson et al. 4104* (holotype, K).

Habenaria nematocaulon sensu King & Pantl., Ann. Roy. Bot. Gard. (Calcutta) 8: 316, t. 416. 1898, non Hook. f., 1890.

Distribution. Assam, Nagaland, Sikkim [Bhutan, Nepal].

Habit. Terrestrial herb.

Remarks. This species is illustrated as *Platanthera nematocaulon* (Hook. f.) Kraenzl. in Swami (2016).

Platanthera urceolata (C. B. Clarke) R. M. Bateman, Ann. Bot. (Oxford) 104: 439. 2009. *Habenaria urceolata* C. B. Clarke, J. Linn. Soc., Bot. 25: 73, t. 30. 1889. *Diphylax urceolata* (C. B. Clarke) Hook. f., Hooker's Icon. Pl. 19: t. 1865. 1889. TYPE: India, Naga Hills, Jakpho, alt. 9000 ft, *C. B. Clarke 41272c* (lectotype, K 247409, designated by Efimov 2016; isotype, CAL [*Clarke 41272a*], K [*Clarke 41272b*]).

Distribution. Arunachal Pradesh, Assam, Nagaland, Sikkim, ?Uttarakhand, West Bengal [Bhutan, China, Myanmar, Nepal].

Habit. Terrestrial herb.

———

Pleione D. Don, Prodr. Fl. Nepal.: 36. 1825.

TYPE: *Pleione praecox* (Sm.) D. Don (*Epidendrum praecox* Sm.).

Pleione arunachalensis Hareesh, P. Kumar & M. Sabu, Phytotaxa 291: 294. 2017. TYPE: India, Arunachal Pradesh, Lower Dibang Valley district, Mayodia, 2100 m, 6 May 2016, *V. S. Hareesh 143761* (holotype, CALI; isotype, CAL, CALI).

Distribution. Arunachal Pradesh. Endemic.

Habit. Terrestrial herb.

Remarks. This species was illustrated as *Pleione saxicola* Tang & F. T. Wang ex S. C. Chen in Singh et al. (2019).

Pleione grandiflora (Rolfe) Rolfe, Orchid Rev. 11: 291. 1903. *Coelogyne grandiflora* Rolfe, J. Linn. Soc., Bot. 36: 22. 1903. TYPE: China: Yunnan, Mengtze, northern mountain forests, alt. 8500 ft, *A. Henry 11116* (holotype, K).

Distribution. Uttarakhand [China].

Habit. Terrestrial herb.

Pleione hookeriana (Lindl.) Rollisson, Nursery Cat. (Rollisson) 1875–1876: 39. 1875. *Coelogyne hook-*

eriana Lindl., Fol. Orchid. 5(*Coelogyne*): 14. 1854. TYPE: India, Sikkim, Darjeeling, 7000–10,000 ft, *J. D. Hooker s.n.* [*74*] (holotype, K-LINDL; isotype, CAL, K).

Coelogyne hookeriana Lindl. var. *brachyglossa* Rchb. f., Gard. Chron., ser. 3, 1: 833. 1887. *Pleione hookeriana* (Lindl.) Rollisson var. *brachyglossa* (Rchb. f.) Rolfe, Orchid Rev. 11: 291. 1903. TYPE: India, Sikkim, *cult. Lawrence s.n.* (holotype, W).

Distribution. Arunachal Pradesh, Assam, Manipur, Nagaland, Sikkim, Uttarakhand, West Bengal [Bhutan, China, Myanmar, Nepal].

Habit. Epiphytic or lithophytic herb.

Pleione humilis (Sm.) D. Don, Prodr. Fl. Nepal.: 37. 1825. *Epidendrum humile* Sm., Exot. Bot. 2: 75, t. 98. 1806. *Dendrobium humile* (Sm.) Sm., Cycl. (Rees) 11: 4. 1808. *Coelogyne humilis* (Sm.) Lindl., Coll. Bot. (Lindley) t. 37. 1826. TYPE: Nepal, *Buchanan-Hamilton s.n.* (holotype, LINN).

Pleione humilis (Sm.) D. Don var. *amitii* R. Pal, Dayama & Medhi, J. Orchid Soc. India 25: 73. 2011, nom. inval. TYPE: India, West Bengal, Chatakhpur, *Ram Pal, M. Dayama & R. P. Medhi s.n.* (NRCO, CAL, no holotype designated).

Distribution. Arunachal Pradesh, Assam, Manipur, Meghalaya, Mizoram, Nagaland, Sikkim, Uttarakhand, West Bengal [Bhutan, China, Myanmar, Nepal].

Habit. Epiphytic or lithophytic herb.

Pleione × lagenaria Lindl. & Paxton, Paxton's Fl. Gard. 2: 5, t. 39, fig. 2. 1851. *Coelogyne × lagenaria* (Lindl. & Paxton) Lindl., Fol. Orchid. 5 (*Coelogyne*): 15. 1854. TYPE: Icon.: Paxton's Fl. Gard. 2: 5, t. 39, fig. 2. 1851.

Distribution. Assam, Meghalaya [?China].

Habit. Epiphytic or lithophytic herb.

Remarks. This taxon is a natural hybrid: *Pleione maculata × P. praecox.*

Pleione maculata (Lindl.) Lindl. & Paxton, Paxton's Fl. Gard. 2: 5, t. 39, fig. 1. 1851. *Coelogyne maculata* Lindl., Gen. Sp. Orchid. Pl.: 43. 1830. TYPE: India, Pundua, Dearung, Mustock Punjee Hill, Oct. 1825, *Wallich 1964 (leg. De Silva 1579)* (holotype, K-WALL).

Pleione diphylla Lindl. & Paxton, Paxton's Fl. Gard. 2: 66. 1851. *Coelogyne diphylla* (Lindl. & Paxton) Lindl., Fol. Orchid. 5(*Coelogyne*): 15. 1854. TYPE: India, Khasi Hills, *Griffith 29/138* (holotype, K).

Distribution. Arunachal Pradesh, Assam, Manipur, Meghalaya, Mizoram, Nagaland, Sikkim, West Bengal [Bhutan, China, Myanmar, Nepal].

Habit. Epiphytic herb.

Pleione praecox (Sm.) D. Don, Prodr. Fl. Nepal.: 37. 1825. *Epidendrum praecox* Sm., Exot. Bot. 2: 73, t. 97. 1806. *Dendrobium praecox* (Sm.) Sm., Cycl. (Rees) 11: 3. 1808. *Coelogyne praecox* (Sm.) Lindl., Coll. Bot. (Lindley) t. 37. 1826. TYPE: Nepal, *Buchanan-Hamilton s.n.* (holotype, LINN).

Coelogyne wallichiana Lindl., Gen. Sp. Orchid. Pl.: 43. 1830. *Pleione wallichiana* (Lindl.) Lindl. & Paxton, Paxton's Fl. Gard. 2: 66. 1851. *Coelogyne praecox* (Sm.) Lindl. var. *wallichiana* (Lindl.) Lindl., Fol. Orchid. 5(*Coelogyne*): 16. 1854. *Pleione praecox* (Sm.) D. Don var. *wallichiana* (Lindl.) E. W. Cooper, Roy. Hort. Soc. Dict. Gard.: 1606. 1951. TYPE: India, Cherrapunji, Oct. 1829, *Wallich 1965 (leg. De Silva 114)* (holotype, K-WALL).

Distribution. Arunachal Pradesh, Assam, Manipur, Meghalaya, Mizoram, Nagaland, Sikkim, Uttarakhand, West Bengal [Bhutan, China, Myanmar, Nepal].

Habit. Epiphytic or lithophytic herb.

Pleione saxicola Tang & F. T. Wang ex S. C. Chen, Acta Phytotax. Sin. 25: 473. 1987. TYPE: China, Yunnan, 2400–2500 m, *K. M. Fang 7914* (holotype, PE; isotype, KUN).

Distribution. Arunachal Pradesh [Bhutan, China].

Habit. Terrestrial or lithophytic herb.

Remarks. The photo labeled as *Pleione saxicola* in Singh et al. (2019) shows *P. arunachalensis*.

Pleione scopulorum W. W. Sm., Notes Roy. Bot. Gard. Edinburgh 113: 218. 1921. TYPE: China, Yunnan, *Forrest 14230* (holotype, E; isotype, K).

Distribution. Arunachal Pradesh, ?Assam [China, Myanmar].

Habit. Terrestrial or lithophytic herb.

———

Plocoglottis Blume, Bijdr. Fl. Ned. Ind. 8: 380. 1825.

TYPE: *Plocoglottis javanica* Blume.

Plocoglottis javanica Blume, Bijdr. Fl. Ned. Ind. 8: 381. 1825. SYNTYPES: Indonesia, Java, Mt. Salak, Mt. Pantjar, etc., *Blume s.n.* (syntype, L [2×]).

Distribution. Andaman & Nicobar Islands [also in Indonesia].

Habit. Terrestrial herb.

Plocoglottis lowii Rchb. f., Gard. Chron. 1865: 434. 1865. TYPE: Borneo, *Hugh Low s.n.* (holotype, W).

Plocoglottis porphyrophylla Ridl., Trans. Linn. Soc. London, Bot. Ser. 2. 3: 368. 1893. TYPE: Malaysia, Pekan, *Ridley s.n.* (holotype, SING).

Distribution. ?Andaman & Nicobar Islands [also in Indonesia].

Habit. Terrestrial herb.

Remarks. This species is listed as "excluded" in Singh et al. (2019).

———

Podochilus Blume, Bijdr. Fl. Ned. Ind. 7: 295. 1825.

TYPE: *Podochilus lucescens* Blume.

Podochilus cultratus Lindl., Gen. Sp. Orchid. Pl.: 234. 1833. TYPE: Nepal, Hetounda, 14 Dec. 1820, *Wallich 7336* (holotype, K-LINDL; isotype, K-WALL).

Distribution. Arunachal Pradesh, Assam, ?Meghalaya, Mizoram, Sikkim, Tripura, West Bengal [Bhutan, China, Myanmar, Nepal].

Habit. Epiphytic herb.

Podochilus falcatus Lindl., Gen. Sp. Orchid. Pl.: 234. 1833. TYPE: Sri Lanka, Mt. Nuera Ellia, *Macrae s.n.* (holotype, K-LINDL).

Distribution. ?Kerala, ?Tamil Nadu [Sri Lanka].

Habit. Epiphytic herb.

Remarks. This species is listed as "excluded" in Singh et al. (2019).

Podochilus khasianus Hook. f., Fl. Brit. India [J. D. Hooker] 6(17): 81. 1890. SYNTYPES: India, Sylhet, Pundua Hills, Aug. 1824, *Wallich 7335B (leg. De Silva 1435)* (syntype, K-WALL); Khasi Hills, Amwee, *Hooker & Thomson s.n.* [4] (syntypes, CAL, K, P).

Distribution. Arunachal Pradesh, Assam, Meghalaya, Sikkim, West Bengal [Bhutan, China, Myanmar].

Habit. Epiphytic herb.

Podochilus malabaricus Wight, Icon. Pl. Ind. Orient. [Wight] 5(1): 20, t. 1748-2. 1851. TYPE: India, Malabar, fl. during the rainy months, *Jerdon s.n.*

Distribution.   Kerala, Tamil Nadu [Sri Lanka].

Habit.   Epiphytic herb.

Podochilus microphyllus Lindl., Gen. Sp. Orchid. Pl.: 234. 1833. TYPE: Malaysia, Penang, *Wallich 7335A (leg. Porter)* (holotype, K-LINDL; isotype, K-WALL).

Distribution.   Andaman & Nicobar Islands [Myanmar].

Habit.   Epiphytic herb.

————

Pogonia Juss., Gen. Pl. [Jussieu]: 65. 1789.

TYPE: *Pogonia ophioglossoides* (L.) Ker Gawl. (*Arethusa ophioglossoides* L.).

Pogonia ?japonica Rchb. f., Linnaea 25: 228. 1852. TYPE: Japan, collector not named.

Distribution.   Arunachal Pradesh [China].

Habit.   Terrestrial herb.

Remarks.   The Indian material illustrated by Bhaumik and Satyanarayana (2014) and also by Singh et al. (2019) as *Pogonia japonica* looks quite different from the species as it is known from Japan, Korea, and eastern China. It may be referable to the Chinese *P. parvula* Schltr., which is currently regarded as a synonym of *P. japonica.*

————

Polystachya Hook., Exot. Fl. 2: t. 103. 1824, nom. cons.

TYPE: *Polystachya luteola* Hook., nom. superfl. (*Epidendrum minutum* Aublet).

Polystachya concreta (Jacq.) Garay & H. R. Sweet, Orquideologia 9: 206. 1974. *Epidendrum concretum* Jacq., Enum. Syst. Pl. 30. 1760. TYPE: Martinique, *Privault 136* (neotype, P, designated by Mytnik-Ejsmont & Baranow, 2010).

Onychium flavescens Blume, Bijdr. Fl. Ned. Ind. 7: 325. 1825. *Polystachya flavescens* (Blume) J. J. Sm., Orch. Java 6: 284, fig. 218. 1905. SYNTYPES: Indonesia, Java, Mt. Salak & Mt. Seribu, *Kuhl & van Hasselt s.n.* (syntypes, L [2×]).
Polystachya purpurea Wight, Icon. Pl. Ind. Orient. [Wight] 5(1): 10, t. 1679. 1851. TYPE: India, Tamil Nadu, top of Iyamally Hill, alt. 3000 ft, June, *Wight s.n.* (holotype, K).

Distribution.   Andaman & Nicobar Islands, Andhra Pradesh, Arunachal Pradesh, Karnataka, Manipur, Odisha, Tamil Nadu [China, Myanmar, Sri Lanka].

Habit.   Epiphytic herb.

Polystachya seidenfadeniana Mytnik & Baranow, Pl. Syst. Evol. 290: 59. 2010. TYPE: India, Tamil Nadu, Salem District, Attur Periakalrayans, alt. 950 m, 1 July 1978, *Matthew 15260* (holotype, C; isotype, RHT).

Distribution.   Tamil Nadu. Endemic.

Habit.   Epiphytic herb.

Polystachya wightii Rchb. f., Ann. Bot. Syst. (Walpers) 6: 640. 1863. *Dendrorkis wightii* (Rchb. f.) Kuntze, Revis. Gen. Pl. 2: 658. 1891 (as "*Dendrorchis*"). TYPE: India, Tamil Nadu, Iyamally Hills near Coimbatore, Sep. 1852, *Wight 2990* (lectotype, K; isotype, C, W, designated by Mytnik-Ejsmont & Baranow, 2010).

Distribution.   Kerala, Tamil Nadu. Endemic.

Habit.   Epiphytic herb.

————

Pomatocalpa Breda, Gen. Sp. Orchid. Asclep. t. [15]. 1829.

TYPE: *Pomatocalpa spicatum* Breda.

Pomatocalpa decipiens (Lindl.) J. J. Sm., Natuurk. Tijdschr. Ned.-Indië 72: 103. 1912. *Cleisostoma decipiens* Lindl., Edwards's Bot. Reg. 30(Misc.): 11. 1844. *Saccolabium decipiens* (Lindl.) Alston in H. Trimen, Handb. Fl. Ceylon 6(Suppl.): 278. 1931. TYPE: Sri Lanka, *Fielding s.n.* (holotype, K-LINDL).

Distribution.   Odisha [Bangladesh, Sri Lanka].

Habit.   Epiphytic herb.

Remarks.   Watthana (2007) suggested that *Pomatocalpa decipiens* falls within the range of variation of *P. spicatum.*

Pomatocalpa maculosum (Lindl.) J. J. Sm. subsp. **andamanicum** (Hook. f.) Watthana, Taiwania 51: 8. 2006. *Cleisostoma andamanicum* Hook. f., Fl. Brit. India [J. D. Hooker] 6(17): 71. 1890. *Pomatocalpa andamanicum* (Hook. f.) J. J. Sm., Natuurk. Tijdschr. Ned.-Indië 72: 103. 1912 (as "*andamanum*"). TYPE: South Andaman Islands, *Kurz s.n.* (holotype, K; isotype, CAL).

Distribution. Andaman & Nicobar Islands [also in Malaysia].

Habit. Epiphytic herb.

Pomatocalpa spicatum Breda, Gen. Sp. Orchid. Asclep. 3: t. [15.] 1829. TYPE: Indonesia, Java, Mt. Barangrang, Breda (1829), pl. [15] (lectotype, designated by Pearce & Cribb, 2002).

Cleisostoma wendlandorum Rchb. f., Allg. Gartenzeitung 24(28): 219. 1856. *Pomatocalpa wendlandorum* (Rchb. f.) J. J. Sm., Natuurk. Tijdschr. Ned.-Indië 72: 108. 1912. TYPE: Indonesia, Java, *cult. Wendland s.n.* (holotype, W).
Cleisostoma mannii Rchb. f., Flora 55: 274. 1872. *Pomatocalpa mannii* (Rchb. f.) J. J. Sm., Natuurk. Tijdschr. Ned.-Indië 72: 105. 1912. TYPE: India, Assam, *G. Mann s.n.* [*11*] (holotype, W).

Distribution. Andaman & Nicobar Islands, Andhra Pradesh, Arunachal Pradesh, Assam, Kerala, Sikkim, Tripura, West Bengal [Bhutan, China, Myanmar].

Habit. Epiphytic herb.

Pomatocalpa undulatum (Lindl.) J. J. Sm., Natuurk. Tijdschr. Ned.-Indië 72: 107. 1912. *Saccolabium undulatum* Lindl., Gen. Sp. Orchid. Pl.: 222. 1833. *Cleisostoma undulatum* (Lindl.) Rchb. f., Flora 55: 274. 1872. *Gastrochilus undulatus* (Lindl.) Kuntze, Revis. Gen. Pl. 2: 661. 1891. *Cleisostoma undulatum* (Lindl.) Tang & F. T. Wang, Acta Phytotax. Sin. 1(1–2): 100. 1951, isonym. TYPE: Bangladesh or India, Sylhet, Mar. 1822, *Wallich 7301 (leg. De Silva, Gomez & Bruce 1031)* (holotype, K-LINDL; isotype, BM, K-WALL, W).

Cleisostoma loratum Rchb. f., Flora 55: 273. 1872. *Pomatocalpa loratum* (Rchb. f.) J. J. Sm., Natuurk. Tijdschr. Ned.-Indië 72: 105. 1912. TYPE: India, Assam, *Mann s.n.* (holotype, W).

Distribution. Assam, Sikkim, West Bengal [Bangladesh].

Habit. Epiphytic herb.

——

Ponerorchis Rchb. f., Linnaea 25: 227. 1852.

TYPE: *Ponerorchis graminifolia* Rchb. f.

Ponerorchis chusua (D. Don) Soó, Acta Bot. Acad. Sci. Hung. 12: 352. 1966. *Orchis chusua* D. Don, Prodr. Fl. Nepal.: 23. 1825. *Gymnadenia puberula* Lindl., Numer. List [Wallich] n. 7059. 1832, nom. inval. *Gymnadenia chusua* (D. Don) Lindl., Numer. List [Wallich] n. 7058. 1832. *Habenaria chusua*

(D. Don) Benth., J. Linn. Soc., Bot. 18: 355. 1881. *Hemipilia chusua* (D. Don) Y. Tang & H. Peng, BioMed Centr. Evol. Biol. 15(96): 27. 2015. TYPE: Nepal, Bagmati Zone, Rashuwa District, Gosainthan [Shishapangma], Aug. 1821, *Wallich s.n.* [*7058*] (holotype, K-WALL).

Distribution. Arunachal Pradesh, Assam, Himachal Pradesh, Sikkim, Uttarakhand [Bhutan, China, Myanmar, Nepal].

Habit. Terrestrial herb.

Ponerorchis cucullata (L.) X. H. Jin, Schuit. & W. T. Jin var. **calcicola** (W. W. Sm.) X. H. Jin, Schuit. & W. T. Jin, Molec. Phylogen. Evol. 77: 51. 2014. *Gymnadenia calcicola* W. W. Sm., Notes Roy. Bot. Gard. Edinburgh 8(38): 188. 1914. *Neottianthe calcicola* (W. W. Sm.) Schltr., Acta Horti Gothob. 1: 136. 1924. *Neottianthe cucullata* (L.) Schltr. var. *calcicola* (W. W. Sm.) Soó, Ann. Hist.-Nat. Mus. Natl. Hung. 26: 353. 1929. *Hemipilia calcicola* (W. W. Sm.) Y. Tang & H. Peng, BioMed Centr. Evol. Biol. 15(96): 28. 2015. SYNTYPES: China, Yunnan, Lichiang Range, eastern flank, 27°35′ N, alt. 13,000 ft, Sep. 1910, *G. Forrest 6536* (syntypes, E, K), *7375* (syntype, E).

Distribution. Sikkim, Uttarakhand [Bhutan, China, Nepal].

Habit. Terrestrial herb.

Ponerorchis nana (King & Pantl.) Soó, Acta Bot. Acad. Sci. Hung. 12: 353. 1966. *Orchis chusua* D. Don var. *nana* King & Pantl., Ann. Roy. Bot. Gard. (Calcutta) 8: 304. 1898. *Gymnadenia chusua* (D. Don) Lindl. var. *nana* (King & Pantl.) Finet, Rev. Gén. Bot. 13: 514. 1901. *Orchis nana* (King & Pantl.) Schltr., Repert. Spec. Nov. Regni Veg. 9: 434. 1911. *Chusua roborowskii* (Maxim.) P. F. Hunt var. *nana* (King & Pantl.) P. F. Hunt, Kew Bull. 26(1): 176. 1971. *Chusua nana* (King & Pantl.) Pradhan, Indian Orchids: Guide Identif. & Cult. 2: 678. 1979. *Ponerorchis chusua* (D. Don) Soó var. *nana* (King & Pantl.) R. C. Srivast., Natl. Acad. Sci. Lett. 18: 61. 1996. TYPE: India, Sikkim, *Pantling 326* (?holotype, K).

Distribution. Arunachal Pradesh, Himachal Pradesh, Sikkim, Uttarakhand [?Bhutan, China, ?Myanmar, ?Nepal].

Habit. Terrestrial herb.

Remarks. This taxon is often considered conspecific with *Ponerorchis chusua*, but may be distinguished by the striped lip which is finely pubescent at the base, with the mid-lobe not surpassing the lateral lobes, and with crenate margins to the lateral lobes. Molecular studies have shown that the two are only distantly related (Jin et al., 2017). The distribution of *P. nana* is uncertain due to confusion with small specimens of *P. chusua* (e.g., the photo in Vij et al., 2013: 137; and in Swami, 2016: 200).

Ponerorchis pathakiana (Av. Bhattacharjee) J. M. H. Shaw, Orchid Rev. 125(1319, Suppl.): 59. 2017. *Amitostigma pathakianum* Av. Bhattacharjee, Phytotaxa 230(3): 268. 2015. TYPE: India, Arunachal Pradesh, Upper Siang, Kamatukut to Kanebango, 2800–3400 m, 5 Aug. 2012, *M. K. Pathak 54472* (holotype, CAL; isotype, CAL).

Distribution. Arunachal Pradesh. Endemic.

Habit. Terrestrial herb.

Ponerorchis puberula (King & Pantl.) Verm., Jahresber. Naturwiss. Vereins Wuppertal 25: 30. 1972. *Orchis puberula* King & Pantl., Ann. Roy. Bot. Gard. (Calcutta) 8: 304, t. 403. 1898. *Amitostigma puberulum* (King & Pantl.) Schltr., Repert. Spec. Nov. Regni Veg. Beih. 4: 92. 1919. *Amitostigma puberulum* (King & Pantl.) Tang & F. T. Wang, Bull. Fan Mem. Inst. Biol. Bot. 7: 6. 1936, isonym. *Chusua puberula* (King & Pantl.) N. Pearce & P. J. Cribb, Fl. Bhutan 3(3): 136. 2002. *Hemipilia puberula* (King & Pantl.) Y. Tang & H. Peng, BioMed Centr. Evol. Biol. 15(96): 28. 2015. TYPE: India, Sikkim, Lachong Valley, 10,000 ft, *Pantling 478* (holotype, CAL; isotype, AMES, K, W).

Distribution. Sikkim [Bhutan].

Habit. Terrestrial herb.

Ponerorchis renzii Deva & H. B. Naithani, Orchid Fl. N. W. Himalaya 199. 1986. *Chusua renzii* (Deva & H. B. Naithani) S. Misra, Orchids India: 258. 2007. *Hemipilia renzii* (Deva & H. B. Naithani) Y. Tang & H. Peng, BioMed Centr. Evol. Biol. 15(96): 28. 2015. TYPE: India, Kumaon, Pithoragarh above Garbyang alt. 10,500 ft, *G. S. Rawat 1225* (holotype, DD).

Distribution. Uttarakhand. Endemic.

Habit. Terrestrial herb.

Ponerorchis secundiflora (Kraenzl.) X. H. Jin, Schuit. & W. T. Jin, Molec. Phylogen. Evol. 77: 51. 2014. *Habenaria secundiflora* Hook. f., Fl. Brit. India [J. D. Hooker] 6(17): 165. 1890, nom. illeg., non Barb. Rodr., 1882. *Peristylus secundiflorus* Kraenzl., Orchid. Gen. Sp. 1: 518. 1898.

Gymnadenia secundiflora (Kraenzl.) Kraenzl., Orchid. Gen. Sp. 1: 936. 1901. *Neottianthe secundiflora* (Kraenzl.) Schltr., Repert. Spec. Nov. Regni Veg. 16: 291. 1919. *Hemipilia secundiflora* (Kraenzl.) Y. Tang & H. Peng, BioMed Centr. Evol. Biol. 15(96): 28. 2015. SYNTYPES: India, Kumaon, alt. 9000–10,000 ft, *Duthie s.n. [3421]* (syntypes, CAL, K); Sikkim, alt. 14,000 ft, *J. D. Hooker 278* (syntype, K-LINDL); Chumbi, *King's collector s.n.* (syntypes, CAL, K).

Distribution. Arunachal Pradesh, Manipur, Meghalaya, Mizoram, Sikkim, Uttar Pradesh, Uttarakhand [Bhutan, China, Nepal].

Habit. Terrestrial herb.

———

Porpax Lindl., Edwards's Bot. Reg. 31(Misc.): 62. 1845.

TYPE: *Porpax reticulata* Lindl.

Porpax albiflora (Rolfe) Schuit., Y. P. Ng & H. A. Pedersen, Bot. J. Linn. Soc. 186: 199. 2018. *Eria albiflora* Rolfe, Bull. Misc. Inform. Kew 1893: 170. 1893. TYPE: India, Nilgiri Hills, *J. O'Brien s.n.* (holotype, K).

Distribution. Karnataka, Kerala, Tamil Nadu. Endemic.

Habit. Epiphytic herb.

Porpax braccata (Lindl.) Schuit., Y. P. Ng & H. A. Pedersen, Bot. J. Linn. Soc. 186: 199. 2018. *Dendrobium braccatum* Lindl., Gen. Sp. Orchid. Pl.: 75. 1830. *Eria braccata* (Lindl.) Lindl., J. Proc. Linn. Soc., Bot. 3: 46. 1859 [1858]. *Pinalia braccata* (Lindl.) Kuntze, Revis. Gen. Pl. 2: 679. 1891. *Conchidium braccatum* (Lindl.) Brieger, Orchideen (Schlechter) 1(11–12): 751. 1981. TYPE: Sri Lanka, *Macrae 53* (holotype, K-LINDL).

Distribution. Kerala [Sri Lanka].

Habit. Epiphytic herb.

Remarks. *Porpax braccata* is quite similar to *P. reticosa* (Wight) Schuit. and is often considered conspecific. Both species show much variation in flower size. As Blatter and McCann (1931) pointed out, *Porpax reticosa* differs in having distinct side-lobes to the lip and also differs in the nature of the sheath covering the pseudobulb. In *P. reticosa*, the sheaths have a dense network of thickened veins that in older pseudobulbs soon detaches from the surface and remains as a persistent, loose net. In *P. braccata*, the veins form a less distinct, more sparsely branched network of hardly thickened veins that does not detach with age.

In India, the endemic *Porpax reticosa* is much more widespread than *P. braccata*, which apparently only occurs in Kerala.

Porpax elwesii (Rchb. f.) Rolfe, Orchid Rev. 16: 8. 1908. *Eria elwesii* Rchb. f., Gard. Chron., n.s., 19: 402. 1883. *Porpax meirax* King & Pantl. var. *elwesii* (Rchb. f.) R. C. Srivast., Natl. Acad. Sci. Lett. 18: 61. 1995. TYPE: Himalaya, *Elwes s.n.* (holotype, K).

Distribution. Andaman & Nicobar Islands, Arunachal Pradesh, Assam, ?Meghalaya, Sikkim, West Bengal [Bhutan, Nepal].

Habit. Epiphytic herb.

Porpax exilis (Hook. f.) Schuit., Y. P. Ng & H. A. Pedersen, Bot. J. Linn. Soc. 186: 199. 2018. *Eria exilis* Hook. f., Hooker's Icon. Pl. 21: t. 2074. 1891. *Pinalia exilis* (Hook. f.) Kuntze, Revis. Gen. Pl. 2: 679. 1891. *Conchidium exile* (Hook. f.) Ormerod, Taiwania 57: 119. 2012. TYPE: India, Travancore, *Johnson s.n.* (holotype, K).

Eria minima Blatt. & McCann, J. Bombay Nat. Hist. Soc. 35(2): 274, fig. 2. 1931. TYPE: India, North Kanara, Anmod, *Sedgwick 3260*.
Porpax chandrasekharanii Bhargavan & C. N. Mohanan, Curr. Sci. 51(20): 990. 1982. *Eria chandrasekharanii* (Bhargavan & C. N. Mohanan) C. S. Kumar & Manilal, Taxon 35(4): 720. 1986. TYPE: India, Kerala, Palghat District, Silent Valley, *Bhargavan 65796* (holotype, CAL).

Distribution. Goa, Karnataka, Kerala, Maharashtra, Tamil Nadu. Endemic.

Habit. Epiphytic herb.

Porpax extinctoria (Lindl.) Schuit., Y. P. Ng & H. A. Pedersen, Bot. J. Linn. Soc. 186: 199. 2018. *Dendrobium extinctorium* Lindl., Edwards's Bot. Reg. 21: t. 1756. 1835. *Eria extinctoria* (Lindl.) Oliv., Bot. Mag. 97: t. 5910. 1871. *Pinalia extinctoria* (Lindl.) Kuntze, Revis. Gen. Pl. 2: 679. 1891. *Conchidium extinctorium* (Lindl.) Y. P. Ng & P. J. Cribb, Orchid Rev. 113: 272. 2005. TYPE: Myanmar, Moulmein [Mawlamyine], *Griffith s.n.* [358] (holotype, K-LINDL).

Distribution. Andaman & Nicobar Islands [Myanmar, ?Nepal].

Habit. Epiphytic herb.

Remarks. Records from Nepal are at least in part referable to *Porpax summerhayesiana* (A. D. Hawkes & A. H. Heller) Schuit., Y. P. Ng & H. A. Pedersen.

Porpax fibuliformis (King & Pantl.) King & Pantl., Ann. Roy. Bot. Gard. (Calcutta) 8: 114, t. 157. 1898. *Eria fibuliformis* King & Pantl., J. Asiat. Soc. Bengal, Pt. 2, Nat. Hist. 64: 336. 1895 [1896]. TYPE: India, Sikkim, Tropical Valleys at the base of the Hills, Sivoke etc., 1000 ft, Oct. 1893, *Pantling s.n.* [278] (holotype, CAL; isotype, K).

Distribution. Assam, Meghalaya, Mizoram, Sikkim, West Bengal [also in Thailand].

Habit. Epiphytic herb.

Porpax filiformis (Wight) Schuit., Y. P. Ng & H. A. Pedersen, Bot. J. Linn. Soc. 186: 199. 2018. *Dendrobium filiforme* Wight, Icon. Pl. Ind. Orient. [Wight] 5(1): 5, t. 1642. 1851. *Eria filiformis* (Wight) Rchb. f., Ann. Bot. Syst. (Walpers) 6: 268. 1861. *Conchidium filiforme* (Wight) Rauschert, Feddes Repert. 94: 444. 1983. TYPE: India, Tamil Nadu, Nilgiri & Iyamally Hills near Coimbatore, *Wight s.n.* (syntype, K); without locality, *J. S. Law s.n.* (syntype, K).

Dendrobium fimbriatum Dalz., Hooker's J. Bot. Kew Gard. Misc. 4: 292. 1852, nom. illeg., non Hook., 1823. *Dendrobium dalzellii* Hook., Hooker's J. Bot. Kew Gard. Misc. 4: 292. 1852. *Eria dalzellii* (Hook.) Lindl., J. Proc. Linn. Soc., Bot. 3: 47. 1859 [1858]. *Eria dalzellii* (Hook.) Lindl. var. *fimbriata* Hook. f., Fl. Brit. India [J. D. Hooker] 5(16): 789. 1890. *Eria conrardii* M. R. Almeida, Fl. Maharashtra 5A: 42. 2009, nom. inval. *Eria conrardii* M. R. Almeida var. *fimbriata* (Dalz.) M. R. Almeida, Fl. Maharashtra 5A: 43. 2009, nom. inval. TYPE: India, Bombay, Ram Ghat, *Dalzell s.n.*

Distribution. Goa, Gujarat, Karnataka, Kerala, Maharashtra, Tamil Nadu. Endemic.

Habit. Epiphytic herb.

Porpax gigantea Deori, Bull. Bot. Surv. India 17(1–4): 174. 1975 [1978]. *Porpax fibuliformis* (King & Pantl.) King & Pantl. var. *gigantea* (Deori) Debta & H. J. Chowdhery, J. Orchid Soc. India 24: 9. 2010. TYPE: India, Meghalaya, Jaintea Hills, Jarain, alt. 4500 ft, 1 July 1972, *N. C. Deori 51757A* (holotype, CAL).

Distribution. Assam, Manipur, Meghalaya, Mizoram [Myanmar].

Habit. Epiphytic herb.

Porpax jerdoniana (Wight) Rolfe, Orchid Rev. 16: 8. 1908. *Lichenora jerdoniana* Wight, Icon. Pl. Ind. Orient. [Wight] 5(1): 18, t. 1738. 1851. *Eria lichenora* Lindl., J. Proc. Linn. Soc., Bot. 3: 46. 1859 [1858], nom. superfl. *Eria jerdoniana* (Wight) Rchb. f., Ann. Bot. Syst. (Walpers) 6: 267. 1861.

Pinalia jerdoniana (Wight) Kuntze, Revis. Gen. Pl. 2: 679. 1891. *Porpax lichenora* (Lindl.) T. Cooke, Fl. Bombay 2(4): 689. 1907. TYPE: India, Malabar Mountains, *Jerdon s.n.* (holotype, K).

Cryptochilus wightii Rchb. f., Bot. Zeitung (Berlin) 20: 214. 1862. TYPE: India, *Wight s.n.* (holotype, W).

Distribution. Andaman & Nicobar Islands, Goa, Karnataka, Kerala, Maharashtra, Tamil Nadu. Endemic.

Habit. Epiphytic herb.

Remarks. *Porpax jerdoniana* is one of the few epiphytic orchids with tesselated leaves.

Porpax lacei (Summerh.) Schuit., Y. P. Ng & H. A. Pedersen, Bot. J. Linn. Soc. 186: 200. 2018. *Eria lacei* Summerh., Bull. Misc. Inform. Kew 1929: 308. 1929. *Conchidium lacei* (Summerh.) Ormerod, Taiwania 57: 119. 2012. TYPE: Myanmar, Amherst District, Lampa Chaung, Dawna Range, *Lace 4751* (holotype, K).

Distribution. Assam, Mizoram [Myanmar].

Habit. Epiphytic herb.

Porpax meirax (C. S. P. Parish & Rchb. f.) King & Pantl., Ann. Roy. Bot. Gard. (Calcutta) 8: 114, t. 158. 1898. *Cryptochilus meirax* C. S. P. Parish & Rchb. f., Trans. Linn. Soc. London 30: 148. 1874. *Eria meirax* (C. S. P. Parish & Rchb. f.) N. E. Br., Gard. Chron., n.s., 14: 603. 1880. TYPE: Myanmar, Tenasserim, Moulmein [Mawlamyine], *C. S. P. Parish s.n.* (holotype, K).

Distribution. ?Andaman & Nicobar Islands, ?Himachal Pradesh, ?Meghalaya, ?Sikkim, ?Uttarakhand, ?West Bengal [Myanmar].

Habit. Epiphytic herb.

Remarks. All records from India may be due to confusion with *Porpax elwesii* and require confirmation.

Porpax microchilos (Dalzell) Schuit., Y. P. Ng & H. A. Pedersen, Bot. J. Linn. Soc. 186: 200. 2018. *Dendrobium microchilos* Dalzell, Hooker's J. Bot. Kew Gard. Misc. 3: 345. 1851. *Eria microchilos* (Dalzell) Lindl., J. Proc. Linn. Soc., Bot. 3: 47. 1859 [1858]. *Conchidium microchilos* (Dalzell) Rauschert, Feddes Repert. 94: 444. 1983 (as "*microchilon*"). TYPE: India, Western Bengal, *Dalzell s.n.* (holotype, K).

Eria tiagii Manilal, C. S. Kumar & J. J. Wood, J. Econ. Taxon. Bot. 5(2): 483, fig. 6. 1984. TYPE: India, Kerala, Silent Valley, *C. Sathish Kumar SV 10736* (holotype, CAL; isotype, K).

Distribution. Goa, Karnataka, Kerala, Maharashtra, Tamil Nadu. Endemic.

Habit. Epiphytic herb.

Porpax nana (A. Rich.) Schuit., Y. P. Ng & H. A. Pedersen, Bot. J. Linn. Soc. 186: 200. 2018. *Eria nana* A. Rich., Ann. Sci. Nat., Bot. sér. 2. 15: 19. 1841. *Pinalia nana* (A. Rich.) Kuntze, Revis. Gen. Pl. 2: 679. 1891. *Conchidium nanum* (A. Rich.) Brieger, Orchideen (Schlechter) 1(11–12): 751. 1981. TYPE: India, Nilgiri Hills, Condas, Oct., *Perrottet s.n.* (holotype, P; isotype, K, L).

Eria muscicola (Lindl.) Lindl. var. *ponmudiana* M. Mohanan & A. N. Henry, J. Econ. Taxon. Bot. 8(2): 425, fig. 10. 1986. TYPE: India, Kerala, Trivandrum District, Ponmudi *M. Mohanan 69213* (holotype, CAL).

Eria muscicola (Lindl.) Lindl. var. *brevilinguis* J. Joseph & V. Chandras., Bull. Bot. Surv. India 15(3–4): 267. 1973 [1976]. *Eria nana* A. Rich. var. *brevilinguis* (J. Joseph & V. Chandras.) Agrawala & H. J. Chowdhery, J. Orchid Soc. India 23: 68. 2009. *Eria brevilinguis* (J. Joseph & V. Chandras.) Bajrach. & K. K. Shrestha, Pleione 3(2): 166. 2009. *Porpax nana* (A. Rich.) Schuit., Y. P. Ng & H. A. Pedersen var. *brevilinguis* (J. Joseph & V. Chandras.) Kottaim., Int. J. Curr. Res. Biosci. Pl. Biol. 6(10): 40. 2019. TYPE: India, Kerala, Trivandrum District, Agastyamalai, *Joseph 44630A* (holotype, CAL).

Distribution. Karnataka, Kerala, Tamil Nadu. Endemic.

Habit. Epiphytic herb.

Porpax parviflora (D. Don) Ormerod & Kurzweil, Phytotaxa 481(1): 215. 2021 (as "*parviflorum*"). *Dendrobium parviflorum* D. Don, Prodr. Fl. Nepal.: 34. 1825. *Callista parviflora* (D. Don) Kuntze, Revis. Gen. Pl. 2: 655. 1891. TYPE: Nepal, Gosainthan [Shishapangma], *Wallich s.n.* (not found).

Dendrobium muscicola Lindl., Gen. Sp. Orchid. Pl.: 75. 1830. *Eria muscicola* (Lindl.) Lindl., J. Proc. Linn. Soc., Bot. 3: 47. 1859 [1858]. *Pinalia muscicola* (Lindl.) Kuntze, Revis. Gen. Pl. 2: 679. 1891. *Conchidium muscicola* (Lindl.) Rauschert, Feddes Repert. 94: 444. 1983. *Porpax muscicola* (Lindl.) Schuit., Y. P. Ng & H. A. Pedersen, Bot. J. Linn. Soc. 186: 200. 2018. TYPE: Nepal, *Wallich 2017* (holotype, K-LINDL; isotype, K-WALL).

Distribution. Andaman & Nicobar Islands, Arunachal Pradesh, Karnataka, Manipur, Meghalaya, Mizoram, Nagaland, Sikkim, Uttarakhand, West Bengal [Bhutan, China, Myanmar, Nepal].

Habit. Epiphytic herb.

Porpax pusilla (Griff.) Schuit., Y. P. Ng & H. A. Pedersen, Bot. J. Linn. Soc. 186: 200. 2018. *Conchidium pusillum* Griff., Not. Pl. Asiat. 3: 321, t. 310. 1851 (as "*pussillum*"). *Eria pusilla* (Griff.) Lindl.,

J. Proc. Linn. Soc., Bot. 3: 48. 1859 [1858]. *Pinalia pusilla* (Griff.) Kuntze, Revis. Gen. Pl. 2: 679. 1891. TYPE: India, Cherrapunji, *Griffith in Assam Herb. 143* (holotype, K-LINDL).

Phreatia uniflora Wight, Icon. Pl. Ind. Orient. [Wight] 5(1): t. 1734. 1851. SYNTYPES: India, Khasi Hills, *Griffith s.n.* (syntype, K-LINDL); Chunassangi, *Griffith s.n.* (not found).

Distribution. Arunachal Pradesh, Meghalaya, Sikkim, West Bengal [China, Myanmar, Nepal].

Habit. Epiphytic herb.

Porpax reticosa (Wight) Schuit., Malesian Orchid J. 24: 107. 2020. *Eria reticosa* Wight, Icon. Pl. Ind. Orient. [Wight] 5(1): 4, t. 1637. 1851. *Eria uniflora* Dalzell, Hooker's J. Bot. Kew Gard. Misc. 4: 111. 1852, nom. superfl. *Pinalia reticosa* (Wight) Kuntze, Revis. Gen. Pl. 2: 679. 1891. TYPE: India, Pycarrah, May–June, *Wight s.n.* (holotype, K K000881644).

Eria rupestris Blatt. & McCann, J. Bombay Nat. Hist. Soc. 35(2): 270, fig. 6. 1931. TYPE: India, Panchgani, Tableland above Convent, *May Langham 231* (holotype, BLAT).

Distribution. Goa, Karnataka, Kerala, Maharashtra, Tamil Nadu, Uttarakhand. Endemic.

Habit. Epiphytic herb.

Remarks. See notes under *Porpax braccata*.

Porpax reticulata Lindl., Edwards's Bot. Reg. 31 (Misc.): 62. 1845. *Cryptochilus reticulatus* (Lindl.) Rchb. f., Bot. Zeitung (Berlin) 20: 214. 1862. *Eria reticulata* (Lindl.) Benth. & Hook. f., Gen. Pl. [Bentham & Hooker f.] 3: 509. 1883 (as "*reticulatum*"). TYPE: East Indies, *Messrs. Loddiges s.n.* (?syntypes, K-LINDL [specimen and colored drawing, without annotation]).

Aggeianthus marchantioides Wight, Icon. Pl. Ind. Orient. [Wight] 5(1): 18, t. 1737. 1851. TYPE: India, Iyamallay Hills towards Paulghat, July–Aug., *Wight s.n.* (holotype, K).

Porpax papillosa Blatt. & McCann, J. Bombay Nat. Hist. Soc. 35(2): 268. 1931. SYNTYPES: India, Concan, *J. E. Stocks s.n.*; *J. S. Law s.n.*; Western Ghats, Belgaum Ghats, *Spooner s.n.* (syntypes, BLAT).

Distribution. Goa, Gujarat, Karnataka, Kerala, Maharashtra, Tamil Nadu [also in Thailand, etc.].

Habit. Epiphytic herb.

Porpax seidenfadenii A. N. Rao, Orchid Memories: 24. 2004. TYPE: India, Arunachal Pradesh,

W. Kameng District, Sissini, 3200 ft, *A. N. Rao 30808* (holotype, Orchid Herbarium, Tipi).

Distribution. Arunachal Pradesh. Endemic.

Habit. Epiphytic herb.

Porpax sikkimensis (Bajrach. & K. K. Shrestha) Schuit., Y. P. Ng & H. A. Pedersen, Bot. J. Linn. Soc. 186: 200. 2018. *Eria sikkimensis* Bajrach. & K. K. Shrestha, Pleione 3(2): 164. 2009. TYPE: India, Namtse (Namchi), 5000 ft, Aug. 1891, *Pantling 163* (holotype, CAL; isotype, K).

Distribution. Sikkim. Endemic.

Habit. Epiphytic herb.

Remarks. *Porpax sikkimensis* is only known from the type collection. Singh et al. (2019) consider this to be a synonym of *Eria muscicola* (Lindl.) Lindl. (= *P. parviflora* (D. Don) Ormerod & Kurzweil). The authors of *E. sikkimensis* stated that it differs from *E. muscicola* in having an ovoid pseudobulb, large oblong-lanceolate, acute floral bracts, and a glabrous lip with serrate margins and without basal calli. Studies on fresh material are desirable.

———

Pteroceras Hasselt ex Hassk., Flora 25 (2, Beibl.): 6. 1842.

TYPE: *Pteroceras radicans* Hassk.

Pteroceras indicum Punekar, Folia Malaysiana 9(2): 119. 2008. TYPE: India, Karnataka, Kannada (North Kanara) District, Anshi National Park along Vaki River near Anshi Nature Camp, *Punekar 187860* (holotype, CAL; isotype, BSI).

Distribution. Karnataka, Tamil Nadu. Endemic.

Habit. Epiphytic herb.

Remarks. This species is doubtfully distinct from *Pteroceras monsooniae* Sasidh. & Sujanapal.

Pteroceras leopardinum (C. S. P. Parish & Rchb. f.) Seidenf. & Smitinand, Orchids Thailand (Prelim. List). 535. 1963. *Thrixspermum leopardinum* C. S. P. Parish & Rchb. f., Trans. Linn. Soc. London 30(1): 145. 1874. *Sarcochilus leopardinus* (C. S. P. Parish & Rchb. f.) Hook. f., Fl. Brit. India [J. D. Hooker] 6(17): 38. 1890. TYPE: Myanmar, Moulmein [Mawlamyine], *C. S. P. Parish s.n.* (holotype, W).

Proteroceras holttumii J. Joseph & Vajr., J. Indian Bot. Soc. 53: 189. 1974. TYPE: India, Tamil Nadu, Coimbatore

District, Vellingiri Hills, alt. 4900 ft, 23 Apr. 1972, *Lakshmanan 40808* (holotype, CAL).

Distribution.   Andaman & Nicobar Islands, Kerala, Tamil Nadu [China, Myanmar].

Habit.   Epiphytic herb.

Pteroceras monsooniae Sasidh. & Sujanapal, Sida 20(3): 923. 2003. *Grosourdya monsooniae* (Sasidh. & Sujanapal) R. Rice, Photo Intro Vandoid Orchid Gen. Asia, Rev. Ed.: 160. 2018. TYPE: India, Kerala, Palakkad District, Parambikulam Wildlife Sanctuary, 3500 ft, 10°32′N, 76°E, 28 May 2000, *P. Sujanapal KFRI 30407* (holotype, KFRI).

Distribution.   Kerala. Endemic.

Habit.   Epiphytic herb.

Remarks.   Although *Pteroceras monsooniae*, *P. indicum*, and *P. muriculatum* (Rchb. f.) P. F. Hunt appear to be misplaced in *Pteroceras*, there is as yet no molecular evidence that they can be accommodated in *Grosourdya* and they differ in plant habit from species known to belong to that genus.

Pteroceras muriculatum (Rchb. f.) P. F. Hunt, Kew Bull. 24(1): 96. 1970. *Thrixspermum muriculatum* Rchb. f., Gard. Chron., n.s., 16: 198. 1881. *Grosourdya muriculata* (Rchb. f.) R. Rice, Photo Intro Vandoid Orchid Gen. Asia, Rev. Ed.: 160. 2018. TYPE: East India, *W. Bull s.n.* (holotype, W).

Distribution.   Andaman & Nicobar Islands. Endemic.

Habit.   Epiphytic herb.

Pteroceras teres (Blume) Holttum, Kew Bull. 14(2): 271. 1960. *Dendrocolla teres* Blume, Bijdr. Fl. Ned. Ind. 7: 289. 1825. *Aerides teres* (Blume) Lindl., Gen. Sp. Orchid. Pl.: 240. 1833. *Sarcochilus teres* (Blume) Rchb. f., Ann. Bot. Syst. (Walpers) 6: 499. 1863. *Thrixspermum teres* (Blume) Rchb. f., Xenia Orchid. 2: 121. 1868. TYPE: Indonesia, Java, Buitenzorg (Bogor) Prov., *Blume s.n.* (holotype, L).

Aerides suaveolens Roxb., Fl. Ind. ed. 1832, 3: 473. 1832. *Sarcochilus suaveolens* (Roxb.) Hook. f., Fl. Brit. India [J. D. Hooker] 6(17): 33. 1890. *Pteroceras suaveolens* (Roxb.) Holtum, Kew Bull. 14(2): 271. 1960. TYPE: Bangladesh, Chittagong, icon. *Aerides suaveolens*, *Roxburgh 2350* (syntypes, CAL, K).

Distribution.   Andaman & Nicobar Islands, Arunachal Pradesh, Assam, ?Meghalaya, Nagaland, Sikkim, Tripura, Uttarakhand, West Bengal [Bangladesh, Myanmar, Nepal].

Habit.   Epiphytic herb.

Renanthera Lour., Fl. Cochinch. 2: 516, 521. 1790.

TYPE: *Renanthera coccinea* Lour.

Renanthera imschootiana Rolfe, Bull. Misc. Inform. Kew 1891: 200. 1891. TYPE: Origin unknown, cult. *Van Imschoot s.n.* (holotype, K).

Renanthera papilio King & Prain, J. Asiat. Soc. Bengal, Pt. 2, Nat. Hist. 64: 328. 1895 [1896]. TYPE: India, Assam, cult. *Chinchona Plantation Sikkim (leg. Chatterton) s.n.*

Distribution.   Arunachal Pradesh, Assam, Manipur, ?Meghalaya, Mizoram, Nagaland [China, Myanmar].

Habit.   Epiphytic herb.

Remarks.   Apart from all species of *Paphiopedilum*, *Renanthera imschootiana* is currently the only Indian orchid species listed on CITES Appendix I.

Rhomboda Lindl., J. Linn. Soc., Bot. 1: 181. 1857.

TYPE: *Rhomboda longifolia* Lindl.

Rhomboda arunachalensis A. N. Rao, J. Econ. Taxon. Bot. 22(2): 426. 1998. TYPE: India, Arunachal Pradesh, E. Kameng District, Kimi, alt. 3520, *A. N. Rao 29426* (holotype, Orchid Herbarium, Tipi).

Distribution.   Arunachal Pradesh. Endemic.

Habit.   Terrestrial herb.

Rhomboda lanceolata (Lindl.) Ormerod, Orchadian 11(7): 329. 1995. *Dossinia lanceolata* Lindl., J. Proc. Linn. Soc., Bot. 1: 186. 1857. *Macodes lanceolata* (Lindl.) Rchb. f., Xenia Orchid. 1: 226. 1858. *Hetaeria lanceolata* (Lindl.) Rchb. f., Trans. Linn. Soc. London 30: 142. 1874. *Anoectochilus pomrangianus* Seidenf., Dansk Bot. Ark. 32(2): 41. 1978. *Odontochilus pomrangianus* (Seidenf.) Szlach., Fragm. Florist. Geobot., Suppl. 3: 115. 1995. TYPE: India, Khasi Hills, Pomrang, 16 Sep. 1850, *Hooker & Thomson s.n.* [2284] (holotype, K; icon. K, K-LINDL).

Distribution.   Arunachal Pradesh, ?Assam, Meghalaya, Nagaland, Sikkim, West Bengal [also in Thailand, etc.].

Habit.   Terrestrial herb.

Remarks.   Bhattacharjee and Chowdhery (2018) pointed out that all verifiable records of *Rhomboda abbreviata* (Lindl.) Ormerod from India are referable to *R. lanceolata*. This species is illustrated as *R. abbreviata* in Swami (2016, 2017).

Rhomboda longifolia Lindl., J. Proc. Linn. Soc., Bot. 1: 181. 1857. *Zeuxine longifolia* (Lindl.) Hook. f., Fl. Brit. India [J. D. Hooker] 6(17): 109. 1890. *Hetaeria longifolia* (Lindl.) Benth., J. Linn. Soc., Bot. 18: 346. 1880. *Odontochilus longifolius* (Lindl.) Tang & F. T. Wang, Acta Phytotax. Sin. 1: 70. 1951. TYPE: India, Sikkim, *J. D. Hooker 335* (holotype, K).

Distribution. Sikkim. Endemic.

Habit. Terrestrial herb.

Remarks. This species is only known from specimens collected by J. D. Hooker in the mid 1800s and is possibly extinct.

Rhomboda monensis Odyuo, D. K. Roy, Av. Bhattacharjee & Ormerod, Phytotaxa 405(1): 61. 2019. TYPE: India, Nagaland, Mon district, Tobu village, 1600–1800 m, 10 August 2017, *N. Odyuo & D. K. Roy 133089* (holotype, ASSAM).

Distribution. Nagaland. Endemic.

Habit. Terrestrial herb.

Rhomboda pulchra (King & Pantl.) Ormerod & Av. Bhattacharjee, Phytotaxa 191: 177. 2014. *Zeuxine pulchra* King & Pantl., J. Asiat. Soc. Bengal, Pt. 2, Nat. Hist. 65: 127. 1896. TYPE: India, Sikkim, Keydung, Lachung Valley, alt. 7500 ft, Aug. 1895, *Pantling 412* (holotype, CAL; isotype, K [7000 ft]).

Distribution. Arunachal Pradesh, Meghalaya, Sikkim, West Bengal. Endemic.

Habit. Terrestrial herb.

———

Rhynchostylis Blume, Bijdr. Fl. Ned. Ind. 7: 285. 434. 1825 (as "*Rynchostylis*").

TYPE: *Rhynchostylis retusa* (L.) Blume (*Epidendrum retusum* L.).

Rhynchostylis gigantea (Lindl.) Ridl., J. Linn. Soc., Bot. 32: 356. 1896. *Saccolabium giganteum* Lindl., Gen. Sp. Orchid. Pl.: 221. 1833. *Gastrochilus giganteus* (Lindl.) Kuntze, Revis. Gen. Pl. 2: 661. 1891. *Anota gigantea* (Lindl.) Fukuy., Trans. Nat. Hist. Soc. Taiwan 34: 111. 1944. TYPE: Myan-

mar, Pegu, Prome, 1826 [a label on the sheet says I-1827], *Wallich 7306* (holotype, K-WALL).

Distribution. ?Arunachal Pradesh, ?Manipur [China, Myanmar].

Habit. Epiphytic herb.

Remarks. This species is listed as "excluded" in Singh et al. (2019).

Rhynchostylis retusa (L.) Blume, Bijdr. Fl. Ned. Ind. 7: 286, t. 49. 1825. *Epidendrum retusum* L., Sp. Pl. 2: 953. 1753. *Aerides retusa* (L.) Sw., J. Bot. (Schrader) 2: 233. 1799. *Limodorum retusum* (L.) Sw., Nova Acta Regiae Soc. Sci. Upsal. 6: 80. 1799. *Saccolabium retusum* (L.) Voigt, Hort. Suburb. Calcutt. 630. 1845. *Gastrochilus retusus* (L.) Kuntze, Revis. Gen. Pl. 2: 661. 1891. TYPE: India, icon. "Ansjeli-maravara" in Rheede, Hort. Malab. 12: 1, t. 1. 1692 (lectotype, designated by Majumdar & Bakshi, 1979).

Aerides praemorsa Willd., Sp. Pl. 4: 130. 1805. *Rhynchostylis praemorsa* (Willd.) Blume, Bijdr. Fl. Ned. Ind.: 286. 1825. *Saccolabium praemorsum* (Willd.) Lindl., Gen. Sp. Orchid. Pl.: 22. 1830. TYPE: Icon. "Biti-marum-maravara" in Rheede, Hort. Malab. 12: t. 2.
Sarcanthus guttatus Lindl., Edwards's Bot. Reg. 17: t. 1443. 1831. *Saccolabium guttatum* (Lindl.) Lindl., Numer. List [Wallich] n. 7308. 1832. *Rhynchostylis guttata* (Lindl.) Rchb. f., Bonplandia (Hannover) 2: 93. 1854. TYPE: Without origin, *cult. Chiswick Garden (leg. Wallich s.n.)* (holotype, K-LINDL).
Aerides guttata Roxb., Fl. Ind. ed. 1832, 3: 471. 1832 (as "*guttatum*"). TYPE: Bangladesh, Dacca [Dhaka], cult. Calcutta s.n. (leg. C. A. Bruce s.n.), icon. *Aerides guttatum* [sic], *Roxburgh 1905* (syntypes, CAL, K).
Saccolabium rheedei Wight, Icon. Pl. Ind. Orient. [Wight] 5(1): 19. 1851 (as "*rheedii*"). *Gastrochilus rheedei* (Wight) Kuntze, Revis. Gen. Pl. 2: 661. 1891. TYPE: icon. "Ansjeli-maravara" in Rheede, Hort. Malab. 12: 1, t. 1. 1692.
Saccolabium garwalicum Lindl., J. Proc. Linn. Soc., Bot. 3: 32. 1859 [1858] (as "*gurwalicum*"). *Rhynchostylis garwalica* (Lindl.) Rchb. f., Ann. Bot. Syst. (Walpers) 6: 888. 1864 (as "*gurwalica*"). *Gastrochilus garwalicus* (Lindl.) Kuntze, Revis. Gen. Pl. 2: 661. 1891 (as "*gurwalicus*"). SYNTYPES: India, Uttarakhand, Garhwal ("Gurwhal"), 3000 ft, *Thomson 181* (syntype, K-LINDL); Kumaon, Gunai Valley, *Thomson 185* (not found).
Saccolabium turneri B. S. Williams, Orch.-Grow. Man., ed. 7: 703. 1894. TYPE: India, not designated.
Rhynchostylis albiflora I. Barua & Bora, J. Econ. Taxon. Bot. 26(1): 251. 2002. *Rhynchostylis retusa* (L.) Blume f. *albiflora* (I. Barua & Bora) Christenson, J. Orchideenfr. 12: 344. 2005. TYPE: India, Assam, Karanga, Jorhat, 26°47′N, 94°12′E, 28 Apr. 1994, *Barua 3286* (holotype, CAL).
Rhynchostylis cymifera Yohannan, J. Mathew & Szlach., Jordan J. Biol. Sci. 11(3): 257. 2018. TYPE: India, Kerala, Malappuram District, Nilambur, Vaniyampuzha Estate, 1200 m, 20 May 2012, *Yohannan 4417* (holotype, TBGT).

Distribution. Andaman & Nicobar Islands, Andhra Pradesh, Arunachal Pradesh, Assam, Bihar, Chhattisgarh, Goa, Gujarat, Haryana, Himachal Pradesh, Jammu & Kashmir, Jharkhand, Karnataka, Kerala, Madhya Pradesh, Maharashtra, Manipur, Meghalaya, Mizoram, Nagaland, Odisha, Sikkim, Tamil Nadu, Tripura, Uttar Pradesh, Uttarakhand, West Bengal [Bangladesh, Bhutan, China, Myanmar, Nepal, Sri Lanka].

Habit. Epiphytic herb.

Remarks. The recently described *Rhynchostylis cymifera* is said to differ from *R. retusa* in having an inflorescence in which the flowers open from the distal towards the basal end, in addition to some other minor differences. However, specimens of otherwise typical *R. retusa* are known in which the flowers open in the same way; therefore, this difference cannot be considered diagnostic.

———

Risleya King & Pantl., Ann. Roy. Bot. Gard. (Calcutta) 8: 246. 1898.

TYPE: *Risleya atropurpurea* King & Pantl.

Risleya atropurpurea King & Pantl., Ann. Roy. Bot. Gard. (Calcutta) 8: 247, t. 328. 1898. TYPE: India, Sikkim, Jongri, alt. 13,000 ft, *Pantling 451* (lectotype, K, designated by Margońska, 2012; isotype, CAL, E, L, P, W).

Distribution. Sikkim [Bhutan, China, Myanmar, Nepal].

Habit. Holomycotrophic terrestrial herb.

———

Robiquetia Gaudich., Voy. Uranie, Bot. 426. 1826 [1829].

TYPE: *Robiquetia ascendens* Gaudich.

Robiquetia andamanica (N. P. Balakr. & N. Bhargava) Kocyan & Schuit., Phytotaxa 161: 68. 2014. *Malleola andamanica* N. P. Balakr. & N. Bhargava, Proc. Indian Acad. Sci., B. 88(4): 317. 1979. TYPE: India, Andaman Island, Hutbay, Mangrove forests, 28 Aug. 1976, *N. Bhargava 4189A* (holotype, CAL).

Distribution. Andaman & Nicobar Islands. Endemic.

Habit. Epiphytic herb.

Robiquetia arunachalensis (A. N. Rao) Kocyan & Schuit. Phytotaxa 161: 68. 2014. *India arunacha-*

lensis A. N. Rao, J. Econ. Taxon. Bot. 22(3): 701. 1998 [1999]. TYPE: India, Arunachal Pradesh, West Kameng, Doimara Reserve Forests, alt. 640 ft, *A. N. Rao 30054* (holotype, Orchid Herbarium, Tipi).

Distribution. Arunachal Pradesh. Endemic.

Habit. Epiphytic herb.

Robiquetia gracilis (Lindl.) Garay, Bot. Mus. Leafl. 23(4): 197. 1972. *Saccolabium gracile* Lindl., Gen. Sp. Orchid. Pl.: 225. 1833. *Malleola gracilis* (Lindl.) Schltr., Repert. Spec. Nov. Regni Veg. Beih. 1: 981. 1913. *Gastrochilus gracilis* (Lindl.) Kuntze, Revis. Gen. Pl. 2: 661. 1891. TYPE: Sri Lanka, *Macrae s.n.* [55] (holotype, K-LINDL).

Distribution. Andaman & Nicobar Islands, Kerala, Tamil Nadu [Sri Lanka].

Habit. Epiphytic herb.

Robiquetia josephiana Manilal & C. S. Kumar, Orchid Rev. 92: 293. 1984. TYPE: India, Kerala, Silent Valley, Palghat District, alt. 4200 ft, 15 Feb. 1982, *C. Sathish Kumar SV 10788* (holotype, CAL).

Distribution. Kerala, Tamil Nadu. Endemic.

Habit. Epiphytic herb.

Robiquetia rosea (Lindl.) Garay, Bot. Mus. Leafl. 23(4): 197. 1972. *Saccolabium roseum* Lindl., Gen. Sp. Orchid. Pl.: 225. 1833. *Malleola rosea* (Lindl.) Schltr., Repert. Spec. Nov. Regni Veg. Beih. 1: 981. 1913. *Gastrochilus roseus* (Lindl.) Kuntze, Revis. Gen. Pl. 2: 661. 1891. TYPE: Sri Lanka, *Macrae s.n.* [64] (holotype, K-LINDL).

Distribution. Karnataka, Tamil Nadu [Sri Lanka].

Habit. Epiphytic herb.

Robiquetia spathulata (Blume) J. J. Sm., Natuurk. Tijdschr. Ned.-Indië 72: 114. 1912. *Cleisostoma spathulatum* Blume, Bijdr. Fl. Ned. Ind. 8: 364. 1825 (as "*spatulata*"). TYPE: Indonesia, Java, Pantjar, *Blume s.n.* (holotype, L; isotype, K).

Cleisostoma spicatum Lindl., Edwards's Bot. Reg. 33: t. 32. 1847. TYPE: Borneo, *Messrs. Rollissons s.n.* (holotype, K-LINDL).
Saccolabium densiflorum Lindl., Gen. Sp. Orchid. Pl.: 220. 1833. *Sarcanthus densiflorus* (Lindl.) C. S. P. Parish & Rchb. f., Trans. Linn. Soc. London 30: 136. 1874. *Aerides densiflora* (Lindl.) Wall. ex Hook. f., Fl. Brit. India 6(17): 72. 1890. *Gastrochilus densiflorus* (Lindl.) Kuntze, Revis. Gen. Pl. 2: 661. 1891. *Rhynchostylis densiflora* (Lindl.) L. O. Williams, Bot. Mus. Leafl. 5: 58. 1937.

Pomatocalpa densiflorum (Lindl.) Tang & F. T. Wang, Acta Phytotax. Sin. 1: 99. 1951. SYNTYPES: Bangladesh or India, Sylhet, *Wallich 7311A* (*leg. De Silva s.n.*) (syntype, K-WALL); Malaysia, Penang, *Wallich 7311B* (*leg. G. Porter s.n.*) (syntype, W).

Distribution. Arunachal Pradesh, Assam, Mizoram [Bangladesh, Bhutan, China, Myanmar].

Habit. Epiphytic herb.

Robiquetia succisa (Lindl.) Seidenf. & Garay, Bot. Tidsskr. 67(1–2): 119. 1972. *Sarcanthus succisus* Lindl., Bot. Reg. 12: t. 1014. 1826. TYPE: China, 1824, *John Damper Parks s.n.* (not found).

Oeceoclades paniculata Lindl., Gen. Sp. Orchid. Pl.: 236. 1833. *Robiquetia paniculata* (Lindl.) J. J. Sm., Natuurk. Tijdschr. Ned.-Indië 72: 114. 1912. SYNTYPES: India, Assam, Gualpara, 1 July 1808, *Wallich 7334A* (*leg. Buchanan-Hamilton s.n.*) (syntypes, K-LINDL, K-WALL); Bangladesh or India, Sylhet, *Wallich s.n.* [*7334B*] (*leg. De Silva s.n.*) (syntypes, K-LINDL, K-WALL).
Saccolabium buccosum Rchb. f., Gard. Chron. 1871: 938. 1871. *Uncifera buccosa* (Rchb. f.) Finet ex Guillaumin, Bull. Soc. Bot. France 77: 333. 1930. TYPE: India, icon. *J. Day s.n.* (holotype, K).

Distribution. Arunachal Pradesh, Assam, Nagaland, Sikkim, West Bengal [Bangladesh, Bhutan, China, Myanmar, Nepal].

Habit. Epiphytic herb.

Robiquetia virescens Ormerod & S. S. Fernando, Rheedea 18: 22. 2008. *Saccolabium virescens* Gardner ex Lindl., J. Proc. Linn. Soc., Bot. 3: 35. 1859 [1858], nom. inval. *Robiquetia virescens* Jayaw., Revised Handb. Fl. Ceylon 2: 255. 1981, nom. inval. TYPE: Locality unknown, *s. coll. Ceylon Plants 488* (holotype, K).

Distribution. Kerala [Sri Lanka].

Habit. Epiphytic herb.

———

Saccolabiopsis J. J. Sm., Bull. Jard. Bot. Buitenzorg, sér. 2, 26: 93. 1918.

TYPE: *Saccolabiopsis bakhuizenii* J. J. Sm.

Saccolabiopsis pusilla (Lindl.) Seidenf. & Garay, Bot. Tidsskr. 67(1–2): 118, fig. 33. 1972. *Oeceoclades pusilla* Lindl., Gen. Sp. Orchid. Pl.: 237. 1833. *Saccolabium pumilio* Rchb. f., Ann. Bot. Syst. (Walpers) 6: 886. 1864. *Saccolabium pusillum* (Lindl.) Lindl. ex Hook. f., Fl. Brit. India [J. D. Hooker] 6(17): 57. 1890, nom. illeg., non Blume, 1825. *Gastrochilus pumilio* (Rchb. f.) Kuntze,

Revis. Gen. Pl. 2: 661. 1891. *Pennilabium pumilio* (Rchb. f.) Pradhan, Indian Orchids: Guide Identif. & Cult. 2: 720. 1979. TYPE: Bangladesh or India, Sylhet, *Wallich 7332A* (*leg. De Silva*) (holotype, K-LINDL; isotype, K-WALL).

Distribution. Andaman & Nicobar Islands, Arunachal Pradesh, Kerala, Manipur, Meghalaya, Sikkim, West Bengal [Bangladesh, Bhutan, Myanmar].

Habit. Epiphytic herb.

———

Sarcoglyphis Garay, Bot. Mus. Leafl. 23(4): 200. 1972.

TYPE: *Sarcoglyphis mirabilis* (Rchb. f.) Garay (*Sarcanthus mirabilis* Rchb. f.).

Sarcoglyphis arunachalensis A. N. Rao, Nordic J. Bot. 10(2): 161, fig. 1. 1990. TYPE: India, Arunachal Pradesh State, West Kameng District, Sessa, alt. 3300 ft, *S. N. Hegde 4199* (holotype, Orchid Herbarium, Tipi).

Distribution. Arunachal Pradesh. Endemic.

Habit. Epiphytic herb.

———

Satyrium Sw., Kongl. Vetensk. Acad. Nya Handl. 21: 214. 1800, nom. cons.

TYPE: *Satyrium bicorne* (L.) Thunb. (*Orchis bicornis* L.) (typ. cons.).

Satyrium nepalense D. Don, Prodr. Fl. Nepal.: 26. 1825. TYPE: Nepal, Bagmati Zone, Rashuwa District, Gosainthan [Shishapangma], *Wallich s.n.* (holotype, BM).

Satyrium albiflorum A. Rich., Ann. Sci. Nat., Bot. sér. 2, 15: 76, t. 5C. 1841. TYPE: India, Nilgiri Hills near Dodabetta., July–Aug., *Perrottet s.n.* (holotype, P).
Satyrium pallidum A. Rich., Ann. Sci. Nat., Bot. sér. 2, 15: 77. 1841. TYPE: India, Nilgiri Hills, Neddoubetta, Sep.–Dec., *Perrottet s.n.* (holotype, P).
Satyrium perrottetianum A. Rich., Ann. Sci. Nat., Bot. sér. 2, 15: 76. 1841. TYPE: India, Nilgiri Hills, Otacamund, July–Aug., *Perrottet s.n.* (holotype, P; isotype, K-LINDL).
Satyrium wightianum Lindl., Gen. Sp. Orchid. Pl.: 340. 1838. *Satyrium nepalense* D. Don var. *wightianum* (Lindl.) Hook. f., Fl. Brit. India [J. D. Hooker] 6(17): 168. 1890 (as "*wightiana*"). TYPE: India, Nilgiri Hills, July–Sep. (fide Wight), *Wight s.n.* (syntypes, K-LINDL).
Satyrium ciliatum Lindl., Gen. Sp. Orchid. Pl.: 341. 1838. *Satyrium nepalense* D. Don var. *ciliatum* (Lindl.) Hook. f., Fl. Brit. India [J. D. Hooker] 6(17): 168. 1890 (as "*ciliata*"). TYPE: Nepal, *Wallich s.n.* (holotype, K-LINDL).

Satyrium neilgherrensis Fyson, Fl. S. Ind. Hill Stations 1: 595.
1932. TYPE: India, Tamil Nadu, Nilgiri Hills, Pulneys,
near Kodaikanal, icon. Wight, *Icon. Pl. Ind. Orient.
(Wight) 5(1) t. 1716. 1851* (iconotype).

Satyrium nepalense D. Don f. *albiflorum* Tuyama in H. Hara,
Fl. E. Himalaya 1: 450. 1966. TYPE: India, West Bengal, Darjeeling, alt. 7000 ft, 15 Sep. 1964, *H. Hara s.n.*

Distribution. Andhra Pradesh, Arunachal Pradesh,
Assam, Himachal Pradesh, Jammu & Kashmir, Karna-
taka, Kerala, Manipur, Meghalaya, Nagaland, Sikkim,
Tamil Nadu, Uttarakhand, West Bengal [Bhutan, China,
Myanmar, Nepal, Pakistan, Sri Lanka].

Habit. Terrestrial herb.

———

Schoenorchis Blume, Bijdr. Fl. Ned. Ind. 8: 361.
1825.

TYPE: *Schoenorchis juncifolia* Blume.

Schoenorchis fragrans (C. S. P. Parish & Rchb. f.)
Seidenf. & Smitinand, Orchids Thailand (Prelim.
List). 611. 1963. *Saccolabium fragrans* C. S. P.
Parish & Rchb. f., J. Bot. 12: 197. 1874. *Gastro-
chilus fragrans* (C. S. P. Parish & Rchb. f.) Kuntze,
Revis. Gen. Pl. 2: 661. 1891. TYPE: Moulmein
[Mawlamyine], 14 May 1873, *C. S. P. Parish &
H. G. Reichenbach s.n.* (holotype, ?W).

Schoenorchis manipurensis Pradhan, Amer. Orchid Soc. Bull.
47: 911. 1978. TYPE: India, Manipur, alt. 2400–3200
ft, July 1975, *R. K. Mohendrajit Singh s.n.* (holotype,
Herbarium U. C. Pradhan).

Distribution. Assam, Manipur [China, Myanmar].

Habit. Epiphytic herb.

Schoenorchis gemmata (Lindl.) J. J. Sm., Natuurk.
Tijdschr. Ned.-Indië 72: 100. 1912. *Saccolabium
gemmatum* Lindl., Edwards's Bot. Reg. 24(Misc.):
50. 1838. *Gastrochilus gemmatus* (Lindl.) Kuntze,
Revis. Gen. Pl. 2: 661. 1891. *Cleisostoma gemma-
tum* (Lindl.) King & Pantl., Ann. Roy. Bot. Gard.
(Calcutta) 8: 234, t. 313. 1898. TYPE: India,
Khasi Hills, 1837, *Gibson s.n.* (not found).

Distribution. Arunachal Pradesh, Assam, Manipur,
Meghalaya, Mizoram, Nagaland, Sikkim, West Bengal
[Bhutan, China, Myanmar, Nepal].

Habit. Epiphytic herb.

Schoenorchis jerdoniana (Wight) Garay, Bot. Mus.
Leafl. 23(4): 202. 1972. *Taeniophyllum jerdonianum* Wight, Icon. Pl. Ind. Orient. [Wight] 5(1): 22,
t. 1756. 1851. *Saccolabium jerdonianum* (Wight)

Rchb. f., Ann. Bot. Syst. (Walpers) 6: 886. 1864.
Gastrochilus jerdonianus (Wight) Kuntze, Revis.
Gen. Pl. 2: 661. 1891. TYPE: India, Malabar, *Jer-
don s.n.* (holotype, K).

Distribution. Karnataka, Kerala, Tamil Nadu.
Endemic.

Habit. Epiphytic herb.

Schoenorchis manilaliana M. Kumar & Sequiera,
Kew Bull. 55(1): 241. 2000. TYPE: India, Kerala,
Palghat District, Siruvani Reserve Forests, 7th
Acre, Siruvani, *Stephen & Michael 008885* (holo-
type, KFRI).

Distribution. Kerala. Endemic.

Habit. Epiphytic herb.

Schoenorchis minutiflora (Ridl.) J. J. Sm., Natuurk.
Tijdschr. Ned.-Indië 72: 101. 1912. *Saccolabium
minutiflorum* Ridl., J. Fed. Malay States Mus. 4:
71. 1909. TYPE: Malay Peninsula, *Ridley 13891.*

Distribution. Andaman & Nicobar Islands [also in
Malaysia].

Habit. Epiphytic herb.

Schoenorchis nivea (Lindl.) Schltr., Repert. Spec.
Nov. Regni Veg. Beih. 1: 986. 1913. *Saccolabium
niveum* Lindl., Gen. Sp. Orchid. Pl.: 224. 1833.
Gastrochilus niveus (Lindl.) Kuntze, Revis. Gen.
Pl. 2: 661. 1891. TYPE: Sri Lanka, Peradenia,
Macrae s.n. (holotype, K-LINDL).

Distribution. Kerala, Tamil Nadu [Sri Lanka].

Habit. Epiphytic herb.

Schoenorchis smeeana (Rchb. f.) Jalal, Jayanthi &
Schuit., Kew Bull. 69(2): art. 9508: 4. 2014. *Sac-
colabium smeeanum* Rchb. f., Gard. Chron., ser. 3,
2: 214. 1887. *Xenikophyton smeeanum* (Rchb. f.)
Garay, Bot. Mus. Leafl. 23(10): 375. 1974. TYPE:
Origin unknown, *cult. Smee s.n.* (holotype, K).

Rhynchostylis latifolia C. E. C. Fisch., Bull. Misc. Inform.
Kew 1927: 358. 1927. *Schoenorchis latifolia* (C. E. C.
Fisch.) C. J. Saldanha, J. Bombay Nat. Hist. Soc. 70:
415. 1973 [1974]. TYPE: India, Cardamonai, Mysore,
Sep. 1903, *C. A. Barber 6093* (holotype, K).

Xenikophyton seidenfadenianum M. Kumar, Sequiera & J. J.
Wood, Kew Bull. 57(1): 227. 2002. TYPE: India, Ker-
ala, Palghat District, Siruvani Forest Reserve, Dam Site,
Stephen 0020621 (holotype, KFRI).

Distribution. Karnataka, Kerala, Tamil Nadu.
Endemic.

Habit. Epiphytic herb.

Seidenfadeniella C. S. Kumar in C. Sathish Kumar & K. S. Manilal, Cat. Indian Orchids 43. 1994.

TYPE: *Sarcanthus roseus* Wight.

Seidenfadeniella filiformis (Rchb. f.) Christenson & Ormerod in K. M. Matthew, Fl. Palni Hills 3: 1258. 1999. *Saccolabium filiforme* Rchb. f., Ann. Bot. Syst. (Walpers) 6: 887. 1864. *Gastrochilus filiformis* (Rchb. f.) Kuntze, Revis. Gen. Pl. 2: 661. 1891. *Schoenorchis filiformis* (Rchb. f.) Schltr., Repert. Spec. Nov. Regni Veg. Beih. 1: 986. 1913. *Cleisostomopsis filiformis* (Rchb. f.) R. Rice, Photo Intro Asian *Bulbophyllum, Coelogyne* & *Dendrobium* Orchids: 184. 2019. TYPE: India, Anamally forests, *Cotton s.n.* (holotype, W).

Saccolabium chrysanthum Alston in Trimen, Handb. Fl. Ceylon 6: 277. 1931. *Seidenfadeniella chrysantha* (Alston) C. S. Kumar in C. Sathish Kumar & K. S. Manilal, Cat. Ind. Orchids 47. 1994. *Schoenorchis chrysantha* (Alston) Garay, Bot. Mus. Leafl. 23(4): 202. 1972. TYPE: Sri Lanka, *Thwaites s.n.* [*Ceylon Plants 633*] (?holotype, CAL).

Distribution. Kerala, Tamil Nadu [Sri Lanka].

Habit. Epiphytic herb.

Seidenfadeniella rosea (Wight) C. S. Kumar in C. Sathish Kumar & K. S. Manilal, Cat. Indian Orchids 46. 1994. *Sarcanthus roseus* Wight, Icon. Pl. Ind. Orient. [Wight] 5(1): 10, t. 1685. 1851. *Schoenorchis roseus* (Wight) Bennett, J. Econ. Taxon. Bot. 6(2): 456. 1985. *Saccolabium roseum* (Wight) Lindl., J. Proc. Linn. Soc., Bot. 3: 36. 1859 [1858], nom. illeg., non Lindl., 1833. *Cleisostomopsis rosea* (Wight) R. Rice, Photo Intro Asian *Bulbophyllum, Coelogyne* & *Dendrobium* Orchids: 184. 2019 (as "*roseus*"). TYPE: India, Nilgiri Hills, Neddawuttim, Aug.–Sep., *Wight s.n.* (holotype, K).

Distribution. ?Arunachal Pradesh, ?Assam, Karnataka, Kerala, Tamil Nadu. Endemic.

Habit. Epiphytic herb.

Sirhookera Kuntze, Revis. Gen. Pl. 2: 681. 1891.

TYPE: *Sirhookera lanceolata* (Wight) Kuntze (*Josephia lanceolata* Wight).

Sirhookera lanceolata (Wight) Kuntze, Revis. Gen. Pl. 2: 681. 1891. *Josephia lanceolata* Wight, Icon. Pl. Ind. Orient. [Wight] 5(1): 19, t. 1742. 1851.

TYPE: India, Nilgiri Hills, Nedawuttin, Aug.–Sep., *Wight s.n.* (holotype, K).

Distribution. Karnataka, Kerala, Tamil Nadu [Sri Lanka].

Habit. Epiphytic herb.

Sirhookera latifolia (Wight) Kuntze, Revis. Gen. Pl. 2: 681. 1891. *Josephia latifolia* Wight, Icon. Pl. Ind. Orient. [Wight] 5(1): 19, t. 1743. 1851. TYPE: India, Pulney Hills, Aug.–Sep., *Wight s.n.* (holotype, K; isotype, MH).

Distribution. Karnataka, Kerala, Tamil Nadu [Sri Lanka].

Habit. Epiphytic herb.

Smithsonia C. J. Saldanha, J. Bombay Nat. Hist. Soc. 71: 73. 1974.

TYPE: *Smithsonia straminea* C. J. Saldanha.

Smithsonia maculata (Dalzell) C. J. Saldanha, J. Bombay Nat. Hist. Soc. 71(1): 74. 1974. *Micropera maculata* Dalzell, Hooker's J. Bot. Kew Gard. Misc. 3: 282. 1851. *Sarcochilus maculatus* (Dalzell) Benth. ex Pfitzer, Grundz. Morph. Orchid. 15. 1881. *Saccolabium maculatum* (Dalzell) Hook. f., Fl. Brit. India [J. D. Hooker] 6(17): 64. 1890. *Gastrochilus maculatus* (Dalzell) Kuntze, Revis. Gen. Pl. 2: 661. 1891. *Loxoma maculatum* (Dalzell) Garay, Bot. Mus. Leafl. 23(4): 184. 1972. *Loxomorchis maculata* (Dalzell) Rauschert, Taxon 31(3): 561. 1982. TYPE: India, Maharashtra, Mt. Syhadree, Tulkut, *Dalzell s.n.* (holotype, K).

Distribution. Karnataka, Kerala, Maharashtra, Tamil Nadu. Endemic.

Habit. Epiphytic herb.

Smithsonia straminea C. J. Saldanha, J. Bombay Nat. Hist. Soc. 71: 73. 1974. *Loxoma straminea* (C. J. Saldanha) Pradhan, Indian Orchids: Guide Identif. & Cult. 2: 718. 1979. *Gastrochilus stramineus* (C. J. Saldanha) R. Rice, Photo Intro Vandoid Orchid Gen. Asia, Rev. Ed.: 159. 2018 (as "*straminea*"). TYPE: India, Karnataka, Mysore, Hassan District, 14 Apr. 1969, *Saldanha 13361* (holotype, St. Joseph's College, Bangalore).

Distribution. ?Arunachal Pradesh, Goa, Karnataka, Kerala, Maharashtra. Endemic.

Habit. Epiphytic herb.

Smithsonia viridiflora (Dalzell) C. J. Saldanha, J. Bombay Nat. Hist. Soc. 71: 75. 1974. *Micropera viridiflora* Dalzell, Hooker's J. Bot. Kew Gard. Misc. 3: 282. 1851. *Saccolabium viridiflorum* (Dalzell) Lindl., J. Linn. Proc. Soc., Bot. 3: 36. 1858. *Gastrochilus viridiflorus* (Dalzell) Kuntze, Revis. Gen. Pl. 2: 661. 1891. *Sarcochilus viridiflorus* (Dalzell) T. Cooke, Fl. Bombay 2(4): 697. 1907, nom. illeg., non (Thwaites) Hook. f., 1890. *Sarcochilus dalzellianus* Santapau, Kew Bull. 1948: 498. 1949. *Gastrochilus dalzellianus* (Santapau) Santapau & Kapadia, J. Bombay Nat. Hist. Soc. 59: 842, t. 52a-c. 1962. *Aerides dalzelliana* (Santapau) Garay, Bot. Mus. Leafl. 23(4): 158. 1972. *Loxoma viridiflora* (Dalzell) Pradhan, Indian Orchids: Guide Identif. & Cult. 2: 718, 522. 1979. TYPE: India, Maharashtra, Mt. Syhadree, Tulkut, *Dalzell s.n.* (holotype, K).

Distribution. Goa, Karnataka, Kerala, Maharashtra, Tamil Nadu. Endemic.

Habit. Epiphytic herb.

———

Smitinandia Holttum, Gard. Bull. Straits Settlem. 25: 105. 1969.

TYPE: *Smitinandia micrantha* (Lindl.) Holttum (*Saccolabium micranthum* Lindl.).

Smitinandia helferi (Hook. f.) Garay, Bot. Mus. Leafl. 23(4): 204. 1972. *Saccolabium helferi* Hook. f., Fl. Brit. India [J. D. Hooker] 6(17): 57. 1890. *Gastrochilus helferi* (Hook. f.) Kuntze, Revis. Gen. Pl. 2: 661. 1891. SYNTYPES: India, Tenasserim or Andaman Islands, *Helfer s.n.* (syntype, CAL); Moulmein [Mawlamyine], *C. S. P. Parish s.n.*; Mergui [Myeik], *Griffith s.n.*

Distribution. Andaman & Nicobar Islands [China, Myanmar].

Habit. Epiphytic herb.

Smitinandia micrantha (Lindl.) Holttum, Gard. Bull. Singapore 25: 106. 1969. *Saccolabium micranthum* Lindl., Gen. Sp. Orchid. Pl.: 220. 1833. *Gastrochilus parviflorus* Kuntze, Revis. Gen. Pl. 2: 661. 1891. *Cleisostoma micranthum* (Lindl.) King & Pantl., Ann. Roy. Bot. Gard. (Calcutta) 8: 234, t. 312. 1898. *Ascocentrum micranthum* (Lindl.) Holttum, Gard. Bull. Singapore 11: 275. 1947. SYNTYPES: Nepal, Bhempeds, May 1821, *Wallich 7300A* (syntype, K-WALL); Bangladesh or

India, Sylhet, *Wallich 7300B (leg. De Silva s.n.)* (syntypes, K-LINDL, K-WALL).

Distribution. Arunachal Pradesh, Assam, Jharkhand, Meghalaya, Nagaland, Odisha, Sikkim, Uttarakhand, West Bengal [Bangladesh, Bhutan, China, Myanmar, Nepal].

Habit. Epiphytic herb.

———

Spathoglottis Blume, Bijdr. Fl. Ned. Ind. 8: 400. 1825.

TYPE: *Spathoglottis plicata* Blume.

Spathoglottis arunachalensis J. Tsering & K. Prasad, Phytotaxa 432(3): 289. 2020. TYPE: India, Arunachal Pradesh, West Kameng district, Sessa Orchid Sanctuary, 1235 m, 4 Oct. 2016, *J. Tsering 40642* (holotype, OHT; isotype, CAL, OHT).

Distribution. Arunachal Pradesh. Endemic.

Habit. Terrestrial herb.

Spathoglottis ixioides (D. Don) Lindl., Gen. Sp. Orchid. Pl.: 120. 1831. *Cymbidium ixioides* D. Don, Prodr. Fl. Nepal.: 36. 1825. TYPE: Nepal, Bagmati Zone, Rashuwa District, Gosainthan [Shishapangma], Aug. 1821, *Wallich 3745* (holotype, K-LINDL; isotype, K-WALL).

Distribution. Arunachal Pradesh, Meghalaya, Nagaland, Sikkim [Bhutan, Nepal].

Habit. Terrestrial or lithophytic herb.

Spathoglottis plicata Blume, Bijdr. Fl. Ned. Ind. 8: 401, t. 76. 1825. TYPE: Indonesia, Java, *Blume 1884* (holotype, L; isotype, L, P).

Distribution. Andaman & Nicobar Islands, Arunachal Pradesh, Assam, Nagaland [Bangladesh, Myanmar, Sri Lanka (introduced)].

Habit. Terrestrial herb.

Spathoglottis pubescens Lindl., Gen. Sp. Orchid. Pl.: 120. 1831. *Spathoglottis plicata* Blume var. *pubescens* (Lindl.) M. Hiroe, Orchid Flowers 2: 89. 1971. SYNTYPES: India, Sylhet, Jentya (Jaintia), July 1820, Cherrapunji, Aug. 1829, Gentea [Jaintia] Hills, July 1830, *Wallich 3744C (leg. H. Bruce s.n., 119, and 184 respectively)* (syntypes, K-WALL [3×]); Myanmar, Pegu, Prome, 1826, *Wallich 3744A* (syntype, K-WALL); Myanmar, Tong Dong, 26 Nov. 1826, *Wallich 3744B* (syntype, K-WALL); *Wallich*

3744 [one of the preceding, but not clear which] (syntype, K-LINDL).

Epipactis graminifolia Roxb., Fl. Ind. ed. 1832, 3: 456. 1832. *Pogonia graminifolia* (Roxb.) Voigt, Hort. Suburb. Calcutt.: 632. 1845. TYPE: Bangladesh or India, Sylhet, icon. *Epipactis graminifolia*, *Roxburgh 2546* (syntypes, CAL, K).

Spathoglottis parvifolia Lindl., Edwards's Bot. Reg. 31: t. 19. 1845. *Spathoglottis pubescens* Lindl. var. *parvifolia* (Lindl.) Hook. f., Fl. Brit. India [J. D. Hooker] 5(16): 814. 1890. TYPE: India, Khasi Hills, *Griffith s.n.* (holotype, K-LINDL; isotype, K).

Spathoglottis khasyana Griff., Not. Pl. Asiat. 3: 323. 1851. TYPE: India, Khasi Hills, Churra, *Griffith s.n.* (not found).

Distribution.   Andaman & Nicobar Islands, Arunachal Pradesh, Manipur, Meghalaya, Mizoram, Nagaland, Tripura [China, Myanmar].

Habit.   Terrestrial herb.

———

Spiranthes Rich., De Orchid. Eur. 20, 28, 36. 1817, nom. cons.

TYPE: *Spiranthes autumnalis* Rich., nom. superfl. (*Ophrys spiralis* L., *Spiranthes spiralis* (L.) Chevall.) (typ. cons.).

Spiranthes australis (R. Br.) Lindl., Bot. Reg. 10: t. 823. 1824. *Neottia australis* R. Br., Prodr. Fl. Nov. Holland. 1: 319. 1810. *Gyrostachys australis* (R. Br.) Blume, Coll. Orchid.: 128. 1859. *Spiranthes sinensis* (Pers.) Ames subsp. *australis* (R. Br.) Kitam., Acta Phytotax. Geobot. 21(1–2): 23. 1964. TYPE: Australia: Port Jackson, May 1802, *Brown s.n.* [*5541*] (lectotype, BM; isotype, AMES, E).

Spiranthes longispicata A. Rich., Ann. Sci. Nat., Bot. sér. 2, 15: 78. 1841. TYPE: India, Nilgiri Hills, Otacamund, Sep.–Nov., *Perrottet s.n.* (holotype, P; isotype, P).

Spiranthes australis (R. Br.) Lindl. var. *wightiana* Hook. f., Fl. Brit. India [J. D. Hooker] 6(17): 102. 1890. *Spiranthes wightiana* Lindl., Numer. List [Wallich] n. 7378. 1832, nom. inval. *Gyrostachys wightiana* (Lindl.) Kuntze, Revis. Gen. Pl. 2: 663. 1891. *Spiranthes sinensis* (Pers.) Ames var. *wightiana* (Hook. f.) C. E. C. Fisch., Fl. Madras 3(8): 1454. 1928. TYPE: *Wallich 7378 (leg. Wight s.n.)* (holotype, K-WALL).

Distribution.   Arunachal Pradesh, Assam, Himachal Pradesh, ?Maharashtra, Manipur, Meghalaya, Mizoram, Nagaland, Odisha, Punjab, Sikkim, Tamil Nadu, Tripura, Uttar Pradesh, Uttarakhand, West Bengal [Bangladesh, Bhutan, China, Myanmar, Nepal, Pakistan, Sri Lanka].

Habit.   Terrestrial herb.

Remarks.   *Spiranthes sinensis* (Pers.) Ames and *S. australis* have long been considered conspecific, but a recent study by Pace et al. (2019) found them to be distinct. The former can be recognized by the glabrous as opposed to pubescent inflorescence rachis. For India, Pace et al. record *S. sinensis* only from Manipur whereas *S. australis* is more widespread. The distribution of both taxa in India needs to be reassessed. This species is illustrated as *S. sinensis* in Swami (2016 [pink-flowered plants only]).

Spiranthes flexuosa (Sm.) Lindl., Bot. Reg. 10: t. 823. 1824. *Neottia flexuosa* Sm. in A. Rees, Cycl. 24: no. 9. 1813. *Gyrostachys australis* (R. Br.) Blume var. *flexuosa* (Sm.) Blume, Fl. Javae: 130. 1859. TYPE: Nepal, Suembu, 1 May 1802, *Buchanan s.n.* (holotype, LINN).

Spiranthes densa A. Rich., Ann. Sci. Nat., Bot. sér. 2, 15: 79. 1841. TYPE: India, Nilgiti Hills, Dodabetta, June, *Perrottet s.n.* (holotype, P; isotype, P).

Spiranthes himalayensis Survesw., Kumar & Mei Sun, Phytokeys 89: 118. 2017. TYPE: India, Manipur, Ukrul district, Imphal-Jessami road, 1 May 2016. S. Surveswaran 1 (HJCB 1001) (holotype, JCB).

Distribution.   Himachal Pradesh, Jammu & Kashmir, Karnataka, Kerala, Manipur, Sikkim, Tamil Nadu [China, Myanmar, Nepal, Sri Lanka].

Habit.   Terrestrial herb.

Remarks.   *Spiranthes flexuosa* differs from *S. australis* mainly in its spring-flowering phenology, whereas *S. australis* and the closely related *S. sinensis* flower in late summer–autumn (Pace et al., 2019). Surveswaran et al. (2020) dispute that *S. flexuosa* is an earlier name for *S. himalayensis* and suggest that more molecular studies are needed to resolve the taxonomy of Asian *Spiranthes*. This species is illustrated as *S. sinensis* in Swami (2016 [white-flowered plant only], 2017).

Spiranthes sinensis (Pers.) Ames, Orchidaceae (Ames) 2: 53. 1908. *Aristotelea spiralis* Lour., Fl. Cochinch. 522. 1790. *Epidendrum aristotelea* Raeusch., Nomencl. Bot., ed. 3: 265. 1797, nom. superfl. *Neottia sinensis* Pers., Syn. Pl. (Persoon) 2: 511. 1807. *Gyrostachys australis* (R. Br.) Blume var. *sinensis* (Pers.) Blume, Fl. Javae Nov. Ser.: 108. 1859. *Spiranthes aristotelea* Merr., Philipp. J. Sci. 15: 230. 1919, nom. superfl. *Spiranthes australis* (R. Br.) Lindl. var. *sinensis* (Pers.) Gagnep. in H. Lecomte, Fl. Indo-Chine 6: 546. 1933, nom. superfl. *Spiranthes spiralis* (Lour.) Makino, J. Jap. Bot. 3: 25. 1926, nom. illeg., non (L.) Chevall., 1827. *Ibidium spirale* (Lour.) Makino, J. Jap. Bot. 6: 37. 1929. TYPE: China, Canton [Guangzhou], *Loureiro s.n.* [*522-1*] (holotype, P).

Distribution.   Manipur [China].

Habit.   Terrestrial herb.

Stereochilus Lindl., J. Proc. Linn. Soc., Bot. 3: 38. 1859 [1858].

TYPE: *Stereochilus hirtus* Lindl.

Stereochilus arunachalensis Chowlu & A. N. Rao, Phytotaxa 433(2): 177. 2020. TYPE: India, Arunachal Pradesh, Lower Dibang Valley District, Hunli, 1000 m, 4 June 2018, *Chowlu 41000* (holotype, ARUN).

Distribution.   Arunachal Pradesh. Endemic.

Habit.   Epiphytic herb.

Stereochilus hirtus Lindl., J. Proc. Linn. Soc., Bot. 3: 38. 1859 [1858]. *Sarcanthus hirtus* (Lindl.) J. J. Sm., Natuurk. Tijdschr. Ned.-Indië 72: 87. 1912. *Sarcochilus hirtus* (Lindl.) Hook. f., Fl. Brit. India [J. D. Hooker] 6(17): 35. 1890. TYPE: India, Khasi Hills, alt. 5000 ft, *Hooker & Thomson 177* (holotype, K-LINDL).

Distribution.   Arunachal Pradesh, Assam, Meghalaya, Sikkim, West Bengal [?Myanmar].

Habit.   Epiphytic herb.

Stereochilus laxus (Rchb. f.) Garay, Bot. Mus. Leafl. 23: 205. 1972. *Sarcanthus laxus* Rchb. f., Bot. Zeitung (Berlin) 24: 378. 1866. TYPE: Myanmar, Moulmein (Mawlamyine), *cult. Royal Botanic Gardens, Kew s.n. (leg. C. S. P. Parish s.n.), October 1865* (holotype, K).

Distribution.   Nagaland [Myanmar].

Habit.   Epiphytic herb.

Remarks.   The Kew specimen, which we assume to be the type, is dated 20 Nov. 1863. In the protologue, Reichenbach gives October 1865. Whether this was a mistake or a reference to a more recent flowering event we cannot say.
      The record from Nagaland is due to J. Kamba et al. (article in prep.).

Stereochilus ringens (Rchb. f.) Garay, Bot. Mus. Leafl. 23(4): 205. 1972. *Sarcanthus ringens* (Rchb. f.) J. J. Sm., Natuurk. Tijdschr. Ned.-Indië 72: 92. 1912. *Cleisostoma ringens* Rchb. f., Gard. Chron., ser. 3, 4: 724. 1888. TYPE: Philippines [doubtful], Sep. 1883, *cult. O' Brien s.n.* (holotype, W).

Stereochilus wattii King & Pantl., J. Asiat. Soc. Bengal, Pt. 2, Nat. Hist. 66: 595. 1898 [1897]. TYPE: India, Assam,

Dikka River, alt. 1000 ft, *G. Watt 542* (?holotype, K [Assam, Aug. 1896, *Watt s.n.*]).

Distribution.   Assam [also, doubtfully, recorded from the Philippines].

Habit.   Epiphytic herb.

Stereosandra Blume, Mus. Bot. 2: 176. 1856.

TYPE: *Stereosandra javanica* Blume.

Stereosandra javanica Blume, Mus. Bot. 2: 176. 1856. TYPE: Indonesia, Java, Bantam Prov. near Harriang, *Kuhl & van Hasselt s.n.* (holotype, L).

Distribution.   Arunachal Pradesh [China, Myanmar].

Habit.   Holomycotrophic terrestrial herb.

Stigmatodactylus Maxim. ex Makino, Ill. Fl. Japan 1(7): 81, t. 43. 1891.

TYPE: *Stigmatodactylus sikokianus* Maxim. ex Makino.

Stigmatodactylus paradoxus (Prain) Schltr., Repert. Spec. Nov. Regni Veg. 10: 4. 1911. *Pantlingia paradoxa* Prain, J. Asiat. Soc. Bengal, Pt. 2, Nat. Hist. 65: 107. 1896. *Stigmatodactylus sikokianus* Maxim. ex Makino var. *paradoxus* (Prain) Maek., Wild Orchids Japan Colour: 187. 1971. TYPE: India, Sikkim Himalaya, Chungtong, alt. 6000 ft, Sep. 1895, *Pantling 420* (holotype, CAL; isotype, AMES, BM, K, L).

Distribution.   Assam, Sikkim [Bhutan].

Habit.   Terrestrial herb.

Stigmatodactylus serratus (Deori) A. N. Rao, J. Econ. Taxon. Bot. 9(2): 255. 1987. *Pantlingia serrata* Deori, Bull. Bot. Surv. India 20(1–4): 175. 1978 [1979]. TYPE: India, Meghalaya, Shillong Peak, alt. 6400 ft, *N. C. Deori 51262A* (holotype, CAL).

Distribution.   Assam, Meghalaya. Endemic.

Habit.   Terrestrial herb.

Strongyleria (Pfitzer) Schuit., Y. P. Ng & H. A. Pedersen, Bot. J. Linn. Soc. 186: 201. 2018.

TYPE: *Strongyleria pannea* (Lindl.) Schuit., Y. P. Ng & H. A. Pedersen (*Eria pannea* Lindl.).

Strongyleria pannea (Lindl.) Schuit., Y. P. Ng & H. A. Pedersen, Bot. J. Linn. Soc. 186: 201. 2018. *Eria pannea* Lindl., Edwards's Bot. Reg. 28(Misc.): 64. 1842. *Pinalia pannea* (Lindl.) Kuntze, Revis. Gen. Pl. 2: 679. 1891. *Mycaranthes pannea* (Lindl.) S. C. Chen & J. J. Wood, Fl. China 25: 348. 2009. TYPE: Singapore, *Messrs. Loddiges 252* (?syntypes, K-LINDL [specimen and colored drawing, without annotation]).

Eria calamifolia Hook. f., Fl. Brit. India [J. D. Hooker] 6(17): 191. 1890. *Pinalia calamifolia* (Hook. f.) Kuntze, Revis. Gen. Pl. 2: 679. 1891. TYPE: India, Upper Assam, Makum forest, *Mann s.n.* (holotype, K).

Distribution. Arunachal Pradesh, Assam, Manipur, Meghalaya, Mizoram, Nagaland, Sikkim, West Bengal [Bhutan, China, Myanmar].

Habit. Epiphytic herb.

———

Taeniophyllum Blume, Bijdr. Fl. Ned. Ind. 8: 355. 1825.

TYPE: *Taeniophyllum obtusum* Blume.

Taeniophyllum alwisii Lindl., J. Proc. Linn. Soc., Bot. 3: 42. 1859 [1858]. TYPE: Sri Lanka, *De Alwiz s.n.* (holotype, K-LINDL, colored drawing only). *Alwisia minuta* Thwaites ex Lindl., J. Proc. Linn. Soc., Bot. 3: 42. 1859 [1858].

Distribution. Kerala, Tamil Nadu [Sri Lanka].

Habit. Epiphytic herb.

Taeniophyllum andamanicum N. P. Balakr. & N. Bhargava, Bull. Bot. Surv. India 20(1–4): 154. 1978 [1979]. TYPE: India, Andaman Island, South Andamans, Baratang Island, Nilambur, 17 July 1977, *N. Bhargava 5932A* (holotype, CAL).

Distribution. Andaman & Nicobar Islands. Endemic.

Habit. Epiphytic herb.

Taeniophyllum arunachalense A. N. Rao & J. Lal, Bioved 2(2): 225. 1991. TYPE: India, Arunachal Pradesh, Pampare Dist., Doimukh forest, 100 m, *J. Lal 437* (holotype, ARUN).

Distribution. Arunachal Pradesh. Endemic.

Habit. Epiphytic herb.

Taeniophyllum campanulatum Carr, Gard. Bull. Straits Settlem. 7: 68. 1932. TYPE: Malaya Penun-sula, Pahang, Fraser Hill, alt. 4000 ft, *Carr s.n.* (holotype, SING).

Distribution. Arunachal Pradesh [also in Malaysia].

Habit. Epiphytic herb.

Taeniophyllum crepidiforme (King & Pantl.) King & Pantl., Ann. Roy. Bot. Gard. (Calcutta) 8: 245, t. 325. 1898. *Sarcochilus crepidiformis* King & Pantl., J. Asiat. Soc. Bengal, Pt. 2, Nat. Hist. 64: 340. 1895 [1896]. TYPE: India, Sikkim, Sep., *Pantling s.n.* [*193*] (holotype, CAL).

Distribution. Arunachal Pradesh, Assam, ?Meghalaya, Sikkim, West Bengal. Endemic.

Habit. Epiphytic herb.

Taeniophyllum filiforme J. J. Sm., Bull. Inst. Bot. Buitenzorg 7: 4. 1900. TYPE: Indonesia, North Sulawesi, Bone near Gorontalo, *J. J. Smith s.n.* (holotype, BO).

Distribution. Andaman & Nicobar Islands [also in Indonesia].

Habit. Epiphytic herb.

Taeniophyllum glandulosum Blume, Bijdr. Fl. Ned. Ind. 8: 356. 1825. TYPE: Indonesia, Java, Mt. Gede/Panggerango, *Blume s.n.* (holotype, L; isotype, L).

Taeniophyllum khasianum J. Joseph & Yogan., J. Indian Bot. Soc. 46: 109. 1967. TYPE: India, Sillong, Laithumkarh, alt. 4500 ft, 21 May 1966, *Joseph 35593* (holotype, CAL).

Distribution. Assam, Meghalaya, Nagaland [China, Myanmar].

Habit. Epiphytic herb.

Taeniophyllum retrospiculatum (King & Pantl.) King & Pantl., Ann. Roy. Bot. Gard. (Calcutta) 8: 244, t. 324. 1898. *Sarcochilus retrospiculatus* King & Pantl., J. Asiat. Soc. Bengal, Pt. 2, Nat. Hist. 64: 340. 1895 [1896] (as "*retro-spiculatus*"). TYPE: India, Sikkim, 5000 [6000] ft, June, *Pantling s.n.* [*165*] (holotype, CAL; isotype, CAL, K [head of Ryang Valley, 6000 ft, May–June 1891]).

Distribution. Sikkim, West Bengal [Bhutan].

Habit. Epiphytic herb.

Taeniophyllum scaberulum Hook. f., Fl. Brit. India [J. D. Hooker] 6(17): 77. 1890. TYPE: India, Travancore, Cottayam, *Johnson s.n.* (holotype, K).

Distribution.   Kerala, Tamil Nadu [Nepal].

Habit.   Epiphytic herb.

Taeniophyllum stella Carr, Gard. Bull. Straits Settlem. 7: 69. 1932. TYPE: Malaya Peninsula, Pahang, Tembeling, *Carr s.n.* (holotype, SING).

Distribution.   Arunachal Pradesh [also in Malaysia].

Habit.   Epiphytic herb.

———

Tainia Blume, Bijdr. Fl. Ned. Ind. 7: 354. 1825.

TYPE: *Tainia speciosa* Blume.

Tainia bicornis (Lindl.) Rchb. f., Bonplandia 5(3): 54. 1857. *Ania bicornis* Lindl. Edwards's Bot. Reg. 28(Misc.): 37. 1842. *Mitopetalum bicorne* (Lindl.) Blume, Mus. Bot. 2: 185. 1856. *Eria bicornis* (Lindl.) Rchb. f., Ann. Bot. Syst. (Walpers) 6: 269. 1861. TYPE: Sri Lanka, *Rev. J. Clowes s.n.* (holotype, K-LINDL).

Distribution.   Karnataka, Kerala, Tamil Nadu [Sri Lanka].

Habit.   Terrestrial herb.

Tainia latifolia (Lindl.) Rchb. f., Bonplandia 5: 54. 1857. *Ania latifolia* Lindl., Gen. Sp. Orchid. Pl.: 130. 1831. *Mitopetalum latifolium* (Lindl.) Blume, Mus. Bot. 2: 185. 1856. TYPE: India, Sylhet, Jentya [Jaintia] Mountains, June 1829, *Wallich 3741 (leg. H. Bruce 241)* (holotype, K-LINDL; isotype, K).

Cymbidium fuscescens Griff., Notul. Pl. Asiat. 3: 405 (nomen), 343 (descr.). 1851. TYPE: Myanmar, Namtuseek, Patkaye [Patkai] Hills, 3500 ft, 17 Mar. 1837, *Griffith s.n.*
Tainia cordata Hook. f., Fl. Brit. India [J. D. Hooker] 6(17): 193. 1890. TYPE: India, Sikkim Himalaya, icon. (holotype, CAL).
Eulophia hastata Lindl., J. Proc. Linn. Soc., Bot. 3: 25. 1859 [1858]. *Tainia hastata* (Lindl.) Hook. f., Fl. Brit. India [J. D. Hooker] 5(16): 821. 1890. TYPE: India, Assam, *Griffith s.n.* (holotype, K-LINDL).
Eria angulata Rchb. f., Flora 55: 275. 1872. *Pinalia angulata* (Rchb. f.) Kuntze, Revis. Gen. Pl. 2: 679. 1891. *Tainia angulata* (Rchb. f.) Benth. ex Kraenzl., Pflanzenr. (Engler) IV 50(50): 165. 1911. TYPE: India, Assam, *Mann s.n.* (holotype, W).
Tainia cordata Hook. f., Fl. Brit. India [J. D. Hooker] 6(17): 193. 1890. TYPE: India, Sikkim, *icon. Hooker f.* (holotype, CAL).
Tainia khasiana Hook. f., Fl. Brit. India [J. D. Hooker] 5(16): 821. 1890. TYPE: India, Khasi Hills, below Churra, alt. 3000 ft, *Hooker & Thomson s.n.* (holotype, K).

Distribution.   Arunachal Pradesh, Assam, Manipur, Meghalaya, Mizoram, Nagaland, Sikkim, West Bengal [Bangladesh, China, Myanmar].

Habit.   Terrestrial herb.

Tainia megalantha (Tang & F. T. Wang) Sushil K. Singh, Agrawala & Jalal, Orchids India: 487. 2019 (as "*megalanthum*"). *Mischobulbum megalanthum* Tang & F. T. Wang, Acta Phytotax. Sin. 1: 76. 1951. *Mischobulbum grandiflorum* Rolfe, Orchid Rev. 20: 127. 1912, nom. illeg., non Schltr., 1911. TYPE: India, Sikkim Himalaya, tropical valleys, Mar. 1896, *Pantling 206* (holotype, CAL; isotype, AMES, FI, K, W).

Distribution.   Arunachal Pradesh, Sikkim, West Bengal. Endemic.

Habit.   Terrestrial herb.

Tainia minor Hook. f., Fl. Brit. India [J. D. Hooker] 5(16): 821. 1890. TYPE: India, Sikkim, Mahalderam, alt. 7000 ft, 28 May 1884, *C. B. Clarke s.n.* [*35517*] (holotype, K; isotype, BM, G, K).

Distribution.   Arunachal Pradesh, Meghalaya, Nagaland, Sikkim, West Bengal [China, Myanmar].

Habit.   Terrestrial herb.

Tainia wrayana (Hook. f.) J. J. Sm., Bull. Jard. Bot. Buitenzorg, sér. 2, 8: 6. 1912. *Ipsea wrayana* Hook. f., Fl. Brit. India [J. D. Hooker] 5(16): 812. 1890. *Mischobulbum wrayanum* (Hook. f.) Rolfe, Orchid Rev. 20: 127. 1912. TYPE: Malaysia: Perak, Gunong Batu Pateh, alt. 4500 ft, *Wray 235* (holotype, K; isotype, CAL).

Nephelaphyllum grandiflorum Hook. f., Fl. Brit. India [J. D. Hooker] 6(17): 192. 1890. *Mischobulbum grandiflorum* (Hook. f.) Schltr., Repert. Spec. Nov. Regni Veg. Beih. 1: 98. 1911. *Tainia grandiflora* (Hook. f.) Gagnep., Bull. Mus. Natl. Hist. Nat., sér. 2, 4: 706. 1932. SYNTYPES: Malaysia: Perak, *Scortechini s.n.*; Malaya, *Kunstler s.n.*

Distribution.   Andhra Pradesh, Arunachal Pradesh, Assam, ?Manipur, West Bengal [also in Indonesia].

Habit.   Terrestrial herb.

———

Taprobanea Christenson, Lindleyana 7(2): 90. 1992.

TYPE: *Taprobanea spathulata* (L.) Christenson (*Epidendrum spathulatum* L.).

Taprobanea spathulata (L.) Christenson, Lindleyana 7(2): 91. 1992. *Epidendrum spathulatum* L., Sp.

Pl. 2: 952. 1753. *Limodorum spathulatum* (L.) Willd., Sp. Pl., ed. 4. 4: 125. 1805. *Cymbidium spathulatum* (L.) Moon, Cat. Pl. Ceylon 60. 1824. *Vanda spathulata* (L.) Spreng., Syst. Veg. (ed. 16) [Sprengel] 3: 719. 1826. TYPE: India, icon. "Ponnampu-maravara" in Rheede, Hort. Malab. 12: 7, t. 3. 1692 (lectotype, designated by Majumdar & Bakshi, 1979).

Aerides maculata Buch.-Ham. ex Sm. in A. Rees, Cycl. 39(1): no. 9. 1818. TYPE: India, Karnataka, Mysore, *Buchanan-Hamilton s.n.*

Distribution. Andhra Pradesh, Karnataka, Kerala, Tamil Nadu [Sri Lanka].

Habit. Climber.

———

Thelasis Blume, Bijdr. Fl. Ned. Ind. 8: 385. 1825.

TYPE: *Thelasis obtusa* Blume.

Thelasis bifolia Hook. f., Fl. Brit. India [J. D. Hooker] 6(17): 86. 1890. TYPE: India, Khasi Hills, *T. Lobb s.n.* (holotype, K).

Distribution. Assam, Meghalaya, Sikkim, West Bengal. Endemic.

Habit. Epiphytic herb.

Thelasis khasiana Hook. f., Fl. Brit. India [J. D. Hooker] 6(17): 87. 1890. *Thelasis pygmaea* (Griff.) Lindl. var. *khasiana* (Hook. f.) Schltr., Mem. Herb. Boissier 8(21): 71. 1900. SYNTYPES: India, Khasi Hills, 3000–4000 ft, *T. Lobb s.n.*; *Hooker & Thomson 29A* (syntypes, K, K-LINDL).

Distribution. Andaman & Nicobar Islands, Arunachal Pradesh, ?Goa, Meghalaya, Tripura [also in Vietnam].

Habit. Epiphytic herb.

Thelasis longifolia Hook. f., Fl. Brit. India [J. D. Hooker] 6(17): 87. 1890. TYPE: India, Khasi Hills, *Hooker & Thomson s.n.* (holotype, K-LINDL; isotype, K).

Distribution. Arunachal Pradesh, Bihar, Madhya Pradesh, Manipur, Meghalaya, Nagaland, Sikkim, Uttarakhand, West Bengal [Nepal].

Habit. Epiphytic herb.

Thelasis pygmaea (Griff.) Lindl., J. Proc. Linn. Soc., Bot. 3: 63. 1859 [1858]. *Euproboscis pygmaea* Griff., Calcutta J. Nat. Hist. 5: 372, t. 26. 1844.

TYPE: Nepal, Apr. 1844, *H. Lawrence s.n.* (not found).

Thelasis pygmaea (Griff.) Lindl. var. *multiflora* Hook. f., Fl. Brit. India [J. D. Hooker] 6(17): 86. 1890. TYPE: India, Sikkim, *Treutler 631* (holotype, K).

Distribution. Andaman & Nicobar Islands, Arunachal Pradesh, Assam, Karnataka, Kerala, Manipur, Meghalaya, Sikkim, Tamil Nadu, West Bengal [China, Myanmar, Nepal].

Habit. Epiphytic herb.

———

Thrixspermum Lour., Fl. Cochinch. 2: 516, 519. 1790.

TYPE: *Thrixspermum centipeda* Lour.

Thrixspermum acuminatissimum (Blume) Rchb. f., Xenia Orchid. 2: 121. 1868. *Dendrocolla acuminatissima* Blume, Bijdr. Fl. Ned. Ind. 7: 288. 1825. *Aerides acuminatissima* (Blume) Lindl., Gen. Sp. Orchid. Pl.: 240. 1833. *Sarcochilus acuminatissimus* (Blume) Rchb. f., Ann. Bot. Syst. (Walpers) 6: 498. 1863. TYPE: Indonesia, Java, Mt. Pantjar, *Blume s.n.* (holotype, L).

Distribution. Assam [also in Indonesia, Thailand, etc.].

Habit. Epiphytic herb.

Thrixspermum centipeda Lour., Fl. Cochinch. 2: 520. 1790. *Sarcochilus centipeda* (Lour.) Náves, Fl. Filip., ed. 3 (Blanco). 13A: 238. 1880. TYPE: Vietnam, Hué, *Loureiro s.n.* (holotype, BM).

Dendrocolla arachnites Blume, Bijdr. Fl. Ned. Ind. 7: 287. 1825. *Aerides arachnites* (Blume) Lindl., Gen. Sp. Orchid. Pl.: 238. 1833, nom. illeg., non Sw., 1800. *Sarcochilus arachnites* (Blume) Rchb. f., Ann. Bot. Syst. (Walpers) 6: 498. 1863. *Thrixspermum arachnites* (Blume) Rchb. f., Xenia Orchid. 2: 121. 1868. TYPE: Indonesia, Java, foot of Mt. Salak, *Blume s.n.*

Distribution. Andaman & Nicobar Islands, Arunachal Pradesh, Assam, Kerala, Manipur, Meghalaya, Mizoram, Sikkim, Tamil Nadu, West Bengal [Bangladesh, China, Myanmar].

Habit. Epiphytic herb.

Thrixspermum crassilabre (King & Pantl.) Ormerod, Orchid Memories: 59. 2004. *Saccolabium crassilabre* King & Pantl., J. Asiat. Soc. Bengal, Pt. 2, Nat. Hist. 66: 593. 1898 [1897]. *Gastrochilus crassilabris* (King & Pantl.) Garay, Bot. Mus. Leafl.

23(4): 179. 1972. TYPE: India, Khasi Hills, alt. 3000 ft, *Pantling drawing 628* (holotype, CAL).

Distribution. Arunachal Pradesh, Assam, ?Meghalaya. Endemic.

Habit. Epiphytic herb.

Thrixspermum hystrix (Blume) Rchb. f., Trans. Linn. Soc. London 30: 145. 1874. *Dendrocolla hystrix* Blume, Bijdr. Fl. Ned. Ind. 7: 291. 1825. *Aerides hystrix* (Blume) Lindl., Gen. Sp. Orchid. Pl.: 242. 1833. *Sarcochilus hystrix* (Blume) Rchb. f., Ann. Bot. Syst. (Walpers) 6: 500. 1863. *Grosourdya hystrix* (Blume) Rchb. f., Xenia Orchid. 2: 123. 1868. TYPE: Indonesia, Java, Buitenzorg (Bogor), *Blume s.n.* (holotype, L; isotype, L).

Distribution. Andaman & Nicobar Islands [also in Indonesia].

Habit. Epiphytic herb.

Thrixspermum indicum Vik. Kumar, D. Verma & A. N. Rao, Phytotaxa 292: 79. 2017. TYPE: India, Meghalaya, East Jaintia Hills, Tuber village, 25°26'19.23"N, 92°16'29.07"E, 1382 m, 7 Dec. 2014, *V. Kumar & D. Verma 113138* (holotype, BSD).

Distribution. Manipur, Meghalaya. Endemic.

Habit. Epiphytic herb.

Thrixspermum merguense (Hook. f.) Kuntze, Revis. Gen. Pl. 2: 682. 1891. *Sarcochilus merguensis* Hook. f., Fl. Brit. India [J. D. Hooker] 6(17): 40. 1890. *Dendrocolla merguensis* (Hook. f.) Ridl., J. Linn. Soc., Bot. 32: 380. 1896. TYPE: Myanmar, Tenasserim, *Griffith 1066* (holotype, K-LINDL).

Distribution. Andaman & Nicobar Islands, Assam [Myanmar].

Habit. Epiphytic herb.

Thrixspermum musciflorum A. S. Rao & J. Joseph, Bull. Bot. Surv. India 11(1–2): 204. 1969 [1971] (as "*muscaeflorum*"). TYPE: India, Meghalaya, Khasi Hills, betw. Umran & Umsaw, beside the Gauhati-Shillong hwy., 23 July 1966, *Joseph 45621A* (holotype, ASSAM; isotype, CAL).

Thrixspermum musciflorum A. S. Rao & J. Joseph var. *nilagiricum* J. Joseph & Vajr., Indian Forester 107(10): 648, fig. 5. 1981. TYPE: India, Tamil Nadu, Nilgiri Hills, Ronning town forests, near Veerakalai Hill tribe settlement on the bank of Kundha River, 25 June 1974, *Vajravelu 44950A* (holotype, MH).

Distribution. Arunachal Pradesh, Assam, Karnataka, Kerala, Manipur, ?Meghalaya, Tamil Nadu. Endemic.

Habit. Epiphytic herb.

Thrixspermum pauciflorum (Hook. f.) Kuntze, Revis. Gen. Pl. 2: 682. 1891. *Sarcochilus pauciflorus* Hook. f., Fl. Brit. India 6: 41. 1890. TYPE: Peninsular Malaysia, Perak, Taiping Hills, *Scortechini s.n.* (holotype, K).

Thrixspermum tsii W. H. Chen & Y. M. Shui, Brittonia 57: 55. 2005. TYPE: China, Yunnan, Malipo County, Jinchang community, Xiao-pingan, ca. 1500 m, 5 June 2002, *Shui Yumin et al. 21607* (holotype, KUN; isotype, NY).

Thrixspermum changlangense K. Gogoi, Pleione 13(1): 167. 2019 (as "*changlangensis*"). TYPE: India, Arunachal Pradesh, Changlang district, Nampong, 500 m, 10 May 2018, *K. Gogoi 00802* (holotype, ASSAM; isotype, DU, TOSEHIM [The Orchid Society of Eastern Himalaya]).

Distribution. Arunachal Pradesh, Meghalaya, Nagaland [China].

Habit. Epiphytic herb.

Remarks. This was first recorded from India by Phukan and Mao (2004). O'Byrne et al. (2015) doubted this identification. However, we are unable to find significant differences between Indian specimens and *Thrixspermum pauciflorum* as illustrated from Malaysia by O'Byrne et al. We consider *T. tsii* to fall within the range of variation (while noting that the published illustration of *T. tsii* is misleading and does not agree with type material that we have seen). The recently described *T. changlangense* is not different from *T. pauciflorum* in any significant character. The record for Nagaland is due to J. Kamba et al. (pers. comm.).

Thrixspermum pulchellum (Thwaites) Schltr., Orchis 5: 57. 1911. *Dendrocolla pulchella* Thwaites, Enum. Pl. Zeyl. [Thwaites]: 430. 1864. *Sarcochilus pulchellus* (Thwaites) Trimen, Syst. Cat. Fl. Pl. Ceylon 89. 1885. TYPE: Sri Lanka, Central Province, Kornegalle, alt. up to 2000 ft, *Thwaites s.n.* [*Ceylon Plants 2354*] (holotype, K; isotype, CAL).

Distribution. Kerala [Sri Lanka].

Habit. Epiphytic herb.

Thrixspermum pygmaeum (King & Pantl.) Holttum, Kew Bull. 14: 275. 1960. *Sarcochilus pygmaeus* King & Pantl., Ann. Roy. Bot. Gard. (Calcutta) 8: 207, t. 277. 1898. TYPE: India, Sikkim, Pemiongsi, alt. 3000 ft, *Pantling 472*.

Distribution. Andhra Pradesh, Arunachal Pradesh, Assam, Karnataka, Sikkim, West Bengal [Bhutan, Nepal].

Habit. Epiphytic herb.

Thrixspermum saruwatarii (Hayata) Schltr., Repert. Spec. Nov. Regni Veg. Beih. 4: 275. 1919. *Sarcochilus saruwatarii* Hayata, Icon. Pl. Formosan. 6: 84. 1916. TYPE: Taiwan, Keitao, Apr. 1916, *B. Hayata s.n.* (holotype, TI).

Distribution. Arunachal Pradesh [Taiwan].

Habit. Epiphytic herb.

Thrixspermum trichoglottis (Hook. f.) Kuntze, Revis. Gen. Pl. 2: 682. 1891. *Sarcochilus trichoglottis* Hook. f., Fl. Brit. India [J. D. Hooker] 6(17): 39. 1890. *Dendrocolla trichoglottis* (Hook. f.) Ridl., J. Linn. Soc., Bot. 32: 381. 1896. SYNTYPES: Malaysia: Perak, *Scortechini s.n.*; *King's collector s.n.*; Singapore, *Ridley s.n.* (syntype, CAL).

Distribution. Andaman & Nicobar Islands, Assam, Kerala, ?Meghalaya [Bangladesh, Myanmar].

Habit. Epiphytic herb.

Thrixspermum walkeri Seidenf. & Ormerod in G. Seidenfaden, Descr. Epidendrorum J. G. König. 1791: 26. 1995. TYPE: Sri Lanka (Ceylon), *Ceylon Plants 3209* (holotype, K).

Distribution. Kerala [Sri Lanka].

Habit. Epiphytic herb.

Remarks. This species was mistakenly identified as *Thrixspermum complanatum* (Koenig) Schltr. by Kumar and Manilal (1994).

———

Thunia Rchb. f., Bot. Zeitung (Berlin) 10: 764. 1852.

TYPE: *Thunia alba* (Lindl.) Rchb. f. (*Phaius albus* Lindl.).

Thunia alba (Lindl.) Rchb. f., Bot. Zeitung (Berlin) 10(44): 764. 1852 var. **alba.** *Phaius albus* Lindl., Pl. Asiat. Rar. (Wallich) 2: 85, t. 198. 1831. TYPE: India, Sylhet, mountains, July 1824, *Wallich 3749B (leg. De Silva 1432)* (lectotype, K-LINDL, designated by Pearce et al., 2001; isotype, K-WALL).

Thunia marshalliana Rchb. f., Linnaea 41: 65. 1877 [1876]. *Phaius marshallianus* (Rchb. f.) N. E. Br., Bull. Misc. Inform. Kew 1889: 101. 1889. *Thunia alba* (Lindl.) Rchb. f. var. *marshalliana* (Rchb. f.) B. Grant, Orch. Burma: 207. 1895. SYNTYPES: Myanmar, Moulmein [Mawlamyine], *cult. Bull s.n.*; origin unknown, *cult. Marshall s.n.*; *cult. Day s.n.*; *cult. Veitch s.n.*; *cult. Wrigley s.n.* (syntypes, W).

Distribution. Andaman & Nicobar Islands, Andhra Pradesh, Arunachal Pradesh, Assam, ?Goa, Karnataka, Kerala, Manipur, Meghalaya, Mizoram, Nagaland, Sikkim, Tamil Nadu, Tripura, Uttarakhand, West Bengal [Bangladesh, Bhutan, China, Myanmar, Nepal].

Habit. Epiphytic or terrestrial herb.

Thunia alba (Lindl.) Rchb. f. var. **bracteata** (Roxb.) N. Pearce & P. J. Cribb, Edinburgh J. Bot. 58(1): 116. 2001. *Limodorum bracteatum* Roxb., Fl. Ind. ed. 1832, 3: 466. 1832. *Thunia bracteata* (Roxb.) Schltr., Repert. Spec. Nov. Regni Veg. Beih. 4: 205. 1919. TYPE: India, Meghalaya, Garrow [Garo] Hills, *M. R. Smith s.n.* (not found). NEOTYPE: Nepal: Kathmandu, Chandaghery [Chandagherry], June 1821, *Wallich 3749A* (neotype, K-LINDL, designated by Seidenfaden, 1986; isoneotype, K-WALL).

Thunia venosa Rolfe, Orchid Rev. 13: 206. 1905. TYPE: India, Sikkim, alt. 3000 ft, *King & Pantling s.n.* [171] (holotype, CAL; isotype, BM, K, W).

Distribution. Andhra Pradesh, Arunachal Pradesh, Assam, Himachal Pradesh, Karnataka, Kerala, Maharashtra, Meghalaya, Odisha, Sikkim, Tamil Nadu, Uttarakhand, West Bengal [Bhutan, China, Myanmar].

Habit. Epiphytic or terrestrial herb.

Thunia bensoniae Hook. f., Bot. Mag. 94: t. 5694. 1868. *Phaius bensoniae* (Hook. f.) Benth., J. Linn. Soc., Bot. 18: 305. 1881. *Phaius albus* Lindl. var. *bensoniae* (Hook. f.) Hook. f., Fl. Brit. India [J. D. Hooker] 5(16): 818. 1890. TYPE: Myanmar, Rangoon, *Benson s.n.*

Thunia winniana L. Linden, Lindenia 10: 43. 1894. TYPE: Origin unknown, *cult. Winn s.n.*

Distribution. ?Andaman & Nicobar Islands, ?Meghalaya [Myanmar].

Habit. Epiphytic or terrestrial herb.

Remarks. This species is listed as "excluded" in Singh et al. (2019).

———

Tipularia Nutt., Gen. N. Amer. Pl. [Nuttall] 2: 195. 1818.

TYPE: *Tipularia discolor* (Pursh) Nutt. (*Orchis discolor* Pursh).

Tipularia cunninghamii (King & Prain) S. C. Chen, S. W. Gale & P. J. Cribb, Fl. China 25: 251. 2009. *Didiciea cunninghamii* King & Prain, J. Asiat. Soc. Bengal, Pt. 2, Nat. Hist. 65: 119. 1896 (as "*cunninghami*"). SYNTYPES: India, Sikkim, Lachen Valley, alt. 12,000 ft, July 1895, *Pantling 396* (syntypes, BM, CAL, K, P); Sikkim, 1889, *Cunningham s.n.* (without locality) (syntypes, CAL, K).

Distribution.   Sikkim, Uttarakhand [also in Taiwan].

Habit.   Terrestrial herb.

Tipularia josephi Rchb. f. ex Lindl., J. Proc. Linn. Soc., Bot. 1: 174. 1857. TYPE: India, Sikkim, alt. 10,000–12,000 ft, *J. D. Hooker s.n.* [*351*] (syntypes, K, K-LINDL).

Distribution.   Sikkim, West Bengal [Bhutan, China, Myanmar, Nepal].

Habit.   Epiphytic herb.

———

Trachoma Garay, Bot. Mus. Leafl. 23: 207. 1972.

TYPE: *Trachoma rhopalorrhachis* (Rchb. f.) Garay (*Dendrocolla rhopalorrhachis* Rchb. f.).

Trachoma coarctatum (King & Pantl.) Garay, Bot. Mus. Leafl. 23(4): 208. 1972. *Saccolabium coarctatum* King & Pantl., J. Asiat. Soc. Bengal, Pt. 2, Nat. Hist. 66: 592. 1898 [1897]. *Tuberolabium coarctatum* (King & Pantl.) J. J. Wood, Nordic J. Bot. 10(5): 481. 1990. TYPE: India, Assam, Jaitea Hills, Amwee, 2900 ft, June, *Pantling 625* (holotype, K).

Distribution.   Assam, Meghalaya [?Myanmar].

Habit.   Epiphytic herb.

———

Trichoglottis Blume, Bijdr. Fl. Ned. Ind. 8: 359. 1825.

TYPE: *Trichoglottis retusa* Blume.

Trichoglottis dawsoniana (Rchb. f.) Rchb. f., Gard. Chron. 1872: 699. 1872. *Cleisostoma dawsonianum* Rchb. f., Gard. Chron. 1868: 815. 1868. *Sarothrochilus dawsonianus* (Rchb. f.) Schltr., Repert. Spec. Nov. Regni Veg. 3: 50. 1906. *Staurochilus dawsonianus* (Rchb. f.) Schltr., Orchideen Beschreib. Kult. Zücht.: 577. 1914. TYPE: Myanmar, Moulmein [Mawlamyine], *Dawson s.n.* (holotype, W).

Distribution.   Andaman & Nicobar Islands [Myanmar].

Habit.   Epiphytic herb.

Trichoglottis orchidea (J. Koenig) Garay, Bot. Mus. Leafl. 23: 209. 1972. *Epidendrum orchideum* J. Koenig, Observ. Bot. (Retzius) 6: 48. 1791. *Limodorum orchideum* (J. Koenig) Willd., Sp. Pl. 4: 126. 1805. *Ceratochilus orchideus* (J. Koenig) Lindl., Gen. Sp. Orchid. Pl.: 232. 1833. TYPE: Not designated.

Trichoglottis cirrhifera Teijsm. & Binn., Natuurk. Tijdschr. Ned.-Indië 5: 493. 1853. TYPE: Indonesia, Java, Mt. Salak, *cult. Bogor s.n.*
Trichoglottis quadricornuta Kurz, J. Asiat. Soc. Bengal, Pt. 2, Nat. Hist. 45: 156. 1876. SYNTYPES: India, Nicobar Islands, Kamorta, *Jelinek s.n.*; *Novara 18.*

Distribution.   Andaman & Nicobar Islands [also in Thailand, etc.].

Habit.   Epiphytic herb.

Trichoglottis ramosa (Lindl.) Senghas, Orchideen (Schlechter) 1(21): 1315. 1988. *Saccolabium ramosum* Lindl., Gen. Sp. Orchid. Pl.: 224. 1833. *Cleisostoma ramosum* (Lindl.) Hook. f., Fl. Brit. India [J. D. Hooker] 6(17): 72. 1890. *Pomatocalpa ramosum* (Lindl.) Summerh., Kew Bull. 1948: 56. 1948. *Gastrochilus ramosus* (Lindl.) Kuntze, Revis. Gen. Pl. 2: 661. 1891. *Sarcanthus ramosus* (Lindl.) J. J. Sm., Natuurk. Tijdschr. Ned.-Indië 72: 92. 1912. *Staurochilus ramosus* (Lindl.) Seidenf., Opera Bot. 95: 95. 1988. TYPE: India, Ganges Delta, icon. *Wallich 654* (not found).

Oeceoclades flexuosa Lindl., Gen. Sp. Orchid. Pl.: 236. 1833. *Saccolabium flexuosum* (Lindl.) Rchb. f., Ann. Bot. Syst. (Walpers) 6: 886. 1864. *Gastrochilus flexuosus* (Lindl.) Kuntze, Revis. Gen. Pl. 2: 661. 1891. TYPE: Myanmar, Attran [Ataran] River, Ripa, 17 May 1827, *Wallich 7333* (syntypes, K-LINDL, K-WALL).

Distribution.   Andhra Pradesh, Assam, Meghalaya, Odisha, Tripura [Bangladesh, Myanmar].

Habit.   Epiphytic herb.

Trichoglottis tenera (Lindl.) Rchb. f., Gard. Chron. 1872: 699. 1872. *Oeceoclades tenera* Lindl., Gen. Sp. Orchid. Pl.: 236. 1833. *Cleisostoma tenerum* (Lindl.) Hook. f., Fl. Brit. India [J. D. Hooker] 6(17): 73. 1890. *Saccolabium tenerum* (Lindl.) Lindl., J. Proc. Linn. Soc., Bot. 3: 36. 1859 [1858]. TYPE: Sri Lanka, *Macrae s.n.* [*66*] (holotype, K-LINDL; isotype, K).

Oeonia alata A. Rich., Ann. Sci. Nat., Bot. sér. 2, 15: 67, t. 11. 1841. TYPE: India, Nilgiri Hills, Neddoubetta, *Perrottet s.n.* (holotype, P).

Distribution.   Karnataka, Kerala, Tamil Nadu [Sri Lanka].

Habit. Epiphytic herb.

———

Trichotosia Blume, Bijdr. Fl. Ned. Ind. 7: 342. 1825.

TYPE: *Trichotosia pauciflora* Blume.

Trichotosia dasyphylla (C. S. P. Parish & Rchb. f.) Kraenzl., Pflanzenr. (Engler) IV, 50 (Heft 50 IIB. 21): 138. 1911. *Eria dasyphylla* C. S. P. Parish & Rchb. f., Trans. Linn. Soc. London 30: 147. 1874. *Pinalia dasyphylla* (C. S. P. Parish & Rchb. f.) Kuntze, Revis. Gen. Pl. 2: 679. 1891. TYPE: Myanmar, Moulmein [Mawlamyine], near Henzai-basin, also road to Meta-Tavoy [Dawei], Mergui [Myeik], 18 July 1818, *C. S. P. Parish s.n.* (holotype, K).

Distribution. Arunachal Pradesh, Assam, Meghalaya, Mizoram, Nagaland, Sikkim, West Bengal [China, Myanmar, Nepal].

Habit. Epiphytic herb.

Trichotosia pulvinata (Lindl.) Kraenzl., Pflanzenr. (Engler) IV, 50 (Heft 50 IIB. 21): 138. 1911. *Eria pulvinata* Lindl., J. Proc. Linn. Soc., Bot. 3: 56. 1859 [1858]. *Pinalia pulvinata* (Lindl.) Kuntze, Revis. Gen. Pl. 2: 679. 1891. TYPE: Myanmar, Mergui [Myeik], 17 Aug. 1834, *Griffith 2* (holotype, K-LINDL).

Eria rufinula Rchb. f., Hamburger Garten- Blumenzeitung 19: 13. 1863. *Pinalia rufinula* (Rchb. f.) Kuntze, Revis. Gen. Pl. 2: 679. 1891. *Trichotosia rufinula* (Rchb. f.) Kraenzl., Pflanzenr. (Engler) IV, 50 (Heft 50): 143. 1911. TYPE: India, Assam, *cult. Day s.n. (leg. Schiller)* (holotype, W).

Distribution. Arunachal Pradesh, Assam, Meghalaya, Sikkim, West Bengal [Bangladesh, China, Myanmar, Nepal].

Habit. Epiphytic herb.

Trichotosia velutina (G. Lodd. ex Lindl.) Kraenzl., Pflanzenr. (Engler), IV, 50(Heft 50 IIB,21): 140. 1911. *Eria velutina* G. Lodd. ex Lindl., Edwards's Bot. Reg. 26(Misc.): 86. 1840. *Pinalia velutina* (G. Lodd. ex Lindl.) Kuntze, Revis. Gen. Pl. 2: 679. 1891. TYPE: Singapore, *cult. Loddiges s.n. (leg. Cuming)* (holotype, K-LINDL).

Distribution. Arunachal Pradesh [Myanmar].

Habit. Epiphytic herb.

———

Tropidia Lindl., Edwards's Bot. Reg. 19: t. 1618. 1833.

TYPE: *Tropidia curculigoides* Lindl.

Tropidia angulosa (Lindl.) Blume, Coll. Orchid. 122. 1859. *Cnemidia angulosa* Lindl., Gen. Sp. Orchid. Pl.: 463. 1840. SYNTYPES: India, Sylhet mountains, *Wallich 7388 (leg. W. Gomez)* (syntypes, K-LINDL, K-WALL); Courtallum, *Wight 928* (syntype, K-LINDL).

Cnemidia semilibera Lindl., Gen. Sp. Orchid. Pl.: 463. 1840. *Tropidia semilibera* (Lindl.) Blume, Coll. Orchid.: 122. 1858. TYPE: "India orientali," without locality, *Wallich s.n.* (holotype, K-LINDL).
Govindooia nervosa Wight, Icon. Pl. Ind. Orient. [Wight] 6: 35, t. 2090. 1853. *Tropidia govindooii* Blume, Coll. Orchid.: 122. 1858 [1859], nom. superfl. *Cnemidia nervosa* (Wight) Bedd. in H. B. Grigg, Man. Nílagiri: 122. 1880. TYPE: India, Courtallum, Aug.–Sep., *Wight s.n.* (holotype, K-LINDL).
Tropidia barbeyana Schltr., Bull. Herb. Boissier, sér. 2, 4: 300. 1906. TYPE: India, Runghee, elev. 1500 m, *native collector s.n. in herb. Barbey-Boissier* (not found).
Tropidia bellii Blatt. & McCann, J. Bombay Nat. Hist. Soc. 35(4): 730. 1932. TYPE: India, North Kanara, Guddehalli, in evergreen forests, alt. 1500 ft, *T. R. Bell 2992* (holotype, BLAT).

Distribution. Andaman & Nicobar Islands, Andhra Pradesh, Arunachal Pradesh, Assam, Goa, Karnataka, Kerala, Meghalaya, Odisha, Sikkim, Tamil Nadu, Tripura, West Bengal [Bangladesh, Bhutan, China, Myanmar].

Habit. Terrestrial herb.

Tropidia bambusifolia (Thwaites) Trimen, Syst. Cat. Fl. Pl. Ceylon: 90. 1885. *Cnemidia bambusifolia* Thwaites, Enum. Pl. Zeyl. [Thwaites]: 314. 1861. TYPE: Sri Lanka, Saffragam Dist., *Thwaites s.n.* [*C. P. 3207*] (syntypes, K, PDA).

Distribution. Andaman & Nicobar Islands [Sri Lanka].

Habit. Terrestrial herb.

Tropidia curculigoides Lindl., Gen. Sp. Orchid. Pl.: 497. 1840. SYNTYPES: Bangladesh or India, Sylhet, 1821, *Wallich 7386A (leg. De Silva & W. Gomez s.n.)* (syntypes, K, K-LINDL, K-WALL); Myanmar, Attran [Ataran] River, Nipa, 29 Jan. 1827, *Wallich 7386* (syntype, K-WALL); Sri Lanka, *Macrae s.n.* (syntype, K-LINDL).

Tropidia assamica Blume, Fl. Javae, n.s. 1: 104. 1858. TYPE: India, Assam, *Griffith s.n.* [144] (holotype, L; isotype, BM, K, K-LINDL, P).

Distribution. Andaman & Nicobar Islands, Arunachal Pradesh, Assam, Kerala, Manipur, Meghalaya,

Nagaland, ?Odisha, Sikkim, West Bengal [Bangladesh, China, Myanmar].

Habit.   Terrestrial herb.

Tropidia hegderaoi S. Misra, Nelumbo 54: 13. 2012. TYPE: India, Tamil Nadu, Namakkal district: Kollimalais, Ariyur shola, Mathikettan, ca. 1300 m, *Misra 2525* (holotype, CAL; isotype, CAL, MH).

Distribution.   Tamil Nadu. Endemic.

Habit.   Terrestrial herb.

Tropidia maxwellii Ormerod, Checkl. Papuasian Orchids: 438. 2017. TYPE: Thailand, Chang Mai Prov., Miuang Dist., Doi Sutep-pui Nat. Park, Palaht Temple area, 720 m, 7 Aug. 1993, *A. Phuakam 43* (holotype, A; isotype, L).

Distribution.   Andaman & Nicobar Islands, Arunachal Pradesh, Assam, Jharkhand, Kerala, Manipur, Odisha, Sikkim, Uttar Pradesh, Uttarakhand [Bangladesh, Myanmar].

Habit.   Terrestrial herb.

Remarks.   Indian specimens of this species have been misidentified as *Tropidia pedunculata* Blume and *T. formosana* Rolfe. These species do not occur in mainland India (Ormerod, 2018a). Misra (2019: 484) suggests that a specimen from the Andaman Islands (*Rashingam 21000*, PBL) could be referable to *T. pedunculata*.

Tropidia namasiae C. K. Liao, T. P. Lin & M. S. Tang, Novon 22: 426. 2013. TYPE: Taiwan, Kaoshing City, Namasia District, 1380 m, 5 May 2009, *C. K. Liao 3594* (holotype, TAI).

Distribution.   Manipur [also in Taiwan, Thailand].

Habit.   Terrestrial herb.

Remarks.   *Tropidia namasiae* seems very close to *T. hegderaoi* and they need to be critically compared.

Tropidia thwaitesii Hook. f., Fl. Brit. India [J. D. Hooker] 6(17): 93. 1890. SYNTYPES: Sri Lanka, *Macrae 58* (syntype, K-LINDL); *Thwaites s.n.* [*Ceylon Plants 3565*] (syntypes, CAL, K).

Distribution.   Andaman & Nicobar Islands, Tamil Nadu [Sri Lanka].

Habit.   Terrestrial herb.

Remarks.   This is considered a synonym of *Tropidia bambusifolia* (Thwaites) Trimen by Singh et al. (2019).

Uncifera Lindl., J. Proc. Linn. Soc., Bot. 3: 39. 1859 [1858].

TYPE: *Uncifera obtusifolia* Lindl. (designated by Christenson, 1986).

Uncifera acuminata Lindl., J. Proc. Linn. Soc., Bot. 3: 40. 1859 [1858]. *Saccolabium acuminatum* (Lindl.) Hook. f., Fl. Brit. India [J. D. Hooker] 6(17): 65. 1890, nom. illeg., non Thwaites, 1861. SYNTYPES: India, Assam and Khasi Hills, *Griffith 18 & s.n.* (syntypes, K, K-LINDL); Khasi Hills, *Hooker & Thomson 193* (syntype, K-LINDL).

Distribution.   Arunachal Pradesh, Assam, Manipur, Meghalaya, Mizoram, Nagaland, Sikkim, West Bengal [Bangladesh, China, Nepal].

Habit.   Epiphytic herb.

Uncifera lancifolia (King & Pantl.) Schltr., Orchideen (Schlechter) 583. 1914. *Saccolabium lancifolium* King & Pantl., J. Asiat. Soc. Bengal, Pt. 2, Nat. Hist. 65: 122. 1896. TYPE: India, Darjeeling, Rississum, 6000 ft, June 1895, *Pantling 152* (holotype, CAL; isotype, BM, K, L, LE, W).

Distribution.   Sikkim, West Bengal. Endemic.

Habit.   Epiphytic herb.

Uncifera obtusifolia Lindl., J. Proc. Linn. Soc., Bot. 3: 40. 1859 [1858]. *Saccolabium obtusifolium* (Lindl.) Hook. f., Fl. Brit. India [J. D. Hooker] 6(17): 65. 1890. *Gastrochilus obtusifolius* (Lindl.) Kuntze, Revis. Gen. Pl. 2: 661. 1891. *Gastrochilus obtusifolius* (Lindl.) A. S. Rao & Mukherji, Rec. Bot. Surv. India 20: 213. 1973, isonym, non (Lindl.) Kuntze, 1891. TYPE: India, Khasi Hills, *Hooker & Thomson 194* (holotype, K-LINDL; isotype, K).

Uncifera heteroglossa Rchb. f., Gard. Chron., n.s., 10: 234. 1878. TYPE: India, Khasi Hills, *cult. Bull 263.* (holotype, W).

Distribution.   Arunachal Pradesh, Assam, Manipur, Meghalaya, Nagaland, Sikkim, West Bengal [Bhutan, Nepal].

Habit.   Epiphytic herb.

———

Vanda Jones ex R. Br., Bot. Reg. 6: t. 506. 1820.

TYPE: *Vanda roxburghii* R. Br.

Vanda alpina (Lindl.) Lindl., Fol. Orchid. 4(*Vanda*): 10. 1853. *Luisia alpina* Lindl., Edwards's Bot.

Reg. 24(Misc.): 56. 1838. *Trudelia alpina* (Lindl.) Garay, Orchid Digest 50(2): 76. 1986. *Stauropsis alpina* (Lindl.) Tang & F. T. Wang, Acta Phytotax. Sin. 1: 93. 1951. SYNTYPES: India, Khasi Hills, Nunklow, *Gibson s.n.* (syntype, K-LINDL.); Khasi Hills, alt. 5000–6000 ft, *Hooker & Thomson s.n.*

Distribution. Arunachal Pradesh, Manipur, Meghalaya, Mizoram, Nagaland, Uttarakhand, West Bengal [Bhutan, China, Nepal].

Habit. Epiphytic herb.

Vanda ampullacea (Roxb.) L. M. Gardiner, Phytotaxa 61: 48. 2012. *Aerides ampullacea* Roxb., Fl. Ind. ed. 1832, 3: 476. 1832 (as "*ampullaceum*"). *Ascocentrum ampullaceum* (Roxb.) Schltr., Repert. Spec. Nov. Regni Veg. Beih. 1: 975. 1913. *Saccolabium ampullaceum* (Roxb.) Lindl., Numer. List [Wallich] n. 7307. 1832. *Oeceoclades ampullacea* (Roxb.) Lindl. ex Voigt, Hort. Suburb. Calcutt. 630. 1845. *Gastrochilus ampullaceus* (Roxb.) Kuntze, Revis. Gen. Pl. 2: 661. 1891. TYPE: Without locality, *M. R. Smith s.n.* (not found); icon. *Aerides ampullaceum* [sic], *Roxburgh 2347* (lectotype, K, selected by Pearce & Cribb, 2009; isolectotype, CAL).

Ascocentrum ampullaceum (Roxb.) Schltr. var. *aurantiacum* Pradhan, Indian Orchids: Guide Identif. & Cult. 2: 561. 1979. TYPE: Manipur, *R. K. Mohendrajit Singh s.n.* (holotype, Herbarium U. C. Pradhan).

Distribution. Andaman & Nicobar Islands, Arunachal Pradesh, Assam, Manipur, Nagaland, Sikkim, Uttarakhand, West Bengal [Bangladesh, Bhutan, China, Myanmar, Nepal].

Habit. Epiphytic herb.

Vanda arbuthnotiana Kraenzl., Gard. Chron., ser. 3, 11: 522. 1892. TYPE: India, Kerala, Malabar, *cult. Sander s.n.*

Distribution. Kerala. Endemic.

Habit. Epiphytic herb.

Remarks. *Vanda arbuthnotiana* is possibly conspecific with *Arachnis labrosa*.

Vanda bicolor Griff., Not. Pl. Asiat. 3: 354. 1851. TYPE: Bhutan: Monass River, alt. 2300 ft, *Griffith 546* (holotype, K-LINDL).

Distribution. Arunachal Pradesh, Assam, Meghalaya, Nagaland [Bhutan, Myanmar, Nepal].

Habit. Epiphytic herb.

Vanda coerulea Griff. ex Lindl., Edwards's Bot. Reg. 33: t. 30. 1847. TYPE: India, Khasi Hills, *Griffith s.n.* (holotype, K-LINDL; isotype, K).

Vanda coerulea Griff. ex Lindl. f. *luwangalba* Kishor, Orchid Rev. 116(1282): 224. 2008. TYPE: India, Manipur, Senapati District, Longa Koireng, *Rajkumar 010* (holotype, K).

Distribution. Arunachal Pradesh, Assam, Bihar, Manipur, Meghalaya, Mizoram, Nagaland [China, Myanmar].

Habit. Epiphytic herb.

Vanda coerulescens Griff., Not. Pl. Asiat. 3: 352. 1851. TYPE: Myanmar, Tsenbo near Bamo, 26 Apr. 1837, *Griffith s.n. (East India Company Herbarium 5187)* (holotype, K).

Distribution. Arunachal Pradesh, Assam, Manipur [Bangladesh, China, Myanmar].

Habit. Epiphytic herb.

Vanda cristata Wall. ex Lindl., Gen. Sp. Orchid. Pl.: 216. 1833. *Aerides cristata* (Wall. ex Lindl.) Wall. ex Hook. f., Fl. Brit. India [J. D. Hooker] 6(17): 53. 1890. *Trudelia cristata* (Wall. ex Lindl.) Senghas, Orchideen (Schlechter) 1(19–20): 1211. 1988. TYPE: Nepal, Naokote, May 1821, *Wallich 7328* (holotype, K-LINDL; isotype, K-WALL).

Vanda striata Rchb. f., Xenia Orchid. 2: 137. 1868. *Luisia striata* (Rchb. f.) Kraenzl. in H. G. Reichenbach, Xenia Orchid. 3: 120. 1893. TYPE: India, Uttarakhand, Dwarahat, Paorie, Ghurwal, *Stewart s.n.* (holotype, W).

Distribution. Arunachal Pradesh, Assam, Bihar, Himachal Pradesh, Manipur, Meghalaya, Mizoram, Nagaland, Sikkim, Uttarakhand, West Bengal [Bangladesh, Bhutan, China, Myanmar, Nepal].

Habit. Epiphytic herb.

Vanda curvifolia (Lindl.) L. M. Gardiner, Phytotaxa 61: 49. 2012. *Saccolabium curvifolium* Lindl., Gen. Sp. Orchid. Pl.: 222. 1833. *Ascocentrum curvifolium* (Lindl.) Schltr., Repert. Spec. Nov. Regni Veg. Beih. 1: 975. 1913. *Gastrochilus curvifolius* (Lindl.) Kuntze, Revis. Gen. Pl. 2: 661. 1891. SYNTYPES: Nepal, *Wallich s.n.* (K-WALL, not found); Sri Lanka, *Macrae s.n.* (syntype, K-LINDL).

Distribution. Andhra Pradesh, Arunachal Pradesh, ?Assam, Nagaland [Myanmar, Sri Lanka].

Habit. Epiphytic herb.

Vanda flavobrunnea Rchb. f., Flora 69: 552. 1886. TYPE: Origin and collector unknown (holotype, W).

Vanda pumila Hook. f., Fl. Brit. India [J. D. Hooker] 6(17): 53. 1890. *Trudelia pumila* (Hook. f.) Senghas, Orchideen (Schlechter) 1(19–20): 1211. 1988. TYPE: Sikkim and Bhutan Himalaya, alt. 2000 ft, icon. *Cathcart s.n.* (holotype, K).

Distribution.   Arunachal Pradesh, Manipur, Meghalaya, Nagaland, Sikkim, Uttarakhand, West Bengal [China, Myanmar, Nepal].

Habit.   Epiphytic herb.

Vanda griffithii Lindl., Paxton's Fl. Gard. 2: 22. 1851. *Luisia griffithii* (Lindl.) Kraenzl., Xenia Orchid. 3: 119. 1893. *Trudelia griffithii* (Lindl.) Garay, Orchid Digest 50(2): 76. 1986. TYPE: Bhutan, Monass River, elev. 2300 ft, *Griffith s.n.* [*846*] (holotype, K-LINDL).

Distribution.   Arunachal Pradesh, Manipur [Bhutan, Nepal].

Habit.   Epiphytic herb.

Vanda jainii A. S. Chauhan, J. Econ. Taxon. Bot. 5(3): 977, fig. 1. 1984. TYPE: India, Meghalaya, West Khasi Hills, Sonapahar, 25 Mar. 1981, *Chauhan 70887A* (holotype, CAL).

Distribution.   Meghalaya. Endemic.

Habit.   Epiphytic herb.

Vanda motesiana Choltco, Orchid Rev. 117: 148. 2009. TYPE: India, Arunachal Pradesh, *Choltco 1005* (holotype, MU).

Distribution.   Arunachal Pradesh, Manipur, Nagaland. Endemic.

Habit.   Epiphytic herb.

Vanda stangeana Rchb. f., Bot. Zeitung (Berlin) 16(47): 351. 1858. TYPE: India, Assam, *Schiller s.n.* (holotype, W).

Distribution.   Arunachal Pradesh, Assam, Manipur, Meghalaya, Nagaland. Endemic.

Habit.   Epiphytic herb.

Remarks.   Singh et al. (2019) suggest that all records from India of *Vanda stangeana* are referable to *V. motesiana*. However, the type of *V. stangeana* supposedly came from Assam and the species has not yet been recorded outside India.

Vanda tessellata (Roxb.) Hook. ex G. Don, Hort. Brit. (Loudon) 372. 1830. *Epidendrum tessellatum* Roxb., Pl. Coromandel 1: 34, t. 42. 1795. *Cymbid-ium tessellatum* (Roxb.) Sw., Nova Acta Regiae Soc. Sci. Upsal. 6: 75. 1799. TYPE: India, Circar Mountains, icon. *Epidendrum tessellatum, Roxburgh 242* (holotype, CAL; isotype, ?G [specimen], K).

Vanda roxburghii R. Br., Bot. Reg. 6: t. 506. 1820. TYPE: India, Bengal, *cult. Banks s.n.*
Cymbidium tesselloides Roxb., Fl. Ind. ed. 2, 3: 463. 1832 (as "*tessaloides*"). *Vanda tesselloides* (Roxb.) Rchb. f., Ann. Bot. Syst. (Walpers) 6: 864. 1864. TYPE: India, Bengal, cult. Calcutta s.n., icon. *Epidendrum tessaloides* [sic], *Roxburgh 1288* (syntypes, CAL, K).
Vanda roxburghii R. Br. var. *spooneri* Gammie, J. Bombay Nat. Hist. Soc. 19: 625. 1909. TYPE: India, Wynaad, *Proudlock s.n.*

Distribution.   Andhra Pradesh, Assam, Bihar, Chhattisgarh, Dadra & Nagar Haveli, Goa, Gujarat, Jharkhand, Karnataka, Kerala, Madhya Pradesh, Maharashtra, Meghalaya, Nagaland, Odisha, Rajasthan, Tamil Nadu, Tripura, Uttar Pradesh, Uttarakhand, West Bengal [Bangladesh, China, ?Myanmar, Nepal, Sri Lanka].

Habit.   Epiphytic herb.

Vanda testacea (Lindl.) Rchb. f., Gard. Chron., n.s., 8: 166. 1877. *Aerides testacea* Lindl., Gen. Sp. Orchid. Pl.: 238. 1833 (as "*testaceum*"). *Vanda parviflora* Lindl. var. *testacea* (Lindl.) Hook. f., Fl. Brit. India [J. D. Hooker] 6(17): 50. 1890. TYPE: Sri Lanka, *Macrae s.n.* (not found).

Aerides wightiana Lindl., Gen. Sp. Orchid. Pl.: 238. 1833 (as "*Wightianum*"). TYPE: India, Madras, *Wallich 7320 (leg. Wight s.n.)* (holotype, K-LINDL; isotype, K-WALL).
Vanda parviflora Lindl., Edwards's Bot. Reg. 30(Misc.): 45. 1844. *Vanda testacea* (Lindl.) Rchb. f. var. *parviflora* (Lindl.) M. R. Almeida, Fl. Maharashtra 5A: 90. 2009. TYPE: India, Bombay, *Messrs. Loddiges s.n.* (holotype, K-LINDL).

Distribution.   Andhra Pradesh, Arunachal Pradesh, Assam, Chhattisgarh, Goa, Himachal Pradesh, Jharkhand, Karnataka, Kerala, Madhya Pradesh, Maharashtra, Manipur, Meghalaya, Mizoram, Nagaland, Odisha, Rajasthan, Sikkim, Tamil Nadu, Uttar Pradesh, Uttarakhand, West Bengal [Bhutan, China, Myanmar, Nepal, Sri Lanka].

Habit.   Epiphytic herb.

Vanda thwaitesii Hook. f. in Trimen, Handb. Fl. Ceylon 4: 193. 1898. TYPE: Sri Lanka, Hunasgiria District, *Alwis s.n.* [*Ceylon Plants 3378*] (not found).

Distribution.   Kerala, Tamil Nadu [Sri Lanka].

Habit.   Epiphytic herb.

Vanda wightii Rchb. f., Ann. Bot. Syst. (Walpers) 6: 932. 1864. TYPE: India, Nilgiri Hills, *Wight s.n.* (holotype, W).

Distribution. Goa, Karnataka, Kerala, Tamil Nadu [Sri Lanka].

Habit. Epiphytic herb.

————

Vanilla Mill., Gard. Dict. Abr. ed. 4. [textus s.n.]. 1754.

TYPE: *Vanilla mexicana* Mill.

Vanilla albida Blume, Cat. Gew. Buitenzorg (Blume): 100. 1823. TYPE: Indonesia, Java, ?*Blume s.n.* (probable syntypes, L).

Distribution. Andaman & Nicobar Islands [also in Indochina, Indonesia, Thailand].

Habit. Climber.

Vanilla andamanica Rolfe, Bull. Misc. Inform. Kew 1918: 237. 1918. TYPE: India, Andaman Islands, Betapur Valley, *C. E. Parkinson 1139* (holotype, K).

Distribution. Andaman & Nicobar Islands. Endemic.

Habit. Climber.

Vanilla aphylla Blume, Bijdr. Fl. Ned. Ind. 8: 422. 1825. SYNTYPES: Indonesia, Java, Bantam Prov., Sadjram, *Blume s.n.*; Nusa Kambangan, *Blume s.n.* (?syntype, L [s. loc.]).

Distribution. Assam [Bangladesh, Myanmar].

Habit. Climber.

Vanilla borneensis Rolfe, J. Linn. Soc., Bot. 32: 460. 1896. TYPE: Indonesia, Borneo, Banjarmasin, *Motley 1248* (holotype, K).

Vanilla pilifera Holttum, Gard. Bull. Singapore 13: 251. 1951. TYPE: Malaya: Johore, Kota Tinggi, Apr. 1951, *J. A. Le Doux s.n.*

Distribution. Assam, Manipur, Mizoram [also in Indonesia, Thailand].

Habit. Climber.

Vanilla sanjappae Rasingam, R. P. Pandey, J. J. Wood & S. K. Srivast., Orchid Rev. 115: 350. 2007. TYPE: India, Little Andaman Island, Krishna nallah, 50 m, 27 Apr. 2006, *L. Rasingam 20990* (holotype, K; isotype, CAL, PBL).

Distribution. Andaman & Nicobar Islands. Endemic.

Habit. Climber.

Vanilla walkerae Wight, Icon. Pl. Ind. Orient. [Wight] 3(3): 12, t. 932. 1845 (as "*walkeriae*"). TYPE: Sri Lanka, *icon. Walker s.n.* (not found).

Distribution. Andhra Pradesh, Karnataka, Kerala, Tamil Nadu [Sri Lanka].

Habit. Climber.

Vanilla wightii Lindl. ex Wight, Icon. Pl. Ind. Orient. [Wight] 3(3): 12, t. 931. 1845. TYPE: India, Kerala, Travancore near Trevandrum [Trivandrum], *Wight s.n.* (?holotype, K-LINDL).

Distribution. Andhra Pradesh, Kerala, Tamil Nadu. Endemic.

Habit. Climber.

————

Vrydagzynea Blume, Coll. Orchid. 71. 1858.

TYPE: *Vrydagzynea albida* (Blume) Blume (*Hetaeria albida* Blume).

Vrydagzynea albida (Blume) Blume, Coll. Orchid. 75, t. 19, fig. 2. 1859. *Hetaeria albida* Blume, Bijdr. Fl. Ned. Ind. 8: 410. 1825 (as "*Etaeria*"). SYNTYPES: Indonesia, Java, Mt. Salak and Mt. Seribu, *Blume s.n.* (?syntype, L, *Blume 812*, s. loc.).

Distribution. Andaman & Nicobar Islands [also in Indonesia].

Habit. Terrestrial herb.

Vrydagzynea viridiflora Hook. f., Fl. Brit. India [J. D. Hooker] 6(17): 96. 1890. TYPE: India, Lower Bengal, Luckempore, on the Megua, Namchung, 1000 ft, 18 Apr. 1885, *C. B. Clarke s.n.* [*37939*] (holotype, K).

Distribution. Assam. Endemic.

Habit. Terrestrial herb.

————

Yoania Maxim., Bull. Acad. Imp. Sci. Saint-Pétersbourg, ser. 3, 18: 68. 1872.

TYPE: *Yoania japonica* Maxim.

Yoania prainii King & Pantl., Ann. Roy. Bot. Gard. (Calcutta) 8: 175. 1898. SYNTYPES: India, Sikkim-Himalaya, Chungthang, 6000 ft, *Pantling 469* (syntypes, BM, CAL, K, P); Naga Hills, 7000 ft, *Prain 15* (syntypes, CAL, K).

Distribution.   Nagaland, Sikkim [China].

Habit.   Holomycotrophic terrestrial herb.

Remarks.   Indian records of *Yoania japonica* Maxim. refer to *Y. prainii*.

———

Zeuxine Lindl., Coll. Bot. (Lindley) Append. [n. 18.]. 1826 ("*Zeuxina*"), nom. et orth. cons.

TYPE: *Zeuxine sulcata* (Roxb.) Lindl. (*Pterygodium sulcatum* Roxb. [as "*sulcata*"]).

Zeuxine affinis (Lindl.) Benth. ex Hook. f., Fl. Brit. India [J. D. Hooker] 6(17): 108. 1890. *Hetaeria affinis* Lindl., Numer. List [Wallich] n. 7383. 1832, nom. inval. (as "*Etaria*"). *Monochilus affinis* Lindl., Gen. Sp. Orchid. Pl.: 487. 1840 (as "*affine*"). *Haplochilus affinis* (Lindl.) D. Dietr., Syn. Pl. (D. Dietrich) 5: 173. 1852. TYPE: Myanmar, Tong Dong, 24 Nov. 1826, *Wallich 7383* (holotype, K-LINDL; isotype, K-WALL).

Hetaeria mollis Lindl., J. Proc. Linn. Soc., Bot. 1: 184. 1857. SYNTYPES: India, Khasi Hills, alt. 3000–4000 ft, *Hooker & Thomson 343* (syntype, K-LINDL); Myanmar, *Griffith s.n.*
Zeuxine grandis Seidenf., Dansk Bot. Ark. 32(2): 90. 1978. TYPE: Thailand, northern part, Omkoi Road at km 13, *Seidenfaden & Smitinand GT 7730* (holotype, CO).
Zeuxine seidenfadenii Deva & H. B. Naithani, Orchid Fl. N. W. Himalaya: 95. 1986. TYPE: India, Dehra Dun, Laxmansidh, 20/04/1978, *Som Deva 9712* (holotype, DD; isotype, DD).

Distribution.   Andaman & Nicobar Islands, Arunachal Pradesh, Assam, Bihar, Kerala, Manipur, Meghalaya, Nagaland, Odisha, Sikkim, Tamil Nadu, Uttar Pradesh, Uttarakhand, West Bengal [Bangladesh, Bhutan, China, Myanmar, Nepal].

Habit.   Terrestrial herb.

Remarks.   We follow Pedersen (in Pedersen et al., 2011) in considering *Zeuxine grandis* conspecific with *Z. affinis*. Bhattacharjee and Chowdhery (2018) regard these as distinct species and report both from India. *Zeuxine affinis* is illustrated as *Z. clandestina* Blume in Swami (2016, 2017).

Zeuxine agyokuana Fukuy., Bot. Mag. (Tokyo) 48: 433. 1934. *Hetaeria agyokuana* (Fukuy.) K. Nakaj., Biol. Mag., Okinawa 8: 82. 1971. *Hetaeria cristata* Blume var. *agyokuana* (Fukuy.) S. S. Ying, Coloured Illustr. Orchid Fl. Taiwan 2: 558. 1990.

TYPE: Taiwan, Taihoku, Mt. Syoagyoku-san near Urai, *Fukuyama 4127* (holotype, lost); Taoyuan County, Mt. Nachieh, 1000–1500 m, 28 Oct. 2010, *T. C. Hsu 3124* (neotype, TAIF, designated by Tian et al., 2014).

Zeuxine pantlingii Av. Bhattacharjee & H. J. Chowdhery, Sida 22(2): 935. 2006. TYPE: India, West Bengal, Darjeeling District, near Mongpoo, alt. 3000 ft, 28 Aug. 2005, *Av. Bhattacharjee 34807* (holotype, CAL).

Distribution.   Meghalaya, West Bengal [China].

Habit.   Terrestrial herb.

Zeuxine andamanica King & Pantl., J. Asiat. Soc. Bengal, Pt. 2, Nat. Hist. 66: 599. 1898 [1897]. TYPE: India, Andaman Islands, *Calcutta Botanical Garden collectors s.n.* (holotype, CAL).

Distribution.   Andaman & Nicobar Islands. Endemic.

Habit.   Terrestrial herb.

Zeuxine blatteri C. E. C. Fisch., Bull. Misc. Inform. Kew 1928: 76. 1928. TYPE: India, Tamil Nadu, Madura Dist., High Wavy Mountain, *Blatter & Hallberg 343* (holotype, K).

Distribution.   Arunachal Pradesh, ?Assam, Karnataka, Kerala, Maharashtra, Meghalaya, Nagaland, ?Odisha, Tamil Nadu [Sri Lanka].

Habit.   Terrestrial herb.

Remarks.   *Zeuxine blatteri* has been synonymized with *Z. gracilis* (Breda) Blume, which Ormerod (2018b) considers to be in error. He also notes that some of the Indian material (e.g., from Odisha) identified as *Z. gracilis* may refer to another species, possibly *Z. pseudogracilis* Ormerod. Clearly, the Indian material needs to be re-evaluated.

Zeuxine chowdheryi Av. Bhattacharjee & Sabap., Taiwania 55(4): 342. 2010 (as "*chowdherii*"). TYPE: India, Tamil Nadu, Coimbatore District, Velliangiri Hills, near Sita Vanam, 22 May 2006, *C. M. Sabapathy 46023* (holotype, CAL).

Distribution.   Tamil Nadu. Endemic.

Habit.   Terrestrial herb.

Zeuxine clandestina Blume, Coll. Orchid. 70, t. 39, fig. 4. 1858. *Monochilus clandestinus* (Blume) Miq., Fl. Ned. Ind. 3: 724. 1859. TYPE: Indonesia, East Java, *A. Waitz s.n.* (holotype, ?L, anonymous).

Distribution.   ?Arunachal Pradesh, ?Assam [also in Thailand, etc.].

Habit.   Terrestrial herb.

Remarks. Records from India need verification (Ormerod, pers. comm.); there is no useful material at Kew.

Zeuxine flava (Wall. ex Lindl.) Trimen, Syst. Cat. Fl. Pl. Ceylon 90. 1885. *Monochilus flavus* Wall. ex Lindl., Gen. Sp. Orchid. Pl.: 487. 1840 (as "*flavum*"). *Hetaeria flava* Lindl., Numer. List [Wallich] n. 7380. 1832, nom. inval. (as "*Etaria*"). *Haplochilus flavus* (Wall. ex Lindl.) D. Dietr., Syn. Pl. (D. Dietrich) 5: 172. 1852. TYPE: Nepal, Gokurrum, May 1821, *Wallich 7380A* (lectotype, K-LINDL, designated by Seidenfaden, 1978; isotype, CAL, K-WALL).

Distribution. Arunachal Pradesh, Assam, Meghalaya, Nagaland, Sikkim, Uttarakhand, West Bengal [Bangladesh, Bhutan, China, Myanmar, Nepal].

Habit. Terrestrial herb.

Zeuxine glandulosa King & Pantl., Ann. Roy. Bot. Gard. (Calcutta) 8: 288, t. 384. 1898. *Heterozeuxine glandulosa* (King & Pantl.) T. Hashim., Proc. World Orchid Conf. 12: 125. 1987. TYPE: Bhutan: Dooar near Jaldacca River, alt. 2000 ft, Mar., *Pantling 434* (holotype, CAL).

Distribution. Assam, Sikkim, West Bengal [Bhutan].

Habit. Terrestrial herb.

Zeuxine goodyeroides Lindl., Gen. Sp. Orchid. Pl.: 486. 1840. *Monochilus goodyeroides* (Lindl.) Lindl., J. Proc. Linn. Soc., Bot. 1: 187. 1857. TYPE: India, Assam, *Griffith in Herb. Ind. Misc. no. 7* (holotype, K-LINDL; isotype, K).

Distribution. Arunachal Pradesh, Assam, Meghalaya, Mizoram, Nagaland, Sikkim, West Bengal [Bhutan, China, Myanmar, Nepal].

Habit. Terrestrial herb.

Zeuxine lindleyana A. N. Rao, Arunachal Forest News 6: 34. 1988. TYPE: India, Kameng District, Tipi, ca. 200 m, 02/02/1983, *A. N. Rao 6496A* (holotype, Orchid Herbarium, Tipi).

Distribution. Arunachal Pradesh, Himachal Pradesh, Odisha, West Bengal. Endemic.

Habit. Terrestrial herb.

Zeuxine longilabris (Lindl.) Trimen, Syst. Cat. Fl. Pl. Ceylon 90. 1885. *Monochilus longilabris* Lindl., Gen. Sp. Orchid. Pl.: 487. 1840 (as "*longilabre*"). *Haplochilus longilabris* (Lindl.) D. Dietr., Syn. Pl.

(D. Dietrich) 5: 173. 1852. TYPE: Sri Lanka, 1829, *Macrae s.n.* [4] (holotype, K-LINDL).

Distribution. Arunachal Pradesh, Assam, ?Bihar, Goa, ?Jharkhand, Karnataka, Kerala, Maharashtra, Meghalaya, Odisha, Tamil Nadu, Tripura, West Bengal [Bangladesh, Myanmar, Nepal, Sri Lanka].

Habit. Terrestrial herb.

Zeuxine membranacea Lindl., Gen. Sp. Orchid. Pl.: 486. 1840. TYPE: Bhutan, *Griffith in Herb. Ind. No. 16* (holotype, K-LINDL; isotype, K).

Zeuxine debrajiana Sud. Chowdhury, Indian Forester 122(1): 87. 1996. TYPE: India, Assam, Sibsagar District (presently Golaghat District), in front of Botany Department of Debraj Roy College, Golaghat, 22 Nov. 1974, *S. Chowdhury 502* (holotype, CAL).

Distribution. Assam, Bihar, Himachal Pradesh, West Bengal [Bangladesh, Bhutan, Myanmar, Nepal].

Habit. Terrestrial herb.

Zeuxine mooneyi S. Misra, Nelumbo 54: 17. 2012. TYPE: India, Odisha, Kendujhar district, Rebana forest, Panasapani, ca. 800 m, 12 July 1982, *Misra 734* (holotype, CAL).

Distribution. Odisha. Endemic.

Habit. Terrestrial herb.

Zeuxine nervosa (Wall. ex Lindl.) Trimen, Syst. Cat. Fl. Pl. Ceylon 90. 1885. *Monochilus nervosus* Wall. ex Lindl., Gen. Sp. Orchid. Pl.: 487. 1840 (as "*nervosus*"). *Hetaeria nervosa* Lindl., Numer. List [Wallich] n. 7381. 1832, nom. inval. (as "*Etaria*"). *Haplochilus nervosus* (Wall. ex Lindl.) D. Dietr., Syn. Pl. (D. Dietrich) 5: 172. 1852. *Heterozeuxine nervosa* (Wall. ex Lindl.) T. Hashim., Ann. Tsukuba Bot. Gard. 5: 21. 1986. TYPE: Bangladesh, Derwani, 11 Mar. 1809, *Wallich 7381A (leg. Buchanan-Hamilton)* (lectotype, K-LINDL, designated by Seidenfaden, 1978, isoype, K-WALL).

Zeuxine assamica I. Barua & K. Barua, J. Econ. Taxon. Bot. 21: 491. 1997. TYPE: India, Assam, Jorhat, 87 m, 26 Mar. 1997, *Iswar Barua 2357* (holotype, CAL; isotype, ASSAM).
Zeuxine dhanikariana Maina, Lalitha & Sreek., J. Econ. Taxon. Bot. 25: 21. 2001. TYPE: India, South Andaman, Goal Tikkiri, near Dhanikari, *Maina & Sreekumar 18433* (holotype, CAL; isotype, PBL).

Distribution. Andaman & Nicobar Islands, Arunachal Pradesh, Assam, Manipur, Nagaland, Odisha, West Bengal [Bangladesh, Bhutan, China, Myanmar, Nepal, Sri Lanka].

Habit. Terrestrial herb.

Zeuxine reflexa King & Pantl., Ann. Roy. Bot. Gard. (Calcutta) 8: 291, t. 388. 1898. TYPE: India, Sikkim, Mungpoo, 3500 ft, July, *Pantling 361* (lectotype, CAL, CAL0000000604, designated by Bhattacharjee et al., 2011; isotype, BM, CAL, K).

Distribution. Arunachal Pradesh, Assam, Sikkim, West Bengal [Bhutan, China, Nepal].

Habit. Terrestrial herb.

Zeuxine rolfeana King & Pantl., J. Asiat. Soc. Bengal, Pt. 2, Nat. Hist. 66: 599. 1898 [1897] (as "*rolfiana*"). TYPE: India, South Andaman, Dhani Khari, 7 Mar. 1891, *G. King s.n.* (lectotype, CAL, no. 459475, designated by Manudev et al., 2014; isotype, CAL, no. 459476).

Distribution. Andaman & Nicobar Islands. Endemic.

Habit. Terrestrial herb.

Zeuxine strateumatica (L.) Schltr., Bot. Jahrb. Syst. 45: 394. 1911. *Orchis strateumatica* L., Sp. Pl. 2: 943. 1753. *Neottia strateumatica* (L.) R. Br., Prodr. Fl. Nov. Holland. 319. 1810. *Spiranthes strateumatica* (L.) Lindl., Bot. Reg. 10: t. 823. 1824. *Adenostylis strateumatica* (L.) Ames, Orchidaceae 2: 58. 1908. TYPE: Sri Lanka, *Herb. Hermann 2: 35, No. 319* (lectotype, BM, designated by Cribb in Cafferty & Jarvis (eds.), 1999).

Pterygodium sulcatum Roxb., Fl. Ind. ed. 1832, 3: 452. 1832 (as "*sulcata*"). *Zeuxine sulcata* (Roxb.) Lindl., Lindl. ex Wight, Cat. Ind. Pl. 123. 1836. *Adenostylis sulcata* (Roxb.) Hayata, Icon. Pl. Formosan. 6(Suppl.): 75. 1917. TYPE: India, Calcutta, icon. *Pterygodium sulcata* [sic], *Roxburgh 1086* (syntypes, CAL, K).
Zeuxine bracteata Wight, Icon. Pl. Ind. Orient. [Wight] 5(1): 16, t. 1724 bis. 1851. TYPE: India, Syndibad, Feb., *J. E. Stocks s.n.* (holotype, K).
Zeuxine brevifolia Wight, Icon. Pl. Ind. Orient. [Wight] 5(1): 16, t. 1725. 1851. TYPE: India, Mysore, Dec.–Jan., *Jerdon s.n.* (holotype, K; isotype, MH).

Zeuxine robusta Wight, Icon. Pl. Ind. Orient. [Wight] 5(1): 16, t. 1726. 1851. TYPE: India, Mysore, Jan., *Jerdon s.n.* (holotype, K).
Zeuxine strateumatica (L.) Schltr. var. *laxiflora* I. Barua, Orchid Fl. Kamrup Distr. Assam: 52. 2001. TYPE: India, Kamrup District, Manas Wild Life Sanctuary, on way to Mathanguri from Bansbari, *Barua 1518* (holotype, CAL; isotype, ASSAM).

Distribution. Andhra Pradesh, Arunachal Pradesh, Assam, Bihar, Chhattisgarh, Delhi, Gujarat, Himachal Pradesh, Jammu & Kashmir, Jharkhand, Karnataka, Kerala, Madhya Pradesh, Maharashtra, Manipur, Meghalaya, Mizoram, Nagaland, Odisha, Punjab, Rajasthan, Sikkim, Tamil Nadu, Tripura, Uttar Pradesh, Uttarakhand, West Bengal [Afghanistan, Bangladesh, Bhutan, China, Myanmar, Nepal, Pakistan, Sri Lanka].

Habit. Terrestrial herb.

EXCLUDED SPECIES

Bulbophyllum chloropterum Rchb. f., Linnaea 22: 835. 1850. TYPE: Brazil, *cult. Hoffmannsegg s.n. (leg. Binus s.n.)* (holotype, W).

Bulbophyllum fallax Rolfe, Gard. Chron., ser. 3, 6: 558. 1889. TYPE: "India, Assam," *cult. Seeger & Tropp s.n. (14 Sep. 1889)* (holotype, K; also icon. K).

Distribution. Brazil. Endemic.

Habit. Epiphytic herb.

Remarks. An examination of the type of *Bulbophyllum fallax*, a species that has never been found again in Assam since its original description, revealed that it cannot be distinguished from *B. chloropterum*, a Brazilian species. There can be no doubt that the type material of *B. fallax*, based on a cultivated specimen, was incorrectly reported from Assam and therefore the species can be excluded from the flora of India.

LITERATURE CITED

Abraham, A. & P. Vatsala. 1981. Introduction to Orchids, with Illustrations and Descriptions of 150 South Indian Orchids. Tropical Botanic Garden and Research Institute, Trivandrum.

Alarcón, M. & C. Aedo. 2002. A revision of the genus *Cephalanthera* (Orchidaceae) in the Iberian Peninsula and Balearic Islands. Anales Jard. Bot. Madrid 69(2): 227–248.

Ansari, R. & R. Balakrishnan. 1990. A revision of the Indian species of *Oberonia* (Orchidaceae). Orchid Monogr. 4: 1–82.

Barbhuiya, H. A., D. Verma, V. Kumar & S. Dey. 2017. Critical notes on some taxa of the genus *Acampe* Lindl. (Orchidaceae). Phytotaxa 303: 271–278.

Baumann, H., S. Kuenkele & R. Lorenz. 1989. Die nomenklatorischen Typen der von Linnaeus veröffentlichten Namen europäischer Orchideen. Mitteilungsbl. Arbeitskreis Heimische Orchid. Baden-Württemberg 21(3): 355–700.

Beentje, H. 2016. The Kew Plant Glossary: An Illustrated Dictionary of Plant Terms. Second Edition. Royal Botanic Gardens, Kew.

Bhattacharjee, A. & R. K. Bhakat. 2010. Lectotypifications in *Goodyera*. Rheedea 20(2): 73–75.

Bhattacharjee, A. & H. J. Chowdhery. 2011. Lecto- and epitypification of *Goodyera hirsuta* Griff. J. Jap. Bot. 86: 162–165.

Bhattacharjee, A. & H. J. Chowdhery. 2013. Epitypification of *Goodyera affinis*. J. Jap. Bot. 88: 286–290.

Bhattacharjee, A. & H. J. Chowdhery. 2014. Lectotypification of the name *Chrysobaphus roxburghii* (Orchidaceae). Taxon 63(5): 1114–1116.

Bhattacharjee, A. & H. J. Chowdhery. 2018. Fascicles of Flora of India. Fascicle 28. Orchidaceae: Orchidoideae, Cranichideae: Subtribe Goodyerinae. Botanical Survey of India, Kolkata.

Bhattacharjee, A., H. J. Chowdhery & R. K. Bhakat. 2011. Lectotypification in *Zeuxine* Lindl. Taiwania 56(2): 153–156.

Bhattacharjee, B., B. K. Dutta & P. K. Hajra. 2018. Orchid Flora of Southern Assam (Barak Valley): Diversity and their Conservation. Bishen Singh Mahendra Pal Singh, Dehra Dun.

Bhaumik, M. & P. Satyanarayana. 2014. *Pterygiella* Oliver (Scrophulariaceae) and *Pogonia* Jussieu (Orchidaceae), two generic records for Indian flora. Indian J. Forest. 37(3): 299–302.

Blatter, E. & C. McCann. 1931. Orchidaceae. *In* E. Blatter, Revision of the Flora of the Bombay Presidency. Part XVI. J. Bombay Nat. Hist. Soc. 35(2): 254–275.

Bone, R. E., P. J. Cribb & S. Buerki. 2015. Phylogenetics of Eulophiinae (Orchidaceae: Epidendroideae): Evolutionary patterns and implications for generic delimitation. Bot. J. Linn. Soc. 179: 43–56.

Cafferty, S. & C. E. Jarvis (editors). 1999. Typification of Linnaean specific and varietal names in the Orchidaceae. Taxon 48: 450–50.

CBD. 2021. Global Strategy. <https://www.cbd.int/gspc/strategy.shtml/>, accessed 1 June 2020.

Chakraborty, S., D. K. Agrawala, G. Aazhivaendhan & J. S. Jalal. 2021. Notes on *Aphyllorchis gollanii* and *Aphyllorchis alpina* (Orchidaceae) and lectotypification of both names. Richardiana, n.s. 5: 34–47.

Chase, M. W., K. M. Cameron, J. V. Freudenstein, A. M. Pridgeon, G. Salazar, C. Van den Berg & A. Schuiteman. 2015. An updated classification of Orchidaceae. Bot. J. Linn. Soc. 177: 151–174.

Chase, M. W., M. J. M. Christenhusz & A. Schuiteman. 2020. Expansion of *Calanthe* to include the species of *Cephalantheropsis*, *Gastrorchis* and *Phaius* (Collabieae; Orchidaceae). Phytotaxa 472(2): 159–168.

Chase, M. W., B. Gravendeel, B. P. Sulistyo, R. K. Wati & A. Schuiteman. 2021. Expansion of the orchid genus *Coelogyne* (Arethuseae; Epidendroideae) to include *Bracisepalum*, *Bulleyia*, *Chelonistele*, *Dendrochilum*, *Dickasonia*, *Entomophobia*, *Geesinkorchis*, *Gynoglottis*, *Ischnogyne*, *Nabaluia*, *Neogyna*, *Otochilus*, *Panisea* and *Pholidota*. Phytotaxa 510(2): 94–134.

Chase, M. W., A. Schuiteman & P. Kumar. 2021. Expansion of the orchid genus *Eulophia* (Eulophiinae; Epidendroideae) to include *Acrolophia*, *Cymbidiella*, *Eulophiella*, *Geodorum*, *Oeceoclades* and *Paralophia*. Phytotaxa 491(1): 47–56.

Chen, S. P., H. Z. Tian, Q. X. Guan, J. W. Zhai, G. Q. Zhang, L. J. Chen, Z. J. Liu, S. R. Lan & M. H. Li. 2019. Molecular systematics of Goodyerinae (Cranichideae, Orchidoideae, Orchidaceae) based on multiple nuclear and plastid regions. Molec. Phylogen. Evol. 139: art. 106542.

Choltco, T. 2009. A new species of *Cleisostoma*. Orchid Rev. 117(1285): 39–41.

Chowlu, K. 2014. A note on the validity of *Bulbophyllum manipurense* Sathish & Suresh (Orchidaceae). Natl. Acad. Sci. Lett. 37(6): 513–515.

Clayton, D. 2017. Charles Parish—Plant Hunter and Botanical Artist in Burma. The Ray Society, London, in association with Royal Botanic Gardens, Kew.

Clayton, D. & P. Cribb. 2013. The Genus *Calanthe*. Natural History Publications (Borneo) in association with Royal Botanic Gardens, Kew.

Cribb, P. J. 1997. The Genus *Cypripedium*. Royal Botanic Gardens, Kew; Timber Press, Portland, Oregon.

Cribb, P. J. 1998. The Genus *Paphiopedilum* (Second Edition). Natural History Publications (Borneo), Kota Kinabalu; Royal Botanic Gardens, Kew.

De Vogel, E. F. 1969. Monograph of the tribe Apostasieae (Orchidaceae). Blumea 17: 313–350.

De Vogel, E. F. 1988. Revisions in Coelogyninae (Orchidaceae) III. The genus *Pholidota*. Orchid Monogr. 3: 1–118.

Deforestation Statistics. 2021. <https://rainforests.mongabay.com/deforestation/archive/India.htm/>, accessed 1 June 2020.

Deori, C., R. Shanpru & S. Phukan. 2007. A *Dendrobium* new for India. Orchid Rev. 115(1273): 44–45.

Deva, S. & H. B. Naithani. 1986. The Orchid Flora of North West Himalaya. Print & Media Associates, New Delhi.

Du Puy, D. J. & P. J. Cribb. 1988. The Genus *Cymbidium*. Timber Press, Portland, Oregon.

Duthie, J. F. 1906. The orchids of the north-western Himalaya. Ann. Roy. Bot. Gard. (Calcutta) 9: 81–211.

Efimov, P. G. 2016. A revision of *Platanthera* (Orchidaceae; Orchidoideae; Orchideae) in Asia. Phytotaxa 254: 1–233.

Efimov, P. G. & X.-H. Jin. 2014. Typification of specific and infraspecific names in Asian *Platanthera* (Orchidaceae). Taxon 63(5): 1117–1121.

Fernando, S. S. & P. Ormerod. 2008. An annotated checklist of the orchids of Sri Lanka. Rheedea 18(1): 1–28.

Gale, S., P. Kumar & T. Phaxaysombath. 2018. A Guide to Orchids of Laos. Natural History Publications (Borneo), Kota Kinabalu.

Garay, L. A. & H. R. Sweet. 1974. Orchids of southern Ryukyu Islands. Botanical Museum, Harvard University, Cambridge.

Geiger, D. L. 2019. Studies on *Oberonia* 5 (Orchidaceae: Malaxideae). Twenty-four new synonyms, and a corrected spelling. Blumea 64: 123–139.

Geiger, D. L. 2020. Studies in *Oberonia* 8 (Orchidaceae: Malaxideae). Additional 24 new synonyms, a corrected spelling, and other nomenclatural matters. Blumea 65: 188–203.

Gogoi, K., R. Das & R. Yonzone. 2015. Orchids of Assam, North East India—An annotated checklist. Int. J. Pharm. Life Sci. 6(1): 4123–4156.

Griffith, W. 1851. Icones Plantarum Asiaticarum, Part 3. Bishop's College Press, Calcutta.

Gyeltshen, N., C. Gyeltshen, K. Tobgay, S. Dalström, D. B. Gurung, N. Gyeltshen & B. B. Ghalley. 2020. Two new spotted *Chiloschista* species (Orchidaceae: Aeridinae) from Bhutan. Lankesteriana 20(3): 281–299.

Hallé, N. 1977. Flore de la Nouvelle-Calédonie et dépendances: 8. Orchidacées. Muséum National d'Histoire Naturelle, Paris.

Hassler, M. 2019. Illustrated World Compendium of Orchids. <https://worldplants.webarchiv.kit.edu/orchids/index.php>, accessed 1 June 2020.

Hooker, J. D. 1890a. Orchideae. Pp. 667–858 *in* The Flora of British India, Vol. 5. L. Reeve & Co., London.

Hooker, J. D. 1890b. Orchideae. Pp. 1–198 *in* The Flora of British India, Vol. 6. L. Reeve & Co., London.

IUCN. 2012. IUCN Red List Categories and Criteria, Version 3.1. Second edition. Prepared by the IUCN Species Survival Commission. IUCN, Gland, Switzerland; Cambridge, United Kingdom.

IUCN Red List. 2021. The IUCN Red List of Threatened Species. <https://www.iucnredlist.org/>, accessed 1 June 2020.

Jain, S. K. & K. Mehrotra. 1984. A Preliminary Inventory of Orchidaceae of India. Botanical Survey of India, Howrah.

Jakha, H. Y., C. R. Deb, N. S. Jamir & S. Dey. 2015. *Arachnis labrosa* var. *zhaoi* (Orchidaceae): A new record for India. Rheedea 25(2): 120–122.

Jalal, J. S. 2018. Orchids of Maharashtra. Botanical Survey of India, Kolkata.

Jalal, J. S. 2019. Diversity and distribution of orchids of Goa, Western Ghats, India. J. Threat. Taxa 11(15): 15015–15042.

Jalal, J. S. & J. Jayanthi. 2018. An updated checklist of the orchids of Maharashtra, India. Lankesteriana 18(1): 23–62.

Jayaweera, D. M. A. 1981. A Revised Handbook to the Flora of Ceylon (Orchidaceae). Balkema, Rotterdam.

Jenny, R. 2021. Orchilibra. <https://orchilibra.com/>, accessed 26 April 2021.

Jin, W. T., A. Schuiteman, M. W. Chase, J. W. Li, S. W. Chung, T. C. Hsu & X. H. Jin. 2017. Phylogenetics of subtribe Orchidinae s.l. (Orchidaceae; Orchidoideae) based on seven markers (plastid *matK*, *psaB*, *rbcL*, *trnL-F*, *trnH-psba*, and nuclear nrITS, *Xdh*): Implications for generic delimitation. BMC Plant Biology 17: art. 222.

Karthikeyan, S., S. K. Jain, M. P. Nayar & M. Sanjappa. 1989. Florae Indicae enumeratio Monocotyledonae (Flora of India, Series 4). Botanical Survey of India, Calcutta.

King, G. & R. Pantling. 1898. The Orchids of Sikkim-Himalaya. Ann. Roy. Bot. Gard. (Calcutta) 8: 1–342, t. 1–448.

Kocyan, A. & A. Schuiteman. 2014. New combinations in Aeridinae (Orchidaceae). Phytotaxa 161(1): 061–085.

Kores, P. J. 1989. A precursory study of Fijian orchids. Allertonia 5: 1–222.

Kumar, C. S. & K. S. Manilal. 1994. A Catalogue of Indian Orchids. Bishen Singh Mahendra Pal Singh, Dehra Dun.

Kumar, P. & S. W. Gale. 2020. Additions to the orchid flora of Laos and taxonomic notes on orchids of the Indo-Burma region—II. Taiwania 65(1): 47–60.

Kumar, P., S. W. Gale, A. Kocyan, G. A. Fischer, L. V. Averyanov, R. Borosova, A. Bhattacharjee, J.-H. Li & K. S. Pang. 2014. *Gastrochilus kadooriei*, a new species from Hong Kong, with notes on allied taxa in section Microphyllae found in the region. Phytotaxa 164: 91–103.

Kumar, P., S. W. Gale, H. Æ. Pedersen, T. Phaxaysombath, S. Bouamanivong & G. A. Fischer. 2018. Additions to the orchid flora of Laos and taxonomic notes on orchids of the Indo-Burma region. Taiwania 63(1): 61–83.

Kumar, V. 2015. Taxonomic Studies on the Tribe Cymbideae Pfitz. (Orchidaceae) in India. Ph.D. Thesis, Kumaun University, Nainital.

Kurzweil, H. 2005. Taxonomic studies in the genus *Disperis* in southeast Asia. Blumea 50: 143–152.

Kurzweil, H. & S. Lwin. 2014. A Guide to Orchids of Myanmar. Natural History Publications (Borneo), Kota Kinabalu.

Lawkush, A. B. & H. J. Chowdhery. 2014. Lectotypification of *Herminium gracile* G.King & R.Pantling (Orchidaceae). Richardiana 14: 19–22.

Lindley, J. 1857. Contributions to the orchidology of India.—No. I. J. Proc. Linn. Soc., Bot. 1(4): 170–190.

Lindley, J. 1858. Contributions to the orchidology of India.—No. II. J. Proc. Linn. Soc., Bot. 3(9): 1–64.

Linnaeus, C. 1753. Species Plantarum. Laurentius Salvius, Holmia [Stockholm].

Lund, I. D. 1987. The genus *Cremastra*, a taxonomic revision. Nordic J. Bot. 8: 197–203.

Mabberley, D. J. 2011. A note on some adulatory botanical plates distributed by Sir Joseph Banks. Kew Bull. 66: 475–477.

Majumdar, N. C. & D. N. Bakshi. 1979. A few Linnaean specific names typified by the illustrations in Rheede's *Hortus Indicus Malabaricus*. Taxon 28(4): 353–354.

Manudev, K. M., A. Bhattacharjee & S. Nampy. 2014. Rediscovery of *Zeuxine rolfiana*, a 'Critically Endangered' endemic Indian orchid from Andaman and Nicobar Islands, with a note on its typification. Ann. Bot. Fenn. 51(6): 409–413.

Margońska, H. B. 2012. Taxonomic Redefinition of the Subtribe Malaxidinae (Orchidales, Malaxideae). Koeltz Scientific Books, Koenigstein.

Matthew, K. M. 1983. The Flora of the Tamilnadu Carnatic, Vol. 3(2). Rapinat Herbarium, Tiruchirapalli.

McNeill, J. 2014. Holotype specimens and type citations: General issues. Taxon 63(5): 1112–1113.

Merrill, E. D. 1938. Critical consideration of Houttuyn's new genera and new species of plants 1773–1783. J. Arnold Arbor. 19: 291–375.

Minderhoud, M. E. & E. F. de Vogel. 1986. A taxonomic revision of the genus *Acriopsis* Reinwardt ex Blume (Orchidaceae). Orchid Monographs 1: 1–16.

Misra, S. 2019. Orchids of India—A Handbook. Bishen Singh Mahendra Pal Singh, Dehra Dun.

Mytnik-Ejsmont, J. & P. Baranow. 2010. Taxonomic study of *Polystachya* Hook. (Orchidaceae) from Asia. Pl. Syst. Evol. 290(1): 57–63.

Ng, Y. P., A. Schuiteman, H. Æ. Pedersen, G. Petersen, S. Watthana, O. Seberg, A. M. Pridgeon, P. J. Cribb & M. W. Chase. 2018. Phylogenetics and systematics of *Eria* and related genera (Orchidaceae: Podochileae). Bot. J. Linn. Soc. 186(2): 179–201.

O'Byrne, P., P. T. Ong & J. J. Vermeulen. 2015. New and little-known species of *Thrixspermum* from Peninsular Malaysia and Sulawesi. Males. Orchid J. 15: 83–108.

Odyuo, N., R. Daimary & C. Deori. 2017. *Liparis formosana* (Orchidaceae)—A new addition to the orchid flora of India. J. Jap. Bot. 92(1): 53–56.

Ormerod, P. 2017. Checklist of Papuasian Orchids. Nature & Travel Books, Lismore.

Ormerod, P. 2018a. Notes on Asiatic *Tropidia* (Orchidaceae: Tropidieae). Harvard Pap. Bot. 23(1): 77–83.

Ormerod, P. 2018b. Notes on *Zeuxine* (Orchidaceae: Goodyerinae). Harvard Pap. Bot. 23(2): 269–277.

Ormerod, P. & C. S. Kumar. 2018. New names in Indian and Sri Lankan orchids. Harvard Pap. Bot. 23(2): 281–284.

Ormerod, P., M. A. Naïve & J. Cootes. 2019. Notes on some Malesian Orchidaceae. Harvard Pap. Bot. 24(2): 281–290.

Ormerod, P., H. Kurzweil & S. Watthana. 2021. Annotated list of Orchidaceae for Myanmar. Phytotaxa 481(1): 1–262.

Pace, M. C. 2020. A recircumscription of *Goodyera* (Orchidaceae), including the description of *Paorchis* gen. nov., and resurrection of *Cionisaccus*, *Eucosia*, and *Salacistis*. Brittonia 72(3): 257–267.

Pace, M. C., G. Giraldo, J. Frericks, C. A. Lehnebach & K. M. Cameron. 2019. Illuminating the systematics of the *Spiranthes sinensis* species complex (Orchidaceae): Ecological speciation with little morphological differentiation. Bot. J. Linn. Soc. 189(1): 36–62.

Pande, H. K. & S. Arora (editors). 2014. India's Fifth National Report to the Convention on Biological Diversity. Ministry of Environment and Forests, Government of India, New Delhi.

Pandey, T. R. & X.-H. Jin. 2021. Taxonomic revision of *Habenaria josephi* group (sect. *Diphyllae* s.l.) in the Pan-Himalaya. Phytokeys 175: 109–135.

Pearce, N. R. & P. J. Cribb. 2002. The Orchids of Bhutan. Charlesworth Group, Huddersfield.

Pearce, N., P. J. Cribb & J. Renz. 2001. Notes relating to the Flora of Bhutan: XLIV. Taxonomic notes, new taxa and additions to the Orchidaceae of Bhutan and Sikkim (India). Edinburgh J. Bot. 58(1): 99–122.

Pedersen, H. Æ., H. Kurzweil, S. Suddee & P. J. Cribb. 2011. Flora of Thailand. Vol. 12, Part 1. The Forest Herbarium, Bangkok.

Pettersson, J. 1990. *Nervilia* (Orchidaceae) in Africa. Acta Univ. Upsal., Compreh. Summ. Uppsala Diss. Fac. Sci. 281.

Phukan, S. & A. A. Mao. 2004. Additions to the Indian orchid flora. Orchid Rev. 112: 114–118.

POWO. 2021. Plants of the World Online. <http://www.plantsoftheworldonline.org/>, accessed 26 April 2021.

Pradhan, U. C. 1976. Indian Orchids: Guide to Identification & Culture, Vol. I. Udai C. Pradhan, Kalimpong.

Pradhan, U. C. 1979. Indian Orchids: Guide to Identification & Culture, Vol. II. Udai C. Pradhan, Kalimpong.

Pridgeon, A. M., P. J. Cribb, M. W. Chase & F. N. Rasmussen (editors). 1999. Genera Orchidacearum [Pridgeon], Vol. 1. General Introduction, Apostasioideae, Cypripedioideae. Oxford University Press, Oxford.

Pridgeon, A. M., P. J. Cribb, M. W. Chase & F. N. Rasmussen (editors). 2001. Genera Orchidacearum [Pridgeon], Vol. 2. Orchidoideae (Part one). Oxford University Press, Oxford.

Pridgeon, A. M., P. J. Cribb, M. W. Chase & F. N. Rasmussen (editors). 2003. Genera Orchidacearum [Pridgeon], Vol. 3. Orchidoideae (Part two), Vanilloideae. Oxford University Press, Oxford.

Pridgeon, A. M., P. J. Cribb, M. W. Chase & F. N. Rasmussen (editors). 2005. Genera Orchidacearum [Pridgeon], Vol. 4. Epidendroideae (Part One). Oxford University Press, Oxford.

Pridgeon, A. M., P. J. Cribb, M. W. Chase & F. N. Rasmussen (editors). 2009. Genera Orchidacearum [Pridgeon], Vol. 5. Epidendroideae (Part Two). Oxford University Press, Oxford.

Pridgeon, A. M., P. J. Cribb, M. W. Chase & F. N. Rasmussen (editors). 2014. Genera Orchidacearum [Pridgeon], Vol. 6. Epidendroideae (Part Three). Oxford University Press, Oxford.

Priyadarshana, T. S., A. G. Atthanagoda, I. H. Wijewardhane, N. Aberathna, I. Peabotuwage & P. Kumar. 2020. *Dendrobium taprobanium* (Orchidaceae): A new species from Sri Lanka with taxonomic notes on some species of the genus. Phytotaxa 432(1): 81–94.

Rao, A. N. & V. Kumar. 2018. Updated checklist of orchid flora of Manipur. Turczaninowia 21(4): 109–134.

Raskoti, B. B., W. T. Jin, X. G. Xiang, A. Schuiteman, D. Z. Li, J. W. Li, W. C. Huang, X. H. Jin & L. Q. Huang. 2016. A phylogenetic analysis of molecular and morphological characters of *Herminium* (Orchidaceae, Orchideae): Evolutionary relationships, taxonomy, and patterns of character evolution. Cladistics 32: 198–210.

Raskoti, B. B., A. Schuiteman, W. T. Jin & X.-H. Jin. 2017. A taxonomic revision of *Herminium* L. (Orchidoideae, Orchidaceae). PhytoKeys 79: 1–74.

Rasmussen, F. N. 1977. The genus *Corymborkis* Thou. (Orchidaceae): A taxonomic revision. Bot. Tidsskr. 71: 161–192.

Ravichandran, V., M. Manikandan & C. Murugan. 2019. *Habenaria richardiana* (Orchidaceae), a little known endemic orchid from Nilgiri Biosphere Reserve (India). Richardiana, n.s. 3: 25–30.

Reichenbach, H. G. 1880. New Garden Plants. P. 776 *in* The Gardeners' Chronicle, new series, Vol. 13. London.

Renz, J. 1984. Orchidaceae. *In* E. Nasir & S. I. Ali (editors), Flora of Pakistan, No. 164. Pan Graphics Ltd., Islamabad.

Renz, J. & G. Taubenheim. 1984. Orchidaceae. Pp. 450–552 & 587–600 *in* P. H. Davis (editor), Flora of Turkey 8. Edinburgh University Press, Edinburgh.

Rheede tot Draakestein, H. van. 1678–1703. Hortus [Indicus] Malabaricus, Vols. 1–12. J. van Someren & J. van Dyck, Amsterdam.

Roy, D. K., H. A. Barbhuiya, A. D. Talukdar & B. K. Sinha. 2014. *Bulbophyllum manabendrae* (Orchidaceae: Epidendroideae), a new species from Meghalaya, India. Phytotaxa 164: 291–295.

Santapau, H. & Z. Kapadia. 1960. Critical notes on the Orchidaceae of Bombay State III. The genus *Oberonia* Lindl. J. Bombay Nat. Hist. Soc. 57(2): 252–269.

Santapau, H. & Z. Kapadia. 1962. Critical notes on the Orchidaceae of Bombay State 8. J. Bombay Nat. Hist. Soc. 59(1): 154–172.

Sarkar, P. K. 1995a. An up-to-date census of Indian orchids. J. Econ. Taxon. Bot., Addit. Ser. 11: 1–32.

Sarkar, P. K. 1995b. Bibliography on Indian Orchidaceae. J. Econ. Taxon. Bot., Addit. Ser. 11: 77–115.

Sarkar, P. K. 1995c. Rare, endangered and endemic orchids in India. J. Econ. Taxon. Bot., Addit. Ser. 11: 33–47.

Sathapattayanon, A. 2008. Taxonomic Revision of Orchids in the Genus *Dendrobium* Sw. section *Formosae* (Benth. & Hook. f.) Hook. f. in Thailand and Adjacent Areas. Thesis, Chulalongkorn University, Bangkok.

Schuiteman, A. 2011a. The strange case of *Dendrobium aphyllum*. Orchid Rev. 119: 104–110.

Schuiteman, A. 2011b. The forgotten identity of *Eria scabrilinguis*. Orchid Rev. 119: 214–221.

Schuiteman, A. 2014. Artificial key to the genera of Aeridinae. Pp. 124–129 *in* A. M. Pridgeon, P. J. Cribb, M. W. Chase & F. N. Rasmussen (editors), Genera Orchidacearum [Pridgeon], Vol. 6. Epidendroideae (Part Three). Oxford University Press, Oxford.

Schuiteman, A., P. Cribb, T. Want & C. King. 2020. *Coelogyne annamensis*. Orchidaceae. Curtis's Bot. Mag. 37(1): 3–12, t. 932.

Seidenfaden, G. 1968. The genus *Oberonia* in mainland Asia. Dansk Bot. Ark. 25(3): 17–125.

Seidenfaden, G. 1976. Contributions to orchid flora of Thailand VII. Bot. Tidsskr. 71: 1–30.

Seidenfaden, G. 1977a. Orchid genera in Thailand V. Orchidoideae. Dansk Bot. Ark. 31(3): 1–149.

Seidenfaden, G. 1977b. Thalia maravara and the rigid airblossom. Bot. Mus. Leafl., Harvard Univ., 25(2), 49–69.

Seidenfaden, G. 1978. Orchid genera in Thailand VI. Neottioideae Lindl. Dansk Bot. Ark. 32(2): 1–195.

Seidenfaden, G. 1979. Orchid genera in Thailand VIII. *Bulbophyllum* Thouars. Dansk Bot. Ark. 33(3): 1–228.

Seidenfaden, G. 1986. Orchid genera in Thailand XIII. Thirty-three epidendroid genera. Opera Bot. 89: 1–216.

Seidenfaden, G. 1988. Orchid genera in Thailand XIV. Fifty-nine vandoid genera. Opera Bot. 95: 1–398.

Seidenfaden, G. 1994. The correct name for *Bulbophyllum uniflorum* Griff., nom. illeg. Nordic J. Bot. 14: 201–203.

Seidenfaden, G. 1995. The Descriptiones Epidendrorum of J. G. König 1791. Olsen & Olsen, Fredensborg.

Seth, C. J. 1982. *Cymbidium aloifolium* and its allies. Kew Bull. 37: 397–402.

Singh, B. 2015. Extended distribution of rare *Anoectochilus papilosus* [sic] L.V. Averyanov (Orchidaceae) to the Indian Himalaya in Asia. Pleione 9(1): 217–221.

Singh, S. K., A. Bhattacharjee, R. Kumar & P. Efimov. 2014. Lectotypification of *Habenaria dyeriana* King & Pantl. (Orchidaceae). Ann. Bot. Fenn. 51(4): 267–271.

Singh, S. S., D. K. Agrawala, J. Jalal, S. S. Dash, A. A. Mao & P. Singh. 2019. Orchids of India: A Pictorial Guide. Botanical Survey of India, Kolkata.

Sourav, M. S. H., R. Halder, P. Kumar & A. Schuiteman. 2017. *Eulophia obtusa* (Orchidaceae: Epidendroideae: Cymbideae) an addition to the flora of Bangladesh, with notes on its ecology and conservation status. Kew Bull. 72: art. 19.

Stone, J. & P. J. Cribb. 2017. Lady Tankerville's Legacy: A Historical and Monographic Review of *Phaius* and *Gastrorchis*. Natural History Publications (Borneo), Kota Kinabalu, in association with Royal Botanic Gardens, Kew.

Suetsugu, K., T.-C. Hsu, H. Fukunaga & S. Sawa. 2016. Epitypification, emendation and synonymy of *Lecanorchis taiwaniana* (Vanilleae, Vanilloideae). Phytotaxa 265: 157–163.

Surveswaran, S., V. Gowda & M. Sun. 2020. Cryptic species and taxonomic troubles: A rebuttal of the systematic treatment of the Asian ladies' tresses orchids (*Spiranthes* species; Orchidaceae) by Pace et al. (2019). Bot. J. Linn. Soc. 194: 375–381.

Swami, N. 2016. Terrestrial Orchids. Naresh Swami, Arunachal Pradesh.

Swami, N. 2017. Orchids of Ziro, Arunachal Pradesh. Naresh Swami, Arunachal Pradesh.

Tang, G. D., G. Q. Zhang, W. J. Hong, Z. J. Liu & X. Y. Zhuang. 2015. Phylogenetic analysis of Malaxideae (Orchidaceae: Epidendroideae): Two new species based on the combined *nr*DNA ITS and chloroplast *mat*K sequence. Guihaia 35: 447–463.

Tetsana, N., H. Æ. Pedersen & K. Sridith. 2013. Five species of *Liparis* (Orchidaceae) newly recorded for Thailand. Thai Forest Bull. (Botany) 41: 48–55.

Tetsana, N., K. Sridith, S. Watthana & H. Æ Pedersen. 2019. A taxonomic revision of *Liparis* (Orchidaceae: Epidendroideae: Malaxideae) in Thailand. Phytotaxa 421(1): 001–065.

Thiers, B. 2021 [continuously updated]. Index Herbariorum: A global directory of public herbaria and associated staff. New York Botanical Garden's Virtual Herbarium. <http://sweetgum.nybg.org/science/ih/>, accessed 26 April 2021.

Tian, H. Z., A. Bhattacharjee, P. Kumar, T. C. Hsu & H. Æ. Pedersen. 2014. Neotypification of *Zeuxine agyokuana* (Goodyerinae, Orchidoideae, Orchidaceae) with a new synonym. Ann. Bot. Fenn. 51: 101–105.

Turland, N. J., J. H. Wiersema, F. R. Barrie, W. Greuter, D. L. Hawksworth, P. S. Herendeen, S. Knapp, W.-H. Kusber, D.-Z. Li, K. Marhold, T. W. May, J. McNeill, A. M. Monro, J. Prado, M. J. Price & G. F. Smith (editors) 2018. International Code of Nomenclature for algae, fungi, and plants (Shenzhen Code). Regnum Veg. 159.

Turner, H. 1992. A revision of the orchid genera *Ania* Lindley, *Hancockia* Rolfe, *Mischobulbum* Schltr. and *Tainia* Blume. Orchid Monogr. 6: 43–100.

Verma, D. & H. A. Barbhuiya. 2014. Neotypification of *Dendrobium darjeelingensis* [sic] and a new combination in *Dendrobium* for *Flickingeria clementsii* (Orchidaceae). Phytotaxa 170(4): 296–296.

Vij, S. P., J. Verma & C. S. Kumar. 2013. Orchids of Himachal Pradesh. Bishen Singh Mahendra Pal Singh, Dehra Dun.

Watthana, S. 2007. The genus *Pomatocalpa* (Orchidaceae) a taxonomic revision. Harvard Pap. Bot. 11(2): 207–256.

WCSP. 2019. World Checklist of Selected Plant Families. Facilitated by the Royal Botanic Gardens, Kew. <http://wcsp.science.kew.org/>, accessed 26 April 2021.

Xiang, X. G., W. T. Jin, D. Z. Li, A. Schuiteman, W. C. Huang, J. W. Li, X. H. Jin & Z. Y. Li. 2014. Phylogenetics of tribe Collabieae (Orchidaceae, Epidendroideae) based on four chloroplast genes with morphological appraisal. PLoS ONE 9(1): art. e87625.

Yukawa, T. & P. J. Cribb. 2014. Nomenclatural changes in the genus *Calanthe* (Orchidaceae). Bull. Natl. Mus. Nat. Sci., Tokyo, B, 40(4): 145–151.

Zhai, J. W., G. Q. Zhang, L. Li, M. Wang, L. J. Chen, S. W. Chung, F. Jiménez Rodríguez, J. Francisco-Ortega, S. R. Lan, F. W. Xing & Z. J. Liu. 2014. A new phylogenetic analysis sheds new light on the relationships in the *Calanthe* alliance (Orchidaceae) in China. Molec. Phylogen. Evol. 77: 216–222.

Zhou, X. X., Z. Q. Cheng, Q. X. Liu, J. L. Zhang, A. Q. Hu, M. Z. Huang, C. Hu & H. Z. Tian. 2016. An updated checklist of Orchidaceae for China, with two new national records. Phytotaxa 276(1): 1–148.

INDEX TO SYNONYMS

retusa (L.) Sw. = **Rhynchostylis retusa** (L.) Blume

rigida Buch.-Ham. ex Sm. = **Acampe praemorsa**
(Roxb.) Blatt. & McCann

rostrata Roxb. = **Micropera rostrata** (Roxb.) N. P.
Balakr.

suaveolens Roxb. = **Pteroceras teres** (Blume) Holttum

taenialis Lindl. = **Phalaenopsis taenialis** (Lindl.)
Christenson & Pradhan

tenuifolia (L.) Moon = **Cleisostoma tenuifolium** (L.)
Garay

teres (Blume) Lindl. = **Pteroceras teres** (Blume)
Holttum

testacea Lindl. = **Vanda testacea** (Lindl.) Rchb. f.

undulata Sm. = **Acampe praemorsa** (Roxb.) Blatt. &
McCann

uniflora (Lindl.) Summerh. = **Papilionanthe uniflora**
(Lindl.) Garay

vandarum Rchb. f. = **Papilionanthe vandarum** (Rchb.
f.) Garay

wightiana Lindl. = **Vanda testacea** (Lindl.) Rchb. f.

williamsii R. Warner = **Aerides rosea** Lodd. ex Lindl. &
Paxton

Aggeianthus

marchantioides Wight = **Porpax reticulata** Lindl.

Agrostophyllum

khasianum Griff. = **Agrostophyllum planicaule** (Wall.
ex Lindl.) Rchb. f.

Alismorkis

alismifolia (Lindl.) Kuntze = **Calanthe alismifolia** Lindl.

alpina (Hook. f. ex Lindl.) Kuntze = **Calanthe alpina**
Hook. f. ex Lindl.

angusta (Lindl.) Kuntze = **Calanthe odora** Griff.

biloba (Lindl.) Kuntze = **Calanthe biloba** Lindl.

brevicornu (Lindl.) Kuntze = **Calanthe brevicornu**
Lindl.

chloroleuca (Lindl.) Kuntze = **Calanthe chloroleuca**
Lindl.

clavata (Lindl.) Kuntze = **Calanthe densiflora** Lindl.

densiflora (Lindl.) Kuntze = **Calanthe densiflora** Lindl.

elytroglossa (Rchb. f. ex Hook. f.) Kuntze = **Calanthe
herbacea** Lindl.

foerstermannii (Rchb. f.) Kuntze = **Calanthe lyroglossa**
Rchb. f.

gracilis (Lindl.) Kuntze = **Calanthe obcordata** (Lindl.)
M. W. Chase, Christenh. & Schuit.

griffithii (Lindl.) Kuntze = **Calanthe griffithii** Lindl.

herbacea (Lindl.) Kuntze = **Calanthe herbacea** Lindl.

lindleyana Kuntze = **Calanthe plantaginea** Lindl.

longipes (Hook. f.) Kuntze = **Calanthe longipes** Hook. f.

lyroglossa (Rchb. f.) Kuntze = **Calanthe lyroglossa**
Rchb. f.

mannii (Hook. f.) Kuntze = **Calanthe mannii** Hook. f.

masuca (D. Don) Kuntze = **Calanthe masuca** (D. Don)
Lindl.

odora (Griff.) Kuntze = **Calanthe odora** Griff.

pachystalix (Rchb. f.) Kuntze = **Calanthe davidii** Franch.

puberula (Lindl.) Kuntze = **Calanthe puberula** Lindl.

purpurea (Lindl.) Kuntze = **Calanthe masuca** (D. Don)
Lindl.

tricarinata (Lindl.) Kuntze = **Calanthe tricarinata** Lindl.

uncata (Lindl.) Kuntze = **Calanthe uncata** Lindl.

vaginata (Lindl.) Kuntze = **Calanthe odora** Griff.

vestita (Wall. ex Lindl.) Kuntze = **Calanthe vestita** Wall.
ex Lindl.

Amesia

royleana (Lindl.) Hu = **Epipactis royleana** Lindl.

Amitostigma

pathakianum Av. Bhattacharjee = **Ponerorchis
pathakiana** (Av. Bhattacharjee) J. M. H. Shaw

puberulum (King & Pantl.) Schltr. = **Ponerorchis
puberula** (King & Pantl.) Verm.

puberulum (King & Pantl.) Tang & F. T. Wang =
Ponerorchis puberula (King & Pantl.) Verm.

Androcorys

angustilabris (King & Pantl.) Agrawala & H. J. Chowdhery
= **Platanthera stenochila** X. H. Jin, Schuit., Raskoti
& Lu Q. Huang

gracilis (King & Pantl.) Schltr. = **Herminium gracile**
King & Pantl.

jaffreyana (King & Pantl.) Lawkush & Vik. Kumar =
Herminium jaffreyanum King & Pantl.

josephi (Rchb. f.) Agrawala & H. J. Chowdhery =
Herminium josephi Rchb. f.

kalimpongensis (Pradhan) Agrawala & H. J. Chowdhery =
Herminium kalimpongensis Pradhan

monophylla (D. Don) Agrawala & H. J. Chowdhery =
Herminium monophyllum (D. Don) P. F. Hunt &
Summerh.

pugioniformis (Lindl. ex Hook. f.) K. Y. Lang = **Herminium pugioniforme** Lindl. ex Hook. f.

Ania

bicornis Lindl. Edwards's Bot. Reg. 28(Misc.): 37. 1842. =
Tainia bicornis (Lindl.) Rchb. f.

hookeriana (King & Pantl.) Tang & F. T. Wang ex
Summerh. = **Ania penangiana** (Hook. f.) Summerh.

latifolia Lindl. = **Tainia latifolia** (Lindl.) Rchb. f.

maculata Thwaites = **Chrysoglossum ornatum** Blume

promensis (Lindl.) Senghas = **Eulophia promensis** Lindl.

Anisopetalon

careyanum Hook. = **Bulbophyllum careyanum** (Hook.)
Spreng.

Anoectochilus

clarkei (Hook. f.) Seidenf. & Smitinand = **Odontochilus
clarkei** Hook. f.

crispus Lindl. = **Odontochilus crispus** (Lindl.) Hook. f.

elwesii (C. B. Clarke ex Hook. f.) King & Pantl. =
Odontochilus elwesii C. B. Clarke ex Hook. f.

flavus Benth. & Hook. f. = **Odontochilus lanceolatus**
(Lindl.) Blume

grandiflorus Lindl. = **Odontochilus grandiflorus**
(Lindl.) Benth. ex Hook. f.

griffithii Hook. f. = **Anoectochilus brevilabris** Lindl.

lanceolatus Lindl. = **Odontochilus lanceolatus** (Lindl.)
Blume

luteus Lindl. = **Odontochilus lanceolatus** (Lindl.)
Blume

pomrangianus Seidenf. = **Rhomboda lanceolata** (Lindl.)
Ormerod

pumilus (Hook. f.) Seidenf. & Smitinand = **Myrmechis
pumila** (Hook. f.) Tang & F. T. Wang

rotundifolius (Blatt.) N. P. Balakr. = **Aenhenrya rotundifolia** (Blatt.) C. S. Kumar & F. N. Rasm.

sikkimensis King & Pantl. = **Anoectochilus brevilabris** Lindl.

tortus (King & Pantl.) King & Pantl. = **Odontochilus tortus** King & Pantl.

Anota

gigantea (Lindl.) Fukuy. = **Rhynchostylis gigantea** (Lindl.) Ridl.

Aopla

reniformis (D. Don) Lindl. = **Habenaria reniformis** (D. Don) Hook. f.

Aorchis

roborovskii (Maxim.) Seidenf. = **Galearis roborovskii** (Maxim.) S. C. Chen, P. J. Cribb & S. W. Gale

spathulata (Lindl.) Verm. = **Galearis spathulata** (Lindl.) P. F. Hunt

Apaturia

lindleyana Wight = **Pachystoma pubescens** Blume

senilis Lindl. = **Pachystoma pubescens** Blume

Apetalon

minutum Wight = **Didymoplexis pallens** Griff.

Aphyllorchis

gollanii Duthie = **Aphyllorchis alpina** King & Pantl.

pantlingii W. W. Sm. = **Neottia pantlingii** (W. W. Sm.) Tang & F. T. Wang

parviflora King & Pantl. = **Neottia acuminata** Schltr.

prainii Hook. f. = **Aphyllorchis montana** Rchb. f.

vaginata Hook. f. = **Chamaegastrodia vaginata** (Hook. f.) Seidenf.

Aplostellis

punctata (Blume) Ridl. = **Nervilia punctata** (Blume) Makino

Aporum

anceps (Sw.) Lindl. = **Dendrobium anceps** Sw.

crumenatum (Sw.) Brieger = **Dendrobium crumenatum** Sw.

cuspidatum Wall. ex Lindl. = **Dendrobium nathanielis** Rchb. f.

jenkinsii Griff. = **Dendrobium parciflorum** Rchb. f. ex Lindl.

keithii (Ridl.) M. A. Clem. = **Dendrobium keithii** Ridl.

kentrophyllum (Hook. f.) Brieger = **Dendrobium kentrophyllum** Hook. f.

mannii (Ridl.) Rauschert = **Dendrobium mannii** Ridl.

nathanielis (Rchb. f.) M. A. Clem. = **Dendrobium nathanielis** Rchb. f.

shompenii (B. K. Sinha & P. S. N. Rao) M. A. Clem. = **Dendrobium shompenii** B. K. Sinha & P. S. N. Rao

spatella (Rchb. f.) M. A. Clem. = **Dendrobium spatella** Rchb. f.

tenuicaule (Hook. f.) Brieger = **Dendrobium tenuicaule** Hook. f.

terminale (C. S. P. Parish & Rchb. f.) M. A. Clem. = **Dendrobium terminale** C. S. P. Parish & Rchb. f.

Appendicula

bifaria Lindl. ex Benth. = **Appendicula cornuta** Blume

teres Griff. = **Ceratostylis subulata** Blume

Arachnanthe

bilinguis (Rchb. f.) Benth. = **Arachnis labrosa** (Lindl. ex Paxton) Rchb. f.

cathcartii (Lindl.) Benth. & Hook. f. = **Arachnis cathcartii** (Lindl.) J. J. Sm.

clarkei (Rchb. f.) Rolfe = **Arachnis clarkei** (Rchb. f.) J. J. Sm.

Arachnis

labrosa (Lindl. ex Paxton) Rchb. f. var. *zhaoi* (Z. J. Liu, S. C. Chen & S. P. Lei) S. C. Chen & J. J. Wood = **Arachnis labrosa** (Lindl. ex Paxton) Rchb. f.

zhaoi Z. J. Liu, S. C. Chen & S. P. Lei = **Arachnis labrosa** (Lindl. ex Paxton) Rchb. f.

Arachnites

monorchis (L.) Hoffm. = **Herminium monorchis** (L.) R. Br.

Archineottia

microglottis (Duthie) S. C. Chen = **Neottia microglottis** (Duthie) Schltr.

pantlingii (W. W. Sm.) S. C. Chen = **Neottia pantlingii** (W. W. Sm.) Tang & F. T. Wang

Arethusa

plicata Andrews = **Nervilia plicata** (Andrews) Schltr.

simplex Thouars = **Nervilia simplex** (Thouars) Schltr.

Aristotelea

spiralis Lour. = **Spiranthes sinensis** (Pers.) Ames

Armodorum

labrosum (Lindl. ex Paxton) Schltr. = **Arachnis labrosa** (Lindl. ex Paxton) Rchb. f.

senapatianum Phukan & A. A. Mao = **Arachnis senapatiana** (Phukan & A. A. Mao) Kocyan & Schuit.

Arrhynchium

labrosum Lindl. ex Paxton = **Arachnis labrosa** (Lindl. ex Paxton) Rchb. f.

Arthrochilium

royleanum (Lindl.) Szlach. = **Epipactis royleana** Lindl.

veratrifolium (Boiss. & Hohen.) Szlach. = **Epipactis veratrifolia** Boiss. & Hohen.

Arundina

affinis Griff. = **Arundina graminifolia** (D. Don) Hochr.

bambusifolia (Roxb.) Lindl. = **Arundina graminifolia** (D. Don) Hochr.

chinensis Blume = **Arundina graminifolia** (D. Don) Hochr.

Ascocentrum

ampullaceum (Roxb.) Schltr. = **Vanda ampullacea** (Roxb.) L. M. Gardiner

ampullaceum (Roxb.) Schltr. var. *aurantiacum* Pradhan = **Vanda ampullacea** (Roxb.) L. M. Gardiner

curvifolium (Lindl.) Schltr. = **Vanda curvifolia** (Lindl.) L. M. Gardiner

himalaicum (Deb, Sengupta & Malick) Christenson = **Holcoglossum himalaicum** (Deb, Sengupta & Malick) Aver.

micranthum (Lindl.) Holttum = **Smitinandia micrantha** (Lindl.) Holttum

Ascotainia

angustifolia (Lindl.) Schltr. = **Ania angustifolia** Lindl.

hookeriana (King & Pantl.) Ridl. = **Ania penangiana** (Hook. f.) Summerh.

penangiana (Hook. f.) Ridl. = **Ania penangiana** (Hook. f.) Summerh.

promensis (Lindl.) Schltr. = **Eulophia promensis** Lindl.

siamensis Rolfe ex Downie = **Ania penangiana** (Hook. f.)
Summerh.

viridifusca (Hook.) Schltr. = **Ania viridifusca** (Hook.)
Tang & F. T. Wang ex Summerh.

Ate

acuminata Thwaites = **Habenaria acuminata** (Thwaites)
Trimen

elwesii (Hook. f.) Szlach. = **Habenaria elwesii** Hook. f.

virens Lindl. = **Habenaria barbata** Wight ex Hook. f.

Bhutanthera

albomarginata (King & Pantl.) Renz = **Herminium
albomarginatum** (King & Pantl.) X. H. Jin, Schuit.,
Raskoti & L. Q. Huang

albovirens Renz = **Herminium albovirens** (Renz) X. H.
Jin, Schuit., Raskoti & L. Q. Huang

alpina (Hand.-Mazz.) Renz = **Herminium handelii** X. H.
Jin, Schuit., Raskoti & L. Q. Huang

Biermannia

decumbens (Griff.) Tang & F. T. Wang ex Merr. & Metcalf =
Phalaenopsis parishii Rchb. f.

taenialis (Lindl.) Tang & F. T. Wang = **Phalaenopsis
taenialis** (Lindl.) Christenson & Pradhan

Birchea

nilgherrensis D. Dietr. = **Luisia tenuifolia** Blume

teretifolia A. Rich. = **Luisia tenuifolia** Blume

Blephariglottis

longicornu (Lindl.) Raf. = **Habenaria longicornu**
Lindl.

Bletia

bicallosa D. Don = **Eulophia bicallosa** (D. Don) P. F.
Hunt & Summerh.

dabia D. Don = **Eulophia dabia** (D. Don) Hochr.

graminifolia D. Don = **Arundina graminifolia** (D. Don)
Hochr.

masuca D. Don = **Calanthe masuca** (D. Don) Lindl.

obcordata Lindl. = **Calanthe obcordata** (Lindl.) M. W.
Chase, Christenh. & Schuit.

tankervilleae (Banks) R. Br. = **Calanthe tankervilleae**
(Banks) M. W. Chase, Christenh. & Schuit.

woodfordii Hook. = **Calanthe woodfordii** (Hook.) M. W.
Chase, Christenh. & Schuit.

Bolbodium

pumilum (Kuntze) Brieger = **Dendrobium pachyphyllum** (Kuntze) Bakh. f.

pusillum (Blume) Rauschert = **Dendrobium pachyphyllum** (Kuntze) Bakh. f.

Bolbophyllaria

biseta (Lindl.) Rchb. f. = **Bulbophyllum bisetum** Lindl.

Bolborchis

crociformis Zoll. & Moritzi = **Nervilia simplex** (Thouars)
Schltr.

Bonatea

benghalensis Griff. = **Habenaria digitata** Lindl.

Brachycorythis

obovalis Summerh. = **Brachycorythis acuta** (Rchb. f.)
Summerh.

Bulbophyllum

agastyamalayanum Gopalan & A. N. Henry =
Bulbophyllum xylophyllum C. S. P. Parish &
Rchb. f.

albidum (Wight) Hook. f. = **Bulbophyllum acutiflorum**
A. Rich.

alopecurum Rchb. f. = **Bulbophyllum triste** Rchb. f.

amplum (Lindl.) Rchb. f. = **Dendrobium amplum** Lindl.

bakhuizenii Steenis = **Bulbophyllum longerepens** Ridl.

berenicis Rchb. f. = **Bulbophyllum caudatum** Lindl.

bicolor (Lindl.) Hook. f. = **Bulbophyllum roseopictum**
J. J. Verm., Schuit. & de Vogel

bootanense (Griff.) C. S. P. Parish & Rchb. f. = **Bulbophyllum umbellatum** Lindl.

brachypodum A. S. Rao & N. P. Balakr. = **Bulbophyllum
yoksunense** J. J. Sm.

brachypodum A. S. Rao & N. P. Balakr. var. *geei* A. S. Rao
& N. P. Balakr. = **Bulbophyllum emarginatum**
(Finet) J. J. Sm.

brachypodum A. S. Rao & N. P. Balakr. var. *parviflorum*
A. S. Rao & N. P. Balakr. = **Bulbophyllum yoksunense** J. J. Sm.

careyanum (Hook.) Spreng. var. *ochracea* Hook. f. =
Bulbophyllum cupreum Lindl.

careyanum Spreng. var. *crassipes* (Hook. f.) Pradhan =
Bulbophyllum crassipes Hook. f.

cauliflorum Hook. f. var. *sikkimense* N. Pearce & P. J.
Cribb = **Bulbophyllum cauliflorum** Hook. f.

cherrapunjeense Barbhuiya & D. Verma = **Bulbophyllum
sarcophyllum** (King & Pantl.) J. J. Sm.

chyrmangense D. Verma, Lavania & Sushil K. Singh =
Bulbophyllum guttulatum (Hook. f.) N. P. Balakr.

cirrhatum (Lindl.) Hook. f. = **Bulbophyllum paleaceum**
(Lindl.) Benth. ex Hemsl.

clarkeanum King & Pantl. = **Bulbophyllum stenobulbon** C. S. P. Parish & Rchb. f.

clarkei Rchb. f. = **Bulbophyllum reptans** (Lindl.) Lindl.

collettii King & Pantl. = **Bulbophyllum cauliflorum**
Hook. f.

conchiferum Rchb. f. = **Bulbophyllum khasyanum**
Griff.

confertum Hook. f. = **Bulbophyllum scabratum** Rchb. f.

congestum Rolfe = **Bulbophyllum odoratissimum** (Sm.)
Lindl.

cylindraceum Lindl. var. *khasyanum* (Griff.) Hook. f. =
Bulbophyllum khasyanum Griff.

densiflorum Rolfe = **Bulbophyllum cariniflorum** Rchb. f.

devangiriense N. P. Balakr. = **Bulbophyllum pteroglossum** Schltr.

dischidiifolium J. J. Sm. subsp. *aberrans* (Schltr.) J. J.
Verm. & P. O'Byrne = **Bulbophyllum aberrans**
Schltr.

dyerianum (King & Pantl.) Seidenf. = **Bulbophyllum
rolfei** (Kuntze) Seidenf.

ebulbum King & Pantl. = **Bulbophyllum apodum**
Hook. f.

fallax Rolfe = **Bulbophyllum chloropterum** Rchb.f.
(excluded species)

flabellum-veneris (J. Koenig) Aver. ?= **Bulbophyllum
lepidum** (Blume) J. J. Sm.

flavidum Lucksom = **Bulbophyllum cariniflorum**
Rchb. f.

fuscescens (Griff.) Rchb. f. = **Dendrobium fuscescens**
Griff.

gamblei (Hook. f.) J. J. Sm. = **Bulbophyllum fischeri** Seidenf.

grandiflorum Griff. = **Bulbophyllum reptans** (Lindl.) Lindl.

hastatum Tang & F. T. Wang = **Bulbophyllum depressum** King & Pantl.

hookeri (Duthie) J. J. Sm. = **Bulbophyllum muscicola** Rchb. f.

imbricatum Griff. = **Bulbophyllum cylindraceum** Lindl.

indicum (C. S. Kumar & Garay) Kottaim. = **Bulbophyllum indicum** (C. S. Kumar & Garay) Sushil K. Singh, Agrawala & Jalal

jejosephii J. J. Verm., Schuit. & de Vogel = **Bulbophyllum moniliforme** C. S. P. Parish & Rchb. f.

josephi M. Kumar & Sequiera = **Bulbophyllum orezii** C. S. Kumar

leopardinum (Wall.) Lindl. ex Wall. var. *tuberculatum* N. P. Balakr. & Sud. Chowdhury = **Bulbophyllum obrienianum** Rolfe

leptanthum Hook. f. var. *gamblei* Hook. f. = **Bulbophyllum gamblei** (Hook. f.) Hook. f.

listeri King & Pantl. = **Bulbophyllum tortuosum** (Blume) Lindl.

longibracteatum Seidenf. = **Bulbophyllum macrocoleum** Seidenf.

maculosum (Lindl.) Garay, Hamer & Siegerist = **Bulbophyllum umbellatum** Lindl.

manabendrae D. K. Roy, Barbhuiya & Talukdar = **Bulbophyllum oblongum** (Lindl.) Rchb. f.

manipurense C. S. Kumar & P. C. S. Kumar = **Bulbophyllum careyanum** (Hook.) Spreng.

mannii Rchb. f. = **Bulbophyllum delitescens** Hance

micranthum Hook. f. = **Bulbophyllum triste** Rchb. f.

mishmeense Hook. f. = **Bulbophyllum paleaceum** (Lindl.) Benth. ex Hemsl.

monanthum (Kuntze) J. J. Sm. = **Bulbophyllum pteroglossum** Schltr.

multiflorum (Breda) Kraenzl. = **Bulbophyllum longerepens** Ridl.

neilgherrense Wight = **Bulbophyllum sterile** (Lam.) Suresh

ochraceum (Ridl.) Ridl. = **Bulbophyllum serratotruncatum** Seidenf.

odoratissimum (Sm.) Lindl. var. *racemosum* N. P. Balakr. = **Bulbophyllum odoratissimum** (Sm.) Lindl.

panigrahianum S. Misra = **Bulbophyllum sarcophyllum** (King & Pantl.) J. J. Sm.

pantlingii Lucksom = **Bulbophyllum cariniflorum** Rchb. f.

paramjitii Agrawala, Sharief & B. K. Singh = **Bulbophyllum moniliforme** C. S. P. Parish & Rchb. f.

parvulum (Hook. f.) J. J. Sm. = **Bulbophyllum rolfei** (Kuntze) Seidenf.

psychoon Rchb. f. = **Bulbophyllum scabratum** Rchb. f.

purpureofuscum J. J. Verm. = **Bulbophyllum paleaceum** (Lindl.) Benth. ex Hemsl.

raui Arora = **Bulbophyllum reptans** (Lindl.) Lindl.

refractoides Seidenf. = **Bulbophyllum wallichii** Rchb. f.

reichenbachianum Kraenzl. = **Bulbophyllum delitescens** Hance

reptans (Lindl.) Lindl. var. *acuta* Malhotra & Balodi = **Bulbophyllum reptans** (Lindl.) Lindl.

reptans (Lindl.) Lindl. var. *subracemosum* Hook. f. = **Bulbophyllum reptans** (Lindl.) Lindl.

rotundatum (Lindl.) Rchb. f. = **Dendrobium rotundatum** (Lindl.) Hook. f.

schmidtianum Rchb. f. = **Bulbophyllum leopardinum** (Wall.) Lindl. ex Wall.

sessile J. J. Sm. = **Bulbophyllum clandestinum** Lindl.

setiferum (Rolfe) J. J. Sm. = **Bulbophyllum delitescens** Hance

sikkimense (King & Pantl.) J. J. Sm. = **Bulbophyllum roxburghii** (Lindl.) Rchb. f.

spectabile Rolfe = **Bulbophyllum pectinatum** Finet

suave Griff. = **Bulbophyllum hirtum** (Sm.) Lindl.

thomsonii (Hook. f.) J. J. Sm. = **Bulbophyllum fischeri** Seidenf.

thomsonii Hook. f. = **Bulbophyllum parviflorum** C. S. P. Parish & Rchb. f.

tiagii A. S. Chauhan = **Bulbophyllum pteroglossum** Schltr.

trichocephalum (Schltr.) Tang & F. T. Wang var. *capitatum* Lucksom = **Bulbophyllum trichocephalum** (Schltr.) Tang & F. T. Wang

trichocephalum (Schltr.) Tang & F. T. Wang var. *racemosum* (N. P. Balakr.) Lucksom = **Bulbophyllum odoratissimum** (Sm.) Lindl.

trichocephalum (Schltr.) Tang & F. T. Wang var. *wallongense* Agrawala, Sabap. & H. J. = **Bulbophyllum trichocephalum** (Schltr.) Tang & F. T. Wang-Chowdhery

umbellatum Lindl. var. *fuscescens* (Hook. f.) P. K. Sarkar = **Bulbophyllum umbellatum** Lindl.

uniflorum Griff. = **Bulbophyllum pteroglossum** Schltr.

virens (Lindl.) Hook. f. = **Bulbophyllum paleaceum** (Lindl.) Benth. ex Hemsl.

wallichii (Lindl.) Merr. & F. P. Metcalf = **Bulbophyllum muscicola** Rchb. f.

yoksunense J. J. Sm var. *geei* (A. S. Rao & N. P. Balakr.) Bennet = **Bulbophyllum emarginatum** (Finet) J. J. Sm.

yoksunense J. J. Sm. var. *parviflorum* (A. S. Rao & N. P. Balakr.) Bennet = **Bulbophyllum yoksunense** J. J. Sm.

Bulleyia

yunnanensis Schltr. = **Coelogyne bulleyia** R. Rice

Calanthe

alpina Hook. f. ex Lindl. var. *keshabii* (Lucksom) R. C. Srivast. = **Calanthe keshabii** Lucksom

angusta Lindl. = **Calanthe odora** Griff.

anjanae Lucksom = **Calanthe griffithii** Lindl.

biloba Lindl. var. *diptera* Hook. f. = **Calanthe biloba** Lindl.

biloba Lindl. var. *treutleri* Hook. f. = **Calanthe biloba** Lindl.

brevicornu Lindl. var. *wattii* Hook. f. = **Calanthe brevicornu** Lindl.

burmanica Rolfe = **Calanthe ceciliae** Low ex Rchb. f.

clavata Lindl. = **Calanthe densiflora** Lindl.

comosa Rchb. f. = **Calanthe perrottetii** A. Rich.

elytroglossa Rchb. f. ex Hook. f. = **Calanthe herbacea** Lindl.

foerstermannii Rchb. f. = **Calanthe lyroglossa** Rchb. f.

fulgens Lindl. = **Calanthe masuca** (D. Don) Lindl.

galeata Lindl. = **Calanthe chloroleuca** Lindl.

gracilis Lindl. = **Calanthe obcordata** (Lindl.) M. W. Chase, Christenh. & Schuit.

masuca (D. Don) Lindl. var. *fulgens* (Lindl.) Hook. f. = **Calanthe masuca** (D. Don) Lindl.

occidentalis Lindl. = **Calanthe tricarinata** Lindl.

pachystalix Rchb. f. ex Hook. f. = **Calanthe davidii** Franch.

pantlingii Schltr. = **Calanthe tricarinata** Lindl.

purpurea Lindl. = **Calanthe masuca** (D. Don) Lindl.

reflexa Maxim. var. *puberula* (Lindl.) Kudô = **Calanthe puberula** Lindl.

vaginata Lindl. = **Calanthe odora** Griff.

veratrifolia R. Br. ex Ker Gawl. = **Calanthe triplicata** (P. Willemet) Ames

viridifusca Hook. = **Ania viridifusca** (Hook.) Tang & F. T. Wang ex Summerh.

wightii Rchb. f. = **Calanthe masuca** (D. Don) Lindl.

wrayi Hook. f. = **Calanthe ceciliae** Low ex Rchb. f.

Calcearia

himalaica (King & Pantl.) M. A. Clem. & D. L. Jones = **Corybas himalaicus** (King & Pantl.) Schltr.

Callista

adunca (Lindl.) Kuntze = **Dendrobium aduncum** Lindl.

aggregata Kuntze = **Dendrobium lindleyi** Steud.

alpestris Kuntze = **Dendrobium monticola** P. F. Hunt & Summerh.

amoena (Wall. ex Lindl.) Kuntze = **Dendrobium amoenum** Wall. ex Lindl.

ampla (Lindl.) Kuntze = **Dendrobium amplum** Lindl.

anceps (Sw.) Kuntze = **Dendrobium anceps** Sw.

aphylla (Roxb.) Kuntze = **Dendrobium aphyllum** (Roxb.) C. E. C. Fisch.

aquea (Lindl.) Kuntze = **Dendrobium aqueum** Lindl.

arachnites Kuntze = **Dendrobium dickasonii** L. O. Williams

aurea (Lindl.) Kuntze = **Dendrobium heterocarpum** Wall. ex Lindl.

barbatula (Lindl.) Kuntze = **Dendrobium barbatulum** Lindl.

bensoniae (Rchb. f.) Kuntze = **Dendrobium bensoniae** Rchb. f.

bicamerata (Lindl.) Kuntze = **Dendrobium bicameratum** Lindl.

brymeriana (Rchb. f.) Kuntze = **Dendrobium brymerianum** Rchb. f.

candida (Wall. ex Lindl.) Kuntze = **Dendrobium moniliforme** (L.) Sw.

capillipes (Rchb. f.) Kuntze = **Dendrobium capillipes** Rchb. f.

carinifera (Rchb. f.) Kuntze = **Dendrobium cariniferum** Rchb. f.

cathcartii (Hook. f.) Kuntze = **Dendrobium salaccense** (Blume) Lindl.

chrysantha (Wall. ex Lindl.) Kuntze = **Dendrobium chrysanthum** Wall ex Lindl.

chrysotoxa (Lindl.) Kuntze = **Dendrobium chrysotoxum** Lindl.

crepidata (Lindl. & Paxton) Kuntze = **Dendrobium crepidatum** Lindl. & Paxton

crumenata (Sw.) Kuntze = **Dendrobium crumenatum** Sw.

cumulata (Lindl.) Kuntze = **Dendrobium cumulatum** Lindl.

densiflora (Lindl.) Kuntze = **Dendrobium densiflorum** Lindl.

denudans (D. Don) Kuntze = **Dendrobium denudans** D. Don

draconis (Rchb. f.) Kuntze = **Dendrobium draconis** Rchb. f.

eriiflora (Griff.) Kuntze = **Dendrobium eriiflorum** Griff.

falconeri (Hook.) Kuntze = **Dendrobium falconeri** Hook.

farmeri (Paxton) Kuntze = **Dendrobium farmeri** Paxton

fimbriata (Hook.) Kuntze = **Dendrobium fimbriatum** Hook.

flabella (Rchb. f.) Kuntze = **Dendrobium plicatile** Lindl.

floribunda (D. Don) Kuntze = **Mycaranthes floribunda** (D. Don) S. C. Chen & J. J. Wood

formosa (Roxb. ex Lindl.) Kuntze = **Dendrobium formosum** Roxb. ex Lindl.

fugax (Rchb. f.) Kuntze = **Dendrobium fugax** Rchb. f.

fuscescens (Griff.) Kuntze = **Dendrobium fuscescens** Griff.

gibsonii (Paxton) Kuntze = **Dendrobium gibsonii** Lindl.

graminifolia Kuntze = **Dendrobium wightii** A. D. Hawkes & A. H. Heller

grandis (Hook. f.) Kuntze = **Dendrobium grande** Hook. f.

gratiosissima (Rchb. f.) Kuntze = **Dendrobium gratiosissimum** Rchb. f.

griffithiana (Lindl.) Kuntze = **Dendrobium griffithianum** Lindl.

haemoglossa (Thwaites) Kuntze = **Dendrobium salaccense** (Blume) Lindl.

herbacea (Lindl.) Kuntze = **Dendrobium herbaceum** Lindl.

heterocarpa (Wall. ex Lindl.) Kuntze = **Dendrobium heterocarpum** Wall. ex Lindl.

heyneana (Lindl.) Kuntze = **Dendrobium heyneanum** Lindl.

hookeriana (Lindl.) Kuntze = **Dendrobium hookerianum** Lindl.

incurva (Lindl.) Kuntze = **Dendrobium incurvum** Lindl.

infundibulum (Lindl.) Kuntze = **Dendrobium infundibulum** Lindl.

jenkinsii (Griff.) Kuntze = **Dendrobium parciflorum** Rchb. f. ex Lindl.

jerdoniana (Wight) Kuntze = **Dendrobium jerdonianum** Wight

kentrophylla (Hook. f.) Kuntze = **Dendrobium kentrophyllum** Hook. f.

kunstleri (Hook. f.) Kuntze = **Dendrobium plicatile** Lindl.

lawiana (Lindl.) Kuntze = **Dendrobium crepidatum** Lindl. & Paxton

lituiflora (Lindl.) Kuntze = **Dendrobium lituiflorum** Lindl.

longicornu (Lindl.) Kuntze = **Dendrobium longicornu** Lindl.

macraei (Lindl.) Kuntze = **Dendrobium macraei** Lindl.

microbulbon (A. Rich.) Kuntze = **Dendrobium microbulbon** A. Rich.

misera (Rchb. f.) Kuntze = **Dendrobium miserum** Rchb. f.

moniliformis (L.) Kuntze = **Dendrobium moniliforme** (L.) Sw.

moschata (Banks) Kuntze = **Dendrobium moschatum** (Banks) Sw.

nana (Hook. f.) Kuntze = **Dendrobium nanum** Hook. f.

nathanielis (Rchb. f.) Kuntze = **Dendrobium nathanielis** Rchb. f.

nobilis (Lindl.) Kuntze = **Dendrobium nobile** Lindl.

normalis (Falc.) Kuntze = **Dendrobium fimbriatum** Hook.

nutans (Lindl.) Kuntze = **Dendrobium nutantiflorum** A. D. Hawkes & A. H. Heller

ochreata (Lindl.) Kuntze = **Dendrobium ochreatum** Lindl.

oculata (Hook.) Kuntze = **Dendrobium fimbriatum** Hook.

ovata (L.) Kuntze = **Dendrobium ovatum** (L.) Kraenzl.

pachyphylla Kuntze = **Dendrobium pachyphyllum** (Kuntze) Bakh. f.

palpebrae (Lindl.) Kuntze = **Dendrobium farmeri** Paxton

parca (Rchb. f.) Kuntze = **Dendrobium parcum** Rchb. f.

parishii (Rchb. f.) Kuntze = **Dendrobium parishii** Rchb. f.

parviflora (D.Don) Kuntze = **Porpax parviflora** (D.Don) Ormerod & Kurzweil

pendula (Roxb.) Kuntze = **Dendrobium pendulum** Roxb.

perula (Rchb. f.) Kuntze = **Dendrobium parciflorum** Rchb. f. ex Lindl.

porphyrochila (Lindl.) Kuntze = **Dendrobium porphyrochilum** Lindl.

praecincta (Rchb. f.) Kuntze = **Dendrobium praecinctum** Rchb. f.

primulina (Lindl.) Kuntze = **Dendrobium polyanthum** Wall. ex Lindl.

pulchella (Roxb. ex Lindl.) Kuntze = **Dendrobium pulchellum** Roxb. ex Lindl.

pumila Kuntze = **Dendrobium pachyphyllum** (Kuntze) Bakh. f.

pusilla (Blume) Kuntze = **Dendrobium pachyphyllum** (Kuntze) Bakh. f.

pycnostachya (Lindl.) Kuntze = **Dendrobium pycnostachyum** Lindl.

rotundata (Lindl.) Kuntze = **Dendrobium rotundatum** (Lindl.) Hook. f.

ruckeri (Lindl.) Kuntze = **Dendrobium ruckeri** Lindl.

salaccensis (Blume) Kuntze = **Dendrobium salaccense** (Blume) Lindl.

secunda (Blume) Kuntze = **Dendrobium secundum** (Blume) Lindl.

spatella (Rchb. f.) Kuntze = **Dendrobium spatella** Rchb. f.

spathacea (Lindl.) Kuntze = **Dendrobium moniliforme** (L.) Sw.

stuposa (Lindl.) Kuntze = **Dendrobium stuposum** Lindl.

sulcata (Lindl.) Kuntze = **Dendrobium sulcatum** Lindl.

tenuicaulis (Hook. f.) Kuntze = **Dendrobium tenuicaule** Hook. f.

terminalis (C. S. P. Parish & Rchb. f.) Kuntze = **Dendrobium terminale** C. S. P. Parish & Rchb. f.

thyrsiflora (Rchb. f. ex André) M. A. Clem. = **Dendrobium thyrsiflorum** Rchb. f. ex André

transparens (Lindl.) Kuntze = **Dendrobium transparens** Lindl.

wardiana (R. Warner) Kuntze = **Dendrobium wardianum** R. Warner

wattii (Hook. f.) Kuntze = **Dendrobium wattii** (Hook. f.) Rchb. f.

williamsonii (J. Day & Rchb. f.) Kuntze = **Dendrobium williamsonii** J. Day & Rchb. f.

Callostylis

bambusifolia (Lindl.) S. C. Chen & J. J. Wood = **Bambuseria bambusifolia** (Lindl.) Schuit., Y. P. Ng & H. A. Pedersen

rigida Blume subsp. *discolor* (Lindl.) Brieger = **Callostylis rigida** Blume

Camarotis

mannii (Hook. f.) King & Pantl. = **Micropera mannii** (Hook. f.) Tang & F. T. Wang

obtusa Lindl. = **Micropera obtusa** (Lindl.) Tang & F. T. Wang

pallida (Roxb.) Lindl. = **Micropera pallida** (Roxb.) Lindl.

purpurea Lindl. = **Micropera rostrata** (Roxb.) N. P. Balakr.

rostrata (Roxb.) Rchb. f. = **Micropera rostrata** (Roxb.) N. P. Balakr.

Carparomorchis

macrantha (Lindl.) M. A. Clem. & D. L. Jones = **Bulbophyllum macranthum** Lindl.

Cephalanthera

royleana (Lindl.) Regel = **Epipactis royleana** Lindl.

thomsonii Rchb. f. = **Cephalanthera longifolia** (L.) Fritsch

Cephalantheropsis

gracilis (Lindl.) S. Y. Hu = **Calanthe obcordata** (Lindl.) M. W. Chase, Christenh. & Schuit.

longipes (Hook. f.) Ormerod = **Calanthe longipes** Hook. f.

obcordata (Lindl.) Ormerod = **Calanthe obcordata** (Lindl.) M. W. Chase, Christenh. & Schuit.

Ceraia

simplicissima Lour. = **Dendrobium simplicissimum** (Lour.) Kraenzl.

Ceratochilus

orchideus (J. Koenig) Lindl. = **Trichoglottis orchidea** (J. Koenig) Garay

Ceratopsis

rosea (D. Don) Lindl. = **Epipogium roseum** (D. Don) Lindl.

Ceratostylis
  teres (Griff.) Rchb. f. = **Ceratostylis subulata** Blume
Cerochilus
  rubens Lindl. = **Hetaeria affinis** (Griff.) Seidenf. &
    Ormerod
Cestichis
  bootanensis (Griff.) T. C. Hsu = **Liparis bootanensis**
    Griff.
  cespitosa (Lam.) Ames = **Liparis cespitosa** (Lam.) Lindl.
  elegans (Lindl.) M. A. Clem. & D. L. Jones = **Liparis
    elegans** Lindl.
  elliptica (Wight) M. A. Clem. & D. L. Jones = **Liparis
    elliptica** Wight
  longipes (Lindl.) Ames = **Liparis viridiflora** (Blume)
    Lindl.
  mannii (Rchb. f.) T. C. Hsu = **Liparis mannii** Rchb. f.
  viridiflora (Blume) T. C. Hsu = **Liparis viridiflora**
    (Blume) Lindl.
Chamaegastrodia
  asraoa (J. Joseph & Abbar.) Seidenf. & A. N. Rao =
    Chamaegastrodia poilanei (Gagnep.) Seidenf. &
    A. N. Rao
Chamorchis
  viridis (L.) Dumort. = **Dactylorhiza viridis** (L.) R. M.
    Bateman, Pridgeon & M. W. Chase
Cheirostylis
  bhotanensis Tang & F. T. Wang = **Cheirostylis monili-
    formis** (Griff.) Seidenf.
  chinensis Rolfe var. *glabra* Bhaumik & M. K. Pathak =
    Cheirostylis moniliformis (Griff.) Seidenf.
  franchetiana King & Pantl. = **Myrmechis pumila** (Hook.
    f.) Tang & F. T. Wang
  kanarensis Blatt. & McCann = **Didymoplexis pallens**
    Griff.
  munnacampensis A. N. Rao = **Cheirostylis yunnanensis**
    Rolfe
  pabongensis Lucksom = **Cheirostylis yunnanensis** Rolfe
  seidenfadeniana C. S. Kumar & F. N. Rasm. = **Cheiro-
    stylis parvifolia** Lindl.
Chelonistele
  apiculata (Lindl.) Pfitzer = **Coelogyne apiculata** (Lindl.)
    Rchb. f.
Chiloschista
  minimifolia (Hook. f.) N. P. Balakr. = **Chiloschista
    fasciata** (F. Muell.) Seidenf. & Ormerod
Chrysobaphus
  roxburghii Wall. = **Anoectochilus roxburghii** (Wall.)
    Lindl.
Chrysoglossum
  erraticum Hook. f. = **Chrysoglossum ornatum** Blume
  hallbergii Blatt. = **Chrysoglossum ornatum** Blume
  latifolium (Blume) Benth. ex Hemsl. = **Diglyphosa
    latifolia** Blume
  macrophyllum King & Pantl. = **Diglyphosa latifolia**
    Blume
  maculatum (Thwaites) Hook. f. = **Chrysoglossum
    ornatum** Blume
Chusua
  chrysea (W. W. Sm.) P. F. Hunt = **Hsenhsua chrysea**
    (W. W. Sm.) X. H. Jin, Schuit., W. T. Jin & L. Q. Huang

nana (King & Pantl.) Pradhan = **Ponerorchis nana**
    (King & Pantl.) Soó
  puberula (King & Pantl.) N. Pearce & P. J. Cribb =
    Ponerorchis puberula (King & Pantl.) Verm.
  renzii (Deva & H. B. Naithani) S. Misra = **Ponerorchis
    renzii** Deva & H. B. Naithani
  roborovskii (Maxim.) P. F. Hunt = **Galearis roborovskii**
    (Maxim.) S. C. Chen, P. J. Cribb & S. W. Gale
  roborowskii (Maxim.) P. F. Hunt var. *nana* (King & Pantl.)
    P. F. Hunt = **Ponerorchis nana** (King & Pantl.) Soó
Cionisaccus
  procerus (Ker Gawl.) M. C. Pace = **Goodyera procera**
    (Ker Gawl.) Hook.
Cirrhopetalum
  acutiflorum (A. Rich.) Hook. f. = **Bulbophyllum
    acutiflorum** A. Rich.
  aemulum W. W. Sm. = **Bulbophyllum forrestii** Seidenf.
  albidum Wight = **Bulbophyllum acutiflorum** A. Rich.
  amplifolium Rolfe = **Bulbophyllum amplifolium** (Rolfe)
    N. P. Balakr. & Sud. Chowdhury
  andersonii Hook. f. = **Bulbophyllum andersonii** (Hook.
    f.) J. J. Sm.
  appendiculatum Rolfe = **Bulbophyllum appendicula-
    tum** (Rolfe) J. J. Sm.
  aureum Hook. f. = **Bulbophyllum aureum** (Hook. f.)
    J. J. Sm.
  blepharistes (Rchb. f.) Hook. f. = **Bulbophyllum
    blepharistes** Rchb. f.
  bootanense Griff. = **Bulbophyllum umbellatum** Lindl.
  brevipes Hook. f. = **Bulbophyllum yoksunense** J. J. Sm.
  caespitosum Wall. ex Lindl. = **Bulbophyllum scabratum**
    Rchb. f.
  caudatum (Lindl.) King & Pantl. = **Bulbophyllum
    caudatum** Lindl.
  cornutum Lindl. = **Bulbophyllum helenae** (Kuntze)
    J. J. Sm.
  delitescens (Hance) Rolfe = **Bulbophyllum delitescens**
    Hance
  dyerianum King & Pantl. = **Bulbophyllum rolfei**
    (Kuntze) Seidenf.
  elatum Hook. f. = **Bulbophyllum elatum** (Hook. f.)
    J. J. Sm.
  elegantulum Rolfe = **Bulbophyllum elegantulum**
    (Rolfe) J. J. Sm.
  emarginatum Finet = **Bulbophyllum emarginatum**
    (Finet) J. J. Sm.
  fimbriatum Lindl. = **Bulbophyllum fimbriatum** (Lindl.)
    Rchb. f.
  flabellum-veneris (J. Koenig) Seidenf. & Ormerod ?=
    Bulbophyllum lepidum (Blume) J. J. Sm.
  gamblei Hook. f. = **Bulbophyllum fischeri** Seidenf.
  gamosepalum Griff. = **Bulbophyllum lepidum** (Blume)
    J. J. Sm.
  guttulatum Hook. f. = **Bulbophyllum guttulatum** (Hook.
    f.) N. P. Balakr.
  hookeri Duthie = **Bulbophyllum muscicola** Rchb. f.
  longiscapum Teijsm. & Binn. = **Bulbophyllum
    blepharistes** Rchb. f.
  macraei Lindl. = **Bulbophyllum macraei** (Lindl.)
    Rchb. f.

maculosum Lindl. = **Bulbophyllum umbellatum** Lindl.

maculosum Lindl. var. *fuscescens* Hook. f. = **Bulbophyllum umbellatum** Lindl.

mannii Mukerjee = **Bulbophyllum delitescens** Hance

maxillare Lindl. = **Bulbophyllum maxillare** (Lindl.) Rchb. f.

mysorense Rolfe = **Bulbophyllum mysorense** (Rolfe) J. J. Sm.

neilgherrense Wight = **Bulbophyllum kaitiense** Rchb. f.

nodosum Rolfe = **Bulbophyllum nodosum** (Rolfe) J. J. Sm.

ochraceum Ridl. = **Bulbophyllum serratotruncatum** Seidenf.

ornatissimum Rchb. f. = **Bulbophyllum ornatissimum** (Rchb. f.) J. J. Sm.

ornatissimum sensu King & Pantl. = **Bulbophyllum appendiculatum** (Rolfe) J. J. Sm.

panigrahianum (S. Misra) S. Misra = **Bulbophyllum sarcophyllum** (King & Pantl.) J. J. Sm.

parvulum Hook. f. = **Bulbophyllum rolfei** (Kuntze) Seidenf.

picturatum Lodd. = **Bulbophyllum picturatum** (Lodd.) Rchb. f.

proudlockii King & Pantl. = **Bulbophyllum proudlockii** (King & Pantl.) J. J. Sm.

restrepia Ridl. = **Bulbophyllum restrepia** (Ridl.) Ridl.

retusiusculum (Rchb. f.) Hook. f. = **Bulbophyllum retusiusculum** Rchb. f.

rothschildianum O'Brien = **Bulbophyllum rothschildianum** (O'Brien) J. J. Sm.

roxburghii Lindl. = **Bulbophyllum roxburghii** (Lindl.) Rchb. f.

sarcophyllum King & Pantl. = **Bulbophyllum sarcophyllum** (King & Pantl.) J. J. Sm.

sarcophyllum King & Pantl. var. *minor* King & Pantl. = **Bulbophyllum sarcophylloides** Garay, Hamer & Siegerist

setiferum Rolfe = **Bulbophyllum delitescens** Hance

sikkimense King & Pantl. = **Bulbophyllum roxburghii** (Lindl.) Rchb. f.

spathulatum Rolfe ex E. W. Cooper = **Bulbophyllum spathulatum** (Rolfe ex E. W. Cooper) Seidenf.

thomsonii Hook. f. = **Bulbophyllum fischeri** Seidenf.

trichocephalum Schltr. = **Bulbophyllum trichocephalum** (Schltr.) Tang & F. T. Wang

viridiflorum Hook. f. = **Bulbophyllum viridiflorum** (Hook. f.) Schltr.

wallichii Lindl. = **Bulbophyllum muscicola** Rchb. f.

wallichii Lindl. = **Bulbophyllum wallichii** Rchb. f.

Cistella

cernua (Willd.) Blume = **Eulophia picta** (R. Br.) Ormerod

Cleisocentron

trichromum (Rchb. f.) Brühl = **Cleisocentron pallens** (Cathcart ex Lindl.) N. Pearce & P. J. Cribb

Cleisostoma

acaule Lindl. = **Gastrochilus acaulis** (Hook. f.) Kuntze

andamanicum Hook. f. = **Pomatocalpa maculosum** (Lindl.) J. J. Sm. subsp. **andamanicum** (Hook. f.) Watthana

auriculatum (Rolfe) Garay = **Cleisostoma discolor** Lindl.

bicuspidatum Hook. f. = **Cleisostoma aspersum** (Rchb. f.) Garay

brevipes Hook. f. = **Cleisostoma striatum** (Rchb. f.) Garay

carinatum (Rolfe ex Downie) Garay = **Cleisostoma duplicilobum** (J. J. Sm.) Garay

dawsonianum Rchb. f. = **Trichoglottis dawsoniana** (Rchb. f.) Rchb. f.

decipiens Lindl. = **Pomatocalpa decipiens** (Lindl.) J. J. Sm.

elegans Seidenf. = **Cleisostoma williamsonii** (Rchb. f.) Garay

gemmatum (Lindl.) King & Pantl. = **Schoenorchis gemmata** (Lindl.) J. J. Sm.

hincksianum (Rchb. f.) Garay = **Cleisostoma appendiculatum** (Lindl.) Benth. & Hook. f. ex B. D. Jacks.

loratum Rchb. f. = **Pomatocalpa undulatum** (Lindl.) J. J. Sm.

mannii Rchb. f. = **Pomatocalpa spicatum** Breda

micranthum (Lindl.) King & Pantl. = **Smitinandia micrantha** (Lindl.) Holttum

pauciflorum (Wight) Senghas = **Cleisostoma tenuifolium** (L.) Garay

pilosulum Gagnep. = **Cleisomeria pilosulum** (Gagnep.) Seidenf. & Garay

ramosum (Lindl.) Hook. f. = **Trichoglottis ramosa** (Lindl.) Senghas

ringens Rchb. f. = **Stereochilus ringens** (Rchb. f.) Garay

sagittiforme Garay = **Cleisostoma linearilobatum** (Seidenf. & Smitinand) Garay

sikkimense Lucksom = **Cleisostoma linearilobatum** (Seidenf. & Smitinand) Garay

spathulatum Blume = **Robiquetia spathulata** (Blume) J. J. Sm.

spicatum Lindl. = **Robiquetia spathulata** (Blume) J. J. Sm.

tenerum (Lindl.) Hook. f. = **Trichoglottis tenera** (Lindl.) Rchb. f.

undulatum (Lindl.) Rchb. f. = **Pomatocalpa undulatum** (Lindl.) J. J. Sm.

undulatum (Lindl.) Tang & F. T. Wang = **Pomatocalpa undulatum** (Lindl.) J. J. Sm.

wendlandorum Rchb. f. = **Pomatocalpa spicatum** Breda

Cleisostomopsis

filiformis (Rchb. f.) R. Rice = **Seidenfadeniella filiformis** (Rchb. f.) Christenson & Ormerod

rosea (Wight) R. Rice = **Seidenfadeniella rosea** (Wight) C. S. Kumar

Cnemidia

angulosa Lindl. = **Tropidia angulosa** (Lindl.) Blume

bambusifolia Thwaites = **Tropidia bambusifolia** (Thwaites) Trimen

nervosa (Wight) Bedd. = **Tropidia angulosa** (Lindl.) Blume

semilibera Lindl. = **Tropidia angulosa** (Lindl.) Blume

Coeloglossum

bracteatum (Muhl. ex Willd.) Parl. var. *kaschmirianum* (Schltr.) Soó = **Dactylorhiza viridis** (L.) R. M. Bateman, Pridgeon & M. W. Chase

densum Lindl. = **Peristylus densus** (Lindl.) Santapau &
Kapadia

kaschmirianum Schltr. = **Dactylorhiza viridis** (L.) R. M.
Bateman, Pridgeon & M. W. Chase

lacertiferum Lindl. = **Peristylus lacertifer** (Lindl.) J. J. Sm.

luteum Dalzell = **Habenaria viridiflora** (Sw.) R. Br. ex
Spreng.

mannii Rchb. f. = **Herminium mannii** (Rchb. f.) Tang &
F. T. Wang

peristyloides Rchb. f. = **Peristylus densus** (Lindl.)
Santapau & Kapadia

secundum Lindl. = **Peristylus secundus** (Lindl.)
Rathakr.

viride (L.) Hartm. = **Dactylorhiza viridis** (L.) R. M.
Bateman, Pridgeon & M. W. Chase

Coelogyne

angustifolia A. Rich. = **Coelogyne odoratissima** Lindl.

arunachalensis H. J. Chowdhery & G. D. Pal = **Coelogyne
ovalis** Lindl.

assamica Linden & Rchb. f. = **Coelogyne fuscescens**
Lindl.

brevifolia Lindl. = **Coelogyne punctulata** Lindl.

calceata (Rchb. f.) Rchb. f. = **Coelogyne pallida** (Lindl.)
Rchb. f.

coronaria Lindl. = **Eria coronaria** (Lindl.) Rchb. f.

corrugata Wight = **Coelogyne nervosa** A. Rich.

diphylla (Lindl. & Paxton) Lindl. = **Pleione maculata**
(Lindl.) Lindl. & Paxton

elata Lindl. = **Coelogyne stricta** (D. Don) Schltr.

flavida Hook. f. ex Lindl. = **Coelogyne prolifera** Lindl.

fuliginosa Lodd. ex Hook. = **Coelogyne ovalis** Lindl.

fuscescens Lindl. var. *assamica* (Linden & Rchb. f.) Pfitzer
= **Coelogyne fuscescens** Lindl.

fuscescens Lindl. var. *brunnea* (Lindl.) Lindl. = **Coelogyne
brunnea** Lindl.

fuscescens Lindl. var. *viridiflorum* Pradhan = **Coelogyne
fuscescens** Lindl.

glandulosa Lindl. = **Coelogyne nervosa** A. Rich.

glandulosa Lindl. var. *bournei* Sandh. Das & S. K. Jain =
Coelogyne mossiae Rolfe

glandulosa Lindl. var. *sathyanarayanae* Sandh. Das &
S. K. Jain = **Coelogyne mossiae** Rolfe

goweri Rchb. f. = **Coelogyne nitida** (Wall. cx D. Don)
Lindl.

graminifolia C. S. P. Parish & Rchb. f. = **Coelogyne
viscosa** Rchb. f.

grandiflora Rolfe = **Pleione grandiflora** (Rolfe) Rolfe

hookeriana Lindl. = **Pleione hookeriana** (Lindl.)
Rollisson

hookeriana Lindl. var. *brachyglossa* Rchb. f. = **Pleione
hookeriana** (Lindl.) Rollisson

humilis (Sm.) Lindl. = **Pleione humilis** (Sm.) D. Don

katakiana (Phukan) R. Rice = **Coelogyne convallariae**
C. S. P. Parish & Rchb. f.

khasyana (Rchb. f.) Rchb. f. = **Coelogyne articulata**
(Lindl.) Rchb. f.

× *lagenaria* (Lindl. & Paxton) Lindl. = **Pleione ×
lagenaria** Lindl. & Paxton

maculata Lindl. = **Pleione maculata** (Lindl.) Lindl. &
Paxton

mishmensis K. Gogoi = **Coelogyne ovalis** Lindl.

occultata Hook. f. var. *uniflora* N. P. Balakr. = **Coelogyne
occultata** Hook. f.

ocellata Lindl. = **Coelogyne punctulata** Lindl.

ochracea Lindl. = **Coelogyne nitida** (Wall. ex D. Don)
Lindl.

odoratissima Lindl. var. *angustifolia* (A. Rich.) Lindl. =
Coelogyne odoratissima Lindl.

ovalis Lindl. var. *latifolia* Hook. f. = **Coelogyne ovalis**
Lindl.

pantlingii Lucksom = **Coelogyne occultata** Hook. f.

parviflora Lindl. = **Coelogyne demissa** (D.Don) M. W.
Chase & Schuit.

praecox (Sm.) Lindl. = **Pleione praecox** (Sm.) D. Don

praecox (Sm.) Lindl. var. *wallichiana* (Lindl.) Lindl. =
Pleione praecox (Sm.) D. Don

punctulata Lindl. f. *brevifolia* (Lindl.) Sandh. Das & S. K.
Jain = **Coelogyne punctulata** Lindl.

rossiana Rchb. f. = **Coelogyne trinervis** Lindl.

thailandica Seidenf. = **Coelogyne quadratiloba**
Gagnep.

thuniana Rchb. f. = **Coelogyne uniflora** Lindl.

treutleri Hook. f. = **Dendrobium treutleri** (Hook. f.)
Schuit. & Peter B. Adams

undulata (Wall. ex Lindl.) Rchb. f. = **Coelogyne rubra**
(Lindl.) Rchb. f.

wallichiana Lindl. = **Pleione praecox** (Sm.) D. Don

Collabiopsis

assamica (Hook. f.) S. S. Ying = **Chrysoglossum
assamicum** Hook. f.

chinensis (Rolfe) S. S. Ying = **Collabium chinense**
(Rolfe) Tang & F. T. Wang

Collabium

assamicum (Hook. f.) Seidenf. = **Chrysoglossum
assamicum** Hook. f.

Conchidium

braccatum (Lindl.) Brieger = **Porpax braccata** (Lindl.)
Schuit., Y. P. Ng & H. A. Pedersen

exile (Hook. f.) Ormerod = **Porpax exilis** (Hook. f.)
Schuit., Y. P. Ng & H. A. Pedersen

extinctorium (Lindl.) Y. P. Ng & P. J. Cribb = **Porpax
extinctoria** (Lindl.) Schuit., Y. P. Ng & H. A.
Pedersen

filiforme (Wight) Rauschert = **Porpax filiformis** (Wight)
Schuit., Y. P. Ng & H. A. Pedersen

lacei (Summerh.) Ormerod = **Porpax lacei** (Summerh.)
Schuit., Y. P. Ng & H. A. Pedersen

microchilos (Dalzell) Rauschert = **Porpax microchilos**
(Dalzell) Schuit., Y. P. Ng & H. A. Pedersen

muscicola (Lindl.) Rauschert = **Porpax parviflora**
(D.Don) Ormerod & Kurzweil

nanum (A. Rich.) Brieger = **Porpax nana** (A. Rich.)
Schuit., Y. P. Ng & H. A. Pedersen

pusillum Griff. = **Porpax pusilla** (Griff.) Schuit., Y. P. Ng
& H. A. Pedersen

Corallorhiza

anandae Malhotra & Balodi = **Corallorhiza trifida** Châtel.

corallorhiza (L.) H. Karst. = **Corallorhiza trifida** Châtel.

foliosa Lindl. = **Oreorchis foliosa** (Lindl.) Lindl.

indica Lindl. = **Oreorchis indica** (Lindl.) Hook. f.

jacquemontii Decne. = **Corallorhiza trifida** Châtel.

patens Lindl. = **Oreorchis patens** (Lindl.) Lindl.

Cordula

charlesworthii (Rolfe) Rolfe = **Paphiopedilum charlesworthii** (Rolfe) Pfitzer

fairrieana (Lindl.) Rolfe = **Paphiopedilum fairrieanum** (Lindl.) Stein

hirsutissima (Lindl. ex Hook.) Rolfe = **Paphiopedilum hirsutissimum** (Lindl. ex Hook.) Stein

insignis (Wall. ex Lindl.) Raf. = **Paphiopedilum insigne** (Wall. ex Lindl.) Pfitzer

spiceriana (Rchb. f.) Rolfe = **Paphiopedilum spicerianum** (Rchb. f.) Pfitzer

venusta (Wall. ex Sims) Rolfe = **Paphiopedilum venustum** (Wall. ex Sims) Pfitzer

villosa (Lindl.) Rolfe = **Paphiopedilum villosum** (Lindl.) Stein

Cordyla

concolor Blume = **Nervilia concolor** (Blume) Schltr.

discolor Blume = **Nervilia plicata** (Andrews) Schltr.

Cordylestylis

foliosa Falc. = **Goodyera procera** (Ker Gawl.) Hook.

himalayensis D. Dietr. = **Goodyera procera** (Ker Gawl.) Hook.

Corybas

purpureus J. Joseph & Yogan. = **Corybas himalaicus** (King & Pantl.) Schltr.

Corymbis

veratrifolia (Reinw.) Rchb. f. = **Corymborkis veratrifolia** (Reinw.) Blume

Corymborkis

acuminata (D. Don) M. R. Almeida = **Crepidium acuminatum** (D. Don) Szlach.

assamica Blume = **Corymborkis veratrifolia** (Reinw.) Blume

densiflora (A. Rich.) M. R. Almeida = **Crepidium densiflorum** (A. Rich.) Sushil K. Singh, Agrawala & Jalal

intermedia (A. Rich.) M. R. Almeida = **Crepidium intermedium** (A. Rich.) Sushil K. Singh, Agrawala & Jalal

latifolia (Sm.) M. R. Almeida = **Crepidium ophrydis** (J. Koenig) M. A. Clem. & D. L. Jones

versicolor (Lindl.) M. R. Almeida = **Crepidium versicolor** (Lindl.) Sushil K. Singh, Agrawala & Jalal

Corysanthes

himalaica King & Pantl. = **Corybas himalaicus** (King & Pantl.) Schltr.

Cottonia

championii Lindl. = **Diploprora championii** (Lindl.) Hook. f.

macrostachya Wight = **Cottonia peduncularis** (Lindl.) Rchb. f. ex Schiller

Cremastra

appendiculata (D. Don) Makino var. *sonamii* Lucksom = **Cremastra appendiculata** (D. Don) Makino

wallichiana Lindl. = **Cremastra appendiculata** (D. Don) Makino

Crepidium

bilobum (Lindl.) Szlach. ex Lucksom = **Crepidium acuminatum** (D. Don) Szlach.

crenulatum (Ridl.) Kottaim. = **Crepidium crenulatum** (Ridl.) Sushil K. Singh, Agrawala & Jalal

nilgiriensis (T. Muthuk., A. Rajendran, Priyadh. & Sarval.) Sushil K. Singh, Agrawala & Jalal = **Liparis atropurpurea** Lindl.

Cryptochilus

carinatus (Gibson ex Lindl.) H. Jiang = **Cryptochilus acuminatus** (Griff.) Schuit., Y. P. Ng & H. A. Pedersen

meirax C. S. P. Parish & Rchb. f. = **Porpax meirax** (C. S. P. Parish & Rchb. f.) King & Pantl.

reticulatus (Lindl.) Rchb. f. = **Porpax reticulata** Lindl.

wightii Rchb. f. = **Porpax jerdoniana** (Wight) Rolfe

Cylindrolobus

bambusifolius (Lindl.) Brieger = **Bambuseria bambusifolia** (Lindl.) Schuit., Y. P. Ng & H. A. Pedersen

crassicaulis (Hook. f.) Brieger = **Bambuseria crassicaulis** (Hook. f.) Schuit., Y. P. Ng & H. A. Pedersen

lohitensis (A. N. Rao, Harid. & S. N. Hedge) A. N. Rao = **Cylindrolobus glandulifer** (Deori & Phukan) A. N. Rao

Cymbidiopsis

lancifolia (Hook.) H. J. Chowdhery = **Cymbidium lancifolium** Hook.

macrorhiza (Lindl.) H. J. Chowdhery = **Cymbidium macrorhizon** Lindl.

Cymbidium

affine Griff. = **Cymbidium mastersii** Griff. ex Lindl.

aloifolium (L.) Sw. var. *pubescens* (Lindl.) Ridl. = **Cymbidium bicolor** Lindl. subsp. **pubescens** (Lindl.) Du Puy & P. J. Cribb

aphyllum (Roxb.) Sw. = **Dendrobium aphyllum** (Roxb.) C. E. C. Fisch.

appendiculatum D. Don = **Cremastra appendiculata** (D. Don) Makino

bambusifolium Roxb. = **Arundina graminifolia** (D. Don) Hochr.

bituberculatum Hook. = **Liparis nervosa** (Thunb.) Lindl. var. **nervosa**

corallorhiza (L.) Sw. = **Corallorhiza trifida** Châtel.

cyperifolium Wall. ex Lindl. var. *szechuanicum* (Y. S. Wu & S. C. Chen) S. C. Chen & Z. J. Liu = **Cymbidium faberi** Rolfe var. **szechuanicum** (Y. S. Wu & S. C. Chen) Y. S. Wu & S. C. Chen

densiflorum Griff. = **Cymbidium elegans** Lindl.

eburneum Lindl. var. *dayanum* (Rchb. f.) Hook. f. = **Cymbidium dayanum** Rchb. f.

elegans Lindl. var. *lutescens* Hook. f. = **Cymbidium elegans** Lindl.

ensifolium (L.) Sw. var. *haematodes* (Lindl.) Trimen = **Cymbidium ensifolium** (L.) Sw. subsp. **haematodes** (Lindl.) Du Puy & P. J. Cribb ex Govaerts

ensifolium (L.) Sw. var. *munroanum* (King & Pantl.) Tang & F. T. Wang = **Cymbidium munronianum** King & Pantl.

erectum Wight = **Cymbidium aloifolium** (L.) Sw.

fuscescens Griff. = **Tainia latifolia** (Lindl.) Rchb. f.

gibsonii Lindl. & Paxton = **Cymbidium lancifolium** Hook.

giganteum Lindl. = **Cymbidium iridioides** D. Don

giganteum Lindl. var. *hookerianum* (Rchb. f.) Desbois = **Cymbidium hookerianum** Rchb. f.

giganteum Lindl. var. *lowianum* Rchb. f. = **Cymbidium lowianum** (Rchb. f.) Rchb. f.

goeringii (Rchb. f.) Rchb. f. var. *mackinnonii* (Duthie) A. N. Rao = **Cymbidium goeringii** (Rchb. f.) Rchb. f.

grandiflorum Griff. = **Cymbidium hookerianum** Rchb. f.

haematodes Lindl. = **Cymbidium ensifolium** (L.) Sw. subsp. **haematodes** (Lindl.) Du Puy & P. J. Cribb ex Govaerts

hookerianum Rchb. f. var. *lowianum* (Rchb. f.) Y. S. Wu & S. C. Chen = **Cymbidium lowianum** (Rchb. f.) Rchb. f.

imbricatum Roxb. = **Coelogyne imbricata** (Hook.) Rchb. f.

intermedium H. G. Jones = **Cymbidium aloifolium** (L.) Sw.

iridifolium Roxb. = **Oberonia mucronata** (D. Don) Ormerod & Seidenf.

ixioides D. Don = **Spathoglottis ixioides** (D. Don) Lindl.

javanicum Blume var. *pantlingii* F. Maek. = **Cymbidium lancifolium** Hook.

longifolium D. Don = **Cymbidium elegans** Lindl.

mackinnonii Duthie = **Cymbidium goeringii** (Rchb. f.) Rchb. f.

mannii Rchb. f. = **Cymbidium bicolor** Lindl. subsp. **obtusum** Du Puy & P. J. Cribb

micromeson Lindl. = **Cymbidium mastersii** Griff. ex Lindl.

moschatum (Banks) Willd. = **Dendrobium moschatum** (Banks) Sw.

nervosum (Thunb.) Sw. = **Liparis nervosa** (Thunb.) Lindl. var. **nervosa**

nitidum Roxb. = **Coelogyne punctulata** Lindl.

nitidum Wall. ex D. Don = **Coelogyne nitida** (Wall. ex D. Don) Lindl.

nutans (Roxb.) Sw. = **Eulophia picta** (R. Br.) Ormerod

ovatum (L.) Willd. = **Dendrobium ovatum** (L.) Kraenzl.

pendulum (Roxb.) Sw. = **Cymbidium aloifolium** (L.) Sw.

pictum R. Br. = **Eulophia picta** (R. Br.) Ormerod

plantaginifolium (J. Koenig) Willd. = **Phreatia plantaginifolia** (J. Koenig) Ormerod

praemorsum (Roxb.) Sw. = **Acampe praemorsa** (Roxb.) Blatt. & McCann

pubescens Lindl. = **Cymbidium bicolor** Lindl. subsp. **pubescens** (Lindl.) Du Puy & P. J. Cribb

sikkimense Hook. f. = **Cymbidium devonianum** Paxton

simonsianum King & Pantl. = **Cymbidium dayanum** Rchb. f.

simulans Rolfe = **Cymbidium aloifolium** (L.) Sw.

sinense Willd. var. *haematodes* (Lindl.) Z. J. Liu & S. C. Chen = **Cymbidium ensifolium** (L.) Sw. subsp. **haematodes** (Lindl.) Du Puy & P. J. Cribb ex Govaerts

spathulatum (L.) Moon = **Taprobanea spathulata** (L.) Christenson

strictum D. Don = **Coelogyne stricta** (D. Don) Schltr.

syringodorum Griff. = **Cymbidium eburneum** Lindl.

szechuanicum Y. S. Wu & S. C. Chen = **Cymbidium faberi** Rolfe var. **szechuanicum** (Y. S. Wu & S. C. Chen) Y. S. Wu & S. C. Chen

tenuifolium (L.) Willd. = **Cleisostoma tenuifolium** (L.) Garay

tenuifolium Lindl. = **Luisia tenuifolia** Blume

tessellatum (Roxb.) Sw. = **Vanda tessellata** (Roxb.) Hook. ex G. Don

tesselloides Roxb. = **Vanda tessellata** (Roxb.) Hook. ex G. Don

Cyperocymbidium

gammieanum (King & Pantl.) A. D. Hawkes = **Cymbidium × gammieanum** King & Pantl.

Cyperorchis

cochlearis (Lindl.) Benth. = **Cymbidium cochleare** Lindl.

eburnea (Lindl.) Schltr. = **Cymbidium eburneum** Lindl.

elegans (Lindl.) Blume = **Cymbidium elegans** Lindl.

× gammieana (King & Pantl.) Schltr. = **Cymbidium × gammieanum** King & Pantl.

gigantea (Blume) Schltr. = **Cymbidium iridioides** D. Don

grandiflora Schltr. = **Cymbidium hookerianum** Rchb. f.

insignis (Rolfe) Schltr. = **Cymbidium insigne** Rolfe

longifolia (D. Don) Schltr. = **Cymbidium elegans** Lindl.

lowiana (Rchb. f.) Schltr. = **Cymbidium lowianum** (Rchb. f.) Rchb. f.

mastersii (Griff. ex Lindl.) Benth. = **Cymbidium mastersii** Griff. ex Lindl.

tigrina (C. S. P. Parish ex Hook.) Schltr. = **Cymbidium tigrinum** C. S. P. Parish ex Hook.

whiteae (King & Pantl.) Schltr. = **Cymbidium whiteae** King & Pantl.

Cypripedium

charlesworthii Rolfe = **Paphiopedilum charlesworthii** (Rolfe) Pfitzer

× crossianum Rchb.f. = **Paphiopedilum × polystigmaticum** (Rchb.f.) Stein

druryi Bedd. = **Paphiopedilum druryi** (Bedd.) Stein

fairrieanum Lindl. = **Paphiopedilum fairrieanum** (Lindl.) Stein

hirsutissimum Lindl. ex Hook. = **Paphiopedilum hirsutissimum** (Lindl. ex Hook.) Stein

insigne Wall. ex Lindl. = **Paphiopedilum insigne** (Wall. ex Lindl.) Pfitzer

macranthos Sw. var. *himalaicum* (Rolfe) Kraenzl. = **Cypripedium himalaicum** Rolfe

macranthos Sw. var. *tibeticum* (King ex Rolfe) Kraenzl. = **Cypripedium tibeticum** King ex Rolfe

× polystigmaticum Rchb.f. = **Paphiopedilum × polystigmaticum** (Rchb.f.) Stein

spicerianum Rchb. f. = **Paphiopedilum spicerianum** (Rchb. f.) Pfitzer

venustum Wall. ex Sims = **Paphiopedilum venustum** (Wall. ex Sims) Pfitzer

villosum Lindl. = **Paphiopedilum villosum** (Lindl.) Stein

Cyrtopera

andamanensis (Rchb. f.) Rolfe = **Eulophia andamanensis** Rchb. f.

bicarinata Lindl. = **Eulophia bicallosa** (D. Don) P. F. Hunt & Summerh.

candida Lindl. = **Eulophia bicallosa** (D. Don) P. F. Hunt & Summerh.

cullenii Wight = **Eulophia flava** (Lindl.) Hook. f.

densiflora (Lindl.) Rchb. f. = **Eulophia densiflora** Lindl.

flava Lindl. = **Eulophia flava** (Lindl.) Hook. f.

fusca Wight = **Eulophia nuda** Lindl.

macrobulbon C. S. P. Parish & Rchb. f. = **Eulophia macrobulbon** (C. S. P. Parish & Rchb. f.) Hook. f.

mannii Rchb. f. = **Eulophia mannii** (Rchb. f.) Hook. f.

mysorensis Lindl. = **Eulophia nuda** Lindl.

nuda (Lindl.) Rchb. f. = **Eulophia nuda** Lindl.

obtusa Lindl. = **Eulophia obtusa** (Lindl.) Hook. f.

plicata Lindl. = **Eulophia nuda** Lindl.

sanguinea Lindl. = **Eulophia zollingeri** (Rchb. f.) J. J. Sm.

zollingeri Rchb. f. = **Eulophia zollingeri** (Rchb. f.) J. J. Sm.

Cyrtopodium

flavum (Lindl.) Benth. = **Eulophia flava** (Lindl.) Hook. f.

Cyrtosia

altissima Blume = **Erythrorchis altissima** (Blume) Blume

Cystopus

clarkei (Hook. f.) Kuntze = **Odontochilus clarkei** Hook. f.

crispus (Lindl.) Kuntze = **Odontochilus crispus** (Lindl.) Hook. f.

elwesii (C. B. Clarke ex Hook. f.) Kuntze = **Odontochilus elwesii** C. B. Clarke ex Hook. f.

flavus (Benth. & Hook. f.) Kuntze = **Odontochilus lanceolatus** (Lindl.) Blume

grandiflorus (Lindl.) Kuntze = **Odontochilus grandiflorus** (Lindl.) Benth. ex Hook. f.

lanceolatus (Lindl.) Kuntze = **Odontochilus lanceolatus** (Lindl.) Blume

pumilus (Hook. f.) Kuntze = **Myrmechis pumila** (Hook. f.) Tang & F. T. Wang

Cystorchis

fusca (Lindl.) Benth. & Hook. f. = **Goodyera fusca** (Lindl.) Hook. f.

Cytheris

cordifolia Lindl. = **Nephelaphyllum cordifolium** (Lindl.) Blume

Dactylorchis

umbrosa (Kar. & Kir.) Wendelbo = **Dactylorhiza umbrosa** (Kar. & Kir.) Nevski

Dactylorhiza

graggeriana (Soó) Soó = **Dactylorhiza hatagirea** (D. Don) Soó

Dendrobium

abhaycharanii (Phukan & A. A. Mao) Schuit. & Peter B. Adams = **Dendrobium calocephalum** (Z. H. Tsi & S. C. Chen) Schuit. & Peter B. Adams

aclinia Lindl. = **Dendrobium incurvum** Lindl.

actinomorphum Blatt. & Hallb. = **Dendrobium crepidatum** Lindl. & Paxton

aggregatum Roxb. = **Dendrobium lindleyi** Steud.

aggregatum Roxb. var. *jenkinsii* (Wall. ex Lindl.) King & Pantl. = **Dendrobium jenkinsii** Wall. ex Lindl.

album Wight = **Dendrobium aqueum** Lindl.

alpestre Royle = **Dendrobium monticola** P. F. Hunt & Summerh.

angulatum Lindl. = **Dendrobium simplicissimum** (Lour.) Kraenzl.

aphyllum (Roxb.) C. E. C. Fisch. var. *cucullatum* (R. Br.) P. K. Sarkar = **Dendrobium aphyllum** (Roxb.) C. E. C. Fisch.

aphyllum (Roxb.) C. E. C. Fisch. var. *katakianum* I. Barua = **Dendrobium aphyllum** (Roxb.) C. E. C. Fisch.

arachnites Rchb. f. = **Dendrobium dickasonii** L. O. Williams

arunachalense C. Deori, S. K. Sarma, Phukan & A. A. Mao = **Dendrobium longicornu** Lindl.

aurantiacum Rchb. f. = **Dendrobium chryseum** Rolfe

aureum Lindl. = **Dendrobium heterocarpum** Wall. ex Lindl.

bambusifolium C. S. P. Parish & Rchb. f. = **Dendrobium salaccense** (Blume) Lindl.

bellatulum Rolfe var. *cleistogamia* Pradhan = **Dendrobium bellatulum** Rolfe

bolboflorum Falc. ex Hook. f. = **Dendrobium bicameratum** Lindl.

braccatum Lindl. = **Porpax braccata** (Lindl.) Schuit., Y. P. Ng & H. A. Pedersen

bulleyi Rolfe = **Dendrobium flexuosum** Griff.

caespitosum King & Pantl. = **Dendrobium porphyrochilum** Lindl.

calceolaria Carey ex Hook. = **Dendrobium moschatum** (Banks) Sw.

cambridgeanum Paxton = **Dendrobium ochreatum** Lindl.

candidum Wall. ex Lindl. = **Dendrobium moniliforme** (L.) Sw.

cariniferum Rchb. f. var. *wattii* Hook. f. = **Dendrobium wattii** (Hook. f.) Rchb. f.

carnosum Teijsm. & Binn. = **Dendrobium pachyphyllum** (Kuntze) Bakh. f.

cathcartii Hook. f. = **Dendrobium salaccense** (Blume) Lindl.

ceraia Lindl. = **Dendrobium simplicissimum** (Lour.) Kraenzl.

chapaense Aver. = **Dendrobium flexuosum** Griff.

chlorops Lindl. = **Dendrobium ovatum** (L.) Kraenzl.

chrysotis Rchb. f. = **Dendrobium hookerianum** Lindl.

ciliatum C. S. P. Parish ex Hook. var. *breve* Rchb. f. = **Dendrobium delacourii** Guillaumin

clavatum Roxb. = **Dendrobium densiflorum** Lindl.

clavatum Wall. ex Lindl. = **Dendrobium chryseum** Rolfe

coerulescens Wall. ex Lindl. = **Dendrobium nobile** Lindl.

crepidatum Griff. = **Eria coronaria** (Lindl.) Rchb. f.

crepidatum Lindl. & Paxton var. *avita* Gammie = **Dendrobium crepidatum** Lindl. & Paxton

cretaceum Lindl. = **Dendrobium polyanthum** Wall. ex Lindl.

cucullatum R. Br. = **Dendrobium aphyllum** (Roxb.) C. E. C. Fisch.

cumulatum Lindl. var. *jenkinsii* Hook. f. = **Dendrobium cumulatum** Lindl.

cuspidatum (Wall. ex Lindl.) Lindl. = **Dendrobium nathanielis** Rchb. f.

dalhousieanum Wall. ex Paxton = **Dendrobium pulchellum** Roxb. ex Lindl.

dalzellii Hook. = **Porpax filiformis** (Wight) Schuit., Y. P. Ng & H. A. Pedersen

demissum D. Don = **Coelogyne demissa** (D.Don) M. W. Chase & Schuit.

denneanum Kerr = **Dendrobium chryseum** Rolfe

densiflorum Lindl. ex Wall. var. *alboluteum* Hook. f. = **Dendrobium thyrsiflorum** Rchb. f. ex André

deuteroarunachalense J. M. H. Shaw = **Dendrobium brunneum** Schuit. & Peter B. Adams

eburneum H. Low = **Dendrobium draconis** Rchb. f.

egertoniae Lindl. = **Dendrobium amoenum** Wall. ex Lindl.

eriiflorum Griff. var. *sikkimense* Lucksom = **Dendrobium eriiflorum** Griff.

extinctorium Lindl. = **Porpax extinctoria** (Lindl.) Schuit., Y. P. Ng & H. A. Pedersen

falconeri Hook. var. *senapatianum* C. Deori, Gogoi & A. A. Mao = **Dendrobium falconeri** Hook.

filiforme Wight = **Porpax filiformis** (Wight) Schuit., Y. P. Ng & H. A. Pedersen

fimbriatum (Blume) Lindl. = **Dendrobium plicatile** Lindl.

fimbriatum Dalz. = **Porpax filiformis** (Wight) Schuit., Y. P. Ng & H. A. Pedersen

fimbriatum Hook. var. *oculatum* Hook. = **Dendrobium fimbriatum** Hook.

flabellum Rchb. f. = **Dendrobium plicatile** Lindl.

floribundum D. Don = **Mycaranthes floribunda** (D. Don) S. C. Chen & J. J. Wood

fuscatum Lindl. = **Dendrobium gibsonii** Lindl.

galliceanum Linden = **Dendrobium thyrsiflorum** Rchb. f. ex André

gamblei King & Pantl. = **Dendrobium macrostachyum** Lindl.

georgei J. Mathew = **Dendrobium herbaceum** Lindl.

graminifolium Wight = **Dendrobium wightii** A. D. Hawkes & A. H. Heller

gunnarii P. S. N. Rao = **Dendrobium trinervium** Ridl.

haemoglossum Thwaites = **Dendrobium salaccense** (Blume) Lindl.

herbaceum Lindl. subsp. *georgei* (J. Mathew) S. Misra = **Dendrobium herbaceum** Lindl.

hexadesmia Rchb. f. = **Dendrobium parcum** Rchb. f.

humile (Sm.) Sm. = **Pleione humilis** (Sm.) D. Don

humile Wight = **Dendrobium crispum** Dalzell

jaintianum Sabap. = **Dendrobium hirsutum** Griff.

javanicum Thunb. ex Sw. = **Eria javanica** (Thunb. ex Sw.) Blume

kallarense J. Mathew, Kad. V. George, Yohannan & K. Madhus. = **Dendrobium barbatulum** Lindl.

kunstleri Hook. f. = **Dendrobium plicatile** Lindl.

lawianum Lindl. = **Dendrobium crepidatum** Lindl. & Paxton

leopardinum Wall. = **Bulbophyllum leopardinum** (Wall.) Lindl. ex Wall.

lindleyanum Griff. = **Dendrobium nobile** Lindl.

listeroglossum Kraenzl. = **Dendrobium parcum** Rchb. f.

longicornu Lindl. var. *hirsutum* (Griff.) Hook. f. = **Dendrobium hirsutum** Griff.

mabeliae Gammie ?= **Dendrobium nanum** Hook. f.

madrasense A. D. Hawkes = **Dendrobium aphyllum** (Roxb.) C. E. C. Fisch.

meghalayense C. Deori, S. K. Sarma, Hynn. & Phukan = **Dendrobium hirsutum** Griff.

meghalayense Y. Kumar & S. Chowdhury = **Dendrobium sulcatum** Lindl.

mesochlorum Lindl. = **Dendrobium amoenum** Wall. ex Lindl.

microchilos Dalzell = **Porpax microchilos** (Dalzell) Schuit., Y. P. Ng & H. A. Pedersen

minutiflorum S. C. Chen & Z. H. Tsi = **Dendrobium sino-minutiflorum** S. C. Chen, J. J. Wood & H. P. Wood

modestum Ridl. = **Dendrobium metrium** Kraenzl.

monile (Thunb.) Kraenzl. = **Dendrobium moniliforme** (L.) Sw.

moschatum (Banks) Sw. var. *unguipetalum* I. Barua = **Dendrobium moschatum** (Banks) Sw.

muscicola Lindl. = **Porpax parviflora** (D.Don) Ormerod & Kurzweil

nageswarayanum Chowlu = **Dendrobium brunneum** Schuit. & Peter B. Adams

nobile Lindl. var. *pallidiflorum* Hook. = **Dendrobium polyanthum** Wall. ex Lindl.

normale Falc. = **Dendrobium fimbriatum** Hook.

nutans Lindl. = **Dendrobium nutantiflorum** A. D. Hawkes & A. H. Heller

nutans Lindl. var. *rubrilabre* Blatt. ex C. E. C. Fisch. = **Dendrobium nutantiflorum** A. D. Hawkes & A. H. Heller

palpebrae Lindl. = **Dendrobium farmeri** Paxton

panduratum Lindl. subsp. *villosum* Gopalan & A. N. Henry = **Dendrobium panduratum** Lindl.

parviflorum D.Don = **Porpax parviflora** (D.Don) Ormerod & Kurzweil

pauciflorum King & Pantl. = **Dendrobium praecinctum** Rchb. f.

paxtonii Paxton = **Dendrobium fimbriatum** Hook.

peguanum Lindl. = **Dendrobium crispum** Dalzell

perpusillum N. P. Balakr. = **Dendrobium pachyphyllum** (Kuntze) Bakh. f.

perula Rchb. f. = **Dendrobium parciflorum** Rchb. f. ex Lindl.

pierardii Roxb. ex Hook. = **Dendrobium aphyllum** (Roxb.) C. E. C. Fisch.

pierardii Roxb. ex R. Br. = **Dendrobium aphyllum** (Roxb.) C. E. C. Fisch.

pierardii Roxb. ex R. Br. var. *cucullatum* (R. Br.) Hook. f. = **Dendrobium aphyllum** (Roxb.) C. E. C. Fisch.

podagraria Hook. f. = **Dendrobium simplicissimum** (Lour.) Kraenzl.

praecox (Sm.) Sm. = **Pleione praecox** (Sm.) D. Don

primulinum Lindl. = **Dendrobium polyanthum** Wall. ex Lindl.

pubescens Hook. = **Dendrolirium lasiopetalum** (Willd.) S. C. Chen & J. J. Wood

pulchellum Roxb. ex Lindl. var. *devonianum* (Paxton) Rchb. f. = **Dendrobium devonianum** Paxton

pumilum Roxb. = **Dendrobium pachyphyllum** (Kuntze) Bakh. f.

nigra (Seidenf.) M. A. Clem. & D. L. Jones = **Liparis gigantea** C. L. Tso

walkerae (Graham) M. A. Clem. & D. L. Jones = **Liparis walkerae** Graham

wightiana (Thwaites) M. A. Clem. & D. L. Jones = **Liparis wightiana** Thwaites

wrayi (Hook. f.) M. A. Clem. & D. L. Jones = **Liparis barbata** Lindl.

Dithrix

decipiens Soó = **Gennaria griffithii** (Hook. f.) X. H. Jin & D. Z. Li

griffithii (Hook. f.) Ormerod & Gandhi = **Gennaria griffithii** (Hook. f.) X. H. Jin & D. Z. Li

Ditulima

anceps (Sw.) Raf. = **Dendrobium anceps** Sw.

Doritis

braceana Hook. f. = **Phalaenopsis taenialis** (Lindl.) Christenson & Pradhan

deliciosa (Rchb. f.) T. Yukawa & K. Kita = **Phalaenopsis deliciosa** Rchb. f.

deliciosa (Rchb. f.) T. Yukawa & K. Kita subsp. hookeriana (O. Gruss & Roellke) T. Yukawa & K. Kita = **Phalaenopsis deliciosa** Rchb. f.

lobbii (Rchb. f.) T. Yukawa & K. Kita = **Phalaenopsis lobbii** (Rchb. f.) H. R. Sweet

mysorensis (C. J. Saldanha) T. Yukawa & K. Kita = **Phalaenopsis mysorensis** C. J. Saldanha

parishii (Rchb. f.) T. Yukawa & K. Kita = **Phalaenopsis parishii** Rchb. f.

pulcherrima Lindl. = **Phalaenopsis pulcherrima** (Lindl.) J. J. Sm.

taenialis (Lindl.) Benth. ex Hook. f. = **Phalaenopsis taenialis** (Lindl.) Christenson & Pradhan

wightii (Rchb. f.) Benth. & Hook. f. = **Phalaenopsis deliciosa** Rchb. f.

Dossinia

lanceolata Lindl. = **Rhomboda lanceolata** (Lindl.) Ormerod

Drymoda

gymnopus (Hook. f.) Garay = **Bulbophyllum gymnopus** Hook. f.

Echioglossum

arietinum (Rchb. f.) Szlach. = **Cleisostoma arietinum** (Rchb. f.) Garay

elegans (Seidenf.) Szlach. = **Cleisostoma williamsonii** (Rchb. f.) Garay

simondii (Gagnep.) Szlach. = **Cleisostoma simondii** (Gagnep.) Seidenf.

striatum Rchb. f. = **Cleisostoma striatum** (Rchb. f.) Garay

williamsonii (Rchb. f.) Szlach. = **Cleisostoma williamsonii** (Rchb. f.) Garay

Elasmatium

repens (L.) Dulac = **Goodyera repens** (L.) R. Br.

Empusa

barbata (Lindl.) T. C. Hsu = **Liparis barbata** Lindl.

nervosa (Thunb.) T. C. Hsu = **Liparis nervosa** (Thunb.) Lindl.

odorata (Willd.) T. C. Hsu = **Liparis odorata** (Willd.) Lindl.

paradoxa Lindl. = **Liparis odorata** (Willd.) Lindl.

Endeisa

flava Raf. = **Dendrobium densiflorum** Lindl.

Entaticus

viridis (L.) Gray = **Dactylorhiza viridis** (L.) R. M. Bateman, Pridgeon & M. W. Chase

Ephemerantha

fimbriata (Blume) P. F. Hunt & Summerh. = **Dendrobium plicatile** Lindl.

kunstleri (Hook. f.) P. F. Hunt & Summerh. = **Dendrobium plicatile** Lindl.

macraei (Lindl.) P. F. Hunt & Summerh. = **Dendrobium macraei** Lindl.

ritaeana (King & Pantl.) P. F. Hunt & Summerh. = **Dendrobium ritaeanum** King & Pantl.

Ephippium

lepidum Blume = **Bulbophyllum lepidum** (Blume) J. J. Sm.

Epicranthes

barbata (Lindl.) Rchb. f. = **Bulbophyllum crabro** (C. S. P. Parish & Rchb. f.) J. J. Verm., Schuit. & de Vogel

Epidendrum

aloifolium L. = **Cymbidium aloifolium** (L.) Sw.

aphyllum (Roxb.) Poir. = **Dendrobium aphyllum** (Roxb.) C. E. C. Fisch.

aristotelea Raeusch. = **Spiranthes sinensis** (Pers.) Ames

calceolare (Buch.-Ham. ex Sm.) D. Don = **Gastrochilus calceolaris** (Buch.-Ham. ex Sm.) D. Don

cespitosum Lam. = **Liparis cespitosa** (Lam.) Lindl.

concretum Jacq. = **Polystachya concreta** (Jacq.) Garay & H. R. Sweet

corallorhizon (L.) Poir. = **Corallorhiza trifida** Châtel.

ensifolium L. = **Cymbidium ensifolium** (L.) Sw. subsp. **ensifolium**

flabellum-veneris J. Koenig, Observ. Bot. (Retzius) 6: 57. 1791.? = **Bulbophyllum lepidum** (Blume) J. J. Sm.

flos-aeris J. Koenig = **Dendrolirium lasiopetalum** (Willd.) S. C. Chen & J. J. Wood

humile Sm. = **Pleione humilis** (Sm.) D. Don

lasiopetalum (Willd.) Poir. = **Dendrolirium lasiopetalum** (Willd.) S. C. Chen & J. J. Wood

liliifolium J. Koenig = **Acriopsis liliifolia** (J. Koenig) Ormerod

monile Thunb. = **Dendrobium moniliforme** (L.) Sw.

moniliforme L. = **Dendrobium moniliforme** (L.) Sw.

moschatum Banks = **Dendrobium moschatum** (Banks) Sw.

nervosum (Thunb.) Thunb. = **Liparis nervosa** (Thunb.) Lindl. var. **nervosa**

odoratum (Lour.) Poir. = **Aerides odorata** Lour.

ophrydis J. Koenig = **Crepidium ophrydis** (J. Koenig) M. A. Clem. & D. L. Jones

orchideum J. Koenig = **Trichoglottis orchidea** (J. Koenig) Garay

ovatum L. = **Dendrobium ovatum** (L.) Kraenzl.

pendulum Roxb. = **Cymbidium aloifolium** (L.) Sw.

plantaginifolium J. Koenig = **Phreatia plantaginifolia** (J. Koenig) Ormerod

praecox Sm. = **Pleione praecox** (Sm.) D. Don

praemorsum Roxb. = **Acampe praemorsa** (Roxb.) Blatt. & McCann

retusum L. = **Rhynchostylis retusa** (L.) Blume

sessile J. Koenig = **Bulbophyllum clandestinum** Lindl.

sinense Andrews = **Cymbidium sinense** (Andrews) Willd.

spathulatum L. = **Taprobanea spathulata** (L.)
Christenson

sterile Lam. = **Bulbophyllum sterile** (Lam.) Suresh

tenuifolium L. = **Cleisostoma tenuifolium** (L.) Garay

tessellatum Roxb. = **Vanda tessellata** (Roxb.) Hook. ex
G. Don

tomentosum J. Koenig = **Dendrolirium tomentosum**
(J. Koenig) S. C. Chen & J. J. Wood

usneoides D. Don. = **Chiloschista usneoides** (D. Don)
Lindl.

Epigeneium

amplum (Lindl.) Summerh., Kew Bull. 12: 260. 1957. =
Dendrobium amplum Lindl.

arunachalense A. N. Rao = **Dendrobium brunneum**
Schuit. & Peter B. Adams

chapaense Gagnep. = **Dendrobium brunneum** Schuit. &
Peter B. Adams

fargesii (Finet) Gagnep. = **Dendrobium fargesii** Finet

fuscescens (Griff.) Summerh. = **Dendrobium fuscescens**
Griff.

naviculare (N. P. Balakr. & Sud. Chowdhury) Hynn. &
Wadhwa = **Dendrobium fuscescens** Griff.

rotundatum (Lindl.) Summerh. = **Dendrobium
rotundatum** (Lindl.) Hook. f.

treutleri (Hook. f.) Ormerod = **Dendrobium treutleri**
(Hook. f.) Schuit. & Peter B. Adams

Epipactis

biflora (Lindl.) A. A. Eaton = **Goodyera biflora** (Lindl.)
Hook. f.

carinata Roxb. = **Nervilia concolor** (Blume) Schltr.

consimilis Wall. ex Hook. f. = **Epipactis veratrifolia**
Boiss. & Hohen.

corallorhiza (L.) Crantz = **Corallorhiza trifida** Châtel.

cordata (Lindl.) A. A. Eaton = **Goodyera viridiflora**
(Blume) Lindl. ex D. Dietr.

foliosa (Lindl.) A. A. Eaton = **Goodyera foliosa** (Lindl.)
Benth. ex C. B. Clarke

fusca (Lindl.) A. A. Eaton = **Goodyera fusca** (Lindl.)
Hook. f.

graminifolia Roxb. = **Spathoglottis pubescens** Lindl.

helleborine (L.) Crantz subsp. *persica* (Soó) H. Sund. =
Epipactis persica (Soó) Hausskn. ex Nannf.

helleborine (L.) Crantz var. *intrusa* (Lindl.) Karthik. =
Epipactis helleborine (L.) Crantz

helleborine (L.) Crantz var. *intrusa* (Lindl.) S. N. Mitra =
Epipactis helleborine (L.) Crantz

hemsleyana (King & Pantl.) A. A. Eaton = **Goodyera
hemsleyana** King & Pantl.

hispida (Lindl.) A. A. Eaton = **Goodyera hispida** Lindl.

intrusa Lindl. = **Epipactis helleborine** (L.) Crantz

juliana Roxb. = **Nervilia juliana** (Roxb.) Schltr.

latifolia (L.) All. var. *intrusa* (Lindl.) Hook. f. = **Epipactis
helleborine** (L.) Crantz

microphylla (Ehrh.) Sw. subsp. *persica* (Soó) Hautz. =
Epipactis persica (Soó) Hausskn. ex Nannf.

monorchis (L.) F. W. Schmidt = **Herminium monorchis**
(L.) R. Br.

ovata (L.) Crantz = **Neottia ovata** (L.) Bluff & Fingerh.

plicata Roxb. = **Nervilia plicata** (Andrews) Schltr.

prainii (Hook. f.) A. A. Eaton = **Goodyera recurva**
Lindl.

procera (Ker Gawl.) A. A. Eaton = **Goodyera procera**
(Ker Gawl.) Hook.

recurva (Lindl.) A. A. Eaton = **Goodyera recurva** Lindl.

repens (L.) Crantz = **Goodyera repens** (L.) R. Br.

robusta (Hook. f.) A. A. Eaton = **Goodyera robusta**
Hook. f.

rubicunda (Blume) A. A. Eaton = **Goodyera rubicunda**
(Blume) Lindl.

schlechtendaliana (Rchb. f.) A. A. Eaton = **Goodyera
schlechtendaliana** Rchb. f.

viridiflora (Blume) Ames = **Goodyera viridiflora**
(Blume) Lindl. ex D. Dietr.

Epiphanes

pallens (Griff.) Rchb. f. = **Didymoplexis pallens** Griff.

Epipogium

indicum H. J. Chowdhery, G. D. Pal & G. S. Giri =
Epipogium roseum (D. Don) Lindl.

nutans (Blume) Rchb. f. = **Epipogium roseum** (D. Don)
Lindl.

sessanum S. N. Hegde & A. N. Rao = **Epipogium
roseum** (D. Don) Lindl.

tuberosum Duthie ?= **Epipogium japonicum** Makino

Eria

acervata Lindl. = **Pinalia acervata** (Lindl.) Kuntze

acutifolia Lindl. = **Pinalia acutifolia** (Lindl.) Kuntze

alba Lindl. = **Pinalia leucantha** Kuntze

albiflora Rolfe = **Porpax albiflora** (Rolfe) Schuit., Y. P.
Ng & H. A. Pedersen

ambrosia Hance = **Bulbophyllum ambrosia** (Hance)
Schltr.

amica Rchb. f. = **Pinalia lineata** (Lindl.) Kuntze

andamanica Hook. f. = **Dendrolirium andamanicum**
(Hook. f.) Schuit., Y. P. Ng & H. A. Pedersen

andersonii Hook. f. = **Pinalia lineata** (Lindl.) Kuntze

angulata Rchb. f. = **Tainia latifolia** (Lindl.) Rchb. f.

ania Rchb. f. = **Ania viridifusca** (Hook.) Tang & F. T.
Wang ex Summerh.

apertiflora Summerh. = **Pinalia apertiflora** (Summerh.)
A. N. Rao

arunachalensis A. N. Rao = **Cylindrolobus arunachalensis** (A. N. Rao) A. N. Rao

bambusifolia Lindl. = **Bambuseria bambusifolia** (Lindl.)
Schuit., Y. P. Ng & H. A. Pedersen

barbata (Lindl.) Rchb. f. = **Eriodes barbata** (Lindl.) Rolfe

bicolor Lindl. = **Pinalia bicolor** (Lindl.) Kuntze

bicornis (Lindl.) Rchb. f. = **Tainia bicornis** (Lindl.)
Rchb. f.

biflora Griff. = **Cylindrolobus biflorus** (Griff.) Rauschert

bilobulata Seidenf. subsp. *meghasaniensis* S. Misra =
Pinalia meghasaniensis (S. Misra) Schuit., Y. P. Ng &
H. A. Pedersen

bilobulata Seidenf. = **Pinalia bilobulata** (Seidenf.)
Schuit., Y. P. Ng & H. A. Pedersen

bipunctata Lindl. = **Pinalia bipunctata** (Lindl.) Kuntze

braccata (Lindl.) Lindl. = **Porpax braccata** (Lindl.)
Schuit., Y. P. Ng & H. A. Pedersen

bractescens Lindl. = **Pinalia bractescens** (Lindl.) Kuntze

bractescens Lindl. var. *kurzii* Hook. f. = **Pinalia bractescens** (Lindl.) Kuntze

brevilinguis (J. Joseph & V. Chandras.) Bajrach. & K. K. Shrestha = **Porpax nana** (A. Rich.) Schuit., Y. P. Ng & H. A. Pedersen

calamifolia Hook. f. = **Strongyleria pannea** (Lindl.) Schuit., Y. P. Ng & H. A. Pedersen

carinata Gibson ex Lindl. = **Cryptochilus acuminatus** (Griff.) Schuit., Y. P. Ng & H. A. Pedersen

chandrasekharanii (Bhargavan & C. N. Mohanan) C. S. Kumar & Manilal = **Porpax exilis** (Hook. f.) Schuit., Y. P. Ng & H. A. Pedersen

confusa Hook. f. = **Pinalia lineata** (Lindl.) Kuntze

connata J. Joseph, S. N. Hegde & Abbar. = **Pinalia connata** (J. Joseph, S. N. Hegde & Abbar.) Ormerod & E. W. Wood

conrardii M. R. Almeida = **Porpax filiformis** (Wight) Schuit., Y. P. Ng & H. A. Pedersen

conrardii M. R. Almeida var. *fimbriata* (Dalz.) M. R. Almeida = **Porpax filiformis** (Wight) Schuit., Y. P. Ng & H. A. Pedersen

convallarioides Lindl. = **Pinalia spicata** (D. Don) S. C. Chen & J. J. Wood

corneri Rchb. f. = **Eria scabrilinguis** Lindl.

corneri Rchb. f. var. *clausa* (King & Pantl.) A. N. Rao = **Eria clausa** King & Pantl.

crassicaulis Hook. f. = **Bambuseria crassicaulis** (Hook. f.) Schuit., Y. P. Ng & H. A. Pedersen

cristata Rolfe = **Cylindrolobus cristatus** (Rolfe) S. C. Chen & J. J. Wood

cylindripoda Griff. = **Eria coronaria** (Lindl.) Rchb. f.

dalzellii (Hook.) Lindl. = **Porpax filiformis** (Wight) Schuit., Y. P. Ng & H. A. Pedersen

dalzellii (Hook.) Lindl. var. *fimbriata* Hook. f. = **Porpax filiformis** (Wight) Schuit., Y. P. Ng & H. A. Pedersen

dasyphylla C. S. P. Parish & Rchb. f. = **Trichotosia dasyphylla** (C. S. P. Parish & Rchb. f.) Kraenzl.

discolor Lindl. = **Callostylis rigida** Blume

elegans (Lindl.) Rchb. f. = **Phreatia elegans** Lindl.

elongata Lindl. = **Dendrolirium laniceps** (Rchb. f.) Schuit., Y. P. Ng & H. A. Pedersen

elwesii Rchb. f. = **Porpax elwesii** (Rchb. f.) Rolfe

excavata Lindl. = **Pinalia excavata** (Lindl.) Kuntze

exilis Hook. f. = **Porpax exilis** (Hook. f.) Schuit., Y. P. Ng & H. A. Pedersen

extinctoria (Lindl.) Oliv. = **Porpax extinctoria** (Lindl.) Schuit., Y. P. Ng & H. A. Pedersen

ferruginea Lindl. = **Dendrolirium ferrugineum** (Lindl.) A. N. Rao

ferruginea Lindl. var. *assamica* Gogoi, Das & R.Yonzone = **Dendrolirium ferrugineum** (Lindl.) A. N. Rao

fibuliformis King & Pantl. = **Porpax fibuliformis** (King & Pantl.) King & Pantl.

filiformis (Wight) Rchb. f. = **Porpax filiformis** (Wight) Schuit., Y. P. Ng & H. A. Pedersen

flava Lindl. = **Dendrolirium lasiopetalum** (Willd.) S. C. Chen & J. J. Wood

flava Lindl. var. *rubida* Lindl. = **Pinalia excavata** (Lindl.) Kuntze

fragrans Rchb. f. = **Eria javanica** (Thunb. ex Sw.) Blume

glandulifera Deori & Phukan = **Cylindrolobus glandulifer** (Deori & Phukan) A. N. Rao

globulifera Seidenf. = **Pinalia globulifera** (Seidenf.) A. N. Rao

gloensis Ormerod & Agrawala = **Cylindrolobus gloensis** (Ormerod & Agrawala) Schuit., Y. P. Ng & H. A. Pedersen

graminifolia Lindl. = **Pinalia graminifolia** (Lindl.) Kuntze

hegdei Agrawala & H. J. Chowdhery = **Cylindrolobus hegdei** (Agrawala & H. J. Chowdhery) A. N. Rao

hindei Summerh. = **Bryobium pudicum** (Ridl.) Y. P. Ng & P. J. Cribb

jengingensis S. N. Hegde = **Cylindrolobus hegdei** (Agrawala & H. J. Chowdhery) A. N. Rao

jerdoniana (Wight) Rchb. f. = **Porpax jerdoniana** (Wight) Rolfe

kamlangensis A. N. Rao = **Dendrolirium kamlangensis** (A. N. Rao) A. N. Rao

khasiana Lindl. = **Cylindrolobus khasianus** (Lindl.) Ormerod & C. S. Kumar

lacei Summerh. = **Porpax lacei** (Summerh.) Schuit., Y. P. Ng & H. A. Pedersen

laniceps Rchb. f. = **Dendrolirium laniceps** (Rchb. f.) Schuit., Y. P. Ng & H. A. Pedersen

lasiopetala (Willd.) Ormerod = **Dendrolirium lasiopetalum** (Willd.) S. C. Chen & J. J. Wood

lichenora Lindl. = **Porpax jerdoniana** (Wight) Rolfe

lineata Lindl. = **Pinalia lineata** (Lindl.) Kuntze

lohitensis A. N. Rao, Harid. & S. N. Hegde = **Cylindrolobus glandulifer** (Deori & Phukan) A. N. Rao

meghasaniensis (S. Misra) S. Misra = **Pinalia meghasaniensis** (S. Misra) Schuit., Y. P. Ng & H. A. Pedersen

meirax (C. S. P. Parish & Rchb. f.) N. E. Br. = **Porpax meirax** (C. S. P. Parish & Rchb. f.) King & Pantl.

merguensis Lindl. = **Pinalia merguensis** (Lindl.) Kuntze

microchilos (Dalzell) Lindl. = **Porpax microchilos** (Dalzell) Schuit., Y. P. Ng & H. A. Pedersen

minima Blatt. & McCann = **Porpax exilis** (Hook. f.) Schuit., Y. P. Ng & H. A. Pedersen

muscicola (Lindl.) Lindl. = **Porpax parviflora** (D.Don) Ormerod & Kurzweil

muscicola (Lindl.) Lindl. var. *brevilinguis* J. Joseph & V. Chandras. = **Porpax nana** (A. Rich.) Schuit., Y. P. Ng & H. A. Pedersen

muscicola (Lindl.) Lindl. var. *ponmudiana* M. Mohanan & A. N. Henry = **Porpax nana** (A. Rich.) Schuit., Y. P. Ng & H. A. Pedersen

mysorensis Lindl. = **Pinalia mysorensis** (Lindl.) Kuntze

nana A. Rich. = **Porpax nana** (A. Rich.) Schuit., Y. P. Ng & H. A. Pedersen

nana A. Rich. var. *brevilinguis* (J. Joseph & V. Chandras.) Agrawala & H. J. Chowdhery = **Porpax nana** (A. Rich.) Schuit., Y. P. Ng & H. A. Pedersen

nepalensis Bajrach. & K. K. Shreshta = **Dendrolirium kamlangensis** (A. N. Rao) A. N. Rao

obesa Lindl. = **Pinalia obesa** (Lindl.) Kuntze

occidentalis Seidenf. = **Pinalia occidentalis** (Seidenf.) Schuit., Y. P. Ng & H. A. Pedersen

paniculata Lindl. = **Mycaranthes floribunda** (D. Don) S. C. Chen & J. J. Wood

pannea Lindl. = **Strongyleria pannea** (Lindl.) Schuit., Y. P. Ng & H. A. Pedersen

pauciflora Wight = **Cylindrolobus pauciflorus** (Wight) Schuit., Y. P. Ng & H. A. Pedersen

planicaulis Wall. ex Lindl. = **Agrostophyllum planicaule** (Wall. ex Lindl.) Rchb. f.

polystachya A. Rich. = **Pinalia polystachya** (A. Rich.) Kuntze

prainii Briq. = **Pinalia obesa** (Lindl.) Kuntze

pseudoclavicaulis Blatt. = **Cylindrolobus pseudoclavicaulis** (Blatt.) Schuit., Y. P. Ng & H. A. Pedersen

pubescens (Hook.) Lindl. ex G.Don = **Dendrolirium lasiopetalum** (Willd.) S. C. Chen & J. J. Wood

pubescens Wight = **Pinalia mysorensis** (Lindl.) Kuntze

pudica Ridl. = **Bryobium pudicum** (Ridl.) Y. P. Ng & P. J. Cribb

pulvinata Lindl. = **Trichotosia pulvinata** (Lindl.) Kraenzl.

pumila Lindl. = **Pinalia pumila** (Lindl.) Kuntze

pusilla (Griff.) Lindl. = **Porpax pusilla** (Griff.) Schuit., Y. P. Ng & H. A. Pedersen

reticosa Wight = **Porpax reticosa** (Wight) Schuit.

reticulata (Lindl.) Benth. & Hook. f. = **Porpax reticulata** Lindl.

rigida (Blume) Rchb. f. = **Callostylis rigida** Blume

rufinula Rchb. f. = **Trichotosia pulvinata** (Lindl.) Kraenzl.

rupestris Blatt. & McCann = **Porpax reticosa** (Wight) Schuit.

secundiflora Griff. = **Cryptochilus strictus** (Lindl.) Schuit., Y. P. Ng & H. A. Pedersen

sharmae H. J. Chowdhury, G. S. Giri & G. D. Pal = **Pinalia sharmae** (H. J. Chowdhery, G. S. Giri & G. D. Pal) A. N. Rao

sikkimensis Bajrach. & K. K. Shrestha = **Porpax sikkimensis** (Bajrach. & K. K. Shrestha) Schuit., Y. P. Ng & H. A. Pedersen

sphaerochila Lindl. = **Pinalia excavata** (Lindl.) Kuntze

spicata (D. Don) Hand.-Mazz. = **Pinalia spicata** (D. Don) S. C. Chen & J. J. Wood

stricta Lindl. = **Cryptochilus strictus** (Lindl.) Schuit., Y. P. Ng & H. A. Pedersen

tiagii Manilal, C. S. Kumar & J. J. Wood = **Porpax microchilos** (Dalzell) Schuit., Y. P. Ng & H. A. Pedersen

tomentosa (J. Koenig) Hook. f. = **Dendrolirium tomentosum** (J. Koenig) S. C. Chen & J. J. Wood

uniflora Dalzell = **Porpax reticosa** (Wight) Schuit.

velutina G. Lodd. ex Lindl. = **Trichotosia velutina** (G. Lodd. ex Lindl.) Kraenzl.

Erythrodes

herpysmoides (King & Pantl.) Schltr. = **Erythrodes hirsuta** (Griff.) Ormerod

seshagiriana A. N. Rao = **Erythrodes blumei** (Lindl.) Schltr.

viridiflora (Blume) Schltr. = **Goodyera viridiflora** (Blume) Lindl. ex D. Dietr.

Erythrorchis

lindleyana (Hook. f. & Thomson) Rchb. f. = **Cyrtosia lindleyana** Hook. f. & Thomson

Esmeralda

cathcartii (Lindl.) Rchb. f. = **Arachnis cathcartii** (Lindl.) J. J. Sm.

clarkei Rchb. f. = **Arachnis clarkei** (Rchb. f.) J. J. Sm.

Eucosia

cordata (Lindl.) T.C. Hsu = **Goodyera viridiflora** (Blume) Lindl. ex D. Dietr.

viridiflora (Blume) M.C.Pace = **Goodyera viridiflora** (Blume) Lindl. ex D. Dietr.

Eulophia

albiflora Edgew. ex Lindl. = **Eulophia herbacea** Lindl.

bicallosa (D. Don) P. F. Hunt & Summerh. var. *major* (King & Pantl.) Pradhan = **Eulophia bicallosa** (D. Don) P. F. Hunt & Summerh.

bicarinata (Lindl.) Hook. f. = **Eulophia bicallosa** (D. Don) P. F. Hunt & Summerh.

bicarinata (Lindl.) Hook. f. var. *major* King & Pantl. = **Eulophia bicallosa** (D. Don) P. F. Hunt & Summerh.

bicolor Dalzell = **Eulophia nuda** Lindl.

brachypetala Lindl. = **Eulophia herbacea** Lindl.

campanulata Duthie = **Eulophia obtusa** (Lindl.) Hook. f.

campestris Wall. ex Lindl. = **Eulophia dabia** (D. Don) Hochr.

candida (Lindl.) Hook. f. = **Eulophia bicallosa** (D. Don) P. F. Hunt & Summerh.

carinata (Willd.) Lindl. = **Eulophia epidendraea** (J. Koenig) C. E. C. Fisch.

cernua (Willd.) M. W. Chase, Kumar & Schuit. = **Eulophia picta** (R. Br.) Ormerod

cernua (Willd.) T. C. Hsu = **Eulophia picta** (R. Br.) Ormerod

cullenii (Wight) Blume = **Eulophia flava** (Lindl.) Hook. f.

cullenii (Wight) Blume var. *minor* C. E. C. Fisch. = **Eulophia flava** (Lindl.) Hook. f.

decipiens Kurz = **Eulophia graminea** Lindl.

emilianae C. J. Saldanha = **Eulophia zollingeri** (Rchb. f.) J. J. Sm.

epidendroides (Willd.) Schltr. = **Eulophia epidendraea** (J. Koenig) C. E. C. Fisch.

geniculata King & Pantl. = **Eulophia promensis** Lindl.

hastata Lindl. = **Tainia latifolia** (Lindl.) Rchb. f.

hemileuca Lindl. = **Eulophia dabia** (D. Don) Hochr.

hirsuta J. Joseph & Vajr. = **Pachystoma hirsuta** (J. Joseph & Vajr.) C. S. Kumar & Manilal

hormusjii Duthie = **Eulophia dabia** (D. Don) Hochr.

macrorhizon Hook. f. = **Eulophia zollingeri** (Rchb. f.) J. J. Sm.

macrostachya Lindl. = **Eulophia pulchra** (Thouars) Lindl.

nuda Lindl. var. *andersonii* Hook. f. = **Eulophia nuda** Lindl.

ramentacea (Roxb.) Lindl. = **Eulophia dabia** (D. Don) Hochr.

rupestris Wall. ex Lindl. = **Eulophia dabia** (D. Don) Hochr.

sanguinea (Lindl.) Hook. f. = **Eulophia zollingeri** (Rchb. f.) J. J. Sm.

spectabilis Suresh = **Eulophia nuda** Lindl.

ucbii Malhotra & Balodi = **Eulophia graminea** Lindl.

vera Royle = **Eulophia herbacea** Lindl.

virens (Roxb.) Spreng. = **Eulophia epidendraea**
(J. Koenig) C. E. C. Fisch.

Eulophidium
pulchrum (Thouars) Summerh. = **Eulophia pulchra**
(Thouars) Lindl.

Eulophus
carinatus (Willd.) R. Br. = **Eulophia epidendraea**
(J. Koenig) C. E. C. Fisch.

Euproboscis
pygmaea Griff. = **Thelasis pygmaea** (Griff.) Lindl.

Eurycaulis
cumulatus (Lindl.) M. A. Clem. = **Dendrobium
cumulatum** Lindl.
perula (Rchb. f.) M. A. Clem. = **Dendrobium parciflorum** Rchb. f. ex Lindl.

Evrardia
asraoa J. Joseph & Abbar. = **Chamaegastrodia poilanei**
(Gagnep.) Seidenf. & A. N. Rao
poilanei Gagnep. = **Chamaegastrodia poilanei**
(Gagnep.) Seidenf. & A. N. Rao

Evrardiana
poilanei (Gagnep.) Aver. = **Chamaegastrodia poilanei**
(Gagnep.) Seidenf. & A. N. Rao

Evrardianthe asraoa (J. Joseph & Abbar.) C. S. Kumar =
Chamaegastrodia poilanei (Gagnep.) Seidenf. &
A. N. Rao

Evrardianthe poilanei (Gagnep.) Rauschert = **Chamaegastrodia poilanei** (Gagnep.) Seidenf. & A. N. Rao

Fimbrorchis
trichosantha (Wall. ex Lindl.) Szlach. = **Habenaria
trichosantha** Wall. ex Lindl.

Flickingeria
abhaycharanii Phukan & A. A. Mao = **Dendrobium
calocephalum** (Z. H. Tsi & S. C. Chen) Schuit. &
Peter B. Adams
calocephala Z. H. Tsi & S. C.Chen = **Dendrobium
calocephalum** (Z. H. Tsi & S. C. Chen) Schuit. &
Peter B. Adams
fimbriata (Blume) A. D. Hawkes = **Dendrobium
plicatile** Lindl.
fugax (Rchb. f.) Seidenf. = **Dendrobium fugax** Rchb. f.
hesperis Seidenf. = **Dendrobium hesperis** (Seidenf.)
Schuit. & Peter B. Adams
kunstleri (Hook. f.) A. D. Hawkes = **Dendrobium
plicatile** Lindl.
macraei (Lindl.) Seidenf. = **Dendrobium macraei**
Lindl.
nodosa (Dalzell) Seidenf. = **Dendrobium nodosum**
Dalzell
pumila (Kuntze) A. D. Hawkes = **Dendrobium
pachyphyllum** (Kuntze) Bakh. f.
rabanii (Lindl.) Seidenf. = **Dendrobium plicatile** Lindl.
ritaeana (King & Pantl.) A. D. Hawkes = **Dendrobium
ritaeanum** King & Pantl.

Froscula
hispida Raf. = **Dendrobium longicornu** Lindl.

Galearis
stracheyi (Hook. f.) P. F. Hunt = **Galearis roborovskii**
(Maxim.) S. C. Chen, P. J. Cribb & S. W. Gale

Galeola
altissima (Blume) Rchb. f. = **Erythrorchis altissima**
(Blume) Blume
falconeri Hook. f. = **Cyrtosia falconeri** (Hook. f.)
Aver.
hydra Rchb. f. = **Galeola nudifolia** Lour.
javanica (Blume) Benth. & Hook. f. = **Cyrtosia javanica**
Blume
lindleyana (Hook. f. & Thomson) Rchb. f. = **Cyrtosia
lindleyana** Hook. f. & Thomson
nana Rolfe ex Downie = **Cyrtosia nana** (Rolfe ex
Downie) Garay

Galeorchis
roborovskii (Maxim.) Nevski = **Galearis roborovskii**
(Maxim.) S. C. Chen, P. J. Cribb & S. W. Gale
spathulata (Lindl.) Soó = **Galearis spathulata** (Lindl.)
P. F. Hunt
stracheyi (Hook. f.) Soó = **Galearis roborovskii** (Maxim.)
S. C. Chen, P. J. Cribb & S. W. Gale

Galera
nutans Blume = **Epipogium roseum** (D. Don) Lindl.
rosea (D. Don) Blume = **Epipogium roseum** (D. Don)
Lindl.

Gamoplexis
orobanchoides Falc. = **Gastrodia falconeri** D. L. Jones &
M. A. Clem.

Garayanthus
carinatus (Rolfe ex Downie) Szlach. = **Cleisostoma
duplicilobum** (J. J. Sm.) Garay
duplicilobus (J. J. Sm.) Szlach. = **Cleisostoma duplicilobum** (J. J. Sm.) Garay
paniculatus (Ker Gawl.) Szlach. = **Cleisostoma paniculatum** (Ker Gawl.) Garay

Gastrochilus
ampullaceus (Roxb.) Kuntze = **Vanda ampullacea**
(Roxb.) L. M. Gardiner
bigibbus (Rchb. f. ex Hook. f.) Kuntze = **Gastrochilus
obliquus** (Lindl.) Kuntze
carinatus (Griff.) Schltr. = **Acampe carinata** (Griff.) S. G.
Panigrahi
cephalotes (Lindl.) Kuntze = **Acampe carinata** (Griff.)
S. G. Panigrahi
congestus (Lindl.) Kuntze = **Acampe praemorsa** (Roxb.)
Blatt. & McCann
crassilabris (King & Pantl.) Garay = **Thrixspermum
crassilabre** (King & Pantl.) Ormerod
curvifolius (Lindl.) Kuntze = **Vanda curvifolia** (Lindl.)
L. M. Gardiner
dalzellianus (Santapau) Santapau & Kapadia = **Smithsonia viridiflora** (Dalzell) C. J. Saldanha
densiflorus (Lindl.) Kuntze = **Robiquetia spathulata**
(Blume) J. J. Sm.
filiformis (Rchb. f.) Kuntze = **Seidenfadeniella filiformis**
(Rchb. f.) Christenson & Ormerod
flexuosus (Lindl.) Kuntze = **Trichoglottis ramosa** (Lindl.)
Senghas
fragrans (C. S. P. Parish & Rchb. f.) Kuntze = **Schoenorchis fragrans** (C. S. P. Parish & Rchb. f.) Seidenf. &
Smitinand

garwalicus (Lindl.) Kuntze = **Rhynchostylis retusa** (L.) Blume

gemmatus (Lindl.) Kuntze = **Schoenorchis gemmata** (Lindl.) J. J. Sm.

giganteus (Lindl.) Kuntze = **Rhynchostylis gigantea** (Lindl.) Ridl.

gracilis (Lindl.) Kuntze = **Robiquetia gracilis** (Lindl.) Garay

helferi (Hook. f.) Kuntze = **Smitinandia helferi** (Hook. f.) Garay

inconspicuus (Hook. f.) Kuntze = **Luisia inconspicua** (Hook. f.) Hook. f. ex King & Pantl.

indicus Garay = **Gastrochilus acaulis** (Hook. f.) Kuntze

jerdonianus (Wight) Kuntze = **Schoenorchis jerdoniana** (Wight) Garay

longifolius (Lindl.) Kuntze = **Acampe praemorsa** (Roxb.) Blatt. & McCann

maculatus (Dalzell) Kuntze = **Smithsonia maculata** (Dalzell) C. J. Saldanha

nilagiricus Kuntze = **Gastrochilus acaulis** (Hook. f.) Kuntze

niveus (Lindl.) Kuntze = **Schoenorchis nivea** (Lindl.) Schltr.

obtusifolius (Lindl.) A. S. Rao & Mukherji = **Uncifera obtusifolia** Lindl.

obtusifolius (Lindl.) Kuntze = **Uncifera obtusifolia** Lindl.

ochraceus (Lindl.) Kuntze = **Acampe ochracea** (Lindl.) Hochr.

papillosus (Lindl.) Kuntze = **Acampe praemorsa** (Roxb.) Blatt. & McCann

parviflorus Kuntze = **Smitinandia micrantha** (Lindl.) Holttum

pseudodistichus (King & Pantl.) Seidenf. = **Gastrochilus pseudodistichus** (King & Pantl.) Schltr.

pulchellus (Wight) Schltr. = **Gastrochilus acaulis** (Hook. f.) Kuntze

pumilio (Rchb. f.) Kuntze = **Saccolabiopsis pusilla** (Lindl.) Seidenf. & Garay

racemifer (Lindl.) Kuntze = **Cleisostoma racemiferum** (Lindl.) Garay

ramosus (Lindl.) Kuntze = **Trichoglottis ramosa** (Lindl.) Senghas

retusus (L.) Kuntze = **Rhynchostylis retusa** (L.) Blume

rheedei (Wight) Kuntze = **Rhynchostylis retusa** (L.) Blume

ringens (Lindl.) Kuntze = **Aerides ringens** (Lindl.) C. E. C. Fisch.

roseus (Lindl.) Kuntze = **Robiquetia rosea** (Lindl.) Garay

rostellatus (Hook. f.) Kuntze = **Cleisostoma discolor** Lindl.

speciosus (Wight) Kuntze = **Aerides maculosa** Lindl.

stramineus (C. J. Saldanha) R. Rice = **Smithsonia straminea** C. J. Saldanha

trichromus (Rchb. f.) Kuntze = **Cleisocentron pallens** (Cathcart ex Lindl.) N. Pearce & P. J. Cribb

undulatus (Lindl.) Kuntze = **Pomatocalpa undulatum** (Lindl.) J. J. Sm.

viridiflorus (Dalzell) Kuntze = **Smithsonia viridiflora** (Dalzell) C. J. Saldanha

wightianus (Lindl.) Kuntze = **Aerides ringens** (Lindl.) C. E. C. Fisch.

Gastrodia

mairei Schltr. = **Gastrodia elata** Blume

orobanchoides (Falc.) Hook. f. = **Gastrodia falconeri** D. L. Jones & M. A. Clem.

pallens (Griff.) F. Muell. = **Didymoplexis pallens** Griff.

shikokiana Makino = **Chamaegastrodia shikokiana** Makino & F. Maek.

Gastroglottis

latifolia (Sm.) Szlach. = **Crepidium ophrydis** (J. Koenig) M. A. Clem. & D. L. Jones

ophrydis (J. Koenig) A. N. Rao = **Crepidium ophrydis** (J. Koenig) M. A. Clem. & D. L. Jones

Gastrorchis

gracilis (Lindl.) Aver. = **Calanthe obcordata** (Lindl.) M. W. Chase, Christenh. & Schuit.

Geodorum

appendiculatum Griff. = **Eulophia picta** (R. Br.) Ormerod

bicolor (Roxb.) Voigt = **Eulophia herbacea** Lindl.

candidum (Roxb.) Lindl. = **Eulophia picta** (R. Br.) Ormerod

densiflorum (Lam.) Schltr. = **Eulophia picta** (R. Br.) Ormerod

densiflorum (Lam.) Schltr. var. *kalimpongense* R. Yonzone, Lama & Bhujel = **Eulophia picta** (R. Br.) Ormerod

dilatatum R. Br. = **Eulophia recurva** (Roxb.) M. W. Chase, Kumar & Schuit.

dilatatum sensu Hook. f. = **Eulophia picta** (R. Br.) Ormerod

laxiflorum Griff. = **Eulophia diffusiflora** M. W. Chase, Kumar & Schuit.

longifolium (Roxb.) Voigt = **Cymbidium cyperifolium** Lindl.

pallidum D. Don = **Eulophia picta** (R. Br.) Ormerod

pictum (R. Br.) Lindl. = **Eulophia picta** (R. Br.) Ormerod

purpureum R. Br. = **Eulophia picta** (R. Br.) Ormerod

ramentaceum (Roxb.) Voigt = **Eulophia dabia** (D. Don) Hochr.

rariflorum Lindl. = **Eulophia picta** (R. Br.) Ormerod

rariflorum Lindl. = **Eulophia diffusiflora** M. W. Chase, Kumar & Schuit.

recurvum (Roxb.) Alston = **Eulophia recurva** (Roxb.) M. W. Chase, Kumar & Schuit.

Georchis

biflora Lindl. = **Goodyera biflora** (Lindl.) Hook. f.

cordata Lindl. = **Goodyera viridiflora** (Blume) Lindl. ex D. Dietr.

foliosa Lindl. = **Goodyera foliosa** (Lindl.) Benth. ex C. B. Clarke

rubicunda (Blume) Rchb. f. = **Goodyera rubicunda** (Blume) Lindl.

schlechtendaliana (Rchb. f.) Rchb. f. = **Goodyera schlechtendaliana** Rchb. f.

viridiflora (Blume) F. Muell. = **Goodyera viridiflora** (Blume) Lindl. ex D. Dietr.

vittata Lindl. = **Goodyera vittata** (Lindl.) Benth. ex
  Hook. f.

Glossaspis
  tentaculata (Lindl.) Spreng. = **Peristylus tentaculatus**
    (Lindl.) J. J. Sm.

Glossula
  tentaculata Lindl. = **Peristylus tentaculatus** (Lindl.)
    J. J. Sm.

Gonogona
  repens (L.) Link = **Goodyera repens** (L.) R. Br.

Goodyera
  affinis Griff. = **Hetaeria affinis** (Griff.) Seidenf. &
    Ormerod
  andersonii King & Pantl. = **Goodyera foliosa** (Lindl.)
    Benth. ex C. B. Clarke
  carnea A. Rich. = **Goodyera procera** (Ker Gawl.) Hook.
  clavata N. Pearce & P. J. Cribb = **Goodyera rubicunda**
    (Blume) Lindl.
  cordata (Lindl.) G. Nicholson = **Goodyera viridiflora**
    (Blume) Lindl. ex D. Dietr.
  dongchenii Lucksom = **Goodyera robusta** Hook. f.
  flabellata A. Rich. = **Cheirostylis flabellata** (A. Rich.)
    Wight
  grandis King & Pantl. = **Goodyera rubicunda** (Blume)
    Lindl.
  hirsuta Griff. = **Erythrodes hirsuta** (Griff.) Ormerod
  macrantha Maxim. = **Goodyera biflora** (Lindl.) Hook. f.
  moniliformis Griff. = **Cheirostylis moniliformis** (Griff.)
    Seidenf.
  ovalifolia Wight = **Hetaeria oblongifolia** Blume
  prainii Hook. f. = **Goodyera recurva** Lindl.
  pubescens R. Br. var. *repens* (L.) Alph. Wood = **Goodyera**
    repens (L.) R. Br.
  recurva Lindl. var. *prainii* (Hook. f.) Pradhan = **Goodyera**
    recurva Lindl.
  repens (L.) R. Br. var. *marginata* (Lindl.) Tang & F. T.
    Wang = **Goodyera marginata** Lindl.
  schlechtendaliana Rchb. f. var. *robusta* (Hook. f.) Av.
    Bhattacharjee & H. J. Chowdhery = **Goodyera**
    robusta Hook. f.
  secundiflora Griff. = **Goodyera foliosa** (Lindl.) Benth. ex
    C. B. Clarke
  viridiflora (Blume) Blume = **Goodyera viridiflora**
    (Blume) Lindl. ex D. Dietr.

Govindooia
  nervosa Wight = **Tropidia angulosa** (Lindl.) Blume

Grafia
  parishii (Rchb. f.) A. D. Hawkes = **Phalaenopsis parishii**
    Rchb. f.

Graphorkis
  andamanensis (Rchb. f.) Kuntze = **Eulophia andaman-**
    ensis Rchb. f.
  bicallosa (D. Don) Kuntze = **Eulophia bicallosa** (D. Don)
    P. F. Hunt & Summerh.
  bicolor (Roxb.) Kuntze = **Eulophia herbacea** Lindl.
  bracteosa (Lindl.) Kuntze = **Eulophia bracteosa** Lindl.
  campestris (Wall. ex Lindl.) Kuntze = **Eulophia dabia**
    (D. Don) Hochr.
  dabia (D. Don) Kuntze = **Eulophia dabia** (D. Don)
    Hochr.

densiflora (Lindl.) Kuntze = **Eulophia densiflora** Lindl.
flava (Lindl.) Kuntze = **Eulophia flava** (Lindl.) Hook. f.
graminea (Lindl.) Kuntze = **Eulophia graminea** Lindl.
herbacea (Lindl.) Lyons = **Eulophia herbacea** Lindl.
macrobulbon (C. S. P. Parish & Rchb. f.) Kuntze =
  Eulophia macrobulbon (C. S. P. Parish & Rchb. f.)
  Hook. f.
mannii (Rchb. f.) Kuntze = **Eulophia mannii** (Rchb. f.)
  Hook. f.
nuda (Lindl.) Kuntze = **Eulophia nuda** Lindl.
obtusa (Lindl.) Kuntze = **Eulophia obtusa** (Lindl.)
  Hook. f.
ochreata (Lindl.) Kuntze = **Eulophia ochreata** Lindl.
pratensis (Lindl.) Kuntze = **Eulophia pratensis** Lindl.
pulchra (Thouars) Kuntze = **Eulophia pulchra** (Thouars)
  Lindl.
rupestris Kuntze = **Eulophia dabia** (D. Don) Hochr.
sanguinea (Lindl.) Kuntze = **Eulophia zollingeri** (Rchb.
  f.) J. J. Sm.

Grastidium
  cathcartii (Hook. f.) M. A. Clem. & D. L. Jones = **Den-**
    drobium salaccense (Blume) Lindl.
  haemoglossum (Thwaites) M. A. Clem. & D. L. Jones =
    Dendrobium salaccense (Blume) Lindl.
  indragiriense (Schltr.) Rauschert = **Dendrobium**
    indragiriense Schltr.
  pensile (Ridl.) Rauschert = **Dendrobium pensile** Ridl.
  salaccense Blume = **Dendrobium salaccense** (Blume)
    Lindl.

Grosourdya
  hystrix (Blume) Rchb. f. = **Thrixspermum hystrix**
    (Blume) Rchb. f.
  monsooniae (Sasidh. & Sujanapal) R. Rice = **Pteroceras**
    monsooniae Sasidh. & Sujanapal
  muriculata (Rchb. f.) R. Rice = **Pteroceras muricula-**
    tum (Rchb. f.) P. F. Hunt

Gymnadenia
  affinis (D. Don) Rchb. f. = **Peristylus affinis** (D. Don)
    Seidenf.
  calcicola W. W. Sm. = **Ponerorchis cucullata** (L.)
    X.H.Jin, Schuit. & W.T.Jin var. **calcicola** (W. W. Sm.)
    X. H. Jin, Schuit. & W. T. Jin
  chusua (D. Don) Lindl. = **Ponerorchis chusua** (D. Don)
    Soó
  chusua (D. Don) Lindl. var. *nana* (King & Pantl.) Finet =
    Ponerorchis nana (King & Pantl.) Soó
  cylindrostachya Lindl. = **Gymnadenia orchidis** Lindl.
    var. **orchidis**
  galeandra (Rchb. f.) Rchb. f. = **Brachycorythis**
    galeandra (Rchb. f.) Summerh.
  helferi Rchb. f. = **Brachycorythis helferi** (Rchb. f.)
    Summerh.
  obcordata (Lindl.) Rchb. f. = **Brachycorythis obcordata**
    (Lindl.) Summerh.
  plantaginea (Roxb.) Lindl. = **Habenaria roxburghii**
    Nicolson
  puberula Lindl. = **Ponerorchis chusua** (D. Don) Soó
  secunda Lindl. = **Peristylus secundus** (Lindl.) Rathakr.
  secundiflora (Kraenzl.) Kraenzl. = **Ponerorchis secun-**
    diflora (Kraenzl.) X. H. Jin, Schuit. & W. T. Jin

gracilis Colebr. = **Peristylus densus** (Lindl.) Santapau & Kapadia

gracillima Hook. f. = **Herminium mannii** (Rchb. f.) Tang & F. T. Wang

graminea A. Rich. = **Habenaria viridiflora** (Sw.) R. Br. ex Spreng.

graminea Lindl. = **Habenaria khasiana** Hook. f.

grandiflora Lindl. = **Habenaria grandifloriformis** Blatt. & McCann

grandiflora Lindl. ex Dalzell & Gibson = **Habenaria grandifloriformis** Blatt. & McCann

grandifloriformis Blatt. & McCann var. *aequiloba* Blatt. & McCann = **Habenaria grandifloriformis** Blatt. & McCann

graveolens Duthie = **Habenaria digitata** Lindl.

griffithii Hook. f. = **Gennaria griffithii** (Hook. f.) X. H. Jin & D. Z. Li

hamigera Griff. = **Habenaria furcifera** Lindl.

hamiltoniana (Lindl.) Hook. f. = **Peristylus hamiltonianus** (Lindl.) Lindl.

helferi (Rchb. f.) Hook. f. = **Brachycorythis helferi** (Rchb. f.) Summerh.

heyneana Lindl. var. *subpubens* (A. Rich.) Pradhan = **Habenaria heyneana** Lindl.

iantha (Wight) Hook. f. = **Brachycorythis iantha** (Wight) Summerh.

indica C. S. Kumar & Manilal = **Habenaria hollandiana** Santapau

intermedia D. Don var. *arietina* (Hook. f.) Finet = **Habenaria arietina** Hook. f.

intrudens Ames = **Peristylus intrudens** (Ames) Ormerod

jerdoniana Wight = **Habenaria diphylla** Dalzell

juncea King & Pantl. = **Platanthera nematocaulon** (Hook. f.) Kraenzl.

lacertifera (Lindl.) Benth. = **Peristylus lacertifer** (Lindl.) J. J. Sm.

latilabris (Lindl.) Hook. f. = **Herminium latilabre** (Lindl.) X. H. Jin, Schuit., Raskoti & L. Q. Huang

lawii (Wight) Hook. f. = **Peristylus lawii** Wight

leptocaulon Hook. f. = **Platanthera leptocaulon** (Hook. f.) Soó

lindleyana Wight = **Habenaria digitata** Lindl.

longecalcarata A. Rich. = **Habenaria longicorniculata** J. Graham

longecalcarata A. Rich. var. *viridis* Blatt. & McCann = **Habenaria longicorniculata** J. Graham

lutea (Wight) Benth. = **Habenaria perrottetiana** A. Rich.

malabarica Hook. f. = **Peristylus brachyphyllus** A. Rich.

malleifera Hook. f. var. *hollandiana* (Santapau) Pradhan = **Habenaria hollandiana** Santapau

marginata Colebr. f. *flavescens* (Hook. f.) Blatt. & McCann = **Habenaria marginata** Colebr.

marginata Colebr. var. *flavescens* (Hook. f.) T. Cooke = **Habenaria marginata** Colebr.

marginata Colebr. var. *fusifera* (Hook. f.) Santapau & Kapadia = **Habenaria marginata** Colebr.

modesta Dalzell = **Habenaria ovalifolia** Wight

montana A. Rich. = **Habenaria longicornu** Lindl.

montana A. Rich. var. *major* A. Rich. = **Habenaria longicornu** Lindl.

monticola Ridl. = **Peristylus monticola** (Ridl.) Seidenf.

neglecta King & Pantl. = **Peristylus densus** (Lindl.) Santapau & Kapadia

nematocaulon Hook. f. = **Platanthera nematocaulon** (Hook. f.) Kraenzl.

obcordata (Lindl.) Fyson = **Brachycorythis obcordata** (Lindl.) Summerh.

oligantha Hook. f. = **Platanthera pachycaulon** (Hook. f.) Soó

orchidis (Lindl.) Hook. f. = **Gymnadenia orchidis** Lindl. var. **orchidis**

pachycaulon Hook. f. = **Platanthera pachycaulon** (Hook. f.) Soó

panchganiensis Santapau & Kapadia = **Habenaria suaveolens** Dalzell

panigrahiana S. Misra var. *parviloba* S. Misra = **Habenaria panigrahiana** S. Misra

parishii (Rchb. f.) Hook. f. = **Peristylus parishii** Rchb. f.

pectinata D. Don subsp. *ensifolia* (Lindl.) Soó = **Habenaria ensifolia** Lindl.

pectinata D. Don var. *arietina* (Hook. f.) Kraenzl. = **Habenaria arietina** Hook. f.

pectinata D. Don var. *gigantea* Pradhan = **Habenaria arietina** Hook. f.

pectinata D. Don var. *khasiensis* Pradhan = **Habenaria arietina** Hook. f.

pelorioides C. S. P. Parish & Rchb. f. = **Habenaria malintana** (Blanco) Merr.

peristyloides Wight = **Peristylus densus** (Lindl.) Santapau & Kapadia

platyphylla Spreng. = **Habenaria roxburghii** Nicolson

platyphylloides M. R. Almeida = **Habenaria roxburghii** Nicolson

polytricha (Hook. f.) Pradhan = **Habenaria pantlingiana** Kraenzl.

polytrichoides Aver. = **Habenaria pantlingiana** Kraenzl.

prainii Hook. f. = **Peristylus prainii** (Hook. f.) Kraenzl.

pseudophrys King & Pantl. = **Peristylus pseudophrys** (King & Pantl.) Kraenzl.

purpureopunctata K. Y. Lang = **Hemipilia purpureopunctata** (K. Y.Lang) X. H. Jin, Schuit. & W. T. Jin

ramayyana Ram. Chary & J. J. Wood = **Habenaria panigrahiana** S. Misra

rariflora A. Rich. var. *latifolia* Blatt. & McCann = **Habenaria rariflora** A. Rich.

robustior (Wight) Hook. f. = **Peristylus secundus** (Lindl.) Rathakr.

rotundifolia Lindl. = **Habenaria grandifloriformis** Blatt. & McCann

schizochilus J. Graham = **Habenaria crinifera** Lindl.

secundiflora Hook. f. = **Ponerorchis secundiflora** (Kraenzl.) X. H. Jin, Schuit. & W. T. Jin

seshagiriana A. N. Rao = **Habenaria pantlingiana** Kraenzl.

sikkimensis Hook. f. = **Platanthera sikkimensis** (Hook. f.) Kraenzl.

spathulata (Lindl.) Benth. = **Galearis spathulata** (Lindl.) P. F. Hunt

agyokuana (Fukuy.) K. Nakaj. = **Zeuxine agyokuana** Fukuy.

albida Blume = **Vrydagzynea albida** (Blume) Blume

asraoa (J. Joseph & Abbar.) Karthik. = **Chamaegastrodia poilanei** (Gagnep.) Seidenf. & A. N. Rao

cristata Blume var. *agyokuana* (Fukuy.) S. S. Ying = **Zeuxine agyokuana** Fukuy.

flava Lindl. = **Zeuxine flava** (Wall. ex Lindl.) Trimen

fusca Lindl. = **Goodyera fusca** (Lindl.) Hook. f.

gardneri (Thwaites) Benth. ex Hook. f. = **Hetaeria oblongifolia** Blume

helferi Hook. f. = **Hetaeria oblongifolia** Blume

lanceolata (Lindl.) Rchb. f. = **Rhomboda lanceolata** (Lindl.) Ormerod

longifolia (Lindl.) Benth. = **Rhomboda longifolia** Lindl.

mollis Lindl. = **Zeuxine affinis** (Lindl.) Benth. ex Hook. f.

nervosa Lindl. = **Zeuxine nervosa** (Wall. ex Lindl.) Trimen

ovalifolia (Wight) Benth. ex Hook. f. = **Hetaeria oblongifolia** Blume

poilanei (Gagnep.) Tang & F. T. Wang = **Chamaegastrodia poilanei** (Gagnep.) Seidenf. & A. N. Rao

pusilla Lindl. = **Cheirostylis pusilla** Lindl.

rubens (Lindl.) Benth. ex Hook. f. = **Hetaeria affinis** (Griff.) Seidenf. & Ormerod

shikokiana (Makino & F. Maek.) Tuyama = **Chamaegastrodia shikokiana** Makino & F. Maek.

Heterozeuxine

glandulosa (King & Pantl.) T. Hashim. = **Zeuxine glandulosa** King & Pantl.

nervosa (Wall. ex Lindl.) T. Hashim. = **Zeuxine nervosa** (Wall. ex Lindl.) Trimen

Himantoglossum

viride (L.) Rchb. = **Dactylorhiza viridis** (L.) R. M. Bateman, Pridgeon & M. W. Chase

Holopogon

microglottis (Duthie) S. C. Chen = **Neottia microglottis** (Duthie) Schltr.

pantlingii (W. W. Sm.) S. C. Chen = **Neottia pantlingii** (W. W. Sm.) Tang & F. T. Wang

Hygrochilus

parishii (Rchb. f.) Pfitzer = **Phalaenopsis marriottiana** (Rchb. f.) Kocyan & Schuit. var. **parishii** (Rchb. f.) Kocyan & Schuit.

Hymeneria

obesa (Lindl.) M. A. Clem. & D. L. Jones = **Pinalia obesa** (Lindl.) Kuntze

Hysteria

veratrifolia Reinw. = **Corymborkis veratrifolia** (Reinw.) Blume

Ibidium

spirale (Lour.) Makino = **Spiranthes sinensis** (Pers.) Ames

Iebine

nervosa (Thunb.) Raf. = **Liparis nervosa** (Thunb.) Lindl. var. **nervosa**

India

arunachalensis A. N. Rao = **Robiquetia arunachalensis** (A. N. Rao) Kocyan & Schuit. Phytotaxa 161: 68. 2014.

Ione

andersonii King & Pantl. = **Bulbophyllum sasakii** (Hayata) J. J. Verm., Schuit. & de Vogel

annamensis Ridl. = **Bulbophyllum medioximum** J. J. Verm., Schuit. & de Vogel

arunachalense A. N. Rao = **Bulbophyllum arunachalense** (A. N. Rao) J. J. Verm., Schuit. & de Vogel

bicolor (Lindl.) Lindl. = **Bulbophyllum roseopictum** J. J. Verm., Schuit. & de Vogel

candida Lindl. = **Bulbophyllum candidum** (Lindl.) Hook. f.

cirrhata Lindl. = **Bulbophyllum paleaceum** (Lindl.) Benth. ex Hemsl.

fuscopurpurea Lindl. = **Bulbophyllum paleaceum** (Lindl.) Benth. ex Hemsl.

intermedia King & Pantl. = **Bulbophyllum interpositum** J. J. Verm., Schuit. & de Vogel

jainii (Hynn. & Malhotra) Seidenf. = **Bulbophyllum jainii** (Hynn. & Malhotra) J. J. Verm., Schuit. & de Vogel

kipgenii Kishor, Chowlu & Vij = **Bulbophyllum kipgenii** (Kishor, Chowlu & Vij) J. J. Verm., Schuit. & de Vogel

paleacea Lindl. = **Bulbophyllum paleaceum** (Lindl.) Benth. ex Hemsl.

racemosa (Sm.) Seidenf. = **Bulbophyllum reptans** (Lindl.) Lindl.

sasakii Hayata = **Bulbophyllum sasakii** (Hayata) J. J. Verm., Schuit. & de Vogel

scariosa (Lindl.) King & Pantl. = **Bulbophyllum sunipia** J. J. Verm.

virens Lindl. = **Bulbophyllum paleaceum** (Lindl.) Benth. ex Hemsl.

Ipsea

wrayana Hook. f. = **Tainia wrayana** (Hook. f.) J. J. Sm.

Iridorchis

gigantea Blume = **Cymbidium iridioides** D. Don

Iridorkis

angustifolia (Lindl.) Kuntze = **Oberonia angustifolia** Lindl.

anthropophora (Lindl.) Kuntze = **Oberonia anthropophora** Lindl.

bicornis (Lindl.) Kuntze = **Oberonia bicornis** Lindl.

brachystachys (Lindl.) Kuntze = **Oberonia brachystachys** Lindl.

brunoniana (Wight) Kuntze = **Oberonia brunoniana** Wight

caulescens (Lindl.) Kuntze = **Oberonia caulescens** Lindl.

clarkei (Hook. f.) Kuntze = **Oberonia clarkei** Hook. f.

ensiformis (Sm.) Kuntze = **Oberonia ensiformis** (Sm.) Lindl.

forcipata (Lindl.) Kuntze = **Oberonia forcipata** Lindl.

griffithiana (Lindl.) Kuntze = **Oberonia griffithiana** Lindl.

iridifolia Kuntze = **Oberonia ensiformis** (Sm.) Lindl.

jenkinsiana (Griff. ex Lindl.) Kuntze = **Oberonia jenkinsiana** Griff. ex Lindl.

lindleyana (Rchb. f.) Kuntze = **Oberonia brunoniana** Wight

longibracteata (Lindl.) Kuntze = **Oberonia longibracteata** Lindl.

stachyurus (Rchb. f.) Kuntze = **Liparis viridiflora**
(Blume) Lindl.

stricklandiana (Rchb. f.) Kuntze = **Liparis stricklandiana** Rchb. f.

torta (Hook. f.) Kuntze = **Liparis torta** Hook. f.

vestita (Rchb. f.) Kuntze = **Liparis vestita** Rchb. f.

viridiflora (Blume) Kuntze = **Liparis viridiflora** (Blume)
Lindl.

walkerae (Graham) Kuntze = **Liparis walkerae** Graham

wightiana (Thwaites) Kuntze = **Liparis wightiana**
Thwaites

wrayi (Hook. f.) Kuntze = **Liparis barbata** Lindl.

Leucostachys
procera (Ker Gawl.) Hoffmanns. = **Goodyera procera**
(Ker Gawl.) Hook.

Lichenora
jerdoniana Wight = **Porpax jerdoniana** (Wight) Rolfe

Limatodis
gracilis (Lindl.) Lindl. = **Calanthe obcordata** (Lindl.)
M. W. Chase, Christenh. & Schuit.

mishmensis Lindl. & Paxton = **Calanthe mishmensis**
(Lindl. & Paxton) M. W. Chase, Christenh. & Schuit.

Limodorum
aphyllum Roxb. = **Dendrobium aphyllum** (Roxb.)
C. E. C. Fisch.

bicallosum (D. Don) Buch.-Ham. ex D. Don = **Eulophia
bicallosa** (D. Don) P. F. Hunt & Summerh.

bicolor Roxb. = **Eulophia herbacea** Lindl.

bracteatum Roxb. = **Thunia alba** Rchb. f. var. **bracteata**
(Roxb.) N. Pearce & P. J. Cribb

candidum Roxb. = **Eulophia picta** (R. Br.) Ormerod

carinatum Willd. = **Eulophia epidendraea** (J. Koenig)
C. E. C. Fisch.

dabia (D. Don) Buch.-Ham. ex D. Don = **Eulophia dabia**
(D. Don) Hochr.

densiflorum Lam. = **Eulophia picta** (R. Br.) Ormerod

epidendroides Willd. = **Eulophia epidendraea** (J.
Koenig) C. E. C. Fisch.

flavum Blume = **Calanthe woodfordii** (Hook.) M. W.
Chase, Christenh. & Schuit.

incarvillei Pers. = **Calanthe tankervilleae** (Banks) M. W.
Chase, Christenh. & Schuit.

laxiflorum Lam. = **Cymbidium ensifolium** (L.) Sw.
subsp. **haematodes** (Lindl.) Du Puy & P. J. Cribb ex
Govaerts

longifolium Roxb. = **Cymbidium cyperifolium** Lindl.

monile Thunb. = **Dendrobium moniliforme** (L.) Sw.

nutans Roxb. = **Eulophia picta** (R. Br.) Ormerod

orchideum (J. Koenig) Willd. = **Trichoglottis orchidea**
(J. Koenig) Garay

pulchrum Thouars = **Eulophia pulchra** (Thouars) Lindl.

ramentaceum Roxb. = **Eulophia dabia** (D. Don) Hochr.

recurvum Roxb. = **Eulophia recurva** (Roxb.) M. W.
Chase, Kumar & Schuit.

retusum (L.) Sw. = **Rhynchostylis retusa** (L.) Blume

roseum D. Don = **Epipogium roseum** (D. Don) Lindl.

royleanum (Lindl.) Kuntze = **Epipactis royleana** Lindl.

spathulatum (L.) Willd. = **Taprobanea spathulata** (L.)
Christenson

tankervilleae Banks = **Calanthe tankervilleae** (Banks)
M. W. Chase, Christenh. & Schuit.

veratrifolium Willd. = **Calanthe triplicata** (P. Willemet)
Ames

virens Roxb. = **Eulophia epidendraea** (J. Koenig)
C. E. C. Fisch.

Liparis
atropurpurea Wight = **Liparis wightiana** Thwaites

auriculata Rchb. f. = **Liparis cespitosa** (Lam.) Lindl.

bicallosa (D. Don) Schltr. = **Eulophia bicallosa** (D. Don)
P. F. Hunt & Summerh.

bidentata Griff. = **Callostylis rigida** Blume

bistriata C. S. P. Parish & Rchb. f. var. *robusta* Hook. f. =
Liparis bistriata C. S. P. Parish & Rchb. f.

bituberculata (Hook.) Lindl. = **Liparis nervosa** (Thunb.)
Lindl. var. **nervosa**

bituberculata (Hook.) Lindl. var. *khasiana* Hook. f. =
Liparis nervosa (Thunb.) Lindl. var. **khasiana**
(Hook. f.) P. K. Sarkar

breviscapa A. P. Das & Lama = **Liparis nervosa** (Thunb.)
Lindl. var. **khasiana** (Hook. f.) P. K. Sarkar

dalzellii Hook. f. = **Liparis odorata** (Willd.) Lindl.

densiflora A. Rich. = **Crepidium densiflorum** (A. Rich.)
Sushil K. Singh, Agrawala & Jalal

diodon Rchb. f. = **Liparis rostrata** Rchb. f.

diphyllos Nimmo ?= **Liparis deflexa** Hook. f.

dolabella Hook. f. = **Liparis stricklandiana** Rchb. f.

duthiei Hook. f. = **Liparis cespitosa** (Lam.) Lindl.

espeevijii S. Misra = **Liparis odorata** (Willd.) Lindl.

flavoviridis Blatt. & McCann = **Liparis deflexa** Hook. f.

griffithii Ridl. = **Liparis stricklandiana** Rchb. f.

hookeri Ridl. = **Liparis bistriata** C. S. P. Parish &
Rchb. f.

indirae Manilal & C. S. Kumar = **Liparis barbata** Lindl.

intermedia A. Rich. = **Crepidium intermedium**
(A. Rich.) Sushil K. Singh, Agrawala & Jalal

khasiana (Hook. f.) Tang & F. T. Wang = **Liparis nervosa**
(Thunb.) Lindl. var. **khasiana** (Hook. f.) P. K. Sarkar

lancifolia Hook. f. = **Liparis bootanensis** Griff.

longipes Lindl. = **Liparis viridiflora** (Blume) Lindl.

longipes Lindl. var. *spathulata* (Lindl.) Ridl. = **Liparis
viridiflora** (Blume) Lindl.

macrantha Hook. f. = **Liparis distans** C. B. Clarke

macrocarpa Hook. f. = **Liparis nervosa** (Thunb.) Lindl.
var. **nervosa**

meniscophora Gagnep. = **Liparis pygmaea** King & Pantl.

nana Rolfe = **Liparis pygmaea** King & Pantl.

nigra Seidenf. = **Liparis gigantea** C. L. Tso

obscura Hook. f. = **Liparis cespitosa** (Lam.) Lindl.

pachypus C. S. P. Parish & Rchb. f. = **Liparis bootanensis** Griff.

paradoxa (Lindl.) Rchb. f. = **Liparis odorata** (Willd.)
Lindl.

pendula Lindl. = **Liparis viridiflora** (Blume) Lindl.

prainii Hook. f. = **Liparis cespitosa** (Lam.) Lindl.

prazeri King & Pantl. = **Liparis deflexa** Hook. f.

pulchella Hook. f. = **Liparis petiolata** (D. Don) P. F. Hunt
& Summerh.

pusilla Ridl. = **Liparis cespitosa** (Lam.) Lindl.

resupinata Ridl. var. *ridleyi* (Hook. f.) King & Pantl. = **Liparis resupinata** Ridl.

ridleyi Hook. f. = **Liparis resupinata** Ridl.

rupestris Griff. var. *purpurascens* Ridl. = **Liparis petiolata** (D. Don) P. F. Hunt & Summerh.

selligera Rchb. f. = **Liparis plantaginea** Lindl.

sikkimensis Lucksom & S. Kumar = **Liparis viridiflora** (Blume) Lindl.

spathulata Lindl. = **Liparis viridiflora** (Blume) Lindl.

stachyurus Rchb. f. = **Liparis viridiflora** (Blume) Lindl.

tenuifolia Hook. f. = **Liparis mannii** Rchb. f.

togashii Tuyama = **Liparis perpusilla** Hook. f.

tortilis P. M. Salim & J. Mathew = **Liparis odorata** (Willd.) Lindl.

udaii S. Misra = **Liparis odorata** (Willd.) Lindl.

vestita Rchb. f. subsp. *seidenfadenii* S. Misra = **Liparis odorata** (Willd.) Lindl.

viridiflora (Blume) Lindl. var. *spathulata* (Lindl.) A. N. Rao = **Liparis viridiflora** (Blume) Lindl.

wightii Rchb. f. = **Liparis elliptica** Wight

wrayi Hook. f. = **Liparis barbata** Lindl.

Lissochilus

flavus (Lindl.) Schltr. = **Eulophia flava** (Lindl.) Hook. f.

obtusus (Lindl.) Schltr. = **Eulophia obtusa** (Lindl.) Hook. f.

pulcher (Thouars) H. Perrier = **Eulophia pulchra** (Thouars) Lindl.

Listera

alternifolia King & Pantl. = **Neottia alternifolia** (King & Pantl.) Szlach.

brevicaulis King & Pantl. = **Neottia brevicaulis** (King & Pantl.) Szlach.

dentata King & Pantl. = **Neottia dentata** (King & Pantl.) Szlach.

divaricata Panigrahi & P. Taylor = **Neottia divaricata** (Panigrahi & P. Taylor) Szlach.

inayatii Duthie = **Neottia inayatii** (Duthie) Schltr.

kashmiriana Duthie = **Neottia inayatii** (Duthie) Schltr.

lindleyana (Decne.) King & Pantl. = **Neottia listeroides** Lindl.

longicaulis King & Pantl. = **Neottia longicaulis** (King & Pantl.) Szlach.

micrantha Lindl. = **Neottia karoana** Szlach.

microglottis Duthie = **Neottia microglottis** (Duthie) Schltr.

mucronata Panigrahi & J. J. Wood = **Neottia mucronata** (Panigrahi & J. J. Wood) Szlach.

nandadeviensis Hajra = **Neottia nandadeviensis** (Hajra) Szlach.

ovata (L.) R. Br. = **Neottia ovata** (L.) Bluff & Fingerh.

pinetorum Lindl. = **Neottia pinetorum** (Lindl.) Szlach.

reniformis D. Don = **Habenaria reniformis** (D. Don) Hook. f.

tenuis Lindl. = **Neottia tenuis** (Lindl.) Szlach.

Loxoma

maculatum (Dalzell) Garay = **Smithsonia maculata** (Dalzell) C. J. Saldanha

straminea (C. J. Saldanha) Pradhan = **Smithsonia straminea** C. J. Saldanha

viridiflora (Dalzell) Pradhan = **Smithsonia viridiflora** (Dalzell) C. J. Saldanha

Loxomorchis

maculata (Dalzell) Rauschert = **Smithsonia maculata** (Dalzell) C. J. Saldanha

Luisia

alpina Lindl. = **Vanda alpina** (Lindl.) Lindl.

birchea Blume = **Luisia tenuifolia** Blume

birchea Blume var. *evangelinae* (Blatt. & McCann) P. K. Sarkar = **Luisia tenuifolia** Blume

evangelinae Blatt. & McCann = **Luisia tenuifolia** Blume

griffithii (Lindl.) Kraenzl. = **Vanda griffithii** Lindl.

grovesii Hook. f. = **Luisia filiformis** Hook. f.

indica Khuraijam & R. K. Roy = **Luisia zeylanica** Lindl.

indivisa King & Pantl. = **Luisia brachystachys** (Lindl.) Blume

laurifolia M. R. Almeida = **Luisia tenuifolia** Blume

laurifolia M. R. Almeida var. *evangelinae* (Blatt. & McCann) M. R. Almeida = **Luisia tenuifolia** Blume

micrantha Hook. f. = **Luisia inconspicua** (Hook. f.) Hook. f. ex King & Pantl.

pseudotenuifolia Blatt. & McCann = **Luisia tenuifolia** Blume

pulniana Vatsala = **Luisia tenuifolia** Blume

striata (Rchb. f.) Kraenzl. = **Vanda cristata** Wall. ex Lindl.

tenuifolia Blume var. *evangelinae* (Blatt. & McCann) Santapau & Kapadia = **Luisia tenuifolia** Blume

trichorrhiza (Hook.) Blume var. *flava* Gogoi = **Luisia trichorrhiza** (Hook.) Blume

truncata Blatt. & McCann = **Luisia zeylanica** Lindl.

uniflora (Lindl.) Blume = **Papilionanthe uniflora** (Lindl.) Garay

Luisiopsis

inconspicua (Hook. f.) C. S. Kumar & P. C. S. Kumar = **Luisia inconspicua** (Hook. f.) Hook. f. ex King & Pantl.

Macodes

lanceolata (Lindl.) Rchb. f. = **Rhomboda lanceolata** (Lindl.) Ormerod

Malaxis

acuminata D. Don = **Crepidium acuminatum** (D. Don) Szlach.

acuminata D. Don f. *biloba* (Lindl.) Tuyama = **Crepidium acuminatum** (D. Don) Szlach.

acuminata D. Don var. *biloba* (Lindl.) Ames = **Crepidium acuminatum** (D. Don) Szlach.

andamanica (King & Pantl.) N. P. Balakr. & Vasudeva Rao = **Crepidium andamanicum** (King & Pantl.) Marg. & Szlach.

angustifolia (Lindl.) Rchb. f. = **Oberonia angustifolia** Lindl.

anthropophora (Lindl.) Rchb. f. = **Oberonia anthropophora** Lindl.

aphylla (King & Pantl.) Tang & F. T. Wang = **Crepidium aphyllum** (King & Pantl.) A. N. Rao

biaurita (Lindl.) Kuntze = **Crepidium biauritum** (Lindl.) Szlach.

bicornis (Lindl.) Rchb. f. = **Oberonia bicornis** Lindl.

biloba (Lindl.) Ames = **Crepidium acuminatum** (D. Don) Szlach.

brunoniana (Wight) Rchb. f. = **Oberonia brunoniana** Wight

calophylla (Rchb. f.) Kuntze = **Crepidium calophyllum** (Rchb. f.) Szlach.

calophylla (Rchb. f.) Kuntze var. *brachycheila* (Hook. f.) Tang & F. T. Wang = **Crepidium calophyllum** (Rchb. f.) Szlach.

caulescens (Lindl.) Rchb. f. = **Oberonia caulescens** Lindl.

cernua Willd. = **Eulophia picta** (R. Br.) Ormerod

cespitosa (Lam.) Thouars = **Liparis cespitosa** (Lam.) Lindl.

congesta (Lindl.) Deb = **Crepidium ophrydis** (J. Koenig) M. A. Clem. & D. L. Jones

cordifolia Sm. = **Liparis petiolata** (D. Don) P. F. Hunt & Summerh.

crenulata (Ridl.) Kuntze = **Crepidium crenulatum** (Ridl.) Sushil K. Singh, Agrawala & Jalal

densiflora (A. Rich.) Kuntze = **Crepidium densiflorum** (A. Rich.) Sushil K. Singh, Agrawala & Jalal

ensiformis Sm. = **Oberonia ensiformis** (Sm.) Lindl.

forcipata (Lindl.) Rchb. f. = **Oberonia forcipata** Lindl.

griffithiana (Lindl.) Rchb. f. = **Oberonia griffithiana** Lindl.

intermedia (A. Rich.) Seidenf. = **Crepidium intermedium** (A. Rich.) Sushil K. Singh, Agrawala & Jalal

iridifolia Rchb. f. = **Oberonia ensiformis** (Sm.) Lindl.

jenkinsiana (Griff. ex Lindl.) Rchb. f. = **Oberonia jenkinsiana** Griff. ex Lindl.

josephiana (Rchb. f.) Kuntze = **Crepidium josephianum** (Rchb. f.) Marg.

khasiana (Hook. f.) Kuntze = **Crepidium khasianum** (Hook. f.) Szlach.

latifolia Sm. = **Crepidium ophrydis** (J. Koenig) M. A. Clem. & D. L. Jones

lindleyana Rchb. f. = **Oberonia brunoniana** Wight

longibracteata (Lindl.) Rchb. f. = **Oberonia longibracteata** Lindl.

mackinnonii (Duthie) Ames = **Crepidium mackinnonii** (Duthie) Szlach.

maximowicziana (King & Pantl.) Tang & F. T. Wang = **Crepidium maximowiczianum** (King & Pantl.) Szlach.

myriantha (Lindl.) Rchb. f. = **Oberonia acaulis** Griff.

nervosa (Thunb.) Sw. = **Liparis nervosa** (Thunb.) Lindl. var. **nervosa**

nilgiriensis T. Muthuk., A. Rajendran, Priyadh. & Sarval. = **Liparis atropurpurea** Lindl.

nutans (Roxb.) Willd. = **Eulophia picta** (R. Br.) Ormerod

obcordata (Lindl.) Rchb. f. = **Oberonia obcordata** Lindl.

odorata Willd. = **Liparis odorata** (Willd.) Lindl.

ophrydis (J. Koenig) Ormerod = **Crepidium ophrydis** (J. Koenig) M. A. Clem. & D. L. Jones

parryae Tang & F. T. Wang = **Crepidium parryae** (Tang & F. T. Wang) Marg.

platycaulon (Wight) Rchb. f. = **Oberonia platycaulon** Wight

pumilio (Rchb. f.) Rchb. f. = **Oberonia pumilio** Rchb. f.

purpurea (Lindl.) Kuntze = **Crepidium purpureum** (Lindl.) Szlach.

pyrulifera (Lindl.) Rchb. f. = **Oberonia pyrulifera** Lindl.

recurva (Lindl.) Rchb. f. = **Oberonia recurva** Lindl.

rufilabris (Lindl.) Rchb. f. = **Oberonia rufilabris** Lindl.

saprophyta (King & Pantl.) Tang & F. T. Wang = **Crepidium saprophytum** (King & Pantl.) A. N. Rao

sikkimensis (Lindl.) Rchb. f. = **Oberonia acaulis** Griff.

stocksii (Hook. f.) Kuntze = **Crepidium intermedium** (A. Rich.) Sushil K. Singh, Agrawala & Jalal

versicolor (Lindl.) Abeyw. = **Crepidium versicolor** (Lindl.) Sushil K. Singh, Agrawala & Jalal

verticillata (Wight) Rchb. f. = **Oberonia verticillata** Wight

verticillata (Wight) Rchb. f. var. *khasiana* (Lindl.) Rchb. f. = **Oberonia pyrulifera** Lindl.

verticillata (Wight) Rchb. f. var. *pubescens* (Lindl.) Rchb. f. = **Oberonia thwaitesii** Hook. f.

viridiflora Blume = **Liparis viridiflora** (Blume) Lindl.

wallichii (Lindl.) Deb = **Crepidium acuminatum** (D. Don) Szlach.

wightiana (Lindl.) Rchb. f. = **Oberonia wightiana** Lindl.

Malleola

andamanica N. P. Balakr. & N. Bhargava = **Robiquetia andamanica** (N. P. Balakr. & N. Bhargava) Kocyan & Schuit.

gracilis (Lindl.) Schltr. = **Robiquetia gracilis** (Lindl.) Garay

rosea (Lindl.) Schltr. = **Robiquetia rosea** (Lindl.) Garay

Mastigion

appendiculatum (Rolfe) Garay = **Bulbophyllum appendiculatum** (Rolfe) J. J. Sm.

ornatissimum (Rchb. f.) Garay, Hamer & Siegerist = **Bulbophyllum ornatissimum** (Rchb. f.) J. J. Sm.

rothschildianum (O'Brien) Lucksom = **Bulbophyllum rothschildianum** (O'Brien) J. J. Sm.

Maxillaria

goeringii Rchb. f. = **Cymbidium goeringii** (Rchb. f.) Rchb. f.

Medusorchis

andamanica (Hook. f.) Szlach. = **Habenaria andamanica** Hook. f.

Mesoclastes

brachystachys Lindl. = **Luisia brachystachys** (Lindl.) Blume

uniflora Lindl. = **Papilionanthe uniflora** (Lindl.) Garay

Mesodactylis

odorata (Blume) Endl. = **Apostasia odorata** Blume

wallichii (R. Br.) Endl. = **Apostasia wallichii** R. Br.

Microchilus

blumei (Lindl.) D. Dietr. = **Erythrodes blumei** (Lindl.) Schltr.

Micropera

maculata Dalzell = **Smithsonia maculata** (Dalzell) C. J. Saldanha

purpurea (Lindl.) Pradhan = **Micropera rostrata** (Roxb.) N. P. Balakr.

viridiflora Dalzell = **Smithsonia viridiflora** (Dalzell) C. J. Saldanha

Microstylis

andamanica King & Pantl. = **Crepidium andamanicum** (King & Pantl.) Marg. & Szlach.

aphylla King & Pantl. = **Crepidium aphyllum** (King & Pantl.) A. N. Rao

biaurita Lindl. = **Crepidium biauritum** (Lindl.) Szlach.

biloba Lindl. = **Crepidium acuminatum** (D. Don) Szlach.

calophylla Rchb. f. = **Crepidium calophyllum** (Rchb. f.) Szlach.

cardonii Prain = **Crepidium mackinnonii** (Duthie) Szlach.

congesta (Lindl.) Rchb. f. = **Crepidium ophrydis** (J. Koenig) M. A. Clem. & D. L. Jones

crenulata Ridl. = **Crepidium crenulatum** (Ridl.) Sushil K. Singh, Agrawala & Jalal

cylindrostachya (Lindl.) Rchb. f. = **Malaxis cylindrostachya** (Lindl.) Kuntze

densiflora (A. Rich.) Alston = **Crepidium densiflorum** (A. Rich.) Sushil K. Singh, Agrawala & Jalal

densiflora (A. Rich.) Kuntze var. *luteola* (Wight) P. K. Sarkar = **Crepidium densiflorum** (A. Rich.) Sushil K. Singh, Agrawala & Jalal

josephiana Rchb. f. = **Crepidium josephianum** (Rchb. f.) Marg.

khasiana Hook. f. = **Crepidium khasianum** (Hook. f.) Szlach.

latifolia (Sm.) J. J. Sm. = **Crepidium ophrydis** (J. Koenig) M. A. Clem. & D. L. Jones

luteola Wight = **Crepidium densiflorum** (A. Rich.) Sushil K. Singh, Agrawala & Jalal

mackinnonii Duthie = **Crepidium mackinnonii** (Duthie) Szlach.

maximowicziana King & Pantl. = **Crepidium maximowiczianum** (King & Pantl.) Szlach.

muscifera (Lindl.) Ridl. = **Malaxis muscifera** (Lindl.) Kuntze

purpurea Lindl. = **Crepidium purpureum** (Lindl.) Szlach.

saprophyta King & Pantl. = **Crepidium saprophytum** (King & Pantl.) A. N. Rao

scottii Hook. f. = **Crepidium calophyllum** (Rchb. f.) Szlach.

stocksii Hook. f. = **Crepidium intermedium** (A. Rich.) Sushil K. Singh, Agrawala & Jalal

versicolor Lindl. = **Crepidium versicolor** (Lindl.) Sushil K. Singh, Agrawala & Jalal

versicolor Lindl. var. *luteola* (Wight) Hook. f. = **Crepidium densiflorum** (A. Rich.) Sushil K. Singh, Agrawala & Jalal

wallichii Lindl. = **Crepidium acuminatum** (D. Don) Szlach.

wallichii Lindl. var. *biloba* (Lindl.) Hook. f. = **Crepidium acuminatum** (D. Don) Szlach.

wallichii Lindl. var. *biloba* King & Pantl. = **Crepidium purpureum** (Lindl.) Szlach.

wallichii Lindl. var. *brachycheila* Hook. f. = **Crepidium calophyllum** (Rchb. f.) Szlach.

Mischobulbum

grandiflorum (Hook. f.) Schltr. = **Tainia wrayana** (Hook. f.) J. J. Sm.

grandiflorum Rolfe = **Tainia megalantha** (Tang & F. T. Wang) Sushil K. Singh, Agrawala & Jalal

megalanthum Tang & F. T. Wang = **Tainia megalantha** (Tang & F. T. Wang) Sushil K. Singh, Agrawala & Jalal

wrayanum (Hook. f.) Rolfe = **Tainia wrayana** (Hook. f.) J. J. Sm.

Mitopetalum

angustifolium (Lindl.) Blume = **Ania angustifolia** Lindl.

bicorne (Lindl.) Blume = **Tainia bicornis** (Lindl.) Rchb. f.

latifolium (Lindl.) Blume = **Tainia latifolia** (Lindl.) Rchb. f.

Monochilus

affinis Lindl. = **Zeuxine affinis** (Lindl.) Benth. ex Hook. f.

clandestinus (Blume) Miq. = **Zeuxine clandestina** Blume

flabellatus (A. Rich.) Wight = **Cheirostylis flabellata** (A. Rich.) Wight

flavus Wall. ex Lindl. = **Zeuxine flava** (Wall. ex Lindl.) Trimen

goodyeroides (Lindl.) Lindl. = **Zeuxine goodyeroides** Lindl.

longilabris Lindl. = **Zeuxine longilabris** (Lindl.) Trimen

nervosus Wall. ex Lindl. = **Zeuxine nervosa** (Wall. ex Lindl.) Trimen

Monomeria

barbata Lindl. = **Bulbophyllum crabro** (C. S. P. Parish & Rchb. f.) J. J. Verm., Schuit. & de Vogel

crabro C. S. P. Parish & Rchb. f. = **Bulbophyllum crabro** (C. S. P. Parish & Rchb. f.) J. J. Verm., Schuit. & de Vogel

gymnopus (Hook. f.) Aver. = **Bulbophyllum gymnopus** Hook. f.

punctata (Lindl.) Schltr. = **Bulbophyllum kingii** Hook. f.

Monorchis

angustilabris (King & Pantl.) O. Schwarz = **Platanthera stenochila** X. H. Jin, Schuit., Raskoti & Lu Q. Huang

duthiei (Hook. f.) O. Schwarz = **Herminium josephi** Rchb. f.

fallax (Lindl.) O. Schwarz = **Herminium fallax** (Lindl.) Hook. f.

gracilis (King & Pantl.) O. Schwarz = **Herminium gracile** King & Pantl.

herminium O. Schwarz = **Herminium monorchis** (L.) R. Br.

jaffreyana (King & Pantl.) O. Schwarz = **Herminium jaffreyanum** King & Pantl.

mackinonii (Duthie) O. Schwarz = **Herminium mackinnonii** Duthie

monophylla (D. Don) O. Schwarz = **Herminium monophyllum** (D. Don) P. F. Hunt & Summerh.

orbicularis (Hook. f.) O. Schwarz = **Platanthera orbicularis** (Hook. f.) X. H. Jin, Schuit. & Raskoti

pugioniformis (Lindl. ex Hook. f.) O. Schwarz = **Herminium pugioniforme** Lindl. ex Hook. f.

quinqueloba (King & Pantl.) O. Schwarz = **Herminium quinquelobum** King & Pantl.

Mycaranthes

merguensis (Lindl.) Rauschert = **Pinalia merguensis** (Lindl.) Kuntze *paniculata* (Lindl.) Schuit., Y. P. Ng & H. A. Pedersen = **Mycaranthes floribunda** (D. Don) S. C. Chen & J. J. Wood

pannea (Lindl.) S. C. Chen & J. J. Wood = **Strongyleria pannea** (Lindl.) Schuit., Y. P. Ng & H. A. Pedersen

stricta (Lindl.) Lindl. = **Cryptochilus strictus** (Lindl.) Schuit., Y. P. Ng & H. A. Pedersen

Myrmechis
franchetiana (King & Pantl.) Schltr. = **Myrmechis pumila** (Hook. f.) Tang & F. T. Wang

Neogyna
gardneriana (Lindl.) Rchb. f. = **Coelogyne gardneriana** Lindl.

Neotainiopsis
barbata (Lindl.) Bennett & Raizada = **Eriodes barbata** (Lindl.) Rolfe

Neottia
australis R. Br. = **Spiranthes australis** (R. Br.) Lindl.

corallorhiza (L.) Kuntze = **Corallorhiza trifida** Châtel.

flexuosa Sm. = **Spiranthes flexuosa** (Sm.) Lindl.

kashmiriana (Duthie) Schltr. = **Neottia inayatii** (Duthie) Schltr.

lindleyana Decne. = **Neottia listeroides** Lindl.

macrophylla D. Don = **Herminium macrophyllum** (D. Don) Dandy

monophylla D. Don = **Herminium monophyllum** (D. Don) P. F. Hunt & Summerh.

parviflora (King & Pantl.) Schltr. = **Neottia acuminata** Schltr.

procera Ker Gawl. = **Goodyera procera** (Ker Gawl.) Hook.

reniformis (D. Don) Spreng. = **Habenaria reniformis** (D. Don) Hook. f.

repens (L.) Sw. = **Goodyera repens** (L.) R. Br.

rubicunda Blume = **Goodyera rubicunda** (Blume) Lindl.

sinensis Pers. = **Spiranthes sinensis** (Pers.) Ames

strateumatica (L.) R. Br. = **Zeuxine strateumatica** (L.) Schltr.

viridiflora Blume = **Goodyera viridiflora** (Blume) Lindl. ex D. Dietr.

Neottianthe
calcicola (W. W. Sm.) Schltr. = **Ponerorchis cucullata** (L.) X.H.Jin, Schuit. & W.T.Jin var. **calcicola** (W. W. Sm.) X. H. Jin, Schuit. & W. T. Jin

cucullata (L.) Schltr. var. *calcicola* (W. W. Sm.) Soó = **Ponerorchis cucullata** (L.) X.H.Jin, Schuit. & W.T.Jin var. **calcicola** (W. W. Sm.) X. H. Jin, Schuit. & W. T. Jin

secundiflora (Kraenzl.) Schltr. = **Ponerorchis secundiflora** (Kraenzl.) X. H. Jin, Schuit. & W. T. Jin

Nephelaphyllum
chinense Rolfe = **Collabium chinense** (Rolfe) Tang & F. T. Wang

grandiflorum Hook. f. = **Tainia wrayana** (Hook. f.) J. J. Sm.

sikkimensis (Hook. f.) Karthik. = **Nephelaphyllum pulchrum** Blume var. **sikkimensis** Hook. f.

Nervilia
aragoana Gaudich. = **Nervilia concolor** (Blume) Schltr.

biflora (Wight) Schltr. = **Nervilia plicata** (Andrews) Schltr.

carinata (Roxb.) Schltr. = **Nervilia concolor** (Blume) Schltr.

crispata (Blume) Schltr. ex K.Schum. & Lauterb. = **Nervilia simplex** (Thouars) Schltr.

crociformis (Zoll. & Moritzi) Seidenf. = **Nervilia simplex** (Thouars) Schltr.

discolor (Blume) Schltr. = **Nervilia plicata** (Andrews) Schltr.

hallbergii Blatt. & McCann = **Nervilia infundibulifolia** Blatt. & McCann

macroglossa (Hook. f.) Schltr. var. *mackinnonii* (Duthie) Pradhan = **Nervilia mackinnonii** (Duthie) Schltr.

monantha Blatt. & McCann = **Nervilia simplex** (Thouars) Schltr.

prainiana (King & Pantl.) Seidenf. & Smitinand = **Nervilia simplex** (Thouars) Schltr.

scottii (Rchb. f.) Schltr. = **Nervilia concolor** (Blume) Schltr.

Nidus
listeroides (Lindl.) Kuntze = **Neottia listeroides** Lindl.

Nujiangia
griffithii (Hook. f.) X. H. Jin & D. Z. Li = **Gennaria griffithii** (Hook. f.) X. H. Jin & D. Z. Li

Oberonia
acaulis Griff. var. *latipetala* Chowlu = **Oberonia acaulis** Griff.

arnottiana Wight = **Oberonia wightiana** Lindl.

arunachalensis A. N. Rao = **Oberonia brachyphylla** Blatt. & McCann

auriculata King & Pantl. = **Oberonia caulescens** Lindl.

bellii Blatt. & McCann = **Oberonia verticillata** Wight

bisaccata Manilal & C. S. Kumar = **Oberonia platycaulon** Wight

croftiana King & Pantl. = **Oberonia recurva** Lindl.

demissa Lindl. = **Oberonia brachystachys** Lindl.

denticulata Wight = **Oberonia mucronata** (D. Don) Ormerod & Seidenf.

denticulata Wight var. *angustifolia* (Lindl.) S. Misra = **Oberonia angustifolia** Lindl.

falcata King & Pantl. = **Oberonia anthropophora** Lindl.

gammiei King & Pantl. = **Oberonia mucronata** (D. Don) Ormerod & Seidenf.

iridifolia Lindl. = **Oberonia ensiformis** (Sm.) Lindl.

iridifolia Lindl. var. *angustifolia* (Lindl.) Hook. f. = **Oberonia angustifolia** Lindl.

iridifolia Lindl. var. *brevifolia* Hook. f. = **Oberonia mucronata** (D. Don) Ormerod & Seidenf.

iridifolia Lindl. var. *denticulata* (Wight) Hook. f. = **Oberonia mucronata** (D. Don) Ormerod & Seidenf.

kamlangensis A. N. Rao = **Oberonia brachyphylla** Blatt. & McCann

katakiana A. N. Rao = **Oberonia caulescens** Lindl.

lindleyana Wight = **Oberonia brunoniana** Wight

lingmalensis Blatt. & McCann = **Oberonia recurva** Lindl.

lobulata King & Pantl. = **Oberonia mucronata** (D. Don) Ormerod & Seidenf.

longifolia M. Kumar & Sequiera = **Oberonia agastyamalayana** C. S. Kumar

repens (L.) Kuntze = **Goodyera repens** (L.) R. Br.

rubicundum (Blume) Kuntze = **Goodyera rubicunda** (Blume) Lindl.

schlechtendalianum (Rchb. f.) Kuntze = **Goodyera schlechtendaliana** Rchb. f.

secundiflorum (Griff.) Kuntze = **Goodyera foliosa** (Lindl.) Benth. ex C. B. Clarke

viridiflorum (Blume) Kuntze = **Goodyera viridiflora** (Blume) Lindl. ex D. Dietr.

vittatum (Lindl.) Kuntze = **Goodyera vittata** (Lindl.) Benth. ex Hook. f.

Orchis

aphylla F. W. Schmidt = **Epipogium aphyllum** Sw.

chrysea (W. W. Sm.) Schltr. = **Hsenhsua chrysea** (W. W. Sm.) X. H. Jin, Schuit., W. T. Jin & L. Q. Huang

chusua D. Don = **Ponerorchis chusua** (D. Don) Soó

chusua var. *nana* King & Pantl. = **Ponerorchis nana** (King & Pantl.) Soó

commelinifolia Roxb. = **Habenaria commelinifolia** (Roxb.) Wall. ex Lindl.

cylindrostachya (Lindl.) Kraenzl. = **Gymnadenia orchidis** Lindl. var. **orchidis**

dentata Sw. = **Habenaria dentata** (Sw.) Schltr.

gigantea Sm. = **Pecteilis gigantea** (Sm.) Raf.

graggeriana Soó = **Dactylorhiza hatagirea** (D. Don) Soó

habenarioides King & Pantl. = **Gymnadenia orchidis** Lindl. var. **orchidis**

hatagirea D. Don = **Dactylorhiza hatagirea** (D. Don) Soó

japonica Thunb. = **Platanthera japonica** (Thunb.) Lindl.

monorchis (L.) Crantz = **Herminium monorchis** (L.) R. Br.

nana (King & Pantl.) Schltr. = **Ponerorchis nana** (King & Pantl.) Soó

obcordata Buch.- Ham. ex D. Don = **Brachycorythis obcordata** (Lindl.) Summerh.

pectinata Sm. = **Habenaria pectinata** D. Don

plantaginea Roxb. = **Habenaria roxburghii** Nicolson

platyphyllos Sw. ex Willd. = **Habenaria roxburghii** Nicolson

puberula King & Pantl. = **Ponerorchis puberula** (King & Pantl.) Verm.

repens (L.) Eyster ex Poir. = **Goodyera repens** (L.) R. Br.

roborovskii Maxim. = **Galearis roborovskii** (Maxim.) S. C. Chen, P. J. Cribb & S. W. Gale

roxburghii Pers. = **Habenaria roxburghii** Nicolson

spathulata (Lindl.) Rchb. f. ex Hook. f. = **Galearis spathulata** (Lindl.) P. F. Hunt

stracheyi Hook. f. = **Galearis roborovskii** (Maxim.) S. C. Chen, P. J. Cribb & S. W. Gale

stracheyi Hook. f. = **Ponerorchis chusua** (D. Don) Soó

strateumatica L. = **Zeuxine strateumatica** (L.) Schltr.

susannae L. = **Pecteilis susannae** (L.) Raf.

triplicata P. Willemet = **Calanthe triplicata** (P. Willemet) Ames

tschiliensis (Schltr.) Soó = **Galearis tschiliensis** (Schltr.) P. J. Cribb, S. W. Gale & R. M. Bateman

umbrosa Kar. & Kir. = **Dactylorhiza umbrosa** (Kar. & Kir.) Nevski

uniflora Roxb. = **Diplomeris pulchella** D. Don

viridiflora Sw. = **Habenaria viridiflora** (Sw.) R. Br. ex Spreng.

viridis (L.) Crantz = **Dactylorhiza viridis** (L.) R. M. Bateman, Pridgeon & M. W. Chase

Oreorchis

foliosa (Lindl.) Lindl. var. *indica* (Lindl.) N. Pearce & P. J. Cribb = **Oreorchis indica** (Lindl.) Hook. f.

rolfei Duthie = **Oreorchis micrantha** Lindl.

Ormerodia

linearilobata (Seidenf. & Smitinand) Szlach. = **Cleisostoma linearilobatum** (Seidenf. & Smitinand) Garay

sagittata (King & Pantl.) Szlach. = **Cleisostoma linearilobatum** (Seidenf. & Smitinand) Garay

Ormostema

albiflora Raf. = **Dendrobium moniliforme** (L.) Sw.

Ornithochilus

cacharensis Barbhuiya, B. K. Dutta & Schuit. = **Phalaenopsis cacharensis** (Barbhuiya, B. K. Dutta & Schuit.) Kocyan & Schuit.

difformis (Wall. ex Lindl.) Schltr. = **Phalaenopsis difformis** (Wall. ex Lindl.) Kocyan & Schuit.

fuscus Wall. ex Lindl. = **Phalaenopsis difformis** (Wall. ex Lindl.) Kocyan & Schuit.

yingjiangensis Z. H. Tsi = **Phalaenopsis yingjiangensis** (Z. H. Tsi) Kocyan & Schuit.

Ortmannia

cernua (Willd.) Opiz = **Eulophia picta** (R. Br.) Ormerod

Otandra

cernua (Willd.) Salisb. = **Eulophia picta** (R. Br.) Ormerod

Otochilus

albus Lindl. = **Coelogyne alba** (Lindl.) Rchb. f.

albus Lindl. var. *lancilabius* (Seidenf.) Pradhan = **Coelogyne lancilabia** (Seidenf.) R. Rice

fuscus Lindl. = **Coelogyne fusca** (Lindl.) Rchb. f.

lancifolius Griff. = **Coelogyne fusca** (Lindl.) Rchb. f.

lancilabius Seidenf. = **Coelogyne lancilabia** (Seidenf.) R. Rice

latifolius Griff. = **Coelogyne porrecta** (Lindl.) Rchb. f.

porrectus Lindl. = **Coelogyne porrecta** (Lindl.) Rchb. f.

Pachychilus

pubescens (Blume) Blume = **Pachystoma pubescens** Blume

senilis (Lindl.) Blume = **Pachystoma pubescens** Blume

Pachyrhizanthe

macrorhizos (Lindl.) Nakai = **Cymbidium macrorhizon** Lindl.

Pachystoma

malabaricum Rchb. f. = **Ipsea malabarica** (Rchb. f.) Hook. f.

senile (Lindl.) Rchb. f. = **Pachystoma pubescens** Blume

Panisea

apiculata Lindl. = **Coelogyne apiculata** (Lindl.) Rchb. f.

demissa (D. Don) Pfitzer = **Coelogyne demissa** (D. Don) M. W. Chase & Schuit.

panchaseensis Subedi = **Coelogyne panchaseensis** (Subedi) M. W. Chase & Schuit.

pantlingii (Pfitzer) Schltr. = **Coelogyne tricallosa** (Rolfe) M. W. Chase & Schuit.

parviflora (Lindl.) Lindl. = **Coelogyne demissa** (D. Don) M. W. Chase & Schuit.

tricallosa Rolfe = **Coelogyne tricallosa** (Rolfe) M. W. Chase & Schuit.

uniflora (Lindl.) Lindl. = **Coelogyne uniflora** Lindl.

Pantlingia

paradoxa Prain = **Stigmatodactylus paradoxus** (Prain) Schltr.

serrata Deori = **Stigmatodactylus serratus** (Deori) A. N. Rao

Paorchis

thailandica (Seidenf.) M. C. Pace = **Goodyera thailandica** Seidenf.

Paphiopedilum

× *spicerovenustum* Pradhan = **Paphiopedilum × polystigmaticum** (Rchb.f.) Stein

× *venustoinsigne* Pradhan = **Paphiopedilum × crossianum** (Rchb.f.) Stein

venustum (Wall. ex Sims) Pfitzer var. *rubrum* Pradhan = **Paphiopedilum venustum** (Wall. ex Sims) Pfitzer

venustum (Wall. ex Sims) Pfitzer var. *teestaensis* Pradhan = **Paphiopedilum venustum** (Wall. ex Sims) Pfitzer

Paracalanthe

gracilis (Lindl.) Kudô = **Calanthe obcordata** (Lindl.) M. W. Chase, Christenh. & Schuit.

reflexa (Maxim.) Kudô var. *puberula* (Lindl.) Kudô = **Calanthe puberula** Lindl.

tricarinata (Lindl.) Kudô = **Calanthe tricarinata** Lindl.

Paragnathis

pulchella (D. Don) Spreng. = **Diplomeris pulchella** D. Don

Paraphaius

flavus (Blume) J.W.Zhai, Z.J.Liu & F.W.Xing = **Calanthe woodfordii** (Hook.) M. W. Chase, Christenh. & Schuit.

Pecteilis

acuifera (Wall. ex Lindl.) M. A. Clem. & D. L. Jones = **Habenaria acuifera** Wall. ex Lindl.

candida Schltr. = **Pecteilis triflora** (D. Don) Tang & F. T. Wang

cephalotes (Lindl.) M. A. Clem. & D. L. Jones = **Habenaria cephalotes** Lindl.

commelinifolia (Roxb.) M. A. Clem. & D. L. Jones = **Habenaria commelinifolia** (Roxb.) Wall. ex Lindl.

crinifera (Lindl.) M. A. Clem. & D. L. Jones = **Habenaria crinifera** Lindl.

dentata (Sw.) M. A. Clem. & D. L. Jones = **Habenaria dentata** (Sw.) Schltr.

elliptica (Wight) M. A. Clem. & D. L. Jones = **Habenaria elliptica** Wight

furcifera (Lindl.) M. A. Clem. & D. L. Jones = **Habenaria furcifera** Lindl.

longicorniculata (J. Graham) M. A. Clem. & D. L. Jones = **Habenaria longicorniculata** J. Graham

longifolia (Buch.- Ham. ex Lindl.) M. A. Clem. & D. L. Jones = **Habenaria longifolia** Buch.- Ham. ex Lindl.

malintana (Blanco) M. A. Clem. & D. L. Jones = **Habenaria malintana** (Blanco) Merr.

marginata (Colebr.) M. A. Clem. & D. L. Jones = **Habenaria marginata** Colebr.

plantaginea (Lindl.) M. A. Clem. & D. L. Jones = **Habenaria plantaginea** Lindl.

rhodocheila (Hance) M. A. Clem. & D. L. Jones = **Habenaria rhodocheila** Hance

roxburghii (Nicolson) M. A. Clem. & D. L. Jones = **Habenaria roxburghii** Nicolson

suaveolens (Dalzell) M. A. Clem. & D. L. Jones = **Habenaria suaveolens** Dalzell

susannae (L.) Raf. subsp. *henryi* (Schltr.) Soó = **Pecteilis henryi** Schltr.

trichosantha (Wall. ex Lindl.) M. A. Clem. & D. L. Jones = **Habenaria trichosantha** Wall. ex Lindl.

Pedilonum

secundum Blume = **Dendrobium secundum** (Blume) Lindl.

Pendulorchis

himalaica (Deb, Sengupta & Malick) Z. J. Liu, K. Wei Liu & X. J. Xiao = **Holcoglossum himalaicum** (Deb, Sengupta & Malick) Aver.

Penkimia

nagalandensis Phukan & Odyuo = **Holcoglossum nagalandense** (Phukan & Odyuo) X. H. Jin

Pennilabium

pumilio (Rchb. f.) Pradhan = **Saccolabiopsis pusilla** (Lindl.) Seidenf. & Garay

Peramium

procerum (Ker Gawl.) Makino = **Goodyera procera** (Ker Gawl.) Hook.

repens (L.) Salisb. = **Goodyera repens** (L.) R. Br.

schlechtendalianum (Rchb. f.) Makino = **Goodyera schlechtendaliana** Rchb. f.

Peristylus

albomarginatus (King & Pantl.) K. Y. Lang = **Herminium albomarginatum** (King & Pantl.) X. H. Jin, Schuit., Raskoti & L. Q. Huang

duthiei (Hook. f.) Deva & H. B. Naithani = **Herminium josephi** Rchb. f.

duthiei (Hook. f.) Deva & H. B. Naithani var. *inayatii* S. Deva & H. B. Naithani = **Herminium macrophyllum** (D. Don) Dandy

elatus Dalzell = **Peristylus plantagineus** (Lindl.) Lindl.

elisabethae (Duthie) R. K. Gupta = **Herminium elisabethae** (Duthie) Tang & F. T. Wang

fallax Lindl. = **Herminium fallax** (Lindl.) Hook. f.

fallax Lindl. var. *dwarikae* Deva & H. B. Naithani = **Herminium fallax** (Lindl.) Hook. f.

goodyeroides Lindl. var. *affinis* (D. Don) T. Cooke = **Peristylus affinis** (D. Don) Seidenf.

gracillimus (Hook. f.) Kraenzl. = **Herminium mannii** (Rchb. f.) Tang & F. T. Wang

kumaonensis Renz = **Herminium elisabethae** (Duthie) Tang & F. T. Wang

lancifolius A. Rich. = **Peristylus secundus** (Lindl.) Rathakr.

macrophyllus (D. Don) Lawkush, Vik. Kumar & Bankoti = **Herminium macrophyllum** (D. Don) Dandy

mannii (Rchb. f.) Mukerjee = **Herminium mannii** (Rchb. f.) Tang & F. T. Wang

neglectus (King & Pantl.) Kraenzl. = **Peristylus densus** (Lindl.) Santapau & Kapadia

nematocaulon (Hook. f.) Banerji & Prabha Pradhan = **Platanthera nematocaulon** (Hook. f.) Kraenzl.

orbicularis (Hook. f.) Agrawala, H. J. Chowdhery &
S. Choudhury = **Platanthera orbicularis** (Hook. f.)
X. H. Jin, Schuit. & Raskoti

orchidis (Lindl.) Kraenzl. = **Gymnadenia orchidis** Lindl.
var. **orchidis**

peristyloides M. R. Almeida = **Peristylus densus** (Lindl.)
Santapau & Kapadia

pseudophrys (King & Pantl.) Pradhan = **Peristylus
pseudophrys** (King & Pantl.) Kraenzl.

robustior Wight = **Peristylus secundus** (Lindl.) Rathakr.

secundiflorus Kraenzl. = **Ponerorchis secundiflora**
(Kraenzl.) X. H. Jin, Schuit. & W. T. Jin

stenostachyus (Lindl.) Kraenzl. = **Peristylus densus**
(Lindl.) Santapau & Kapadia

stocksii (Hook. f.) Kraenzl. = **Peristylus caranjensis**
(Dalzell) Ormerod & C. S. Kumar

superanthus J. J. Wood = **Platanthera superantha** (J. J.
Wood) X. H. Jin, Schuit., Raskoti & Lu Q. Huang

viridis (L.) Lindl. = **Dactylorhiza viridis** (L.) R. M.
Bateman, Pridgeon & M. W. Chase

xanthochlorus Blatt. & McCann = **Peristylus densus**
(Lindl.) Santapau & Kapadia

Phaius

albus Lindl. = **Thunia alba** (Lindl.) Rchb. f. var. **alba**

albus Lindl. var. *bensoniae* (Hook. f.) Hook. f. = **Thunia
bensoniae** Hook. f.

bensoniae (Hook. f.) Benth. = **Thunia bensoniae** Hook. f.

blumei Lindl. var. *pulchra* King & Pantl. = **Calanthe
tankervilleae** (Banks) M. W. Chase, Christenh. &
Schuit.

epiphyticus Seidenf. = **Calanthe densiflora** Lindl.

flavus (Blume) Lindl. = **Calanthe woodfordii** (Hook.)
M. W. Chase, Christenh. & Schuit.

gracilis (Lindl.) S. S. Ying = **Calanthe obcordata**
(Lindl.) M. W. Chase, Christenh. & Schuit.

incarvillei (Pers.) Kuntze = **Calanthe tankervilleae**
(Banks) M. W. Chase, Christenh. & Schuit.

longipes (Hook. f.) Holttum = **Calanthe longipes** Hook. f.

luridus Thwaites = **Calanthe testacea** M. W. Chase,
Christenh. & Schuit.

marshallianus (Rchb. f.) N. E. Br. = **Thunia alba** (Lindl.)
Rchb. f. var. **alba**

mishmensis (Lindl. & Paxton) Rchb. f. = **Calanthe
mishmensis** (Lindl. & Paxton) M. W. Chase,
Christenh. & Schuit.

nanus Hook. f. = **Calanthe nana** (Hook. f.) M. W. Chase,
Christenh. & Schuit.

roeblingii O'Brien = **Calanthe wallichii** (Lindl.) M. W.
Chase, Christenh. & Schuit.

tankervilleae (Banks) Blume = **Calanthe tankervilleae**
(Banks) M. W. Chase, Christenh. & Schuit.

tankervilleae (Banks) Blume var. *pulchra* (King & Pantl.)
Karthik. = **Calanthe tankervilleae** (Banks) M. W.
Chase, Christenh. & Schuit.

veratrifolius Lindl. = **Calanthe tankervilleae** (Banks)
M. W. Chase, Christenh. & Schuit.

vestitus (Wall. ex Lindl.) Rchb. f. = **Calanthe vestita**
Wall. ex Lindl.

wallichii Lindl. = **Calanthe wallichii** (Lindl.) M. W.
Chase, Christenh. & Schuit.

woodfordii (Hook.) Merr. = **Calanthe woodfordii** (Hook.)
M. W. Chase, Christenh. & Schuit.

Phalaenopsis

braceana (Hook. f.) Christenson = **Phalaenopsis
taenialis** (Lindl.) Christenson & Pradhan

decumbens (Griff.) Holttum = **Phalaenopsis parishii**
Rchb. f.

decumbens (Griff.) Holttum var. *lobbii* (Rchb. f.) P. F. Hunt
= **Phalaenopsis lobbii** (Rchb. f.) H. R. Sweet

deliciosa Rchb. f. subsp. *hookeriana* (O. Gruss & Roellke)
Christenson = **Phalaenopsis deliciosa** Rchb. f.

hygrochila J. M. H. Shaw = **Phalaenopsis marriottiana**
(Rchb. f.) Kocyan & Schuit. var. **parishii** (Rchb. f.)
Kocyan & Schuit.

mastersii King & Pantl. = **Phalaenopsis pulcherrima**
(Lindl.) J. J. Sm.

parishii Rchb. f. var. *lobbii* Rchb. f. = **Phalaenopsis
lobbii** (Rchb. f.) H. R. Sweet

speciosa Rchb. f. = **Phalaenopsis tetraspis** Rchb. f.

speciosa Rchb. f. var. *tetraspis* (Rchb. f.) H. R. Sweet =
Phalaenopsis tetraspis Rchb. f.

tigrina Ming H. Li = **Phalaenopsis marriottiana** (Rchb.
f.) Kocyan & Schuit. var. **parishii** (Rchb. f.) Kocyan &
Schuit.

wightii Rchb. f. = **Phalaenopsis deliciosa** Rchb. f.

Pholidota

articulata Lindl. = **Coelogyne articulata** (Lindl.)
Rchb. f.

articulata Lindl. var. *griffithii* (Hook. f.) King & Pantl. =
Coelogyne articulata (Lindl.) Rchb. f.

assamica Regel = **Coelogyne imbricata** (Hook.) Rchb. f.

bracteata (D. Don) Seidenf. = **Coelogyne imbricata**
(Hook.) Rchb. f.

calceata Rchb. f. = **Coelogyne pallida** (Lindl.) Rchb. f.

convallariae (C. S. P. Parish & Rchb. f.) Hook. f. =
Coelogyne convallariae C. S. P. Parish & Rchb. f.

convallariae (C. S. P. Parish & Rchb. f.) Hook. f. var.
breviscapa Deori & J. Joseph = **Coelogyne convallariae** C. S. P. Parish & Rchb. f.

griffithii Hook. f. = **Coelogyne articulata** (Lindl.) Rchb. f.

imbricata Hook. = **Coelogyne imbricata** (Hook.) Rchb. f.

imbricata Hook. var. *sessilis* Hook. f. = **Coelogyne
pallida** (Lindl.) Rchb. f.

katakiana Phukan = **Coelogyne convallariae** C. S. P.
Parish & Rchb. f.

khasyana Rchb. f. = **Coelogyne articulata** (Lindl.)
Rchb. f.

missionariorum Gagnep. = **Coelogyne missionariorum**
(Gagnep.) R. Rice

obovata Hook. f. = **Coelogyne articulata** (Lindl.) Rchb. f.

pallida Lindl. Edwards's Bot. Reg. 21: t. 1777. 1835. =
Coelogyne pallida (Lindl.) Rchb. f.

pallida Lindl. var. *sessilis* (Hook. f.) P. K. Sarkar =
Coelogyne pallida (Lindl.) Rchb. f.

protracta Hook. f. = **Coelogyne protracta** (Hook. f.)
R. Rice

pygmaea H. J. Chowdhery & G. D. Pal = **Coelogyne
imbricata** (Hook.) Rchb. f.

recurva Lindl. = **Coelogyne recurva** (Lindl.) Rchb. f.

rubra Lindl. = **Coelogyne rubra** (Lindl.) Rchb. f.

suaveolens Lindl. = **Coelogyne suaveolens** (Lindl.) Hook. f.

undulata Wall. ex Lindl. = **Coelogyne rubra** (Lindl.) Rchb. f.

wattii King & Pantl. = **Coelogyne wattii** (King & Pantl.) M. W. Chase & Schuit.

Phreatia

secunda (Blume) Lindl. = **Phreatia plantaginifolia** (J. Koenig) Ormerod

uniflora Wight = **Porpax pusilla** (Griff.) Schuit., Y. P. Ng & H. A. Pedersen

Phyllomphax

acuta (Rchb. f.) Schltr. = **Brachycorythis acuta** (Rchb. f.) Summerh.

affinis (D. Don) Schltr. = **Peristylus affinis** (D. Don) Seidenf.

galeandra (Rchb. f.) Schltr. = **Brachycorythis galeandra** (Rchb. f.) Summerh.

helferi (Rchb. f.) Schltr. = **Brachycorythis helferi** (Rchb. f.) Summerh.

obcordata (Lindl.) Schltr. = **Brachycorythis obcordata** (Lindl.) Summerh.

obovalis (Summerh.) Szlach. = **Brachycorythis acuta** (Rchb. f.) Summerh.

splendida (Summerh.) Szlach. = **Brachycorythis splendida** Summerh.

wightii (Summerh.) Szlach. = **Brachycorythis wightii** Summerh.

Phyllorkis

acutiflora (A. Rich.) Kuntze = **Bulbophyllum acutiflorum** A. Rich.

affinis (Wall. ex Lindl.) Kuntze = **Bulbophyllum affine** Lindl.

albida (Wight) Kuntze = **Bulbophyllum acutiflorum** A. Rich.

andersonii (Hook. f.) Kuntze = **Bulbophyllum andersonii** (Hook. f.) J. J. Sm.

apoda (Hook. f.) Kuntze = **Bulbophyllum apodum** Hook. f.

aurea (Hook. f.) Kuntze = **Bulbophyllum aureum** (Hook. f.) J. J. Sm.

biseta (Lindl.) Kuntze = **Bulbophyllum bisetum** Lindl.

blepharistes (Rchb. f.) Kuntze = **Bulbophyllum blepharistes** Rchb. f.

bootanensis (Griff.) Kuntze = **Bulbophyllum umbellatum** Lindl.

brevipes (Hook. f.) Kuntze = **Bulbophyllum yoksunense** J. J. Sm.

candida (Lindl.) Kuntze = **Bulbophyllum candidum** (Lindl.) Hook. f.

capillipes (C. S. P. Parish & Rchb. f.) Kuntze = **Bulbophyllum capillipes** C. S. P. Parish & Rchb. f.

caudata (Lindl.) Kuntze = **Bulbophyllum caudatum** Lindl.

cirrhata (Hook. f.) Kuntze = **Bulbophyllum paleaceum** (Lindl.) Benth. ex Hemsl.

conchifera (Rchb. f.) Kuntze = **Bulbophyllum khasyanum** Griff.

conferta (Hook. f.) Kuntze = **Bulbophyllum scabratum** Rchb. f.

crassipes (Hook. f.) Kuntze = **Bulbophyllum crassipes** Hook. f.

cuprea (Lindl.) Kuntze = **Bulbophyllum cupreum** Lindl.

cylindracea (Lindl.) Kuntze = **Bulbophyllum cylindraceum** Lindl.

elata (Hook. f.) Kuntze = **Bulbophyllum elatum** (Hook. f.) J. J. Sm.

eublephara (Rchb. f.) Kuntze = **Bulbophyllum eublepharum** Rchb. f.

fimbriata (Lindl.) Kuntze = **Bulbophyllum fimbriatum** (Lindl.) Rchb. f.

fuscopurpurea (Wight) Kuntze = **Bulbophyllum fuscopurpureum** Wight

gamblei (Hook. f.) Kuntze = **Bulbophyllum fischeri** Seidenf.

guttulata (Hook. f.) Kuntze = **Bulbophyllum guttulatum** (Hook. f.) N. P. Balakr.

gymnopus (Hook. f.) Kuntze = **Bulbophyllum gymnopus** Hook. f.

helenae Kuntze = **Bulbophyllum helenae** (Kuntze) J. J. Sm.

hirta (Sm.) Kuntze = **Bulbophyllum hirtum** (Sm.) Lindl.

hymenantha (Hook. f.) Kuntze = **Bulbophyllum hymenanthum** Hook. f.

iners (Rchb. f.) Kuntze = **Bulbophyllum iners** Rchb. f.

kaitiensis (Rchb. f.) Kuntze = **Bulbophyllum kaitiense** Rchb. f.

kingii (Hook. f.) Kuntze = **Bulbophyllum kingii** Hook. f.

leopardina (Wall.) Kuntze = **Bulbophyllum leopardinum** (Wall.) Lindl. ex Wall.

leptantha (Hook. f.) Kuntze = **Bulbophyllum leptanthum** Hook. f.

lobbii (Lindl.) Kuntze = **Bulbophyllum lobbii** Lindl.

longiscapa (Teijsm. & Binn.) Kuntze = **Bulbophyllum blepharistes** Rchb. f.

macraei (Lindl.) Kuntze = **Bulbophyllum macraei** (Lindl.) Rchb. f.

macrantha (Lindl.) Kuntze = **Bulbophyllum macranthum** Lindl.

maculosa (Lindl.) Kuntze = **Bulbophyllum umbellatum** Lindl.

mishmeensis (Hook. f.) Kuntze = **Bulbophyllum paleaceum** (Lindl.) Benth. ex Hemsl.

monantha Kuntze = **Bulbophyllum pteroglossum** Schltr.

moniliformis (C. S. P. Parish & Rchb. f.) Kuntze = **Bulbophyllum moniliforme** C. S. P. Parish & Rchb. f.

nasuta (Rchb. f.) Kuntze = **Bulbophyllum nasutum** Rchb. f.

neilgherrensis (Wight) Kuntze = **Bulbophyllum sterile** (Lam.) Suresh

odoratissima (Sm.) Kuntze = **Bulbophyllum odoratissimum** (Sm.) Lindl.

ornatissima (Rchb. f.) Kuntze = **Bulbophyllum ornatissimum** (Rchb. f.) J. J. Sm.

parviflora (C. S. P. Parish & Rchb. f.) Kuntze = **Bulbophyllum parviflorum** C. S. P. Parish & Rchb. f.

penicillium (C. S. P. Parish & Rchb. f.) Kuntze = **Bulbophyllum penicillium** C. S. P. Parish & Rchb. f.

picturata (Lodd.) Kuntze = **Bulbophyllum picturatum** (Lodd.) Rchb. f.

polyrhiza (Lindl.) Kuntze = **Bulbophyllum polyrhizum** Lindl.

protracta (Hook. f.) Kuntze = **Bulbophyllum protractum** Hook. f.

psychoon (Rchb. f.) Kuntze = **Bulbophyllum scabratum** Rchb. f.

repens (Griff.) Kuntze = **Bulbophyllum repens** Griff.

reptans (Lindl.) Kuntze = **Bulbophyllum reptans** (Lindl.) Lindl.

retusiuscula (Rchb. f.) Kuntze = **Bulbophyllum retusiusculum** Rchb. f.

rolfei Kuntze = **Bulbophyllum rolfei** (Kuntze) Seidenf.

roxburghii (Lindl.) Kuntze = **Bulbophyllum roxburghii** (Lindl.) Rchb. f.

rufina (Rchb. f.) Kuntze = **Bulbophyllum rufinum** Rchb. f.

schmidtiana (Rchb. f.) Kuntze = **Bulbophyllum leopardinum** (Wall.) Lindl. ex Wall.

secunda (Hook. f.) Kuntze = **Bulbophyllum secundum** Hook. f.

sessilis Kuntze = **Bulbophyllum clandestinum** Lindl.

stenobulbon (C. S. P. Parish & Rchb. f.) Kuntze = **Bulbophyllum stenobulbon** C. S. P. Parish & Rchb. f.

striata (Griff.) Kuntze = **Bulbophyllum striatum** (Griff.) Rchb. f.

tenuifolia (Blume) Kuntze = **Bulbophyllum tenuifolium** (Blume) Lindl.

thomsonii (Hook. f.) Kuntze = **Bulbophyllum parviflorum** C. S. P. Parish & Rchb. f.

tortuosa (Blume) Kuntze = **Bulbophyllum tortuosum** (Blume) Lindl.

tremula (Wight) Kuntze = **Bulbophyllum tremulum** Wight

umbellata (Lindl.) Kuntze = **Bulbophyllum umbellatum** Lindl.

virens (Lindl.) Kuntze = **Bulbophyllum paleaceum** (Lindl.) Benth. ex Hemsl.

viridiflora (Hook. f.) Kuntze = **Bulbophyllum viridiflorum** (Hook. f.) Schltr.

wallichii (Lindl.) Kuntze = **Bulbophyllum muscicola** Rchb. f.

xylophylla (C. S. P. Parish & Rchb. f.) Kuntze = **Bulbophyllum xylophyllum** C. S. P. Parish & Rchb. f.

Physurus

blumei Lindl. = **Erythrodes blumei** (Lindl.) Schltr.

herpysmoides King & Pantl. = **Erythrodes hirsuta** (Griff.) Ormerod

hirsutus (Griff.) Lindl. = **Erythrodes hirsuta** (Griff.) Ormerod

viridiflorus (Blume) Lindl. = **Goodyera viridiflora** (Blume) Lindl. ex D. Dietr.

Pinalia

alba Buch.-Ham. ex Lindl. = **Pinalia spicata** (D. Don) S. C. Chen & J. J. Wood

amica (Rchb. f.) Kuntze = **Pinalia lineata** (Lindl.) Kuntze

andamanica (Hook. f.) Kuntze = **Dendrolirium andamanicum** (Hook. f.) Schuit., Y. P. Ng & H. A. Pedersen

andersonii (Hook. f.) Kuntze = **Pinalia lineata** (Lindl.) Kuntze

angulata (Rchb. f.) Kuntze = **Tainia latifolia** (Lindl.) Rchb. f.

bambusifolia (Lindl.) Kuntze = **Bambuseria bambusifolia** (Lindl.) Schuit., Y. P. Ng & H. A. Pedersen

barbata (Lindl.) Kuntze = **Eriodes barbata** (Lindl.) Rolfe

biflora (Griff.) Kuntze = **Cylindrolobus biflorus** (Griff.) Rauschert

braccata (Lindl.) Kuntze = **Porpax braccata** (Lindl.) Schuit., Y. P. Ng & H. A. Pedersen

calamifolia (Hook. f.) Kuntze = **Strongyleria pannea** (Lindl.) Schuit., Y. P. Ng & H. A. Pedersen

carinata (Gibson ex Lindl.) Kuntze = **Cryptochilus acuminatus** (Griff.) Schuit., Y. P. Ng & H. A. Pedersen

confusa (Hook. f.) Kuntze = **Pinalia lineata** (Lindl.) Kuntze

crassicaulis (Hook. f.) Kuntze = **Bambuseria crassicaulis** (Hook. f.) Schuit., Y. P. Ng & H. A. Pedersen

dasyphylla (C. S. P. Parish & Rchb. f.) Kuntze = **Trichotosia dasyphylla** (C. S. P. Parish & Rchb. f.) Kraenzl.

exilis (Hook. f.) Kuntze = **Porpax exilis** (Hook. f.) Schuit., Y. P. Ng & H. A. Pedersen

extinctoria (Lindl.) Kuntze = **Porpax extinctoria** (Lindl.) Schuit., Y. P. Ng & H. A. Pedersen

ferruginea (Lindl.) Kuntze = **Dendrolirium ferrugineum** (Lindl.) A. N. Rao

jerdoniana (Wight) Kuntze = **Porpax jerdoniana** (Wight) Rolfe

lobbii Kuntze = **Dendrolirium laniceps** (Rchb. f.) Schuit., Y. P. Ng & H. A. Pedersen

muscicola (Lindl.) Kuntze = **Porpax parviflora** (D.Don) Ormerod & Kurzweil

nana (A. Rich.) Kuntze = **Porpax nana** (A. Rich.) Schuit., Y. P. Ng & H. A. Pedersen

nilgherensis Kuntze = **Cylindrolobus pauciflorus** (Wight) Schuit., Y. P. Ng & H. A. Pedersen

paniculata (Lindl.) Kuntze = **Mycaranthes floribunda** (D. Don) S. C. Chen & J. J. Wood

pannea (Lindl.) Kuntze = **Strongyleria pannea** (Lindl.) Schuit., Y. P. Ng & H. A. Pedersen

pulvinata (Lindl.) Kuntze = **Trichotosia pulvinata** (Lindl.) Kraenzl.

pusilla (Griff.) Kuntze = **Porpax pusilla** (Griff.) Schuit., Y. P. Ng & H. A. Pedersen

reticosa (Wight) Kuntze = **Porpax reticosa** (Wight) Schuit.

rufinula (Rchb. f.) Kuntze = **Trichotosia pulvinata** (Lindl.) Kraenzl.

scabrilinguis (Lindl.) Kuntze = **Eria scabrilinguis** Lindl.

stricta (Lindl.) Kuntze = **Cryptochilus strictus** (Lindl.) Schuit., Y. P. Ng & H. A. Pedersen

tomentosa (J. Koenig) Kuntze = **Dendrolirium tomentosum** (J. Koenig) S. C. Chen & J. J. Wood

velutina (G. Lodd. ex Lindl.) Kuntze = **Trichotosia velutina** (G. Lodd. ex Lindl.) Kraenzl.

vittata (Lindl.) Kuntze = **Eria vittata** Lindl.

Plantaginorchis

cephalotes (Lindl.) Szlach. = **Habenaria cephalotes** Lindl.

Plectoglossa
  perrottetiana (A. Rich.) K. Prasad & Venu = **Habenaria perrottetiana** A. Rich.
Pleione
  barbata (Lindl. ex Griff.) Kuntze = **Coelogyne barbata** Lindl. ex Griff.
  brevifolia (Lindl.) Kuntze = **Coelogyne punctulata** Lindl.
  breviscapa (Lindl.) Kuntze = **Coelogyne breviscapa** Lindl.
  corymbosa (Lindl.) Kuntze = **Coelogyne corymbosa** Lindl.
  diphylla Lindl. & Paxton = **Pleione maculata** (Lindl.) Lindl. & Paxton
  elata (Lindl.) Kuntze = **Coelogyne stricta** (D. Don) Schltr.
  fimbriata (Lindl.) Kuntze = **Coelogyne fimbriata** Lindl.
  flaccida (Lindl.) Kuntze = **Coelogyne flaccida** Lindl.
  flavida (Hook. f. ex Lindl.) Kuntze = **Coelogyne prolifera** Lindl.
  fuliginosa (Lodd. ex Hook.) Kuntze = **Coelogyne ovalis** Lindl.
  fuscescens (Lindl.) Kuntze = **Coelogyne fuscescens** Lindl.
  gardneriana (Lindl.) Kuntze = **Coelogyne gardneriana** Lindl.
  glandulosa (Lindl.) Kuntze = **Coelogyne nervosa** A. Rich.
  graminifolia (C. S. P. Parish & Rchb. f.) Kuntze = **Coelogyne viscosa** Rchb. f.
  griffithii (Hook. f.) Kuntze = **Coelogyne griffithii** Hook. f.
  hookeriana (Lindl.) Rollisson var. *brachyglossa* (Rchb. f.) Rolfe = **Pleione hookeriana** (Lindl.) Rollisson
  humilis (Sm.) D. Don var. *amitii* R. Pal, Dayama & Medhi = **Pleione humilis** (Sm.) D. Don
  longipes (Lindl.) Kuntze = **Coelogyne longipes** Lindl.
  micrantha (Lindl.) Kuntze = **Coelogyne micrantha** Lindl.
  nervosa (A. Rich.) Kuntze = **Coelogyne nervosa** A. Rich.
  nitida (Lindl.) Kuntze = **Coelogyne nitida** (Wall. ex D. Don) Lindl.
  occultata (Hook. f.) Kuntze = **Coelogyne occultata** Hook. f.
  ochracea (Lindl.) Kuntze = **Coelogyne nitida** (Wall. ex D. Don) Lindl.
  odoratissima (Lindl.) Kuntze = **Coelogyne odoratissima** Lindl.
  praecox (Sm.) D. Don var. *wallichiana* (Lindl.) E. W. Cooper = **Pleione praecox** (Sm.) D. Don
  prolifera (Lindl.) Kuntze = **Coelogyne prolifera** Lindl.
  rigida (C. S. P. Parish & Rchb. f.) Kuntze = **Coelogyne rigida** C. S. P. Parish & Rchb. f.
  rossiana (Rchb. f.) Kuntze = **Coelogyne trinervis** Lindl.
  suaveolens (Lindl.) Kuntze = **Coelogyne suaveolens** (Lindl.) Hook. f.
  thuniana (Rchb. f.) Kuntze = **Coelogyne uniflora** Lindl.
  treutleri (Hook. f.) Kuntze = **Dendrobium treutleri** (Hook. f.) Schuit. & Peter B. Adams
  trinervis (Lindl.) Kuntze = **Coelogyne trinervis** Lindl.
  uniflora (Lindl.) Kuntze = **Coelogyne uniflora** Lindl.

  viscosa (Rchb. f.) Kuntze = **Coelogyne viscosa** Rchb. f.
  wallichiana (Lindl.) Lindl. & Paxton = **Pleione praecox** (Sm.) D. Don
Plocoglottis
  porphyrophylla Ridl. = **Plocoglottis lowii** Rchb. f.
Podanthera
  pallida Wight = **Epipogium roseum** (D. Don) Lindl.
Podochilus
  cornutus (Blume) Schltr. = **Appendicula cornuta** Blume
  reflexus (Blume) Schltr. = **Appendicula reflexa** Blume
Pogonia
  biflora Wight = **Nervilia plicata** (Andrews) Schltr.
  carinata (Roxb.) Lindl. = **Nervilia concolor** (Blume) Schltr.
  concolor (Blume) Blume = **Nervilia concolor** (Blume) Schltr.
  crispata Blume = **Nervilia simplex** (Thouars) Schltr.
  falcata King & Pantl. = **Nervilia falcata** (King & Pantl.) Schltr.
  flabelliformis Lindl. = **Nervilia concolor** (Blume) Schltr.
  gammieana Hook. f. = **Nervilia gammieana** (Hook. f.) Pfitzer
  graminifolia (Roxb.) Voigt = **Spathoglottis pubescens** Lindl.
  hookeriana King & Pantl. = **Nervilia hookeriana** (King & Pantl.) Schltr.
  juliana (Roxb.) Wall ex Lindl. = **Nervilia juliana** (Roxb.) Schltr.
  khasiana King & Pantl. = **Nervilia khasiana** (King & Pantl.) Schltr.
  mackinnonii Duthie = **Nervilia mackinnonii** (Duthie) Schltr.
  macroglossa Hook. f. = **Nervilia macroglossa** (Hook. f.) Schltr.
  plicata (Andrews) Lindl. = **Nervilia plicata** (Andrews) Schltr.
  prainiana King & Pantl. = **Nervilia simplex** (Thouars) Schltr.
  punctata Blume = **Nervilia punctata** (Blume) Makino
  scottii Rchb. f. = **Nervilia concolor** (Blume) Schltr.
  simplex (Thouars) Rchb. f. = **Nervilia simplex** (Thouars) Schltr.
Polychilos
  cornu-cervi Breda = **Phalaenopsis cornu-cervi** (Breda) Blume & Rchb. f.
  lobbii (Rchb. f.) Shim = **Phalaenopsis lobbii** (Rchb. f.) H. R. Sweet
  mannii (Rchb. f.) Shim = **Phalaenopsis mannii** Rchb. f.
  mysorensis (C. J. Saldanha) Shim = **Phalaenopsis mysorensis** C. J. Saldanha
  parishii (Rchb. f.) Shim = **Phalaenopsis parishii** Rchb. f.
  speciosus (Rchb. f.) Shim = **Phalaenopsis tetraspis** Rchb. f.
  taenialis (Lindl.) Shim = **Phalaenopsis taenialis** (Lindl.) Christenson & Pradhan
Polystachya
  flavescens (Blume) J. J. Sm. = **Polystachya concreta** (Jacq.) Garay & H. R. Sweet
  pumila (Kuntze) Kraenzl. = **Dendrobium pachyphyllum** (Kuntze) Bakh. f.

purpurea Wight = **Polystachya concreta** (Jacq.) Garay & H. R. Sweet

Polystylus
  cornu-cervi (Breda) Hasselt ex Hassk. = **Phalaenopsis cornu-cervi** (Breda) Blume & Rchb. f.

Pomatocalpa
  andamanicum (Hook. f.) J. J. Sm. = **Pomatocalpa maculosum** (Lindl.) J. J. Sm. subsp. **andamanicum** (Hook. f.) Watthana
  armigerum (King & Pantl.) Tang & F. T. Wang = **Cleisostoma armigerum** King & Pantl.
  bambusarum (King & Pantl.) Garay = **Cleisostoma bambusarum** (King & Pantl.) King & Pantl.
  densiflorum (Lindl.) Tang & F. T. Wang = **Robiquetia spathulata** (Blume) J. J. Sm.
  loratum (Rchb. f.) J. J. Sm. = **Pomatocalpa undulatum** (Lindl.) J. J. Sm.
  mannii (Rchb. f.) J. J. Sm. = **Pomatocalpa spicatum** Breda
  ramosum (Lindl.) Summerh. = **Trichoglottis ramosa** (Lindl.) Senghas
  wendlandorum (Rchb. f.) J. J. Sm. = **Pomatocalpa spicatum** Breda

Ponerorchis
  chrysea (W. W. Sm.) Soó = **Hsenhsua chrysea** (W. W. Sm.) X. H. Jin, Schuit., W. T. Jin & L. Q. Huang
  chusua var. *nana* (King & Pantl.) R. C. Srivast. = **Ponerorchis nana** (King & Pantl.) Soó

Porpax
  chandrasekharanii Bhargavan & C. N. Mohanan = **Porpax exilis** (Hook. f.) Schuit., Y. P. Ng & H. A. Pedersen
  fibuliformis (King & Pantl.) King & Pantl. var. *gigantea* (Deori) Debta & H. J. Chowdhery = **Porpax gigantea** Deori
  lichenora (Lindl.) T. Cooke = **Porpax jerdoniana** (Wight) Rolfe
  meirax King & Pantl. var. *elwesii* (Rchb. f.) R. C. Srivast. = **Porpax elwesii** (Rchb. f.) Rolfe
  muscicola (Lindl.) Schuit., Y. P. Ng & H. A. Pedersen = **Porpax parviflora** (D.Don) Ormerod & Kurzweil
  nana (A. Rich.) Schuit., Y. P. Ng & H. A. Pedersen var. *brevilinguis* (J. Joseph & V. Chandras.) Kottaim. = **Porpax nana** (A. Rich.) Schuit., Y. P. Ng & H. A. Pedersen
  papillosa Blatt. & McCann = **Porpax reticulata** Lindl.

Preptanthe
  vestita (Wall. ex Lindl.) Rchb. f. = **Calanthe vestita** Wall. ex Lindl.

Pristiglottis
  torta (King & Pantl.) Aver. = **Odontochilus tortus** King & Pantl.

Proteroceras
  holttumii J. Joseph & Vajr. = **Pteroceras leopardinum** (C. S. P. Parish & Rchb. f.) Seidenf. & Smitinand

Pteroceras
  alatum (Holttum) Holttum = **Macropodanthus alatus** (Holttum) Seidenf. & Garay
  appendiculatum (Blume) Holttum = **Grosourdya appendiculata** (Blume) Rchb. f.

berkeleyi (Rchb. f.) Holttum = **Macropodanthus berkeleyi** (Rchb. f.) Seidenf. & Garay
carrii (L. O. Williams) Holttum = **Grosourdya muscosa** (Rolfe) Garay
suaveolens (Roxb.) Holtum = **Pteroceras teres** (Blume) Holttum
unguiculatum (Lindl.) H. A. Pedersen = **Brachypeza unguiculata** (Lindl.) Kocyan & Schuit.

Pterygodium
  sulcatum Roxb. = **Zeuxine strateumatica** (L.) Schltr.

Ptilocnema
  bracteata D. Don = **Coelogyne imbricata** (Hook.) Rchb. f.

Raciborskanthos
  striatus (Rchb. f.) Szlach. = **Cleisostoma striatum** (Rchb. f.) Garay

Renanthera
  bilinguis Rchb. f. = **Arachnis labrosa** (Lindl. ex Paxton) Rchb. f.
  papilio King & Prain = **Renanthera imschootiana** Rolfe

Rhamphidia
  gardneri Thwaites = **Hetaeria oblongifolia** Blume
  ovalifolia (Wight) Lindl. = **Hetaeria oblongifolia** Blume
  rubens (Lindl.) Lindl. = **Hetaeria affinis** (Griff.) Seidenf. & Ormerod
  rubicunda (Blume) F. Muell. = **Goodyera rubicunda** (Blume) Lindl.

Rhynchostylis
  albiflora I. Barua & Bora = **Rhynchostylis retusa** (L.) Blume
  cymifera Yohannan, J. Mathew & Szlach. = **Rhynchostylis retusa** (L.) Blume
  densiflora (Lindl.) L. O. Williams = **Robiquetia spathulata** (Blume) J. J. Sm.
  garwalica (Lindl.) Rchb. f. = **Rhynchostylis retusa** (L.) Blume
  guttata (Lindl.) Rchb. f. = **Rhynchostylis retusa** (L.) Blume
  latifolia C. E. C. Fisch. = **Schoenorchis smeeana** (Rchb. f.) Jalal, Jayanthi & Schuit.
  papillosa (Lindl.) Heynh. = **Acampe praemorsa** (Roxb.) Blatt. & McCann
  praemorsa (Willd.) Blume = **Rhynchostylis retusa** (L.) Blume
  retusa (L.) Blume f. *albiflora* (I. Barua & Bora) Christenson = **Rhynchostylis retusa** (L.) Blume

Rhytionanthos
  aemulus (W. W. Sm.) Garay, Hamer & Siegerist = **Bulbophyllum forrestii** Seidenf.
  balaeniceps (Rchb. f.) C. S. Kumar & Garay = **Bulbophyllum balaeniceps** Rchb. f.
  bootanensis (Griff.) Garay, Hamer & Siegerist = **Bulbophyllum umbellatum** Lindl.
  cornutus (Lindl.) Garay, Hamer & Siegerist = **Bulbophyllum helenae** (Kuntze) J. J. Sm.
  indicus C. S. Kumar & Garay = **Bulbophyllum indicum** (C. S. Kumar & Garay) Sushil K. Singh, Agrawala & Jalal
  nodosus (Rolfe) Garay, Hamer & Siegerist = **Bulbophyllum nodosum** (Rolfe) J. J. Sm.

rheedei (Manilal & C. S. Kumar) Garay, Hamer & Siegerist = **Bulbophyllum rheedei** Manilal & C. S. Kumar

spathulatus (Rolfe ex E. W. Cooper) Garay, Hamer & Siegerist = **Bulbophyllum spathulatum** (Rolfe ex E. W. Cooper) Seidenf.

Ritaia

    himalaica (Hook. f.) King & Pantl. = **Ceratostylis himalaica** Hook. f.

Robiquetia

    bambusarum (King & Pantl.) R. Rice = **Cleisostoma bambusarum** (King & Pantl.) King & Pantl.

    paniculata (Lindl.) J. J. Sm. = **Robiquetia succisa** (Lindl.) Seidenf. & Garay

    virescens Jayaw. = **Robiquetia virescens** Ormerod & S. S. Fernando

Roptrostemon

    concolor (Blume) Lindl. = **Nervilia concolor** (Blume) Schltr.

    discolor (Blume) Lindl. = **Nervilia plicata** (Andrews) Schltr.

Saccolabium

    acaule (Lindl.) Hook. f. = **Gastrochilus acaulis** (Hook. f.) Kuntze

    acuminatum (Lindl.) Hook. f. = **Uncifera acuminata** Lindl.

    acutifolium Lindl. = **Gastrochilus acutifolius** (Lindl.) Kuntze

    affine King & Pantl. = **Gastrochilus affinis** (King & Pantl.) Schltr.

    ampullaceum (Roxb.) Lindl. = **Vanda ampullacea** (Roxb.) L. M. Gardiner

    bambusarum (King & Pantl.) Tang & F. T. Wang = **Cleisostoma bambusarum** (King & Pantl.) King & Pantl.

    bellinum Rchb. f. = **Gastrochilus bellinus** (Rchb. f.) Kuntze

    bigibbum Rchb. f. ex Hook. f. = **Gastrochilus obliquus** (Lindl.) Kuntze

    buccosum Rchb. f. = **Robiquetia succisa** (Lindl.) Seidenf. & Garay

    calceolare (Buch.- Ham. ex Sm.) Lindl. = **Gastrochilus calceolaris** (Buch.-Ham. ex Sm.) D. Don

    carinatum Griff. = **Acampe carinata** (Criff.) S. G. Panigrahi

    cephalotes (Lindl.) Hook. f. = **Acampe carinata** (Griff.) S. G. Panigrahi

    chrysanthum Alston = **Seidenfadeniella filiformis** (Rchb. f.) Christenson & Ormerod

    coarctatum King & Pantl. = **Trachoma coarctatum** (King & Pantl.) Garay

    congestum (Lindl.) Hook. f. = **Acampe praemorsa** (Roxb.) Blatt. & McCann

    crassilabre King & Pantl. = **Thrixspermum crassilabre** (King & Pantl.) Ormerod

    curvifolium Lindl. = **Vanda curvifolia** (Lindl.) L. M. Gardiner

    dasypogon (Sm.) Lindl. = **Gastrochilus dasypogon** (Sm.) Kuntze

    decipiens (Lindl.) Alston = **Pomatocalpa decipiens** (Lindl.) J. J. Sm.

densiflorum Lindl. = **Robiquetia spathulata** (Blume) J. J. Sm.

distichum Lindl. = **Gastrochilus distichus** (Lindl.) Kuntze

filiforme Rchb. f. = **Seidenfadeniella filiformis** (Rchb. f.) Christenson & Ormerod

flabelliforme Blatt. & McCann = **Gastrochilus flabelliformis** (Blatt. & McCann) C. J. Saldanha

flexuosum (Lindl.) Rchb. f. = **Trichoglottis ramosa** (Lindl.) Senghas

fragrans C. S. P. Parish & Rchb. f. = **Schoenorchis fragrans** (C. S. P. Parish & Rchb. f.) Seidenf. & Smitinand

garwalicum Lindl. = **Rhynchostylis retusa** (L.) Blume

gemmatum Lindl. = **Schoenorchis gemmata** (Lindl.) J. J. Sm.

giganteum Lindl. = **Rhynchostylis gigantea** (Lindl.) Ridl.

gracile Lindl. = **Robiquetia gracilis** (Lindl.) Garay

guttatum (Lindl.) Lindl. = **Rhynchostylis retusa** (L.) Blume

helferi Hook. f. = **Smitinandia helferi** (Hook. f.) Garay

himalaicum Deb, Sengupta & Malick = **Holcoglossum himalaicum** (Deb, Sengupta & Malick) Aver.

inconspicuum Hook. f. = **Luisia inconspicua** (Hook. f.) Hook. f. ex King & Pantl.

intermedium Griff. ex Lindl. = **Gastrochilus intermedius** (Griff. ex Lindl.) Kuntze

jerdonianum (Wight) Rchb. f. = **Schoenorchis jerdoniana** (Wight) Garay

lancifolium King & Pantl. = **Uncifera lancifolia** (King & Pantl.) Schltr.

lineare Lindl. = **Aerides ringens** (Lindl.) C. E. C. Fisch.

longifolium (Lindl.) Hook. f. = **Acampe praemorsa** (Roxb.) Blatt. & McCann

maculatum (Dalzell) Hook. f. = **Smithsonia maculata** (Dalzell) C. J. Saldanha

micranthum Lindl. = **Smitinandia micrantha** (Lindl.) Holttum

minutiflorum Ridl. = **Schoenorchis minutiflora** (Ridl.) J. J. Sm.

nilagiricum Hook. f. = **Gastrochilus acaulis** (Hook. f.) Kuntze

niveum Lindl. = **Schoenorchis nivea** (Lindl.) Schltr.

obliquum Lindl. = **Gastrochilus obliquus** (Lindl.) Kuntze

obtusifolium (Lindl.) Hook. f. = **Uncifera obtusifolia** Lindl.

ochraceum Lindl. = **Acampe ochracea** (Lindl.) Hochr.

pallens Cathcart ex Lindl. = **Cleisocentron pallens** (Cathcart ex Lindl.) N. Pearce & P. J. Cribb

paniculatum Wight = **Aerides ringens** (Lindl.) C. E. C. Fisch.

papillosum Lindl. = **Acampe praemorsa** (Roxb.) Blatt. & McCann

peninsulare (Dalzell) Alston = **Cleisostoma tenuifolium** (L.) Garay

pilosulum (Gagnep.) Tang & F.T.Wang = **Cleisomeria pilosulum** (Gagnep.) Seidenf. & Garay

praemorsum (Roxb.) Hook. f. = **Acampe praemorsa** (Roxb.) Blatt. & McCann

praemorsum (Willd.) Lindl. = **Rhynchostylis retusa** (L.)
Blume

pseudodistichum King & Pantl. = **Gastrochilus pseudo-distichus** (King & Pantl.) Schltr.

pumilio Rchb. f. = **Saccolabiopsis pusilla** (Lindl.)
Seidenf. & Garay

pusillum (Lindl.) Lindl. ex Hook. f. = **Saccolabiopsis pusilla** (Lindl.) Seidenf. & Garay

racemiferum Lindl. = **Cleisostoma racemiferum** (Lindl.)
Garay

ramosum Lindl. = **Trichoglottis ramosa** (Lindl.)
Senghas

retusum (L.) Voigt = **Rhynchostylis retusa** (L.) Blume

rheedei Wight = **Rhynchostylis retusa** (L.) Blume

ringens Lindl. = **Aerides ringens** (Lindl.) C. E. C. Fisch.

roseum (Wight) Lindl. = **Seidenfadeniella rosea** (Wight)
C. S. Kumar

roseum Lindl. = **Robiquetia rosea** (Lindl.) Garay

rostellatum Hook. f. = **Cleisostoma discolor** Lindl.

rubrum Wight = **Aerides ringens** (Lindl.) C. E. C. Fisch.

smeeanum Rchb. f. = **Schoenorchis smeeana** (Rchb. f.)
Jalal, Jayanthi & Schuit.

speciosum Wight = **Aerides maculosa** Lindl.

tenerum (Lindl.) Lindl. = **Trichoglottis tenera** (Lindl.)
Rchb. f.

tenuifolium (L.) Alston = **Cleisostoma tenuifolium** (L.)
Garay

trichromum Rchb. f. = **Cleisocentron pallens** (Cathcart
ex Lindl.) N. Pearce & P. J. Cribb

turneri B. S. Williams = **Rhynchostylis retusa** (L.)
Blume

undulatum Lindl. = **Pomatocalpa undulatum** (Lindl.)
J. J. Sm.

virescens Gardner ex Lindl. = **Robiquetia virescens**
Ormerod & S. S. Fernando

viridiflorum (Dalzell) Lindl. = **Smithsonia viridiflora**
(Dalzell) C. J. Saldanha

wightianum (Lindl. ex Wight) Hook. f. = **Acampe
praemorsa** (Roxb.) Blatt. & McCann

wightianum Lindl. = **Aerides ringens** (Lindl.) C. E. C.
Fisch.

Salacistis

clavata (N. Pearce & P. J. Cribb) M. C. Pace = **Goodyera
rubicunda** (Blume) Lindl.

fumata (Thwaites) M. C. Pace = **Goodyera fumata**
Thwaites

fumata (Thwaites) T. C. Hsu = **Goodyera fumata**
Thwaites

rubicunda (Blume) M. C. Pace = **Goodyera rubicunda**
(Blume) Lindl.

rubicunda (Blume) T. C. Hsu = **Goodyera rubicunda**
(Blume) Lindl.

Sarcanthus

appendiculatus (Lindl.) E. C. Parish = **Cleisostoma
appendiculatum** (Lindl.) Benth. & Hook. f. ex B. D.
Jacks.

arietinus Rchb. f. = **Cleisostoma arietinum** (Rchb. f.)
Garay

armiger (King & Pantl.) J. J. Sm. = **Cleisostoma
armigerum** King & Pantl.

aspersus Rchb. f. = **Cleisostoma aspersum** (Rchb. f.)
Garay

auriculatus Rolfe = **Cleisostoma discolor** Lindl.

bambusarum King & Pantl. = **Cleisostoma bambusarum**
(King & Pantl.) King & Pantl.

brevipes (Hook. f.) J. J. Sm. = **Cleisostoma striatum**
(Rchb. f.) Garay

carinatus Rolfe ex Downie = **Cleisostoma duplicilobum**
(J. J. Sm.) Garay

densiflorus (Lindl.) C. S. P. Parish & Rchb. f. = **Robique-tia spathulata** (Blume) J. J. Sm.

discolor (Lindl.) J. J. Sm. = **Cleisostoma discolor** Lindl.

duplicilobus J. J. Sm. = **Cleisostoma duplicilobum** (J. J.
Sm.) Garay

filiformis Lindl. = **Cleisostoma filiforme** (Lindl.) Garay

guttatus Lindl. = **Rhynchostylis retusa** (L.) Blume

hincksianus Rchb. f. = **Cleisostoma appendiculatum**
(Lindl.) Benth. & Hook. f. ex B. D. Jacks.

hirtus (Lindl.) J. J. Sm. = **Stereochilus hirtus** Lindl.

insectifer Rchb. f. = **Pelatantheria insectifera** (Rchb. f.)
Ridl.

khasiaensis Tang & F. T. Wang = **Cleisostoma aspersum**
(Rchb. f.) Garay

laxus Rchb.f. = **Stereochilus laxus** (Rchb.f.) Garay

linearilobatus Seidenf. & Smitinand = **Cleisostoma
linearilobatum** (Seidenf. & Smitinand) Garay

lorifolius C. S. P. Parish ex Hook. f. = **Cleisostoma
racemiferum** (Lindl.) Garay

macrodon Rchb. f. = **Cleisostoma discolor** Lindl.

pallidus Lindl. = **Cleisostoma racemiferum** (Lindl.)
Garay

paniculatus (Ker Gawl.) Lindl. = **Cleisostoma panicula-tum** (Ker Gawl.) Garay

parishii Hook. f. = **Cleisostoma parishii** (Hook. f.) Garay

pauciflorus Wight = **Cleisostoma tenuifolium** (L.) Garay

peninsularis Dalzell = **Cleisostoma tenuifolium** (L.)
Garay

praemorsus (Roxb.) Lindl. ex Spreng. = **Acampe
praemorsa** (Roxb.) Blatt. & McCann

racemifer (Lindl.) Rchb. f. = **Cleisostoma racemiferum**
(Lindl.) Garay

ramosus (Lindl.) J. J. Sm. = **Trichoglottis ramosa**
(Lindl.) Senghas

ringens (Rchb.) J. J. Sm. = **Stereochilus ringens**
(Rchb. f.) Garay

rolfeanus King & Pantl. = **Cleisostoma rolfeanum** (King
& Pantl.) Garay

roseus Wight = **Seidenfadeniella rosea** (Wight) C. S.
Kumar

sagittatus King & Pantl. = **Cleisostoma linearilobatum**
(Seidenf. & Smitinand) Garay

secundus Griff. = **Cleisostoma subulatum** Blume

striatus (Rchb. f.) J. J. Sm. = **Cleisostoma striatum**
(Rchb. f.) Garay

subulatus (Blume) Rchb. f. = **Cleisostoma subulatum**
Blume

succisus Lindl. = **Robiquetia succisa** (Lindl.) Seidenf. &
Garay

tenuifolius (L.) Seidenf. = **Cleisostoma tenuifolium** (L.)
Garay

uraiensis Hayata = **Cleisostoma uraiense** (Hayata)
Garay & H. R. Sweet

williamsonii Rchb. f. = **Cleisostoma williamsonii** (Rchb.
f.) Garay

Sarcochilus

acuminatissimus (Blume) Rchb. f. = **Thrixspermum
acuminatissimum** (Blume) Rchb. f.

alatus Holttum = **Macropodanthus alatus** (Holttum)
Seidenf. & Garay

appendiculatus (Blume) J. J. Sm. = **Grosourdya
appendiculata** (Blume) Rchb. f.

arachnites (Blume) Rchb. f. = **Thrixspermum centipeda**
Lour.

berkeleyi (Rchb. f.) Hook. f. = **Macropodanthus
berkeleyi** (Rchb. f.) Seidenf. & Garay

bimaculatus King & Pantl. = **Biermannia bimaculata**
(King & Pantl.) King & Pantl.

carrii L. O. Williams = **Grosourdya muscosa** (Rolfe)
Garay

centipeda (Lour.) Náves = **Thrixspermum centipeda**
Lour.

crepidiformis King & Pantl. = **Taeniophyllum crepidiforme** (King & Pantl.) King & Pantl.

dalzellianus Santapau = **Smithsonia viridiflora** (Dalzell)
C. J. Saldanha

difformis (Wall. ex Lindl.) Tang & F. T. Wang = **Phalaenopsis difformis** (Wall. ex Lindl.) Kocyan &
Schuit.

fasciatus F. Muell. = **Chiloschista fasciata** (F. Muell.)
Seidenf. & Ormerod

hystrix (Blume) Rchb. f. = **Thrixspermum hystrix**
(Blume) Rchb. f.

leopardinus (C. S. P. Parish & Rchb. f.) Hook. f. =
Pteroceras leopardinum (C. S. P. Parish & Rchb. f.)
Seidenf. & Smitinand

maculatus (Dalzell) Benth. ex Pfitzer = **Smithsonia
maculata** (Dalzell) C. J. Saldanha

maculatus Carr = **Grosourdya muscosa** (Rolfe) Garay

mannii Hook. f. = **Micropera mannii** (Hook. f.) Tang &
F. T. Wang

merguensis Hook. f. = **Thrixspermum merguense**
(Hook. f.) Kuntze

minimifolius Hook. f. = **Chiloschista fasciata** (F. Muell.)
Seidenf. & Ormerod

muscosus Rolfe = **Grosourdya muscosa** (Rolfe) Garay

obtusus (Lindl.) Benth. ex Hook. f. = **Micropera obtusa**
(Lindl.) Tang & F. T. Wang

pauciflorus Hook. f. = **Thrixspermum pauciflorum**
(Hook. f.) Kuntze

praemorsus (Roxb.) Spreng. = **Acampe praemorsa**
(Roxb.) Blatt. & McCann

pulchellus (Thwaites) Trimen = **Thrixspermum
pulchellum** (Thwaites) Schltr.

purpureus (Lindl.) Benth. ex Hook. f. = **Micropera
rostrata** (Roxb.) N. P. Balakr.

pygmaeus King & Pantl. = **Thrixspermum pygmaeum**
(King & Pantl.) Holttum

retrospiculatus King & Pantl. = **Taeniophyllum
retrospiculatum** (King & Pantl.) King & Pantl.

roxburghii Hook. f. = **Micropera pallida** (Roxb.) Lindl.

saruwatarii Hayata = **Thrixspermum saruwatarii**
(Hayata) Schltr.

suaveolens (Roxb.) Hook. f. = **Pteroceras teres** (Blume)
Holttum

tenuifolius (L.) Náves = **Cleisostoma tenuifolium** (L.)
Garay

teres (Blume) Rchb. f. = **Pteroceras teres** (Blume)
Holttum

trichoglottis Hook. f. = **Thrixspermum trichoglottis**
(Hook. f.) Kuntze

unguiculatus Lindl. = **Brachypeza unguiculata** (Lindl.)
Kocyan & Schuit.

usneoides (D. Don) Rchb. f. = **Chiloschista usneoides**
(D. Don) Lindl.

viridiflorus (Dalzell) T. Cooke = **Smithsonia viridiflora**
(Dalzell) C. J. Saldanha

wightii Hook. f. = **Chiloschista fasciata** (F. Muell.)
Seidenf. & Ormerod

Sarcoglyphis

manipurensis A. N. Rao, Vik. Kumar & H. B. Sharma =
Cleisostoma parishii (Hook. f.) Garay

Sarcopodium

affine (Wall. ex Lindl.) Lindl. & Paxton = **Bulbophyllum
affine** Lindl.

amplum (Lindl.) Lindl. = **Dendrobium amplum** Lindl.

chapaense (Gagnep.) Tang & F. T. Wang = **Dendrobium
brunneum** Schuit. & Peter B. Adams

fargesii (Finet) Tang & F. T. Wang = **Dendrobium
fargesii** Finet

fuscescens (Griff.) Lindl. = **Dendrobium fuscescens**
Griff.

griffithii Lindl. = **Bulbophyllum griffithii** (Lindl.)
Rchb. f.

leopardinum (Wall.) Lindl. & Paxton = **Bulbophyllum
leopardinum** (Wall.) Lindl. ex Wall.

lobbii (Lindl.) Lindl. & Paxton = **Bulbophyllum lobbii**
Lindl.

macranthum (Lindl.) Lindl. & Paxton = **Bulbophyllum
macranthum** Lindl.

rotundatum Lindl. = **Dendrobium rotundatum** (Lindl.)
Hook. f.

striatum (Griff.) Lindl. = **Bulbophyllum striatum** (Griff.)
Rchb. f.

uniflorum Lindl. = **Bulbophyllum pteroglossum** Schltr.

Sarothrochilus

dawsonianus (Rchb. f.) Schltr. = **Trichoglottis dawsoniana** (Rchb. f.) Rchb. f.

Satyrium

albiflorum A. Rich. = **Satyrium nepalense** D. Don

ciliatum Lindl. = **Satyrium nepalense** D. Don

epipogium L. = **Epipogium aphyllum** Sw.

lanceum (Thunb. ex Sw.) Pers. = **Herminium lanceum**
(Thunb. ex Sw.) Vuijk

monorchis (L.) Pers. = **Herminium monorchis** (L.)
R. Br.

neilgherrensis Fyson = **Satyrium nepalense** D. Don

nepalense D. Don f. *albiflorum* Tuyama = **Satyrium
nepalense** D. Don

nepalense D. Don var. *ciliatum* (Lindl.) Hook. f. =
Satyrium nepalense D. Don

mucronata D. Don = **Oberonia mucronata** (D. Don)
Ormerod & Seidenf.

odoratissima Sm. = **Bulbophyllum odoratissimum**
(Sm.) Lindl.

racemosa Sm. = **Bulbophyllum reptans** (Lindl.) Lindl.

Stereochilus

bicuspidatus (Hook. f.) King & Pantl. = **Cleisostoma
aspersum** (Rchb. f.) Garay

wattii King & Pantl. = **Stereochilus ringens** (Rchb. f.)
Garay

Stichorkis

bistriata (C. S. P. Parish & Rchb. f.) Marg., Szlach. &
Kulak, = **Liparis bistriata** C. S. P. Parish & Rchb. f.

bootanensis (Griff.) Marg., Szlach. & Kulak = **Liparis
bootanensis** Griff.

cespitosa (Lam.) Thouars ex Marg. = **Liparis cespitosa**
(Lam.) Lindl.

distans (C. B. Clarke) Marg., Szlach. & Kulak = **Liparis
distans** C. B. Clarke

elegans (Lindl.) Marg., Szlach. & Kulak = **Liparis elegans** Lindl.

elliptica (Wight) Marg., Szlach. & Kulak = **Liparis
elliptica** Wight

luteola (Lindl.) Marg., Szlach. & Kułak = **Liparis luteola**
Lindl.

mannii (Rchb. f.) Marg., Szlach. & Kulak = **Liparis
mannii** Rchb. f.

plantaginea (Lindl.) Marg., Szlach. & Kulak = **Liparis
plantaginea** Lindl.

pusilla (Ridl.) Marg., Szlach. & Kulak = **Liparis
cespitosa** (Lam.) Lindl.

stricklandiana (Rchb. f.) Marg., Szlach. & Kułak =
Liparis stricklandiana Rchb. f.

tenuifolia (Hook. f.) Marg., Szlach. & Kulak = **Liparis
mannii** Rchb. f.

torta (Hook. f.) Marg., Szlach. & Kulak = **Liparis torta**
Hook. f.

vestita (Rchb. f.) Marg., Szlach. & Kulak = **Liparis
vestita** Rchb. f.

viridiflora (Blume) Marg., Szlach. & Kulak = **Liparis
viridiflora** (Blume) Lindl.

Stigmatodactylus

sikokianus Maxim. ex Makino var. *paradoxus* (Prain)
Maek. = **Stigmatodactylus paradoxus** (Prain)
Schltr.

Stimegas

venustum (Wall. ex Sims) Raf. = **Paphiopedilum
venustum** (Wall. ex Sims) Pfitzer

Sturmia

bituberculata (Hook.) Rchb. f. = **Liparis nervosa**
(Thunb.) Lindl. var. **nervosa**

longipes (Lindl.) Rchb. f. = **Liparis viridiflora** (Blume)
Lindl.

nervosa (Thunb.) Rchb. f. = **Liparis nervosa** (Thunb.)
Lindl. var. **nervosa**

Styloglossum

clavatum (Lindl.) T. Yukawa & P. J. Cribb = **Calanthe
densiflora** Lindl.

densiflorum (Lindl.) T. Yukawa & P. J. Cribb = **Calanthe
densiflora** Lindl.

lyroglossum (Rchb. f.) T. Yukawa & P. J. Cribb = **Calanthe
lyroglossa** Rchb. f.

Sunipia

andersonii (King & Pantl.) P. F. Hunt = **Bulbophyllum
sasakii** (Hayata) J. J. Verm., Schuit. & de Vogel

annamensis (Ridl.) P. F.Hunt = **Bulbophyllum medioximum** J. J.Verm., Schuit. & de Vogel

arunachalensis (A. N. Rao) J. M. H. Shaw = **Bulbophyllum arunachalense** (A. N. Rao) J. J. Verm., Schuit. &
de Vogel

bicolor Lindl. = **Bulbophyllum roseopictum** J. J. Verm.,
Schuit. & de Vogel

candida (Lindl.) P. F. Hunt = **Bulbophyllum candidum**
(Lindl.) Hook. f.

cirrhata (Lindl.) P. F. Hunt = **Bulbophyllum paleaceum**
(Lindl.) Benth. ex Hemsl.

fuscopurpurea (Lindl.) P. F. Hunt = **Bulbophyllum
paleaceum** (Lindl.) Benth. ex Hemsl.

intermedia (King & Pantl.) P. F. Hunt = **Bulbophyllum
interpositum** J. J. Verm., Schuit. & de Vogel

jainii Hynn. & Malhotra = **Bulbophyllum jainii** (Hynn.
& Malhotra) J. J. Verm., Schuit. & de Vogel

kipgenii (Kishor, Chowlu & Vij) J. M. H. Shaw = **Bulbophyllum kipgenii** (Kishor, Chowlu & Vij) J. J. Verm.,
Schuit. & de Vogel

paleacea (Lindl.) P. F. Hunt = **Bulbophyllum paleaceum**
(Lindl.) Benth. ex Hemsl.

racemosa (Sm.) Tang & F. T. Wang = **Bulbophyllum
reptans** (Lindl.) Lindl.

scariosa Lindl. = **Bulbophyllum sunipia** J. J. Verm.

virens (Lindl.) P. F. Hunt = **Bulbophyllum paleaceum**
(Lindl.) Benth. ex Hemsl.

Synmeria

crinifera (Lindl.) Szlach. = **Habenaria crinifera** Lindl.

Taeniophyllum

jerdonianum Wight = **Schoenorchis jerdoniana** (Wight)
Garay

khasianum J. Joseph & Yogan. = **Taeniophyllum
glandulosum** Blume

Tainia

angulata (Rchb. f.) Benth. ex Kraenzl. = **Tainia latifolia**
(Lindl.) Rchb. f.

angustifolia (Lindl) Benth. & Hook. f. = **Ania angustifolia**
Lindl.

barbata Lindl. = **Eriodes barbata** (Lindl.) Rolfe

chinensis (Rolfe) Gagnep. = **Collabium chinense** (Rolfe)
Tang & F. T. Wang

cordata Hook. f. = **Tainia latifolia** (Lindl.) Rchb. f.

cordata Hook. f. = **Tainia latifolia** (Lindl.) Rchb. f.

cordifolia (Lindl.) Gagnep. = **Nephelaphyllum
cordifolium** (Lindl.) Blume

grandiflora (Hook. f.) Gagnep. = **Tainia wrayana** (Hook.
f.) J. J. Sm.

hastata (Lindl.) Hook. f. = **Tainia latifolia** (Lindl.)
Rchb. f.

hookeriana King & Pantl. = **Ania penangiana** (Hook. f.)
Summerh.

khasiana Hook. f. = **Tainia latifolia** (Lindl.) Rchb. f.

maculata (Thwaites) Hook. f. = **Chrysoglossum
ornatum** Blume

penangiana Hook. f. = **Ania penangiana** (Hook. f.)
Summerh.

promensis (Lindl.) Hook. f. = **Eulophia promensis** Lindl.

viridifusca (Hook.) Benth. ex Hook. f. = **Ania viridifusca**
(Hook.) Tang & F. T. Wang ex Summerh.

Tainiopsis

barbata (Lindl.) Schltr. = **Eriodes barbata** (Lindl.) Rolfe

Thelasis

elegans (Lindl.) Blume = **Phreatia elegans** Lindl.

pygmaea (Griff.) Lindl. var. *khasiana* (Hook. f.) Schltr. =
Thelasis khasiana Hook. f.

pygmaea (Griff.) Lindl. var. *multiflora* Hook. f. = **Thelasis
pygmaea** (Griff.) Lindl.

Thelymitra

malintana Blanco = **Habenaria malintana** (Blanco) Merr.

Thicuania

moschata (Banks) Raf. = **Dendrobium moschatum**
(Banks) Sw.

Thrixspermum

appendiculatum (Blume) Kuntze = **Grosourdya
appendiculata** (Blume) Rchb. f.

arachnites (Blume) Rchb. f. = **Thrixspermum centipeda**
Lour.

berkeleyi Rchb. f. = **Macropodanthus berkeleyi** (Rchb.
f.) Seidenf. & Garay

changlangense K. Gogoi = **Thrixspermum pauciflorum**
(Hook. f.) Kuntze

leopardinum C. S. P. Parish & Rchb. f. = **Pteroceras
leopardinum** (C. S. P. Parish & Rchb. f.) Seidenf. &
Smitinand

mannii (Hook. f.) Kuntze = **Micropera mannii** (Hook. f.)
Tang & F. T. Wang

minimifolium (Hook. f.) Kuntze = **Chiloschista fasciata**
(F. Muell.) Seidenf. & Ormerod

muriculatum Rchb. f. = **Pteroceras muriculatum**
(Rchb. f.) P. F. Hunt

musciflorum A. S. Rao & J. Joseph var. *nilagiricum* J. Joseph & Vajr. = **Thrixspermum musciflorum** A. S.
Rao & J. Joseph

teres (Blume) Rchb. f. = **Pteroceras teres** (Blume)
Holttum

tsii W. H. Chen & Y. M. Shui = **Thrixspermum pauciflorum** (Hook. f.) Kuntze

unguiculatum (Lindl.) Rchb. f. = **Brachypeza unguiculata** (Lindl.) Kocyan & Schuit.

usneoides (D. Don) Rchb. f. = **Chiloschista usneoides**
(D. Don) Lindl.

Thunia

alba (Lindl.) Rchb. f. var. *marshalliana* (Rchb. f.) B. Grant
= **Thunia alba** (Lindl.) Rchb. f. var. **alba**

bracteata (Roxb.) Schltr. = **Thunia alba** Rchb. f. var.
bracteata (Roxb.) N. Pearce & P. J. Cribb

marshalliana Rchb. f. = **Thunia alba** (Lindl.) Rchb. f. var.
alba

venosa Rolfe = **Thunia alba** Rchb. f. var. **bracteata**
(Roxb.) N. Pearce & P. J. Cribb

winniana L. Linden = **Thunia bensoniae** Hook. f.

Trias

bonaccordensis C. S. Kumar = **Bulbophyllum bonaccordense** (C. S. Kumar) J. J. Verm., Schuit. & de Vogel

disciflora (Rolfe) Rolfe = **Bulbophyllum disciflorum**
Rolfe

manabendrae (D. K. Roy, Barbhuiya & Talukdar) S. Misra
= **Bulbophyllum oblongum** (Lindl.) Rchb. f.

nasuta (Rchb. f.) Stapf = **Bulbophyllum nasutum**
Rchb. f.

obesa (Lindl.) Mason = **Pinalia obesa** (Lindl.) Kuntze

oblonga Lindl. = **Bulbophyllum oblongum** (Lindl.)
Rchb. f.

ovata Lindl. = **Bulbophyllum oblongum** (Lindl.) Rchb. f.

pusilla J. Joseph & H. Deka = **Bulbophyllum moniliforme** C. S. P. Parish & Rchb. f.

stocksii Benth. ex Hook. f. = **Bulbophyllum stocksii**
(Benth. ex Hook. f.) J. J. Verm., Schuit. & de Vogel

vitrina Rolfe = **Bulbophyllum nasutum** Rchb. f.

Tribrachia

hirta (Sm.) Lindl. = **Bulbophyllum hirtum** (Sm.) Lindl.

odoratissima (Sm.) Lindl. = **Bulbophyllum odoratissimum** (Sm.) Lindl.

reptans Lindl. = **Bulbophyllum reptans** (Lindl.) Lindl.

Trichoglottis

cirrhifera Teijsm. & Binn. = **Trichoglottis orchidea**
(J. Koenig) Garay

difformis (Wall. ex Lindl.) T. B. Nguyen & D. H. Duong =
Phalaenopsis difformis (Wall. ex Lindl.) Kocyan &
Schuit.

quadricornuta Kurz = **Trichoglottis orchidea** (J. Koenig)
Garay

Trichosma

coronaria (Lindl.) Kuntze = **Eria coronaria** (Lindl.)
Rchb. f.

suavis Lindl. = **Eria coronaria** (Lindl.) Rchb. f.

Trichotosia

crassicaulis (Hook. f.) Kraenzl. = **Bambuseria crassicaulis** (Hook. f.) Schuit., Y. P. Ng & H. A. Pedersen

ferruginea (Lindl.) Kraenzl. = **Dendrolirium ferrugineum** (Lindl.) A. N. Rao

rufinula (Rchb. f.) Kraenzl. = **Trichotosia pulvinata**
(Lindl.) Kraenzl.

Tripudianthes

blepharistes (Rchb. f.) Szlach. & Kras = **Bulbophyllum
blepharistes** Rchb. f.

dickasonii (Seidenf.) Szlach. & Kras = **Bulbophyllum
dickasonii** Seidenf.

proudlockii (King & Pantl.) Szlach. & Kras = **Bulbophyllum proudlockii** (King & Pantl.) J. J. Sm.

viridiflora (Hook. f.) Szlach. & Kras = **Bulbophyllum
viridiflorum** (Hook. f.) Schltr.

wallichii (Rchb. f.) Szlach. & Kras = **Bulbophyllum
wallichii** Rchb. f.

Tropidia

assamica Blume = **Tropidia curculigoides** Lindl.

barbeyana Schltr. = **Tropidia angulosa** (Lindl.) Blume

bellii Blatt. & McCann = **Tropidia angulosa** (Lindl.)
Blume

govindooii Blume = **Tropidia angulosa** (Lindl.) Blume

semilibera (Lindl.) Blume = **Tropidia angulosa** (Lindl.)
Blume

Trudelia

alpina (Lindl.) Garay = **Vanda alpina** (Lindl.) Lindl.

cristata (Wall. ex Lindl.) Senghas = **Vanda cristata** Wall. ex Lindl.

griffithii (Lindl.) Garay = **Vanda griffithii** Lindl.

pumila (Hook. f.) Senghas = **Vanda flavobrunnea** Rchb. f.

Tuberolabium

coarctatum (King & Pantl.) J. J. Wood = **Trachoma coarctatum** (King & Pantl.) Garay

Tylostylis

discolor (Lindl.) Hook. f. = **Callostylis rigida** Blume

rigida (Blume) Blume = **Callostylis rigida** Blume

Uncifera

buccosa (Rchb. f.) Finet ex Guillaumin = **Robiquetia succisa** (Lindl.) Seidenf. & Garay

heteroglossa Rchb. f. = **Uncifera obtusifolia** Lindl.

Vanda

amesiana Rchb. f. = **Holcoglossum amesianum** (Rchb. f.) Christenson

cathcartii Lindl. = **Arachnis cathcartii** (Lindl.) J. J. Sm.

clarkei (Rchb. f.) N. E. Br. = **Arachnis clarkei** (Rchb. f.) J. J. Sm.

coerulea Griff. ex Lindl. f. luwangalba Kishor = **Vanda coerulea** Griff. ex Lindl.

congesta Lindl. = **Acampe praemorsa** (Roxb.) Blatt. & McCann

goverae Gower = **Cymbilabia undulata** (Lindl.) D. K. Liu & Ming H. Li

himalaica (Deb, Sengupta & Malick) L. M. Gardiner = **Holcoglossum himalaicum** (Deb, Sengupta & Malick) Aver.

longifolia Lindl. = **Acampe praemorsa** (Roxb.) Blatt. & McCann

multiflora Lindl. = **Acampe praemorsa** (Roxb.) Blatt. & McCann

paniculata (Ker Gawl.) R. Br. = **Cleisostoma paniculatum** (Ker Gawl.) Garay

parishii Rchb. f. = **Phalaenopsis marriottiana** (Rchb. f.) Kocyan & Schuit. var. **parishii** (Rchb. f.) Kocyan & Schuit.

parviflora Lindl. = **Vanda testacea** (Lindl.) Rchb. f.

parviflora Lindl. var. testacea (Lindl.) Hook. f. = **Vanda testacea** (Lindl.) Rchb. f.

peduncularis Lindl. = **Cottonia peduncularis** (Lindl.) Rchb. f. ex Schiller

pulchella Wight = **Gastrochilus acaulis** (Hook. f.) Kuntze

pumila Hook. f. = **Vanda flavobrunnea** Rchb. f.

roxburghii R. Br. = **Vanda tessellata** (Roxb.) Hook. ex G. Don

roxburghii R. Br. var. spooneri Gammie = **Vanda tessellata** (Roxb.) Hook. ex G. Don

simondii Gagnep. = **Cleisostoma simondii** (Gagnep.) Seidenf.

spathulata (L.) Spreng. = **Taprobanea spathulata** (L.) Christenson

striata Rchb. f. = **Vanda cristata** Wall. ex Lindl.

teres (Roxb.) Lindl. = **Papilionanthe teres** (Roxb.) Schltr.

tesselloides (Roxb.) Rchb. f. = **Vanda tessellata** (Roxb.) Hook. ex G. Don

testacea (Lindl.) Rchb. f. var. parviflora (Lindl.) M. R. Almeida = **Vanda testacea** (Lindl.) Rchb. f.

trichorrhiza Hook. = **Luisia trichorrhiza** (Hook.) Blume

undulata Lindl. = **Cymbilabia undulata** (Lindl.) D. K. Liu & Ming H. Li

vandarum (Rchb. f.) K. Karas. = **Papilionanthe vandarum** (Rchb. f.) Garay

wightiana Lindl. ex Wight = **Acampe praemorsa** (Roxb.) Blatt. & McCann

Vandopsis

parishii (Rchb. f.) Schltr. = **Phalaenopsis marriottiana** (Rchb. f.) Kocyan & Schuit. var. **parishii** (Rchb. f.) Kocyan & Schuit.

undulata (Lindl.) J. J. Sm. = **Cymbilabia undulata** (Lindl.) D. K. Liu & Ming H. Li

Vanilla

pilifera Holttum = **Vanilla borneensis** Rolfe

Wolfia

spectabilis Dennst. = **Eulophia nuda** Lindl.

Xenikophyton

seidenfadenianum M. Kumar = **Schoenorchis smeeana** (Rchb. f.) Jalal, Jayanthi & Schuit.

smeeanum (Rchb. f.) Garay = **Schoenorchis smeeana** (Rchb. f.) Jalal, Jayanthi & Schuit.

Xiphosium

acuminatum Griff. = **Cryptochilus acuminatus** (Griff.) Schuit., Y. P. Ng & H. A. Pedersen

Zeuxine

assamica I. Barua & K. Barua = **Zeuxine nervosa** (Wall. ex Lindl.) Trimen

bracteata Wight = **Zeuxine strateumatica** (L.) Schltr.

brevifolia Wight = **Zeuxine strateumatica** (L.) Schltr.

debrajiana Sud. Chowdhury = **Zeuxine membranacea** Lindl.

dhanikariana Maina, Lalitha & Sreek. = **Zeuxine nervosa** (Wall. ex Lindl.) Trimen

franchetiana (King & Pantl.) King & Pantl. = **Myrmechis pumila** (Hook. f.) Tang & F. T. Wang

grandis Seidenf. = **Zeuxine affinis** (Lindl.) Benth. ex Hook. f.

longifolia (Lindl.) Hook. f. = **Rhomboda longifolia** Lindl.

moniliformis (Griff.) Griff. = **Cheirostylis moniliformis** (Griff.) Seidenf.

pantlingii Av. Bhattacharjee & H. J. Chowdhery = **Zeuxine agyokuana** Fukuy.

pulchra King & Pantl. = **Rhomboda pulchra** (King & Pantl.) Ormerod & Av. Bhattacharjee

pumila (Hook. f.) King & Pantl. = **Myrmechis pumila** (Hook. f.) Tang & F. T. Wang

robusta Wight = **Zeuxine strateumatica** (L.) Schltr.

seidenfadenii Deva & H. B. Naithani = **Zeuxine affinis** (Lindl.) Benth. ex Hook. f.

strateumatica (L.) Schltr. var. laxiflora I. Barua = **Zeuxine strateumatica** (L.) Schltr.

sulcata (Roxb.) Lindl. = **Zeuxine strateumatica** (L.) Schltr.

Zosterostylis

arachnites Blume = **Cryptostylis arachnites** (Blume) Hassk.

Contributors

B. Ramamurthy Kailash is a Senior Research Associate at Ashoka Trust for Research in Ecology and the Environment (ATREE), Bangalore, India. He is a contributor to the India Biodiversity Portal and the Plant Checklist of India, for which he is involved in compiling plant names from the published sources; curating nomenclatural issues in collaboration with global taxonomic and nomenclatural experts; and compiling species pages.

André Schuiteman is research leader at the Royal Botanic Gardens, Kew. He has published numerous papers on the taxonomy and evolution of Orchidaceae with an emphasis on tropical Asia and is currently focusing on the orchid flora of New Guinea.

Uttam Babu Shrestha is director at the Global Institute for Interdisciplinary Studies, Nepal. He is involved in compiling, cross-checking, and curating the plant names for the Plant Checklist of India.